Advances in Science, Technology & Innovation

IEREK Interdisciplinary Series for Sustainable Development

Advances in Science, Technology & Innovation (ASTI) is a series of peer-reviewed books based on important emerging research that redefines the current disciplinary boundaries in science, technology and innovation (STI) in order to develop integrated concepts for sustainable development. It not only discusses the progress made towards securing more resources, allocating smarter solutions, and rebalancing the relationship between nature and people, but also provides in-depth insights from comprehensive research that addresses the **17 sustainable development goals (SDGs)** as set out by the UN for 2030.

The series draws on the best research papers from various IEREK and other international conferences to promote the creation and development of viable solutions for a **sustainable future and a positive societal** transformation with the help of integrated and innovative science-based approaches. Including interdisciplinary contributions, it presents innovative approaches and highlights how they can best support both economic and sustainable development, through better use of data, more effective institutions, and global, local and individual action, for the welfare of all societies.

The series particularly features conceptual and empirical contributions from various interrelated fields of science, technology and innovation, with an emphasis on digital transformation, that focus on providing practical solutions to **ensure food, water and energy security to achieve the SDGs.** It also presents new case studies offering concrete examples of how to resolve sustainable urbanization and environmental issues in different regions of the world.

The series is intended for professionals in research and teaching, consultancies and industry, and government and international organizations. Published in collaboration with IEREK, the Springer ASTI series will acquaint readers with essential new studies in STI for sustainable development.

ASTI series has now been accepted for Scopus (September 2020). All content published in this series will start appearing on the Scopus site in early 2021.

More information about this series at http://www.springer.com/series/15883

Mohamed Ben Ahmed • Sehl Mellouli •
Luis Braganca • Boudhir Anouar Abdelhakim •
Kwintiana Ane Bernadetta
Editors

Emerging Trends in ICT for Sustainable Development

The Proceedings of NICE2020 International Conference

 Springer

Editors
Mohamed Ben Ahmed
Computer Engineering Department
Faculty of Sciences and Techniques of Tangier
Abdelmalek Essaadi University
Tangier, Morocco

Sehl Mellouli
Systèmes d'Information
Organisationnels
Université Laval
Québec, QC, Canada

Luis Braganca
Civil Engineering Department
School of Engineering, University of Minho
Minho, Portugal

Boudhir Anouar Abdelhakim
Computer Engineering
Faculty of Sciences and Techniques of Tangier
Abdelmalek Essaadi University
Tangier, Morocco

Kwintiana Ane Bernadetta
Department of Computer Application
School of Computer Information
and Mathematics
Abdur Rahman Crescent
Institute of Science & Technology
Chennai, India

ISSN 2522-8714   ISSN 2522-8722   (electronic)
Advances in Science, Technology & Innovation
IEREK Interdisciplinary Series for Sustainable Development
ISBN 978-3-030-53442-4   ISBN 978-3-030-53440-0   (eBook)
https://doi.org/10.1007/978-3-030-53440-0

This Springer imprint is published by the registered company Springer Nature Switzerland AG
The registered company address is: Gewerbestrasse 11, 6330 Cham, Switzerland

Committee

Steering Committee

Prof. Ben Ahmed Mohamed, FSTT, Abdelmalek Essaadi University, Morocco
Prof. Boudhir Anouar Abdelhakim, FSTT, Abdelmalek Essaadi University, Morocco
Prof. Wassila Mtalaa, Luxembourg Institute of Science and Technology, Luxembourg
Prof. Bernadetta Kwintiana Ane, University of Stuttgart, Germany.

Local Chair

Prof Said Raghay, FSTG, Cadi Ayaad University, Marrakech, Morocco
Prof Moulay Driss Laanaoui, ENS, Cadi Ayaad University, Marrakech, Morocco.

Technical Program Committee Chairs

Prof. Olga Sergeyeva Saint-Petersburg State University, Russia.

Publications Chairs:

Topical Editor of IA and Sustainability

Prof. Bernadetta Kwintiana Ane, University of Stuttgart, Germany.

Topical Editor of Green Networks

Prof. Dr. İsmail Rakıp Karaş, Karabuk University, Turkey.

Topical Editor of Environmental Informatics for Sustainable Development

Prof. Luis Bragança, University of Minho, Portugal.

Topical Editor of Computing for Sustainable Development

Prof. Sehl Mellouli, Laval University, Canada.

Special Issue Co-Chairs

Prof. Domingos Santos, Polytechnic Institute Castelo Branco, Portugal
Prof. Senthil Kumar, Hindustan College of Arts and Science, India.

Keynote and Panels Chairs:

Prof. Joel Rodrigues, National Institute of Telecommunications , Brazil.

Tutorials Chair

Prof. Kashif Saleem, King Saud University, Riyadh, Saudi Arabia.

Web Chair

Prof. Ben Ahmed Mohamed, FSTT, Abdelmalek Essaadi University, Morocco.

Publicity and Social Media Co-Chairs:

Abdellaoui Alaoui El Arbi, EIGSI , Casablanca, Morocco.

Technical Program Committee

A. Souissi, Mohammed V University, Rabat, Morocco
Abdel-Badeeh M. Salem, Ain shams university, Egypt
Abdelkrim Haqiq, University Hassan I of settat, Morocco
Abderrahim El Mhouti, Mohammed Premier University, Morocco
Abderrahmane Lakas, United Arab Emirates University, UAE
Abeer Mohamed ElKorany, Faculty of Computers and Information Cairo University, Egypt
Abik Mounia, ENSIAS, UM5, Morocco
Abtoy Anouar, ENSATE, UAE, Morocco
Accorsi, Riccardo, Bologna University, Italy

Adel Alti, University of SETIF-1, Algeria
Ahmad S. Almogren, King Saud University, Saudi Arabia
Ahmed Kadhim Hussein, Babylon University, Iraq
Ahouzi Esmail, INPT-Rabat, Morocco
Alabdulkarim Lamya, King Saud University, Saudi Arabia
Alghamdi Jarallah, Prince Sultan University, Saudi Arabia
Anabtawi Mahasen, Al-Quds University, Palestine
Arsalane Zarghili, USMBA, Fez, Morocco
Assaghir Zainab , Lebanese University, Lebanon
Aziz Darouichi, FST Marrakech, Morocco
Ben Yahya Sadok, Faculty of Sciences of Tunis, Tunisia
Benaouicha Said, UAE, Morocco
Bernadetta Kwintiana Ane, University of Stuttgart, Germany
Bessai-Mechmach Fatma Zohra, CERIST, Algeria
Bouhorma Mohamed, FSTT, UAE University, Morocco
Boulmalf Mohammed, UIR, Morocco
Chadli Lala Saadia, University Sultan Moulay Slimane, Beni-Mellal, Morocco
Damir Žarko, Zagreb University, Croatia
Dominique Groux, UPJV, France
Dousset Bernard UPS, Toulouse, France
El Alami Ali , Moulay Ismail, University, Morocco
EL Amrani Chaker, FSTT, UAE University, Morocco
El Bouchti Abdelali, Bournemouth University, United Kingdom
El Kafhali Said, Hassan 1st University, Settat, Morocco
El Mamoun Souidi, FSR, Mohammed V University, Morocco
Elhaddadi Anass, Mohammed Premier University, Morocco
El-Hebeary Mohamed Rashad,Cairo University, Egypt
El-Mahdi El-Guarmah, ERA, Morocco
EN-Naimi EL Mokhtar, UAE, Morocco
Enrique Arias Castilla-La Mancha University, Spain
Ensari Tolga, Istanbul University, Turkey
Faisal Shah Khan, Khalifa University, Abu Dhabi, UAE
Fennan Abdelhadi, Abdelmalek Essaadi University, Morocco
Ghadi Abderrahim, FSTT, UAE, Morocco
Haddadi Kamel IEMN Lille University, France
Hanaa Hachimi, ENSAK, Ibn Tofail University, Morocco
Harmouzi Mustapha, FS, UAE, Morocco
Harroud Hamid, University AlAkhawayn of Ifrane, Morocco
Hazim Tawfik Cairo University, Egypt
Hioual Ouassila, Abbes laghrour khenchela, Algeria
Jaime Lioret Mauri Polytechnic University of Valencia, Spain
Joel Rodrigues, National Institute of Telecommunications , Brazil
Josep Lluis de la Rosa, Girone université, Spain
Jun Wu, Shanghai Jiao Tong University, China
Jus Kocijan, Nova Gorica University, Slovenia
Kamel Hussein Rahouma, Faculty of Engineering, Minia University, Egypt
Kashif Saleem , King Saud University, Kingdom of Saudi Arabia
Khoudeir Majdi IUT, Poitiers university, France
Labib Arafeh, Al-Quds University, Palestine
Laila Moussaid, ENSEM, Hassan II University, Casablanca, Morocco
Lalam Mustapha, Mouloud Mammeri University of Tizi Ouzou, Algeria
Lamya Alabdulkarim, King Saud University, Saudi Arabia
Loncaric Sven, Zagreb University, Croatia

Preface

Sustainable development is a relevant topic nowadays. It attracts government and deciders, industrials and researchers of many disciplines. Several works are conducted in order to understand and improve the environmental, social and economic domains in which humans interact with around the world. Generally, the drive towards sustainable development cannot ignore the import role of ICT (Information and Communication technologies) advances.

This book is one of the important series which encloses recent and advanced researches on Emerging Trends in ICT for Sustainable Development. It's an interesting manuscript that will help newer and advanced researchers, industrials and deciders to understand, to attract and conclude new ideas, innovative solutions, incorporating the concepts of sustainable development. It's also an opportunity to know the exiting works and scientific contributions of the literature in order to develop and think to new ones.

In its edition, this book lists original researches in new directions and advances focused on multidisciplinary fields and closely related to the fields of information systems, communication and technologies and their applications. This edition is the result of a reviewed, evaluated and presented work in more than 15 sessions opened and listed in NICE20 Conference as follows:

- Artificial Intelligence and Sustainability
- Green Networks and Intelligent Transportation Systems
- Environmental Informatics and Sustainable Environment
- Computing Technologies for Sustainable Development

The present book contains selected and extended papers of the 3rd international conference on Networks, Intelligent systems, Computing and Environmental informatics for sustainable development.

Co-organized conjointly by Medi-Ast Association and Faculty of sciences and techniques held from 31 March to 02 April, 2020, in Marrakech—Morocco.

We thank all authors from across the globe for choosing NICE20 to submit their manuscripts. A sincere gratitude to all keynotes speakers for offering their valuable time and sharing their knowledge with the conference attendees. Specials thanks are addressed to all organizing committee members, to local chairs and local committee in Faculty of Sciences and technologies of Marrakech, to all programme committee members, to all chairs of sessions for their efforts and the time spent in order to success this event.

Many thanks to the Springer staff for their support and guidance. In particular, our special thanks to Dr. Nabil Khelifi, Reyhaneh Majidi and Nareshkumar Mani for their help, support and guidance.

Tangier, Morocco   Mohamed Ben Ahmed
Québec, Canada   Sehl Mellouli
Minho, Portugal   Luis Braganca
Tangier, Morocco   Boudhir Anouar Abdelhakim
Chennai, India   Kwintiana Ane Bernadetta

Keynote Speakers

Prof. Steven Furnell

Centre for Security, Communications & Network Research, University of Plymouth, UK

Biography: Steven Furnell is a Professor of information security and leads the Centre for Security, Communications & Network Research at the University of Plymouth. He is also an Adjunct Professor with Edith Cowan University in Western Australia and an Honorary Professor with Nelson Mandela University in South Africa. His research interests include usability of security and privacy, security management and culture and technologies for user authentication and intrusion detection. He has authored over 320 papers in refereed international journals and conference proceedings, as well as books including Cybercrime: Vandalizing the Information Society and Computer Insecurity: Risking the System. He is the current Chair of Technical Committee 11 (security and privacy) within the International Federation for Information Processing, and a member of related working groups on security management, security education and human aspects of security. He is also a board member of the Chartered Institute of Information Security and chairs the academic partnership committee and southwest branch.

Passwords: A Lesson in Cybersecurity Failure?

Abstract: Passwords being dead is not actually true, but what can we expect in this era of fake news? Imagine for a moment that passwords were dead. What would they be remembered for? What would be their legacy if their oft predicted passing finally came to pass?

For many users, they would likely be thought of as the things that they were forced to have but struggled to remember. Ironically, they could not remember the passwords themselves, but they would remember they were difficult to use. Meanwhile, in the Cybersecurity community, few would mourn their passing. Passwords have hardly distinguished themselves when it comes to upholding protection and, in some cases, they have become a byword for vulnerability. They will be remembered as the things the users got wrong. And for those who see people as the weakest link in Cybersecurity, passwords could be the archetype of why.

The technology has the potential to be used well, and it is not acceptable to simply blame the users either. Much of the responsibility rests with those offering passwords—or other forms of security—in the first place. If people are expected to use security, we cannot just assume that they already know how to do so.

Prof. Mohammed Bouhorma

Abdelmalek Essaadi University, Tangier, Morocco

Biography: Bouhorma is an experienced academic who has more than 22 years of teaching and tutoring experience in the areas of Information Security, Security Protocols, AI, Big Data and Digital Forensics at Abdelmalek Essaadi University. He received his M.S. and Ph.D. degrees in Electronics and Telecommunications from INPT in France. Previous roles included Head of Department of Computing at FST of Tangier and Director of LIST at FSTT. He is also responsible for the Master degree programme of Computer Sciences and Networks. He has held a Visiting Professor position at many Universities (France, Spain, Egypt and Riyadh). He was head (and founder) of the Computer and Communication Systems Laboratory from 2013 to 2017. He has directed and supervised more than 30 doctoral theses in Morocco and in other countries and graduated over 15 Ph.D. Students. His research interests include computer security, wireless sensors network, Cybersecurity, Big Data Analytics, AI, Smart Cities technology and serious games. He is an editorial board member for over dozens of international journal and has published more than 100 research papers in journals and conferences. He is a member of the experts' committee for the evaluation of Master's courses in computer networks and security.

The IoT Architectures and Applications: From Technology to the Smart Revolution

Abstract: The Internet of Things (IoT) is defined as an integrated part of the future Internet, it is the key to the next phase of the industrial revolution. The recent deployment of IoT in Smart City infrastructures has led to very large amounts of data being generated each day across a variety of domains, with applications including healthcare, smart transport, smart home, smart energy and smart security. In this talk, we propose a novel architecture for IoT technologies, highlights some of the most important technologies and applications that have the potential to make a striking difference in human life, especially for the VANET (Vehicular ad hoc networks), Smart Home Design for Smart Medical Surveillance and Monitoring System for Elderly.

Prof. Serge Miranda

University of Nice Sophia Antipolis, France

Biography: Serge Miranda is a full-time Professor of Computer Science at the University of Nice Sophia Antipolis (UNS), France, a position he has held since October 1983 after a Ph.D. in Toulouse University (France) and a Master thesis at UCLA (the University of California, Los Angeles with an INRIA Scholarship). He has authored or coauthored more than 100 publications and published 6 successful French books on databases (15 Editions) with many invited conferences around the world (in French, English and Spanish). He has been running an MBDS master degree and innovation laboratory (since 1992) devoted to database, Big Data and mobiquitous information systems with an important financial involvement of industry partners to prototype information services of the future. A genuine MBDS business model was set up between the University of Nice and key IT industry partners based upon joined tutored contracts. Several innovative wireless information cell phone-centric applications with communicating objects (RFID, NFC, QR Code, captors, sensors,..) were prototyped at MBDS in the area of m-payment, digital campus, health (elderly people), travel/tourism, sustainable economy (fair trade, NFC posters for illiterate, etc.). MBDS is de facto an INNOVATION laboratory which gave the key initial impetus of Nice becoming the first NFC city in Europe in 2010. MBDS has been successfully delocalized in Haiti (since 1998), Morocco, Madagascar and Russia. Serge Miranda founded and became the first president (until 2012) of a multidisciplinary University foundation DreamIT on December 2009 around the MBDS kernel. DreamIT was key in rebuilding MBDS facility in Haiti: on April 2011, was inaugurated the "first mobiquitous NFC building in America" for MBDS degree in the Science Campus of the University of Haiti in Port of Prince (the initial MBDS building was destroyed during the devastating 2010 earthquake) On 21 March 1998, he was decorated ("Chevalier Ordre du Merite") by Senator Pierre Laffitte (founder of Sophia Antipolis science park) on behalf of the Ministry of Industry of France for recognition of his original contribution between higher education and industry in the science park of Sophia Antipolis.

Towards a Unifying Theory for Big Data Management (and data lakes) Based Upon Categories

Abstract: We entered a new data-centric economy era with the widespread use of supporting Big Data infrastructures to deliver predictive real-time analysis and augmented intelligence in the three "P" sectors (Public, Private and Professional). Every economic sector will be drastically impacted. Such Big Data infrastructure involves two major scientific fields:

- Computer science for data management (i.e. building a virtual or real Data Lake)
- Mathematics for machine learning-ML- or AI for deep learning-DL-.

From a computing point of view, three types of Data are involved in big data architecture:

- Structured data (with predefined schema),
- Semi-structured data (around XML with meta data) and
- Unstructured data (no schema, no metadata).

A Data Lake is a generalization of data warehouse to semi-structured and non-structured data. The data lake could be real (with ETL pumping systems like in most data warehouses) or virtual with distributed large data sets and the polystore approach (1) (2).

Today, there exists no present SQL standard to manage a data lake with many proprietary proposals encompassing new key features like "external tables" within From-SQL clauses referring to N.O.SQL files. Expected use of a data lake is a predictive real-time analysis by data scientists using a large variety of ML and DL methods generally in supervised, unsupervised or reinforced modes; no interactivity exists among these methods.

This conference encompasses two parts:

- A presentation of the three formal models underlying the three types of data: set theory, graph theory and linear algebra (matrices)
- A functional comparison of the two basic formal approaches to offer a unique interface to these three models based upon either the "category theory" (3), (4), (5), (6), or "associative arrays' (7), (8), the former being the most promising from an implementation point of view (Microsoft and Oracle are working on them) and teachability

Interesting research is still to be conducted to make the category theory the formal foundation of big data and polystores as Codd's relational model dit it for structured data and SQL.

Contents

Artificial Intelligence for Sustainability

An Intelligent Chatbot Using NLP and TF-IDF Algorithm for Text Understanding Applied to the Medical Field

Ayanouz Soufyane⬛, Boudhir Anouar Abdelhakim⬛, and Mohamed Ben Ahmed⬛

Abstract

This is the era of intelligent machines. Due to the progress and advancements of machine learning, deep learning, and artificial intelligence, machines are impersonating humans. Natural language processing-based conversational software agents are known as chatbots. These chatbots are the best example of such machines. To spend a healthy life, health care is much important. But it is not easy to have a consultation with the appropriate doctor while having health problems. Our proposed solution for this problem is to create an artificial intelligence-based medical chatbot, which will be able to detect the ailment and also to give the necessary details about any condition. The purpose of this chatbot is to minimize the healthcare cost and to improve the approachability to medical knowledge. This chatbot will act as a medical reference book that will help the patients to diagnose their sickness and also to recover their health. A patient can only gain the actual benefit of the chatbot only when a chatbot can diagnose the accurate disease and also provide the necessary information about the condition. A text-to-text verdict bot involves people in the discussion about their health problems and also provides a personalized diagnosis depending on the symptoms. Hence, people will get ideas related to their health conditions and will get the right safety. In this work, we performed a detailed survey on recent literature. We examined many publications from the last 5 years, which are related to chatbots. We also presented a hybrid architecture based on deep learning models like NLP and the TF-IDF algorithm.

Keywords

NLP • TF-IDF • Deep learning • Artificial intelligence • Chatbots • Medical reference • Health care • Conversational agent

1 Introduction

People are extremely busy with the Internet nowadays that they have forgotten how important it is to stay healthy. They ignore small diseases and think they will recover with time, but sometimes these trivial diseases become a significant disease and can cause death in the worst scenario (Ayanouz et al. 2020). So, there should be a way to overcome this problem, and hence people nowadays tend to engage more in question answer forums to get health-related answers instead of going to a doctor or waiting for some expert to answer their question later on. So, primary purpose of this research is to establish a system that can reduce this gap, and users can get their answers in no time. Also, already available systems have many issues like telephonic calls are never picked soon, and they charge for each call. Also, people have to wait for a long time to get the answers they seek.

The proposed Chatbot system uses Natural Language Processing (NLP) techniques to provide communication between humans and computers and vice versa. This system uses three primary analyses to perform natural language processing which are Identification of the linguistic relations, parsing of subjects into objects, and semantic interpretation of word meanings.

A chatbot uses a combination of texts and vocal language techniques to imitate the discussion of humans in its own, particularly accepted format. Copying of human conversation is the primary aim of this system. In order to achieve this

A. Soufyane (✉) · B. A. Abdelhakim · M. B. Ahmed
Faculty of Sciences and Techniques, University of Abdelmalek
Essaadi, Tangier, Morocco
e-mail: Ayanouz.soufiane@gmail.com

B. A. Abdelhakim
e-mail: boudhir.anouar@gmail.com

M. B. Ahmed
e-mail: m.benahmed@gmail.com

particular setup, an interface is made that can send input and receive a response on a specific device. The chatbot interacts with a human by storing the preceding commands and keeping track of the human–computer interaction to generate the correct answer and perform the required functionality. Medical Chatbots use artificial intelligence algorithms to examine user queries, recognize them, and give a proper reply to the asked question. Sometimes, a significant disease starts from a small problem. Let's take into account a minor problem that begins with a headache and leads to a brain tumor that is a significant disease. So, this problem can be overcome with the periodic use of a medical chatbot that keeps the record of previous diseases of a human and can predict this problem in advance using its medical record database. So, most of the conditions can be identified in advance and can be dealt with before they become a big problem.

2 Related Work

Most commonly, we humans have to go to a hospital or a doctor for a routine checkup personally or worse for even a small disease. It is quite hectic and time consuming. So regarding this issue, Users don't know the treatment or symptoms of every disease; hence, they have to see the doctor. Also, if the doctors start to deal with their patients on telephonic calls, it is not possible. Such problems can only be solved by using some technology. So, medical Chatbots are introduced that have high learning abilities and excellent problem-solving skills. These Chatbots have been found very helpful in providing guidance on daily routine to patients who have minor diseases. Using medical Chatbot Natural Language Processing, health-related problems of people can be quickly solved. Any novice user can easily ask any question regarding their health using Google's voice to text conversion and vice versa from the Chatbot. The question is sent to a Medical Chatbot, and hence, the bot replies with the solution on the given screen of an android device using the Android App (Ayanouz et al. 2020).

Virtual assistants have been found to be very helpful to patients and medical-field tasks. Some of the main Chatbots, Florence, Molly, and Ada, are very useful Chatbots with powerful algorithms that allow them to take care of human health by providing medical assistance anytime and anywhere. These bots are developed by keeping in mind the basic rules of artificial intelligence to perform a dialogue with humans. Users don't need to personally go to the hospital anymore as they can now talk directly with Medical Chatbots from their home. These Chatbots have a unique ability to assist humans via applications, text messages, and instant messages. Chatbots have already been in use in other industries like making cars, robotic hands, retail, etc. to expedite, support, and improve processes. This technology is gaining grip in health care also, where these Chatbots are helping patients to perform countless tasks (Bates 2019).

Many jobs that were traditionally performed by humans have been replaced by Chatbots in the form of Customer Service Agents and Online Educators. Chatbots have gone through a variety of modifications and improvements since the start of rule-based bots to the advanced artificial intelligence Chatbots. The work of an advanced chatbot is improved significantly than older Chatbots. Nowadays, these bots can interact and chat with humans and can also learn from them. The primary purpose of this research paper was to go through past research on Chatbots, mostly which are called Conversational Agents, by the use of Bibliometric Analysis. This research can help other researchers significantly to find areas for future research in modern Chatbots. The result was quite shocking as an excellent new research opportunity was found in Chatbots that will lead to the development of new learning technologies. This new technology will redirect future research in Chatbots to a new development phase. Many different research references are provided that are based on outcomes that were achieved from this analysis (Bayu and Ferry 2016).

As humans interact with each other using natural language, Chatbots also use human language to communicate with social users. Chatbots are software programs that are made to interact with humans using natural words. These software programs aimed to make socially interacting humans think that they are interacting with the human on the other side. The purpose of this research paper is to study existing Chatbots named ALICE and ELIZA that interact with humans. Also, to explain that, it is easy to make Chatbots using ALICE because ALICE is based on simple pattern matching techniques, while ELIZA is based on rules, so it is complicated to make a Chatbot using ELIZA. Then, the proposed solution is to implement ALICE as a domain-specific chatterbox, which is a Student Information System that will help students in solving various queries related to universities (Dharwadkar and Deshpande 2018).

Chatbots are also known as chatter robots, and are software agents that simulate human conversation via text or voice messages. One of the first and main goals of a Chatbot had always been to resemble an intelligent human and make it hard for others to understand their real nature. With the development of more Chatbots of various architectures and capabilities, their usage has widely expanded [10].

These conversational agents can go into a point of fooling the users and making them believe they are talking to a human, but are very limited in improving their knowledge base at runtime.

In order to understand the user input and provide a meaningful response, the chatbot uses artificial intelligence and deep learning methods. Moreover, they interact with humans, using natural language and different applications of Chatbots such as medical Chatbots and call centers.

Fig. 1 Classification of Chatbot
Application

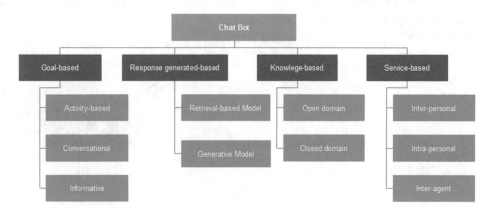

A chatbot could help doctors, nurses, patients, or their families. Better organization of patient information, medication management, helping in emergencies or with first aid, and offering a solution for superficial medical issues, these are all possible situations for chatbots to step in and reduce the burden on medical professionals (Schuetzler et al. 2018).

3 Taxonomy of Chatbot

Two major developments are associated with the latest concern about chatbots. First, the growth of messaging service has tremendously increased in the last few years. It supports features, like booking, ordering and payments, which usually need a different website or application. Thus, directly from their preferred messaging apps, users can carry out tasks like asking questions, booking a restaurant, and buying goods without having to download a series of different applications. Line, WeChat, WhatsApp, and Facebook Messenger are some of the common applications. Second, a combination of deep learning techniques and advanced AI techniques has led to an improvement in the quality of decision-making and understanding on affordable processing power. It manages and processes a large amount of data to produce outcomes that supersede human performance.

Advisory, entertainment, commercial, and service chatbots are the four main categories of chatbot applications (Higashinaka 2014). Advisory chatbots are aimed at repairing goods, offering maintenance, giving recommendations on service, or providing suggestions. Chatbots in this category can offer advice and support, as well as contact people, when necessary. The aim of entertainment chatbots is to engage customers with movies, favorite band, sports, or other events. These chatbots offer ticket deals, details on upcoming events, and bet-placing options. The basic function of commercial chatbots is to simplify customers' purchases. For instance, a pizza company can use a messaging interface to notify promotions or take delivery orders. Service chatbots are concerned with providing facilities to

customers. For instance, instead of using phone calls or emails, a logistics firm can make use of an instant messaging channel to respond to delivery queries and provide copies of delivery documents. As displayed in Fig. 1, we identified four types of chatbot applications in this paper, which include response generated based, service based, knowledge based, and goal based.

Response generated-based Chatbot
These chatbots are classified in relation with their role in response generation. The input and output of the response models are in the form of natural language text. One of the tasks of the dialogue manager is to decide on the combination of response models. The dialogue manager generates a response in three steps (Lo and Lee 2017). First, the dialogue manager utilizes all response models to create a set of responses. Second, it sends a response on the basis of priority. Third, in the absence of any priority response, the model selection policy is applied in selecting the response.

Service-based Chatbot
The facilities provided to the customer are the classification criteria for service-based chatbots. They are designed for a commercial or personal purpose. A logistics firm, for example, can use a chatbot to send copies of delivery documents instead of making phone calls, or people can use the app to order a meal from McDonald's.

Knowledge-based Chatbot
This category of chatbots is characterized by the knowledge obtained from a training dataset or primary data sources. Closed-domain and open-domain are the two major data sources. Closed-domain data sources deal with a specified knowledge domain. The dataset contains all the relevant information for providing an answer to the question, and examples include bAbI, MCTest, and Daily Mail. Open-domain data sources are concerned with general topics, and the questions determine the answers to be provided. Quiz bowl and Allen Al Science are typical examples of open-domain.

Fig. 2 Representation of the
generative model

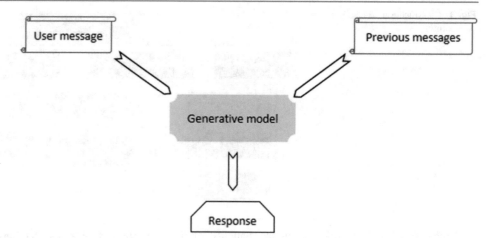

Goal-based Chatbot

The main goal to achieve, is the underlying factor for classifying goal-based chatbots. They are created to accomplish a given task and help users complete the task by getting useful information from them via short conversations (Lo and Lee 2017). For example, customers can get answers to their questions or solutions to their problems by engaging with the Chatbot on a company's website.

4 Chatbot Models

This is the era of artificial intelligence and chatbots. The development and advancement in this field have proved that in customer services, chatbots are replacing humans. The advancement in the field of artificial intelligence is grown up to be something more than just a simple dream of fiction of science. Have you ever gotten a thought that humans will ever interact with the machines? But nowadays, chatbots have proved that this idealistic thought is possible by their human-like behavior, their intelligence, and their capability of learning with different experiences using machine learning.

4.1 Different Types of Architecture Models of Chatbots

For the development of a chatbot, it is necessary to define its architecture. The response of a chatbot can be of two types: either it can produce a response from scratch according to the models of machine learning, or in the second type of response, chatbots can use heuristics for the selection of suitable response from a library where predefined responses are stored.

Generative Models

Generative models are used to develop smart chatbots. These chatbots are more advanced in nature. Chatbots of this type

are rarely used because the implementation of complex algorithms is required for this purpose. That's why they are challenging to develop and build as compared to other models. Millions of examples and a lot of time are required to train these types of chatbots. This is the point where deep learning models can involve in the conversation. But after all this, you cannot make sure that the model will generate what type of response. Figure 2 is a representation of this model.

Retrieval-Based Models

The Retrieval-Based Model of a chatbot (in Fig. 3) is much reliable and easier to implement. Although we cannot get 100% accurate responses by the chatbot, but we can know the possible responses, and we can also ensure that no incorrect or inappropriate response will be generated by the chatbot (Higashinaka 2014).

Nowadays, this type of model is more in use as compared to other models. Several APIs and algorithms are easily accessible by the developers for the development of chatbots on this model. These types of chatbots consider the conversational context and predefined list of messages to generate the most reliable response.

4.2 Response Generation Mechanism of Chatbots

Chatbots use two different ways to understand human messages and to determine the intent of messages.

Pattern-Based Heuristics

Chatbots can generate a response in two different ways:

1. By using machine learning classifiers.
2. By using the logic of if-else conditions.

The most straightforward way is to define some set of predefined rules that will act as conditions. Artificial Intelligence Language is the main language that is used in writing response patterns in the development of chatbots.

Fig. 3 Representation of the retrieval-based model

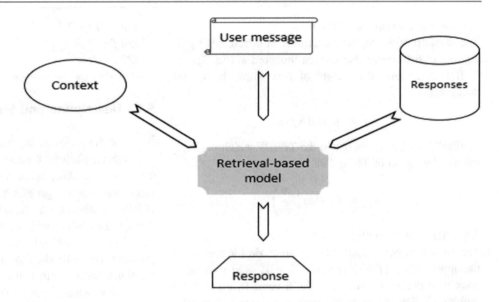

Chatbots use rich patterns consisting of a predefined set of rules with natural language processing pipelines to build a smart bot using AIML. These smart Chatbots are able to parse human messages, find their synonyms, find concepts used in a sentence, find parts of speech, and find which rule matches the human question. But, these smart Chatbots do not use any APIs or Machine learning concepts until we specifically program them that way.

Intent Classification Using Machine Learning

Pattern-based searching can generate good results but, to achieve this, we have to write all the patterns manually. So, this is a very tiring and time-consuming task. Let's suppose if there is only one intent, then it works very fast, but if there are hundreds of intents, then determining different intents is a challenging task.

For the training of Chatbots, we are wholly based on machine learning technology that uses intent classification. We have sets of thousands of predefined patterns from which a Chatbot can quickly pickup data patterns and learn from them.

Response Generation

After understanding the human message, the next thing a Chatbot has to do is to generate a proper response. So, there are two main ways by which we can make a response in Chatbots:

1. Generate response based upon the template of the intent and put some variables in it to make it complete.
2. Generate a simple response that is static.

Which method would be taken to generate the response depends on the development company. These companies choose different techniques for different Chatbots, depending on the nature of the work in which Chatbots are intended to be used.

Let's take an example to clarify this. A Chatbot that is being employed by different weather forecast companies will have many different options to use this Chatbot. Some will program it to say, "it's a rainy day today" or "it will rain today." Some companies will try to make it more persuasive, "Probability of rain today is 90%, so you should carry your umbrella to work today."

Also, as you can see, the style of response varies depending on the nature of the chat. So, these Chatbots analyze and study previous conversations to generate customized reactions according to the intended nature of a chat. The given diagram will explain further how to separate a response that is created and also which modules are responsible for the selection of these different modules

5 Algorithm and Architecture of proposed system

5.1 Text understanding algorithm (TF-IDF)

When building a model with the goal of understanding text, we need a strategy to score the relative importance of words using TF-IDF.

Term Frequency (TF)

The division of the number of times a word appears in a document by the total number of words (Eq. 1).

NB: Every document has its own term frequency

$$tf_{i,j} = \frac{n_{i,j}}{\sum_k n_{i,j}} \qquad (1)$$

Inverse Data Frequency (IDF)

The division of the log of the number of documents by the number of documents that contain the word **w** (Eq. 2).

IDF determines the weight of rare words in the full document

$$idf(w) = \log(N/df_t) \tag{2}$$

Finally, in Eq. 3, we obtain the TF-IDF which is simply the multiplication of TF by IDF:

$$w_{i,j} = tf_{i,j} * \log\left(\frac{N}{df_i}\right) \tag{3}$$

Algorithm implementation

The algorithm below shows the pseudocode for computing the approximate TF-IDF measure. The TF-IDF measure is created in buffer B and returns the k items (t, id) with the highest TF-IDF values in the given data stream. The algorithm has two steps:

1. The first, it computes $A - tf$: execute the term frequency part of the alogrithm
2. The second phase $A - idf$: execute the inverse document frequency part of the alogrithm

The last results are used to calculate $A - tf - idf$. The pseudocode assumes a parallel implementation (Sharma et al. 2017).

TF-IDF Algorithm

Initialization
V ← Sort by key(B.id)
C ← Reduce by key(V.id)
Begin
for i ← 1 to k do
id = Bi.id
$T^F i = {}^B i.counter$ / Get total(C, id)
end for
for i ← 1 to k do
Ci.total = 1
end for
D ← Count(C.id)
V ← Sort by key(B.t)
for i ← 1 to k do
$Ci.t = {}^V i.t$
$^C i.total = 1$
end for
C ← Reduce by key(C.t)
for i ← 1 to k do
$t = {}^B i.t$
$^{IDF} i = \log(D$ / Get total(C, t))
end for
for i ← 1 to k do

$T F {}^{IDF} i ← T {}^F i * {}^{IDF} i$
end for
END

5.2 Description and Working Mechanism

Some of the chatbots are just like comprehensive medical books that are helpful for doctors, patients, and for those who are curious to learn about health. The user feels that he is incorporated in the procedure of his health (Ayanouz et al. 2020). The patient, who feels indulged, who is intermingling through the chatbot with the system of health care, will stay with the system, which is essential for them and the healthcare provider. The older chatbots were just like the client communication system. For this purpose, a question and answer page is available on their website. The chatbots facilitate the user by getting common questions related to health and then predict the disease without providing any interface to the human.

Our system in Fig. 4, allows users to submit their queries and complaints about their health. The primary concern of our system is customer satisfaction. As many people don't have awareness about underlying physical conditions, our system will help the user by giving them proper information and guidance to live a healthy and good life. Moreover, many people spend their lives with debility, but they never bother about their symptoms because they always think that they do not require any doctor. So our proposed system will help such people.

The internal architecture of our medical chatbot is explained in Fig. 5.

In Fig. 5, you can see that we have tried to display how our Chatbot system works internally. We have explained it briefly in the below section.

5.3 Environnement and Deployment

Additional Plugins

Medical Chatbots among other bots use components and plugins to help them work efficiently. These plugins and components provide Smart Automation tools and Solution APIs to Chatbots to enhance their workforce.

Front-End Platforms

We can use many different platforms that are already working with a vast list of clients to develop the front-end of Chatbots. Some of these systems are given below:

- Google Hangouts
- Microsoft Teams
- Slack
- Skype for Business
- Facebook

Fig. 4 General architecture of
the medical chatbot

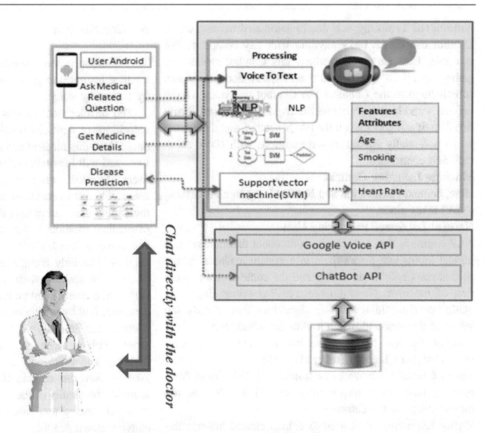

Fig. 5 Representation of the
internal architecture

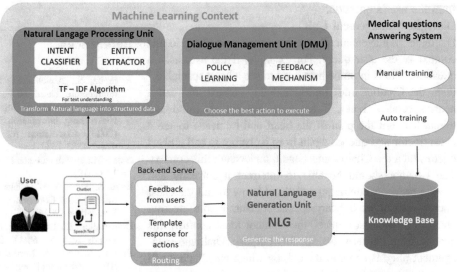

Back-End Server

A back-end server is responsible for handling the user's
request and then routing it to the components accordingly
and vice versa.

Natural language generation unit

This unit is responsible for the generation of an appropriate
response to the question according to the data stored in the
knowledge base, in other words, the process that automatically transforms data into plain-English content.

Medical Question and Answer System

FAQs is a very important component of any system. So, a
Chatbot has stores FAQ questions in its database so that it
can quickly give an answer to the user's common problems:

Manual Training: As the name says, domain experts have
to map frequently asked questions with answers manually,
such that if any user asks any questions that come from the
FAQs section then, the Chatbot must be able to identify it
and hence, answer it accordingly quickly.

Automated Training: It is due to automated training that a Chatbot can do its tasks anytime with any company. For example, Company A for which this Chatbot works can gather up all its questionnaire and Q&A documents and submit them to the Chatbot. The Chatbot can study all the materials and hence, gather all the required data itself. From this data, it can generate all the FAQs needed, and thus, later, it can potentially answer all the questions with 100% accuracy and confidence.

Machine Learning Context

Here, context clarification and Natural Language Processing (NLP) takes place.

Natural Language Processing Unit

NLP engine is the most crucial component that understands human inputs and parses them into system readable format so that the Chatbot can understand the context of the question of the user. Natural Language Processing Engine is made up of machine learning algorithms that identify the intent of the user and match it with the saved intents of the Chatbot database to determine which intentions are supported by that Chatbot (Erra et al. 2015).

Intent Classifier: It works as follows: 1) Take input from user, 2) Understand its meaning, and 3) Relate it to the intents stored in the database.

Entity Extractor: It is used to extract critical information from the user's question.

Dialogue Management Unit

Dialogue Management Agent can work according to the context of the user query. For example, if the user asks a question, "He is in immediate need to see a heart specialist," then, the Chatbot must understand this context and call a heart specialist. Later, if the user changes his mind and thinks it is not the pain of the heart and he needs to see a chest doctor and queries again from a chatbot to call a chest doctor, then the Chatbot must again understand this context. The Chatbot should be able to interpret it correctly and rewrite previous command and should now to call the chest doctor instead of a heart specialist. But, before making changes to the last request, the Chatbot must confirm this change from the user. So, for this purpose, Dialogue management plugins come to the aid, by which the Chatbot can confirm from the user before making some changes.

Feedback Mechanism: In this module, an agent is assigned to periodically take feedback from a user whether the Chatbot is working as intended or not, and if the user is satisfied with the dialogue and response of the bot.

Policy Learning: This technique enables the bot to learn from the experience of the chat with the user and to save the path that it followed previously. It is done so that the user can get maximum satisfaction. The Chatbot learns from past chat and deducts the happiest flow from the previous conversation and in the new chat, it uses its prior learning and deduction to perform best in a new session.

6 Conclusion

The Chatbot is very helpful for personal use or for the medical institutes to get information about any disease either by text or by voice.

The development of this smart chatbot is based on the algorithms of Artificial Intelligence as well as on the trained data. Thus, this medical assistant will fruitfully protect many lives and will inevitably raise medical awareness among the users. As said earlier, the era of the future is all about messaging apps because people spend most of their time on messaging apps rather than the others. That's why the future of medical chatbots is very bright. In distant places, people can have a comfortable conversation with this medical chatbot. The only requirement for this chatbot is a smartphone or a desktop with an Internet connection. The efficiency of a chatbot can be increased by increasing the use of databases and by raising more combinations of words so that it could handle all types of diseases. Moreover, a conversation with a voice can also be added to improve the usability and efficiency of this system. In this work, we assume that the combination of NLP and TF-IDF will improve the quality of the proposed Chatbot system and can be an interesting text-based recommender system for smart conversational agents.

References

Ayanouz, S., Abdelhakim, B.A., Benhmed, M.: A smart chatbot architecture based NLP and machine learning for health care assistance. In: Proceedings of the 3rd International Conference on Networking, Information Systems & Security, Marrakech, Morocco (2020) Association for Computing Machinery. ACM ISBN 978-1-4503-7634-1

Bates, M.: Health care chatbots are here to help. IEEE Pulse **10**(3), 12–14 (2019)

Bayu, S., Ferry, W.W.: Chatbot using a knowledge in database. In: 7th International Conference on Intelligent Systems Modelling and Simulation, (2016).

Dharwadkar, R., Deshpande, N.A.: A medical chatbot. Int. J. Comput. Trends Technol. **60**(1), (2018).

Erra, U., Senatore, S., Minnella, F., Caggianesec, G.: Approximate TF–IDF based on topic extraction from massive message stream using the GPU. Elsivier, Information Sciences **292**(20), 143–161 (2015)

Higashinaka, R.: Towards an open domain conversational system fully based on natural language processing. In: COLING'14, pp. 928–939 (2014).

Lo, H.N., Lee, C.B.: Chatbots and conversational agents: A bibliometric analysis. In: IEEE International Conference on Industrial Engineering and Engineering Management (IEEM), pp. 215–219, Singapore (2017).

Schuetzler, R.M., Grimes, G.M., Giboney, J.S.: An investigation of conversational agent relevance, presence, and engagement. In: AMCIS 2018 Proceedings (2018).

Sharma, V., Goyal, M., Malik. D.: An intelligent behaviour shown by chatbot system. Int. J. New Technol. Res. (IJNTR) **3**, 52–54 (2017)

Artificial Intelligence in Predicting the Spread of Coronavirus to Ensure Healthy Living for All Age Groups

Stitini Oumaima, Kaloun Soulaimane, and Bencharef Omar

Abstract

Ensure healthy lives and promote well being for all at all ages represents the third goal of sustainable development goals. Actually CoronaVirus epidemic temporary named 2019-nCoV, is accelerating and spreading very quickly, and the number of cases of infection from this infectious disease is changing enormously day by day, which threatens many lives and damages the economy. Researchers are turning to artificial intelligence to track the virus as it spreads. This paper presents a systematic review of the literature that analyze the use of Artificial Intelligence algorithms in order to predict the spread of the Coronavirus.

Keywords

Artificial intelligence • Medical imaging • Deep learning • Sustainable development • Convolutional neural network

1 Introduction

Artificial intelligence not only creates new opportunities, but also poses greater challenges to sustainable development in the medical field, so the relationship between artificial intelligence and sustainability in this field is worth studying (Razzak et al. 2018). The deep integration of artificial intelligence technology and medical practices is one of the important characteristics of computerization in this area and more specifically in medical imaging (Yasmin et al. 2013; Wang 2019; Lewis et al. 2019).

Covid-19 has become a pandemic today's the virus that causes COVID-19 is infecting people and spreading easily from person-to-person. Cases have been detected in most countries worldwide and community spread is being detected in a growing number of countries which has pushed most researchers has thought how to react to this pandemic.

Artificial intelligence (AI) algorithms, particularly deep learning, have demonstrated remarkable progress in image-recognition tasks (Maier et al. 2018). Methods ranging from convolutional neural networks to variational auto encoders have found myriad applications in the medical image analysis field (Pesapane et al. 2018), propelling it forward at a rapid pace. Historically, in radiology practice, trained physicians visually assessed medical images for the detection, characterization and monitoring of diseases. AI methods excel at automatically recognizing complex patterns in imaging data and providing quantitative (Shen et al. 2017), rather than qualitative, assessments of radiographic characteristics (Huawei Cloud 1042). This paper provides a systematic review to investigate how artificial intelligence can predict the spread of coronavirus using medical imaging. The goals of this study are to identify newest trends in the use or research of AI algorithms for predicting the evolution of the corona virus. Identify open questions in the use or research of AI algorithms. This paper is organized as follows: Sect. 2 presents a theoretical background Sect. 3 explains the related work, Sect. 4 explains the systematic literature review flow of this study. Section 5 shows the systematic literature review result and before finishing we presents a discussion and then conclusion.

2 Theoretical Background

This section gives an overview of two main research fields related to this article, namely machine learning algorithms and deep learning algorithms.

S. Oumaima (✉) · K. Soulaimane · B. Omar
Cadi Ayyad University, Marraakech, Morocco
e-mail: oumaima.stitini@ced.uca.ma

K. Soulaimane
e-mail: so.kaloun@uca.ac.ma

B. Omar
e-mail: Bencharefomar@gmail.com

Table 1 Table Research questions for SLR

Research Question	
RQ1	How to predict the spread of the coronavirus using artificial intelligence?
RQ2	How artificial intelligence can help to diagnose the virus?
RQ3	Which technique to use to predict the end time of an epidemic or how to predict the end time of an epidemic using deep learning?
RQ4	How artificial intelligence will help identify the risks of infection from this pandemic?
RQ5	How is it done interpreting radiological images using artificial intelligence to obtain intervention recommendations?

Table 2 Search strings for SLR

Search strings	
S1	Artificial intelligence in medical imaging OR artificial intelligence in medical imaging and sustainability.
S2	Machine Learning in Medical Imaging OR deep learning in Medical Imaging AND Machine Learning and deep learning in Medical Imaging.
S3	The use of Artificial Intelligence algorithm in health-care domain to detect diseases from medical imaging AND review paper.
S4	Neural Networks in Medical Imaging Applications AND A Survey

Table 3 Including and excluding criteria

Including Criteria	Excluding Criteria
Paper was published in a journal as scientific article.	Paper is not written in English
Paper proposes a comparison between approaches to check what the best in medical imaging is.	Papers that just illustrate the pandemic or the disease and no relation to computer technology.
Chinese scientific articles because they are the only ones who have shown how can we detect the virus.	

Table 4 Number of papers found in each research databases

Journals				
Research Database	IEEE	ELSEVIER	SPRINGER	OTHER
S1	0	0	1	2
S2	1	2	2	2
S3	0	0	1	3
S4	0	0	0	1
TOTAL	1	2	4	8

2.1 Artificial Intelligence Methods in Medical Imaging

Today there are two levels of artificial intelligence: weak AI and strong AI. Weak AI consists of reproducing a specific task, or "reproducing identically" human intelligence using a computer program. The latter is designed to perform on the activity for which it was programmed, without the possibility of evolving. This level of artificial intelligence is already present in daily radiology, such as voice recognition for automatic dictation of reports, but also optical character recognition (OCR) technology.

The second level of artificial intelligence is capable of "producing" human intelligence, that is to say of analyzing a situation, the different possible solutions depending on the context, and finally to act to meet the needs for which it was designed. In what is called strong intelligence, there are still two sub-fields which are: "Machine Learning" and "Deep Learning" (Suzuk 2017).

2.2 Machine Learning Algorithms

The use of machine learning (ML) has grown rapidly in the field of medical imaging, including computer-aided detection and diagnosis (CADe), radiomics and analysis of medical images, because objects such as lesions and organs in medical images may be too complex to be represented accurately by a simple equation or model (Morra et al. 2019).

One of the most popular uses of ML in computer-aided diagnosis (CAD), including CADe and CADx, and medical image analysis is the classification of objects such as lesions into certain classes (e.g., lesions or non-lesions, and malignant or benign) based on input features (e.g., contrast, area, and circularity) obtained from segmented objects. This class of ML is referred to as ML with feature input or feature based ML.

Machine Learning (ML) and Artificial Intelligence (AI) have progressed rapidly in recent years. Techniques of ML and AI have played important role in medical field like medical image processing, computer-aided diagnosis, image interpretation, image fusion, image registration, image segmentation, image-guided therapy, image retrieval and analysis Techniques of ML extract information from the images and represents information effectively and efficiently (Morra et al. 2019).

The ML facilitate and assist doctors that they can diagnose and predict accurate and faster the risk of diseases and prevent them in time. These techniques enhance the abilities of doctors and researchers to understand that how to analyze the generic variations which will lead to disease.

These techniques composed of conventional algorithms without learning like Support Vector Machine (SVM) (Currie et al. 2019), Neural Network (NN), KNN etc. and deep learning algorithms such as Convolutional Neural Network (CNN) (Pesapane et al. 2018).

Support Vector Machine:

Support vector machine (SVM) is a proven success and a state-of-the-art method in many areas, and a promising machine learning technique.

The experimental results show that this SVM classifier can get 96.56% accuracy which is higher about 3.42% than 92.94% using SVM, and the error recognition rates are close to 100% averagely. (An Improved SVM Classifier for Medical Image Classification) (Currie et al. 2019).

Neural Network

Healthcare and medical imaging applications can be used to deliver even better performance, efficiency and well-focused treatment when neural networks become part of such applications.

Over the past twenty to thirty years clinical applications are habitually utilizing medical imaging in different forms and helping in better disease diagnostic and treatment. The usage of Neural Networks in applications of Medical Imaging opened new doors for researchers, stirring them to excel in this domain (Pesapane et al. 2018).

Deep learning algorithms

Recently, an ML area called deep learning emerged in the computer vision field and became very popular in virtually all fields. It started from an event in late 2012. A deep-learning approach based on a CNN won an overwhelming victory in the best-known worldwide computer vision competition, ImageNet Classification, with the error rate smaller by 11% than that in the 2nd place of 26% (Maier et al. 2018).

Deep learning is one extensively applied techniques that provides state of the aft accuracy. It opened new doors in medical image analysis that have not been before. Applications of deep learning in healthcare covers a broad range of problems ranging from cancer screening and disease monitoring to personalized treatment suggestions. Various sources of data today - radiological imaging (X-Ray, CT and MRI scans), pathology imaging and recently, genomic sequences have brought an immense amount of data at the physicians disposal (Shen et al. 2017).

2.3 Examples

Radiology-based

Thoracic imaging
Lung cancer is one of the most common and deadly tumors. Lung cancer screening can help identify pulmonary nodules, with early detection being lifesaving in many patients. Artificial intelligence (AI) can help in automatically identifying these nodules and categorizing them as benign or malignant.

Abdominal and pelvic imaging

With the rapid growth in medical imaging, especially computed tomography (CT) and magnetic resonance imaging (MRI), more incidental findings, including liver lesions, are identified. AI may aid in characterizing these lesions as benign or malignant and prioritizing follow-up evaluation for patients with these lesions.

Colonoscopy

Colonic polyps that are undetected or misclassified pose a potential risk of colorectal cancer. Although most polyps are initially benign, they can become malignant over time. Hence, early detection and consistent monitoring with robust AI-based tools are critical.

Mammography

Screening mammography is technically challenging to expertly interpret. AI can assist in the interpretation, in part by identifying and characterizing micro calcifications (small deposits of calcium in the breast).

Brain imaging

Brain tumors are characterized by abnormal growth of tissue and can be benign, malignant, primary or metastatic; AI could be used to make diagnostic predictions.

Radiation oncology

Radiation treatment planning can be automated by segmenting tumors for radiation dose optimization. Furthermore, assessing response to treatment by monitoring over time is essential for evaluating the success of radiation therapy efforts. AI is able to perform these assessments, thereby improving accuracy and speed.

Non-radiology-based

Dermatology

Diagnosing skin cancer requires trained dermatologists to visually inspect suspicious areas. With the large variability in sizes, shades and textures, skin lesions are rather challenging to interpret. The massive learning capacity of deep learning algorithms qualifies them to handle such variance and detect characteristics well beyond those considered by humans.

Pathology

The quantification of digital whole-slide images of biopsy samples is vital in the accurate diagnosis of many types of cancers. With the large variation in imaging hardware, slide preparation, magnification and staining techniques, traditional AI methods often require considerable tuning to address this problem. More robust AI is able to more accurately perform mitosis detection, segment histologic primitives (such as nuclei, tubules and epithelium), count events and characterize and classify tissue.

DNA and RNA sequencing

The ever-increasing amount of available sequencing data continues to provide opportunities for utilizing genomic end points in cancer diagnosis and care. AI-based tools are able to identify and extract high-level features correlating somatic point mutations and cancer types as well as predict the effect of mutations on sequence specificities of RNA-binding and DNA-binding proteins.

3 Related Work

This article offers different approaches that we can use in medical imaging to ensure a healthy lifestyle for all that is essential to sustainable development, each approach emits problems and analyzes the main aspect. However, this research shows the steps followed for sorting articles, such as the definition of keywords, inclusion and exclusion criteria and the databases searched. It presented a systematic review of artificial intelligence in medical imaging, classifying the articles according to the field applied and the techniques used.

3.1 Procedure for Systematic Literature Review

In the field of knowledge and especially in Computer Engineering, a SLR has great importance to increase the quality of the research. SLR consists of creating an analytical methodology to identify, select, evaluate and synthesize the main scientific research that allows elaborating a more objective bibliographic review.

4 SLR Flow

This systematic analysis displays a flow diagram summarizing the different steps taken to carry out the SLR, which are:

Stage 1: Definition of the research questions and keywords: at this stage the research questions (Table 1) and the keywords (Table 2) were defined.

Stage 2: Elaboration of search strings and choice of the search databases: at this step the search strings were defined according to the format of the search databases selected. IEEE, Elsevier and British Institute of Radiology, were selected due to their credibility, adequacy to the computing area.

Stage 3: Elaboration of inclusion and exclusion criteria: The inclusion and exclusion criteria for gathering documents are addressed in Table 3.

Stage 4: Search of papers: at this step, we searched the papers using the search strings (Table 2) elaborated based on the research questions and the keywords.

Stage 5 Pre-selection: only the title and abstract of the whole papers retrieved in stage4

were read at this step. Following the inclusion and exclusion criteria, the papers were either selected for full reading or discarded.

Stage 6: Complete reading of selected papers: the papers selected in step 5 were thoroughly read and evaluated for their relevance within the scope of the research. Criteria were defined to evaluate the relevance of papers. For this evaluation, metrics (Table 6) were defined based on the research questions (Table 1), which allowed the classification of the papers.

Stage 7 Classification: Finally, the papers were evaluated and ranked as highly, partially, or not relevant to the established research questions.

5 SLR Results

The Table 4 shows the number of papers found by each string S1, S2, S3 et S4 (Table 2) executed in each databases. The total number of papers per databases corresponds to the sum of all papers found with all search strings as shown in Eq. 1 (Fig. 1)

$$Total = s1 + s2 + s3 + s4 \qquad (1)$$

5.1 Systematic Review Result

Question answers:

RQ1- artificial intelligence has plunged into Big Data in order to analyze its data. Artificial intelligence shows great promise in determining where and how quickly diseases spread. Scientists are increasingly using it to predict the spread of disease. Researchers are also consulting AI to model the movement of people in cities, this vehicle may be linked to the spread of pathogens.

RQ2-the use of natural language processing and machine learning algorithms to peruse information from hundreds of sources for early signs of infectious epidemics. A lot of Chinese researches show that computer vision is the best to detect coronavirus infection (Towards data science 2020).

RQ3-To answer to this question we summarize the major techniques that are described in relation to the field of

computer vision, and with the aim of improving the performance of the algorithm (Maier et al. 2018).

Machine learning can be divided into several different strategies:

- Rote learning.
- Learning by instruction.
- Deductive learning
- Learning by analogy
- Inductive learning.

Computer vision algorithms can now show symptoms of fever, coughing, and respiratory problems on a wide scale at the same time. The Chinese technology developed by "giant Baidu' is an artificial intelligent system that uses cameras fitted with machine vision and infrared sensors to predict the temperatures of individuals in public places. The system can screen up to 200 people per minute and detect their temperature within a range of 0.5 degrees Celsius. The AI flags anyone who has a temperature above 37.3 degrees (Liu et al. 2019).

RQ4-Some apps even give specific instructions, such as how to take precautions, how to do self-isolation for 14 days, and where to find a fever clinic or designated hospital. Self-screening features like this have also been created on smartphone apps and are a helpful tool for early infection detection and care.

RQ5-Artificial intelligence is finding its place in the fight against the coronavirus. AI-assisted medical image reading programs have been tested and launched to help diagnose COVID-19 patients more quickly. A cloud-based coronavirus pneumonia diagnosis service using AI-assisted CT image interpretation was launched by Huawei Cloud, a healthcare business branch of Huawei (Jiang et al. 2007).

6 Selection of Publications Reviewed in This Study

The procedure involved searching and downloading journal papers from different top level electronic journal databases. These electronic journal databases were chosen because they contained published papers on the topic under review. Three types of articles were downloaded and reviewed in this study that are part of artificial intelligence to solve many problems in medical imaging.

6.1 Machine Learning Category

The purpose of this category of papers was to establish what review studies have been carried out by other authors previously as well as give a brief background of the field of recommender systems.

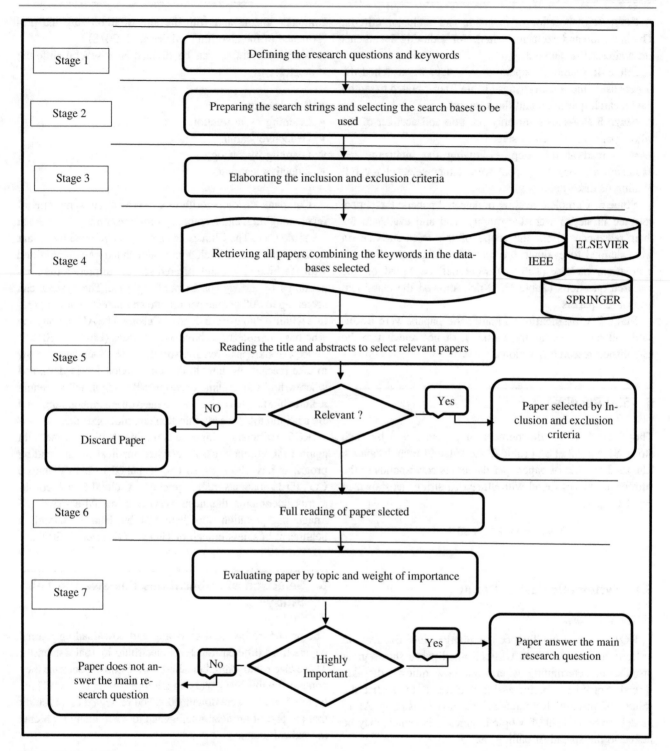

Fig. 1 The flow diagram for systematic literature review

Methodology	Study	Goal	Limitations
Machine Learning	Liu et al. 2019	In this paper, they propose a joint regression and classification method for PD diagnosis upon structural and diffusion magnetic resonance images. The unified multitask model can explore relationships among features, samples, and clinical scores. Four clinical variables of depression, sleep, olfaction, and cognition scores are regressed out, while the classification of PD is also conducted from the multimodal imaging data.	No Limitations

Deep learning category

Methodology	Study	Goal	Limitations
Deep Learning	Currie et al. 2019	In this paper, the area of deep learning in medical imaging is overviewed, including what was changed in machine learning before and after the introduction of deep learning, what is the source of the power of deep learning, two major deep-learning models: a massive-training artificial neural network (MTANN) and a convolutional neural network (CNN), similarities and differences between the two models, and their applications to medical imaging.	Limitations of "deep" CNNs (in ML with image input) include a very high computational cost for training because of the high dimensionality of input data and the required large number of training images.
	Dutta 2020	In this paper, they have highlighted the barriers that are reducing the growth of deep learning in the health care sector and highlighted state-of-the-art applications of deep learning in medical image analysis.	
	Jiang et al. 2007	Computational modeling for medical image analysis has had a significant impact on both clinical applications and scientific research. Recent progress in deep learning has shed new light on medical image analysis by enabling the discovery of morphological and/or textural patterns in images solely from data. Deep learning methods have achieved state-of-the-art performance across different medical applications; however, there is still room for improvement.	An important limitation of typical deep CNNs arises from the fixed architecture of the models themselves
	Lee et al. 2019	In this short paper on deep learning in manipulating medical images. At the same time, they are based on two goals. The first aim is to introduce the field and the theory involved with deep learning. And the second purpose is to provide a description of the sector and, in general, future alternative implementations.	Deep learning is extremely data hungry. This is one of the main limitations that the field is currently facing, and performance grows only logarithmically with the amount of data used.

(continued)

Neural networks category

Methodology	Study	Goal	Limitations
Neural Networks	Huawei Cloud 2020	In this paper, they propose a joint regression and classification method for PD diagnosis upon structural and diffusion magnetic resonance images. The unified multitask model can explore relationships among features, samples, and clinical scores. Four clinical variables of depression, sleep, olfaction, and cognition scores are regressed out, while the classification of PD is also conducted from the multimodal imaging data.	No limitations

7 Discussion

The process of literature review was started with research of leading academic databases (Science Direct, IEEE, ELSEVIER, WILEY, SPRINGER) about the technique used to spread a coronavirus pandemic especially medical imaging in the medical

domain and its practice. The keywords were composed of « medical imaging », « artificial intelligence in medical imaging » and « machine learning medical imaging ».

Initial refinement was made considering two criteria:

- Quality of journals (by evaluating impact factor and citation rates).
- The article found should use artificial intelligence algorithm in medical imaging.

Then, the studies were retrieved and refined by title and abstract basis. In total, 15 papers were retrieved. It was refined to 25 papers by title basis elimination and by abstract basis elimination. In the next phase, 15 papers were found meeting the following criteria of quality: reliability of the source, integrity in the content and providing applicable studies. In the final phase, finding were synthesized and reported.

8 Conclusion

In this short research review we have tried to enlighten the role of machine learning, deep learning and neural networks in the advancement of medical imaging. We have discussed some of the existing techniques that make use of neural networks in order to deliver better medical imaging services. Understanding the principles and applications of AI, ML, and DL in medical imaging will facilitate assimilation and expedite advantages to practice.

References

Currie, G., Hawk, K.E., Vial, A.: Machine learning and deep learning in medical imaging: Intelligent imaging. Elsevier, 1–11 (2019)

Dutta, S.: A 2020 guide to deep learning for medical imaging and the healthcare industry. (2020)

Huawei Cloud, https://blog.csdn.net/devcloud/article/details/104263141, last accessed 2020/03/26

Imaging technology news, https://www.itnonline.com/article/deployment-health-it-china's-fight-against-covid-19-epidemic, last accessed 2020/03/25

Jiang, Y., Li, Z., Zhang, L., Sun, P.: An Improved SVM Classifier for Medical Image Classification. Springer, 1–2 (2007)

Lee, L.I., Kanthasamy, S., Ayyalaraju, R.S., Ganatra, R.: The current state of artificial intelligence in medical imaging and nuclear medicine. Brit. Inst. Radiol. 1–8 (2019)

Lewis, S.J., Gandomkar, Z., Brennan, P.C.: Artificial intelligence in medical imaging practice: Looking to the future. J. Med. Radiat. Sci. 1–4 (2019)

Liu, X., Faes, L., Kale, A.U., Wagner, S.K., Fu, D.J., Fu, D.J.: A comparison of deep learning performance against health-care professionals in detecting diseases from medical imaging: A systematic review and meta-analysis. Elsevier, 1–24 (2019)

Maier, A., Syben, C., Lasser, T., Riess, C.: A gentle introduction to deep learning in medical image processing. Elsevier, 1–16(2018)

Morra, L., Delsanto, S., Correale, L.: Artificial intelligence in medical imaging: From theory to clinical oractice. CRC, 1–58 (2019)

Pesapane, F., Codari, M., Sardanelli, F.: Artificial intelligence in medical imaging: Threat or opportunity? Radiologists again at the forefront of innovation in medicine.Springer, 1–10 (2018)

Razzak, M.I., Naz, S., Zaib, A.: Deep Learning for Medical Image Processing: Overview, Challenges and the Future. Springer, 1–28 (2018)

Shen, D., Wu, G., Suk, H.: Deep learning in medical image analysis. Nat. Cent. Biotechnol. Inf. 1–28 (2017)

Suzuk, K.: Overview of deep learning in medical imaging. Springer, 4–20 (2017)

Towards data science, https://towardsdatascience.com/using-kalman-filter-to-predict-corona-virus-spread-72d91b74cc8, last accessed 2020/03/21

Towardsdatascience, https://towardsdatascience.com/analyzing-coronavirus-covid-19-data-using-pandas-and-plotly-2e34fe2c4edc, last accessed 2020/03/23

Wang, Q., Yinghuan, S., Dinggang, S.: Machine Learning in Medical Imaging. IEEE, 1–2 (2019)

Yasmin, M., Sharif, M., Mohsin, S.: Neural networks in medical imaging applications: A survey. World Appl. Sci. J.1–13 (2013)

Sustainability of Artificial Intelligence and Deep Learning Algorithms for Medical Image Classification: Case of Cancer Pathology

Dahdouh Yousra, Anouar Boudhir Abdelhakim, and Ben Ahmed Mohamed

Abstract

In the context of Artificial Intelligence's Sustainability (AI), deep learning has sparked tremendous global interest in recent years. Deep Learning has been widely adopted in image recognition, speech recognition, and natural language processing, but is only beginning to impact on health care. In pathology, artificial intelligence and especially deep learning algorithms has been applied to pathology image analysis tasks such as tumor region identification, prognosis prediction, and detection of cancer areas and classification. This chapter presents different deep learning methods such as convolutional networks (ConvNets) for automatic classification of medical images—cancer pathology. We have proposed different models and different architectures. The first proposed model consists of feature extraction and classification with simple CNN; the second one is composed of two parts: we have used CNN as feature extractor by removing the last classification layers and we have passed the features to Random Forest; and the last one is by using transfer learning–Fine-Tuning–pre-trained CNN "DenseNet201" as classifier. Finally, we have evaluated our models using three metrics: accuracy, precision, and F1 score.

Keywords

Medical domain • Pathological images • Deep learning • Convolutional neural networks • Classification • Cancer • Random forest • DenseNet

D. Yousra (✉) · A. B. Abdelhakim · B. A. Mohamed
Faculty of Sciences and Techniques, Abdelmalek Essaâdi University, Tangier, Morocco
e-mail: dahdouhyousra@gmail.com

A. B. Abdelhakim
e-mail: aboudhir@uae.ac.ma

B. A. Mohamed
e-mail: mbenahmed@uae.ac.ma

1 Introduction

The emergence of Artificial Intelligence (AI) and its progressively wider impact on many sectors across the society requires an assessment of its effect on sustainable development, where it is a wide-ranging branch of computer science concerned with building smart machines capable of performing tasks that typically require human intelligence. As in other domains, artificial intelligence is becoming increasingly important in medicine and health care. In particular, deep learning has been primarily applied to medical imaging analysis, in which deep learning systems have shown robust diagnostic performance in detecting various medical conditions, accurately diagnosing diseases, and predicting patient prognoses (Kumar and Rao 2018).

The unwanted growth of cells causing cancer is a serious threat to humans; the recent statistics shows that millions of people all over the world suffer from various types of cancer (Nahid et al. 2018). Cancer identification is largely depending on digital biomedical image analysis such as histopathological images by doctors and physicians. As it is an Automated classification of high-resolution, histopathology slides are one of the most popular yet challenging problems in medical image analysis. With the fast development of computer and medical imaging technologies, image understanding by computer programs has become an attractive and active topic in the machine learning field and in application-specific studies (Cruzroa et al. 2011). Toward building an intelligent Computer-Aided Diagnosis (CAD) system, fast and accurate annotation or the grading of medical images has become a key technique in CAD systems in most medical fields.

Deep learning has produced a set of image analysis techniques, which automatically extract relevant features, transforming the field of computer vision, has done remarkably well in image classification and processing tasks, and offers fascinating perspectives to address medical images, mainly owing to convolutional neural networks (CNN) (LeCun 2015). A convolutional neural network is a

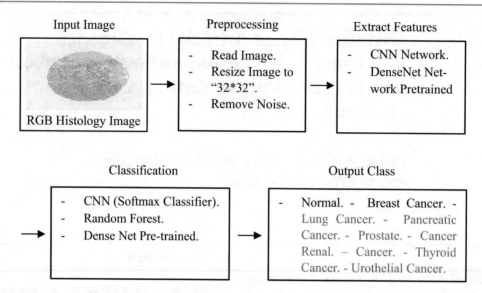

Fig. 1 Histopatholgical image classification model architecture

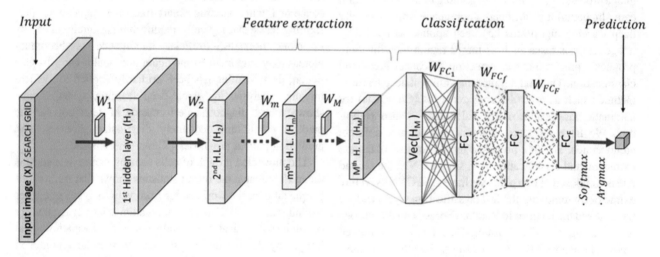

Fig. 2 Typical CNN architecture. (Arun and Katiyar 2013)

Fig. 3 The proposed CNN
model

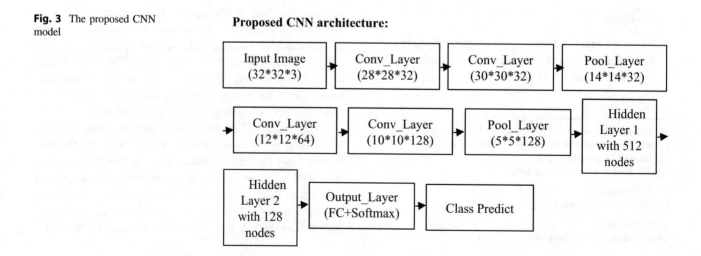

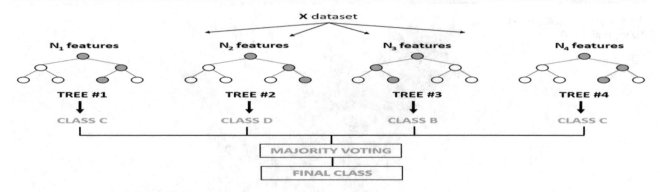

Fig. 4 Random forest architecture. (Breiman 2001)

Fig. 5 The CNN-RF architecture

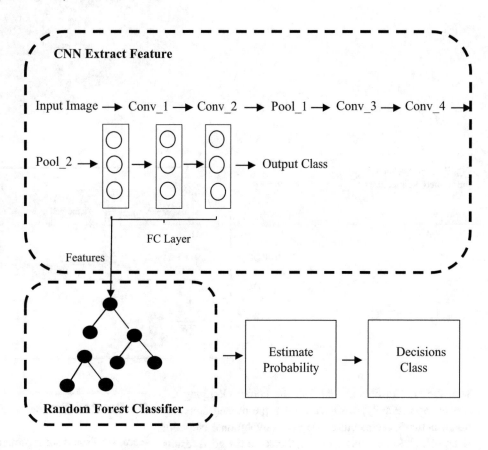

type of artificial neural network usually designed to extract features from data and to classify given high-dimensional data. CNN is designed specifically to reorganize two-dimensional shapes with a high degree of invariance to translation, scaling, skewing, and other forms of distortion. The structure includes feature extraction, feature mapping, and subsampling layers. Convolutional neural networks (CNNs) use a data-driven approach to automatically learn feature representations for images, achieving super-human performance on benchmark image classification datasets, those networks are the current state-of-the-art architecture for medical image analysis.

In this paper, we have treated the classification task using pathological medical images in the context of deep learning and computer vision; we have described accurate and efficient algorithms for this challenging problem, where we have proposed different models and their different architectures. First, we built a model that consists of feature extraction and the classification with simple CNN (Multi-Label Classification) as an initial experience, and for the second model we have used CNN for extracting image features from preprocessed visual inputs and then passed those features to random forest—a shallow classifier, which is used to classify the extracted features (CNN outputs)—

Fig. 6 DenseNet with 5 layers (Huang et al. 2017)

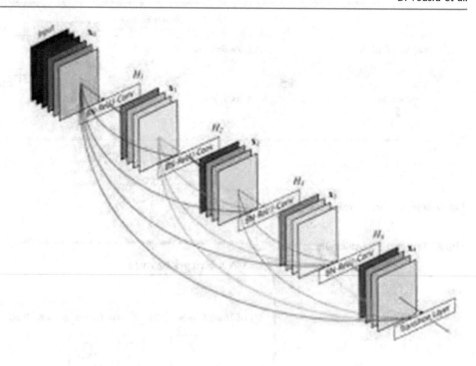

Fig. 7 DensNet model pre-trained architecture

Image ➤ 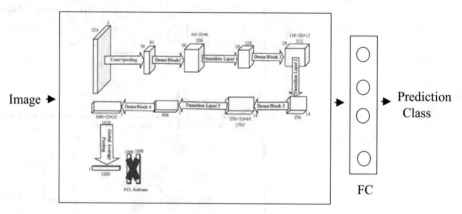 ➤ Prediction Class

FC

with the goal to obtain a label in the output clarifying the cancer type. Finally, we have used a pre-trained convolutional network architecture "Dense Convolutional Network (DenseNet)" to treat this problem, thanks to the good results that DenseNet shows in different classification problems.

This document is organized as follows: in Sect. 2, we discuss the related works, in Sect. 3, we give a detailed description of the proposed models for treating the pathological image classification challenge. And in Sect. 4, we present our experiment results, including the introduction to the dataset used, and the experiment settings; we have presented the experimental evaluation of the algorithms in the same section. Finally, in Sect. 5, we finished our paper by a conclusion and point to future challenges.

2 Related Works

Here, we discuss some previous works, including state-of-the-art methods, aiming to automate the diagnostic procedure in the context of cancer histopathology. We also discuss various works that have been carried out in the context of transferability of the pre-trained deep models (CNNs). Medical image classification and especially Histopathological image classification have become more and more important for disease diagnosis, treatment (Wei et al. 2017; Wei et al. 2017; Su 2016), clinical practice, and the discovery of diseases such as cancer. In previous, most existing papers on pathological image classification have used manually

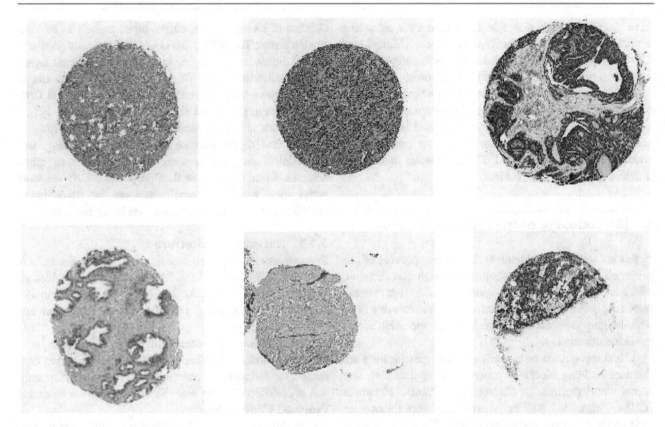

Fig. 8 Example of images used

Table 1 Results of our proposed CNN Model

Method	F1 Score	Precision	Accuracy	Optimizer
CNN	0.36	0.39	0.40	SGD
	0.48	0.50	0.51	RMSprop
	0.50	**0.52**	**0.55**	**Adam**

designed features are extracted to classify the tumor/normal tissue, such as fractal features (Huang and Lee 2009) and textural features (Sertel et al. 2009). But, due to the complexity of medical images, traditional methods have been unable to meet actual application needs.

In recent years, the development of deep learning theory has provided a technical approach for solving medical image classification tasks owing to breakthrough of CNN in image recognition. Many current researches use CNN as a classifier (Hou et al. 2016; Wang and Khosla 2016). In Ref. (Li et al. 2014), the authors designed a customized Convolutional Neural Networks (CNN) with a shallow convolution layer to classify lung image patches with interstitial lung disease (ILD). For avoid the over-fitting problem, they applied different technologies like intensive dropout and input distortion.

Byungjae Lee and Kyunghyun Paeng proposed in (Lee and Paeng 2018) a robust and effective method for metastasis detection and pN-stage classification in breast cancer from multiple gigapixel pathology images using CNN-based metastasis detection and random forest-based lymph node classification. And (Xu et al. 2017) dealt with the image classification problem in histopathology domain with the task as a transfer learning. They have used deep convolutional activation features trained with ImageNet knowledge and applied a CNN model "AlexNet architecture" to the extraction of features from brain tumor and colon cancer digital histopathology datasets. In the same context was used more advanced form of transfer learning, called fine-tuning, to adapt a pre-trained CNN to a different dataset. Fine-tuning has been applied to a variety of different medical imaging modality or disease-specific classification tasks using classification forests and an augmented set of image miniatures of cardiac MRI (Margeta et al. 2016).

This article describes an accurate and efficient algorithms for this challenging problem, and aims to present different convolutional neural networks to classify tissue images. We have proposed to use a simple CNN model for extracting features and classifying the type of cancer (Multi-Label Classification). Or CNN to extract images features and then random forest is used to classify those ones. And other model pre-trained CNN "DenseNet" to gain time and extract image's features and evaluate the histopathological images classification problem. The systems proposed in this paper shows superior performance for classifying cancer types, where we have tried to build different system to improve different results for this problem.

3 Methodology of the Proposed Models

In this section, we will highlight the key components of our proposed model for histopathological image classification, which are based on computer vision; the last one has become increasingly important and effective in recent years due to its wide-ranging applications especially in the health and medicine domains.

Visual recognition tasks, such as image classification are the core building blocks of many of the applications, and recent developments in Convolutional Neural Networks (CNNs) have led to the outstanding performance in state-of-the-art visual recognition tasks. CNN usually contains convolutional layers, pooling layers, one or more fully connected layers, and a softmax layer. The convolutional layers combined with the pooling layers are used for extracting features. The softmax layer is regarded as the classifier (Khan et al. 2018).

The main design principles of our model are as follows: (1) To perform the image preprocessing, such as read images, resize images to 32*32, and remove noise; (2) For extracting prominent features from images, we have used the CNN or the pre-trained DensNet network; (3) Choosing the proper activation function (Nair and Hinton 2010); (4) The initial weights are also important. If the initial weights are too small, then the deep network would not be able to learn, and if they are too large, then the initial weights would undergo divergence (Mishkinand and Matas 2015]; (5) Using dropout to reduce overfitting and local response normalization (Krizhevsky et al. 2017) to reduce error rates is equally important; (6) Choose the performance classifier to classify the type of cancer (Multi-Label Classification); and (7) Choose the proper learning rate (Fig. 1).

3.1 Convolutional Neural Network

Convolutional Neural Networks (CNN) have completely dominated the machine vision space in recent years. CNN consists of an input layer, output layer, as well as multiple hidden layers. The hidden layers of CNN typically consist of convolutional layers, pooling layers, fully connected layers, and normalization layers (ReLU). Additional layers can be used for more complex models. Examples of typical CNN can be seen in (Arun and Katiyar 2013) (Fig. 2).

CNN has also been incorporated into medical imaging analysis and digital microscopy tissue images, diagnostic and clinical prognostics, such as classification of tumor samples. In this paper, we thus proposed a CNN network architecture which is specially adapted for tissue image classification and turned to avoid overfitting problem.

3.1.1 Network Architecture

Preprocessing The original images are resized from $3 \times 300 \times 300$ to $3 \times 32 \times 32$. Then RGB pixel values are extracted and $3 \times 32 \times 32$ data is formed for each image. Then the whole dataset is normalized using zero mean and unit standard deviation.

Model Selection Preprocessed images are given as input to the CNN model for training. In the experiments we conducted, the following architecture gives the best performance and is considered as the best model for the given dataset.

Proposed CNN architecture

The model is composed of an input layer, followed by four layers of convolution and two layers of MaxPooling and three layers of a fully connected output layer to extract features. The last layer of fully connected is used for classification. The input image is of size 32 * 32; the image passes first to the first convolution layer. This layer is composed of 32 filters of size 3 * 3; the activation function ReLU is used; this activation function forces the neurons to return positive values; after this convolution 32 feature maps of size 30* 30 will be created. Then, the 32 feature maps that are obtained are given as input to the second convolution layer which is also made up of 32 filters. The ReLU activation function is applied to the convolutional layers. MaxPooling is applied afterwards to reduce the image size, and parameter pooling layer window size is fixed as 2×2. At the exit of this layer, we will have 32 feature maps of size 14 * 14 (Fig. 3).

We repeat the same thing with convolution layers three and four (layer three is composed of 64 filters and layer four is composed of 128 filters); the activation function ReLU is always applied to each convolution. The Max Pooling layer is applied after layers of convolutions four. At the end of the last MaxPooling layer, we will have 128 feature maps of size 5 * 5. After these four convolutional layers, we use a neural network composed of three fully connected layers. The first layer has 512 neurons and the second layer has 128 neurons, where the activation function used is ReLU, and the third layer is a softmax layer which calculates the probability distribution of the 8 classes (number of classes in the dataset used).

3.1.2 Training

In order to train and evaluate our CNN model, we implement a cross-validation technique Train_Test Split on the whole dataset, where 80% of the images are used for training, and 20% are used for testing.

In model training, 80% of the images are separated from the training set for the actual training, and the remaining 20% are used for back-propagation validation. The performance evaluation criteria are accuracy, precision, and F1 score.

We trained our model conducting several experiments for different optimizers, such as RMSprop (Ruder 2017), Adam (Ruder 2017), and SGD, learning rates, and other hyperparameters, where our results are summarized in Table 1.

3.2 Convolutional Neural Network with Random Forest

Random Forest (RF) (Breiman 2001) is a supervised statistical classification method that has been successfully applied to a variety of problems. Some examples include image processing (Lepetit and Fua 2006), pose recognition (Fanelli et al. 2011), and medical diagnosis (Malof et al. 2016).

Random Forest (Breiman 2001) is a general term for ensemble methods using tree-type classifiers {h (x, βk), k = 1...} for classification and regression, where the {βk} are independent identically distributed random vectors and x is an input pattern. In training, the Random Forest algorithm creates multiple CART-like trees each trained on a bootstrapped sample of the original training data, and searches only across a randomly selected subset of the input variables to determine a split (for each node).

Each tree is grown as follows: sample N (the number of cases in the training set) cases at random with replacement from the original data. This sample will be the training set for growing the tree; at each node, m predictors are randomly selected out of the M input variables (m < M) and the best split on these m predictors is used to split the node. Each tree is grown to the largest extent possible; there is no pruning. For classification, each tree in the Random Forest casts a unit vote for the most popular class at input x. The output of the classifier is determined by a majority vote of the trees (Fig. 4).

At the end, we have fusioned the simple CNN and RF classifier for obtaining good results of multi-label pathological image classification, where we have passed features (CNN last FC layer result) as input to the random forest classifier for obtaining the cancer type.

3.2.1 Network Architecture

The proposed approach consists of preprocessing, feature extractor training based on CNN, and classification based on Random Forest (Fig. 5).

Preprocessing: It is a common practice to perform several simple preprocessing steps before attempting to generate features from data. In this work, images were preprocessed by resizing them from $3 \times 300 \times 300$ to $3 \times 32 \times 32$. Then, RGB pixel values are extracted and $3 \times 32 \times 32$ data is formed for each image. Then the whole dataset is normalized using zero mean and unit standard deviation.

Feature extractor training based on CNN: The CNN network is composed of 11 layers, counting the input and output layers, all of which contain trainable parameters (weights). The input layer is a matrix with size 32 by 32 raw pixel images. Layer C1 is a convolutional layer with 32 feature maps. Layer C2 is a convolutional layer with 32 feature maps. Layer C3 is a convolutional layer with 64 feature maps. Layer C4 is a convolutional layer with 128 feature maps. Layer S2 is a sub-sampling layer with size 2*2. The layer F5 is fully connected with 512 nodes, implementing a general-purpose classifier over the features extracted by the earlier layers. Classification based on random forest: Once the features are extracted from the CNN's fully connected layer, it will be passed to a Random Forest classifier for predicting a label for the input patterns.

The values from the trained CNN network were used as a new feature vector to represent each input pattern, and were fed into the Random Forest for learning and testing. Once the Random Forest classifier is well trained, it performs the recognition task and makes new decisions on testing images with such automatically extracted features. In the experiments, the Random Forests was trained with default parameters.

3.2.2 Training

Many classifiers were applied on the pathological image data such as random forest, LinearSVC. But in our case, we find Random Forest works efficiently on large datasets, carries a very low risk of overfitting, and is a robust classifier for noisy data(medical). Generating an ensemble of trees using random vectors, which control the growth of each tree in the ensemble, significantly increases the classification accuracy.

In order to improve the performance of the classifier in our experiments, we started from 100 trees. The optimal number of trees is chosen by considering the error rate. By setting the number of trees to 210, the error rate is low, almost close to the minimum error rate, and fewer number of

Table 2 Results of our CNN-RF Model

Method	F1 Score	Precision	Accuracy
CNN-SVC	0.45	0.46	0.48
CNN-RF	0.45	0.48	0.50
CNN-RF	**0.66**	**0.68**	**0.70**

trees reduce the computational burden; so, the classifier performance is faster. The number of randomly selected predictors is set to 8. The performance evaluation criteria are accuracy, precision, and F1 score. Our results are summarized in Table 2.

3.3 Densely Connected Convolutional Networks

DenseNet, Dense Convolutional Network, was firstly introduced in 2016 by Huang et al. (Kingma and Ba 2015). It is another very deep neural network with dense blocks. The name of the network came from the fact of its dense connectivity to ensure maximum information flow between layers in the network. DenseNet connects all layers directly with each other. To preserve the feed-forward nature, each layer obtains additional inputs from all preceding layers and passes on its own feature maps to all subsequent layers. The layer is designed in such a way that the activation maps of all preceding layers are considered as separate inputs whereas its own activation maps are passed on as inputs to all subsequent layers. DenseNet is mainly composed of a convolution layer, a Dense block, a transition layer, and a classifier after the global average pooling at the input end (Fig. 6).

DenseNet has many benefits that evoke interest among researchers. It can moderate the vanishing-gradient issue, fortify feature propagation, promote feature reutilization, and reduce the number of parameters. In this case, we proposed a pre-trained DenseNet model, specifically DenseNet201 that consists of 201 layers and was trained using the ImageNet dataset, then it will be followed by a fully connected layer of our own.

3.3.1 Network Architecture

The transfer learning technique is commonly used to elevate the problems of the large amount of data, and powerful computer hardware (GPUs) are required in order to train CNN from scratch, where a CNN model is trained on a huge dataset then the features learned from this model are transferred to a model at hand. In the transfer learning technique, the fully connected layer of the model is removed and the remaining layers of the architecture are used as a feature extractor for the new task. Therefore, only the fully connected layers of the new model are trained in the current model.

In our case, we have used the pre-trained network DenseNet201; it more efficient, accurate, and deeper. The architecture starts with convolution and pooling layers followed by four dense blocks and transition layers. After a final dense block, a global average pooling is included followed by a softmax classifier (Fig. 7).

First, we have resized all the images to 224 × 224x3 RGB color space. After that, we initialized weights of different layers of our proposed network by using a pre-trained model on ImageNet. Then, we employed that last layer fine-tuning on our histopathology cancer image dataset. Therefore, the ImageNet pre-trained weights were preserved while the last fully connected layer was updated continuously. The first convolutional layer of the network is then un-frozen, and the entire network is fine-tuned on the histopathological training data for obtaining the result of the classification.

The advantage of DenseNet is feature concatenation that helps us to learn the features in any stage without the need to compress them and the ability to control and manipulate those features. This technique allows to reduce the training process complexity and eliminated the over fitting problem.

3.3.2 Training

DenseNet-201 pre-trained CNN architectures are employed to classify histopathology images into 8 classes. The implementation was done using Keras and TensorFlow libraries. In order to train and evaluate our model, we have used Adam optimizer to minimize the loss function. DenseNet training was pretty fast to take, and the performance evaluation criteria are accuracy, precision, and F1 score. Our results are summarized in Table 3.

4 Experiments and Evaluation

4.1 Dataset

The HPA (http://www.proteinatlas.org/) dataset was used for training and testing classifiers. It contains 7053 images including 3297 normal images and 3756 cancer images from Seven types of cancer, namely breast cancer, lung cancer, pancreatic cancer, prostate cancer, renal cancer, thyroid cancer, and urothelial cancer (Fig. 8).

Table 3 Results of our DenseNet Model

Method	F1 Score	Precision	Accuracy	Optimizer
DenseNet201 Pre-trained	0.44	0.54	0.58	Adam
	0.58	0.62	0.65	RMSprop
	0.62	**0.65**	**0.78**	**Adam**

Table 4 Results of the performance models

Method	F1 Score	Precision	Accuracy
CNN	0.50	0.52	0.55
CNN-RF	0.66	0.68	0.70
DENSENET201	**0.62**	**0.65**	**0.78**

The performance evaluation of the proposed models is a very important task, where we have used in this section different performance measures, accuracy, F1-score, and precision.

Precision: the relation between the true matches (True Positives) among all the re-identified matches (True Positives and False Positives).

$$Precision = \frac{True\ Positives}{(True\ Positives\ + False\ Positives)} \quad (1)$$

Recall: the relation between all the true matches (True Positives) among all the real matches (True Positives and False Negatives).

$$Recall = \frac{TruePositives}{(TruePositives + FalseNegatives)} \quad (2)$$

F1 score: it represents the harmonic mean between precision and recall.

$$F1\ Score = 2 \times \frac{(Recall \times Precision)}{(Recall + Precision)} \quad (3)$$

Accuracy it is defined as the number of times that the model predicts the correct answer for inputs.

$$Accuracy = \frac{TP + TN}{TP + TN + FP + FN} \quad (4)$$

We trained our models conducting several experiments for different optimizers, learning rates, and other hyperparameters (Table 4). The following table shows the results obtained on the test set.

5 Conclusion

AI is now increasingly ubiquitous in different fields especially the Medical field. It has the potential to revolutionize the way we discover, learn, live, communicate, and work. It has tremendous potential for the society. The intelligence of machines and robotics with deep learning capabilities have created profound disrupting and enabling impacts on business, governments, and Medicine. They are also influencing the larger trends in global sustainability.

In this paper, we proposed different CNN models for medical image classification in the context of AI's sustainability. The first model is composed of four convolution layers, two of MaxPooling, and three fully connected layers, which helps to extract features; the last fully connected layer is used for classification. The second one is composed of two parts; we have used CNN as a feature extractor by removing the classification layers and we have passed the output to train Random Forest. Finally, in the last proposed model we have used Fine-Tuning pre-trained CNN "DenseNet201" as classifier. Through various experiments, results demonstrate that the proposed last model DenseNet achieves high performance in histopathology image classification, where it achieves about 78% of accuracy.

References

Arun, P., Katiyar, S.: A CNN based Hybrid approach towards automatic age registration. Geodesy and Cartography, **62**1 (2013)

Breiman, L.: Random forests. Mach. Learn. **45**(1), 5–32 (2001)

Cruzroa, A., Caicedo, J.C., Gonzalez, F.A.: Visual pattern mining in histology image collections using bag of features. Artif. Intell. Med. **52**(2), 91–106 (2011)

Fanelli, G., Gall, J., Van Gool, L.: "Real time head pose estimation with random regression forests," IEEE Conf. Comput. Vis. Pattern Recognit. pp. 617–624 (2011)

Hou, L., Samaras, D., Kurc, T.M., Gao, Y., Davis, J.E., Saltz, J.H.: Patch-based convolutional neural network for whole slide tissue image classification. In: CVPR, (2016)

Huang, G., Liu, Z., Van Der Maaten, L., Weinberger, K.Q.: Densely connected convolutional networks. In: Proceedings of the IEEE conference on computer vision and pattern recognition, pp. 4700–4708, (2017)

Huang, P.W., Lee, C.H.: Automatic classification for pathological prostate images based on fractal analysis. IEEE Trans. Pattern Anal. Mach. Intell. **28**, 1037–1050 (2009)

Khan, S., Rahmani, H., Shah, S.A.A., Bennamoun, M.: A Guide to Convolutional Neural Networks for Computer Vision, Morgan & Claypool publishers (2018)

Kingma, D., Ba, J.L.: "Adam: a Method for Stochastic Optimization". International Conference on Learning Representations, 1–13 (2015)

Krizhevsky, A., Sutskever, I., Hinton, G.E.: "Imagenet classification with deep convolutional neural networks," Communications of the ACM, **606**, pp. 1097–1105

Kumar, K., Rao, A.C.S.: Breast cancer classification of image using convolutional neural network. In: 2018 4th International Conference on Recent Advances in Information Technology (RAIT). IEEE, (2018) p. 1–6

LeCun, Y., Bengio, Y., Hinton, G.: Deep learning. Nature **521**(7553), 436–444 (2015)

Lee, B., Paeng, K.: A robust and effective approach towards accurate metastasis detection and pn-stage classification in breast cancer. arXiv preprint arXiv:1805.12067, 2018

Lepetit, V., Fua, P.: Keypoint recognition using randomized trees. IEEE Trans. Pattern Anal. Mach, Intell (2006)

Li, Q., Cai, W., Wang, X., Zhou, Y., Feng, D.D., Chen, M.: "Medical image classification with convolutional neural network," In: Proceedings of 13th International Conference on Control Automation Robotics & Vision (ICARCV), IEEE, pp. 844–848, Singapore, December 2014

Malof, J.M. Collins, L.M., Bradbul, K., Newell, R.G.: A deep convolutional neural network and random forest classifier for solar photovoltaic array detection in aerial imagery, pp. 20–23 (2016) Burmingham, UK

Margeta, J., Criminisi, A., Lozoya, R.C., Lee, D., Ayache, N.: Fine-tuned convolutional neural nets for cardiac MRI acquisition plane recognition. Comput. Methods Biomech. (2016). https://doi.org/10.1080/21681163.2015.1061448

Mishkinand, D., Matas, J.: All you need is a good in it," arXiv:1511.06422http:// arxiv.org/abs/1511.06422"

Nahid, A.A., Mehrabi, M.A., Kong, Y.: Histopathological breast cancer image classification by deep neural network techniques guided by local clustering, Published 7 March 2018

Nair, V., Hinton, G.E.: "Rectified linear units improve restricted boltzmann machines," In: Proceedings of International Conference on Machine Learning, pp. 807–814. Haifa, Israel (2010)

Ruder, S.: An overview of gradient descent optimization algorithms, arXiv:1609.04747v2 [cs. LG] 15 Jun 2017

Sertel, O., Kong, J., Catalyurek, U.V., Lozanski, G., Saltz, J.H., Gurcan, M.N.: Histopathological image analysis using model-based intermediate representations and color texture: Follicular lymphoma grading. J. Signal Process. Syst. **55**, 169 (2009)

Su, R., Zhang, C., Pham, T.D., Davey, R., Bischof, L., Vallotton, P., Sun, C.: Detection of tubule boundaries based on circular shortest path and polar-transformation of arbitrary shapes[J]. J. Microsc. **264** (2), 127–42 (2016)

Wang, D., Khosla, A., Gargeya, R., Irshad, H., Beck, A.H.: Deep learning for identifying metastatic breast cancer. arXiv preprint arXiv:1606.05718, 2016

Wei L, Xing P, Tang J, Zou Q. PhosPred-RF: A novel sequence-based predictor for phosphorylation sites using sequential information only[J]. IEEE Trans. Nano Biosci. (2017)

Wei, L., Tang, J., Zou, Q.: Local-DPP: An improved DNAbinding protein prediction method by exploring local evolutionary information[J]. Inf. Sci. **384**, 135–44 (2017)

Xu, Y., Jia, Z., Wang, L.B., Ai, Y., Zhang, F., Lai, M., Eric, I., Chang, C.: Large scale tissue histopathology image classification, segmentation, and visualization via deep convolutional activation featuresy. BMC Bioinformatics, **18**,281 (2017)

An Intelligent Strategy for Developing Scientific Learning Skills

Okacha Diyer, Naceur Achtaich, and Khalid Najib

Abstract

The enormous evolution and the offers proposed by technology and scientific researches in several fields are relevant to the requirements of development. So, education, like other sectors, exploits this tendency in various actions. The objective of this technological development is to improve the quality of learning, namely knowledge reception and learning assessment operations. In our research, we are interested in evaluating learning skills which are dedicated to scientific and technological disciplines through an intelligent system. In fact, these disciplines are rich in the five skills: Appropriate, Analyze and Reason, Achieve, Validate and Communicate. Through well-adapted teaching situations, students can acquire interesting cognitive skills to challenge future needs. The proposed evaluation strategy in our research helps the teacher make good decisions in an appropriate time to reduce the difficulties encountered. This approach also allows students to acquire knowledge according to their learning ability. The evaluation operation is implicitly done and each pupil progresses in his learning following educational activities according to his learning level. This strategy for evaluating learning skills allows students to apply all the skills mentioned above, in order to be able to face the different situations in their learning process. The intelligent system also allows a very precise summative evaluation of each student taking into consideration each stage of his performance. Therefore, the presence of a formative, judicious and a motivating evaluation is necessary.

Keywords

Innovative assessment approaches • Learning skills evaluation • Digital assistance systems • Teaching performance

1 Introduction

Several researchers have focused on learning skills which have significantly been enriched and evolved (Baafi et al. 2017; Bandaranaike 2018; Sullivan and Bandaranaike 2017). The main objective is to follow the immense progress that exists in the learner's environment. They treated the improved and expected skills in this great technological revolution. The transmission of skills in the educational field takes place through different pedagogical approaches. Researchers have been interested in specifying the different skills to be developed (Ceylan et al. 2018; Harshvardhan 2014; Jayaprakash and Karthik 2015; Nuray and Bülent 2018). In their works, they argued that their choices are driven by the importance of these skills in the development of the learning process. The interaction between advanced didactic means (software) and learners is of great importance for the success of an educational objective (Arena et al. 2018; Ganeshan et al. 2018; Tvenge and Ogorodnyk 2018). This research focused on the evaluation of the relevant means for a judicious contribution of technological tools and procedures in the learning operation. Researchers who have carried out work on competence assessments have been able to characterize them by measurable quantities so that they can be easily evaluated (Abrahams et al. 2019; Kennedy et al. 2014). Thanks to the amazing development in the world of scientific and technological research, artificial intelligence has undergone a huge evolution in terms of software and technological tools. Thus, different

O. Diyer (✉) · N. Achtaich
Faculty of Sciences Ben M'Sik, University Hassan II, Casablanca, Morocco
e-mail: odiyer@yahoo.fr

N. Achtaich
e-mail: nachtaich@gmail.com

K. Najib
Superior National School of Mines, Rabat, Morocco
e-mail: najibkhalid@gmail.com

sectors have considerably improved, and they offer more efficient services than before. Researches that are conducted, using artificial intelligence in the educational field (Popenici and Kerr 2017; Tuomi 2018), focus on the learners' personalized and dynamic learning paths. The enormous development in the fields of scientific research, computer programming, technological tools, didactic means and pedagogical procedures allowed the learners to obtain a qualified teaching which answers the needs of each one (Ganeshan et al. 2018; Tvenge and Ogorodnyk 2018; Popenici and Kerr 2017; Gottfredsonk and Mosher 2011; Karsenti 2019; Mäkiö et al. 2016). Recent researches (Chattopadhyay et al. 2018; Gerardo et al. 2018; Ogunlade and Bello 2019) are interested in the use of artificial intelligence in the student's evaluation operation at the end of the session. This operation makes it possible to reduce the difficult situations encountered by learners in the appropriate time. This allows equity between learners in that they acquire the necessary skills and allow everyone to progress according to their abilities. In our research, we were interested in dynamic evaluation through artificial intelligence to allow the teacher to instantly adjust the students learning. As a part of improving assessment in the educational system, we proposed skills assessment strategy using an intelligent system that helps the teacher make good decisions. The researches which evoked the evaluation of different skills (Brundiers and Wiek 2017; Sudana et al. 2019) focused on approaches to carry out the assessment of some skills. We cite our research (Diyer et al. 2020), which aims at using artificial intelligence in learning skills evaluation approach for scientific and technological fields. We present an extension of this research by adopting an algorithm to determine the final mark of a student. This paper is organized as follows. In Sect. 2, we present the research method that we have adopted in which we characterize the five learning skills to be evaluated. Then, we describe the algorithm approach by focusing on three levels of skills acquisition. In Sect. 3, we give the implicit notation algorithm described by our method. Section 4 is reserved for the description of the results obtained by adopting this intelligent evaluation system. Also, it speaks about the necessary interpretations which allow the teacher to make the appropriate decision. Finally, we end this paper with a conclusion.

2 Research Method

We define the learning skills that are dedicated to scientific and technical subjects. Then, we present the used research method and the approach that allows us to achieve our objective.

2.1 Learning Skills Characterization

Researchers have investigated several types of learning skills (Baafi et al. 2017; Jayaprakash and Karthik 2015; Nuray and Bülent 2018; Abrahams et al. 2019; Kennedy et al. 2014). In our work, we are interested in five skills that are widely answered in scientific and technological knowledge, namely Appropriate, Analyze and reason, Achieve, Validate and Communicate, see Fig. 1.

We present in Table 1 the capacities that characterize each of the skills mentioned above. From these capabilities, all stakeholders will be able to identify these skills and assess them. To acquire scientific or technological information, the student must go through these five skills by implementing the capacities listed in Table 1.

2.2 Approach Used

We present the approach adopted by the teacher, see Fig. 2, using artificial intelligence to assess the different skills C1, C2, C3, C4 and C5. Thus, the pedagogical approach considered helps the teacher make decisions to remedy the skills difficulties encountered by the learners. The teacher assesses the learners' performance on a technological support (computer, tablet, etc.) as the tests given are corrected. The system interprets the skills' result obtained for each student. It then suggests suitable educational situations in order to be able to remedy the difficulties encountered. The teacher monitors the progression of support offered by the system, see Fig. 2. The result of the performance assessment for each learner is displayed progressively.

We assume that the test composition is in the form of a number of exercises, and each exercise is composed of skills percentage C1, C2, C3, C4 and C5. The teacher must determine the size of each skill in each exercise. To assess the performance of students' skills, the teacher corrects a student's sheet by prescribing in a tablet, the value of each skill which breaks down from 0 to 4, see Table 2.

The students' skills result in a test which includes four exercises is given in Table 3. From these results, the teacher can know an overall assessment on the skill level of each student.

The system interprets the results obtained by a student in a specific test and it proposes the corresponding level for each skill, see Table 4. We divide the student's results into four levels as indicated in Table 4. For each $1 \leq i \leq 5$;

For a given skill Ci, $(1 \leq i \leq 5)$, and based on the student's result, the system offers pedagogical situations in the form of exercises to improve the acquisition of a given skill. Indeed, for $1 \leq i \leq 5$,

Fig. 1 The five learning skills for scientific and technological knowledge

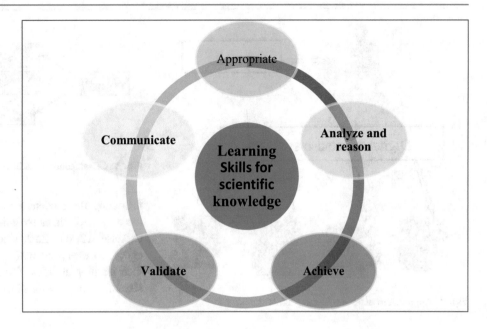

Table 1 Capacities of the learning skills for scientific and technological knowledge

Symbols	Learning skills	Capacities
C1	Appropriate	Identify data
		Find information
		Extract information
		Organize information
C2	Analyze and reason	Find relationships between data
		Detect lines of investigation
		Suggest a resolution method
C3	Achieve	Execute a resolution method
		Progress according to a logic
		Apply other information
		Experiment
C4	Validate	confirm
		Criticize
		Argue
C5	Communicate	Report the result in a written test
		Organize the work
		Write logically

- Level i1: the system offers exercises, concerning the same concept studied, of a very clear and very simple level to which the existence of the competence Ci is preponderant.
- Level i2: the system offers exercises, concerning the same concept studied, of a fairly clear and fairly simple level to which the existence of the competence Ci is preponderant.
- Level i3: the system offers exercises, concerning the same concept studied, of a reflection considerable level to which the existence of the competence Ci is preponderant.
- Level i4: the skill Ci is well acquired, and the teacher is satisfied with the result obtained.

For a given skill Ci, the system assigns each student a given skill level and offers the corresponding exercises. Depending on the pupil's performance, the system offers progression in the levels, see Fig. 3. In each level, there are situations that improve the level of appropriation of the skill Ci.

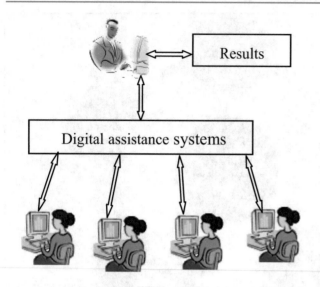

Fig. 2 Approach to assessing learning skills

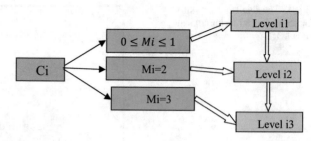

Fig. 3 Classification of skill acquisition levels

following the student's response that the system proposes the new step, either by going to level i2 or it offers another exercise i12, etc. Each student belongs to a skill level category, so everyone will have exercises to do according to their learning abilities. The system evaluates and marks each step of the algorithm. Thus, it offers at the end an additional

Table 2 Quantifications of learning skills appreciations

Appreciations	Ratings
Very well	4
Well	3
medium	2
insufficient	1
Very insufficient	0

Table 3 Results of a student's learning skills assessment

Student 1	EX 1	EX 2	EX 3	EX 4	Means
C1	N11	N21	N31	N41	M1
C2	N12	N22	N32	N42	M2
C3	N13	N23	N33	N43	M3
C4	N14	N24	N34	N44	M4
C5	N15	N25	N35	N45	M4

Table 4 Interventions levels according to the obtained results

Skill	Mean	Classifications	Diagnostic	Levels
Ci	Mi	$0 \leq Mi \leq 1$	Insufficient	Level i1
		Mi = 2	Medium	Level i2
		Mi = 3	Pretty good	Level i3
		Mi = 4	Property acquired	Level i4

We propose the algorithm used by the system to attenuate the difficulty encountered for a given skill, see Figs. 4, 5 and 6. Depending on the student's result, the system offers pedagogical situations to remedy the difficulties encountered. Level i1 is the most basic for acquiring the skills. This level begins with an exercise i11, as shown in Fig. 4, and

note which helps the teacher know more precisely the degree and speed of each pupil comprehension. The teacher, using the results obtained by the system, can suggest group work during learning operations later. Supports are given for students who really have huge problems in one type of skill, see Figs. 4, 5 and 6, and the teacher can suggest to do it at home. Depending on the number of students who have passed to the support stage, the teacher must evaluate his approach in the learning operation.

- Support i1: pedagogical situations which focus on the skill concerned Ci of a basic level which is very clear and very simple.
- Support i2: educational situations that focus on the skill concerned Ci at a fairly clear and fairly simple level.
- Support i3: educational situations that focus on the skill concerned Ci requiring a considerable level of reflection.

We propose the system operating algorithm, and we describe the level progression adopted by the intelligent system. According to the result obtained by the student, the system assigns it in a level, see Table 4. Thus the student's support activity begins with exercise i11 (in the case of level i1) or with exercise i21 (in the case of level i2) or by exercise i31 (in the case of level i3), see Figs. 4, 5 and 6. Next, the steps taken by the students depend on their response to the exercises provided by the system. The notes which are inserted in Figs. 4, 5 and 6 corresponds to a scale which will be introduced in the algorithm. The explanation of how a student's final mark is calculated will be presented in paragraph 3.

To take stock of the year, we can assume that three tests have been completed in a given class. The same procedure is repeated, and the system records the services of each student in each stage. The results for a given student will be in the form of Table 5.

For $1 \leq i \leq 5$ and $0 \leq j \leq 3$, Mij denotes the average received by the student in competence Ci and the test Tj. Ni indicate the average of the competence Ci in the three tests. From these results, we will be able to know the evolution of student in accordance with the acquisition of a given competence. The information obtained can offer us a global vision on the level of each learner skills and his progress.

3 Implicit Notation Algorithm

We are interested in measuring the performance of each student according to their interaction with the proposed tests. In this regard, the system proposes, firstly, the appropriate level for each student and the corresponding exercises. Secondly, it presents a counter that accumulates the scores of the correct answers and determines the final result of each learning skill. Indeed, the intelligent system takes into consideration each step of the algorithm by assigning a score for the correct answers. Thus, we present the scoring algorithm. See Figs. 4, 5 and 6, and Table 6, for all $1 \leq i \leq 5$ and for all $1 \leq j \leq 3$,

Let us set for all $1 \leq i \leq 5$ and for all $1 \leq j \leq 3$, nij the result of level j. Let us set, for all $1 \leq i \leq 5$, Ri the final result of a pupil for the learning skill Ci. The value of Ri is determined as follows in Table 7.

In Table 8, we present the interpretation of the result obtained by the intelligent system. It allows storing each student's information for later use. According to the previous table, we can deduce the following relationship:

for all $1 \leq i \leq 5$, $0 \leq Ri \leq 12$.

The system records the information of each student for all the tests proposed by the teacher. It monitors the performance of each learner by interpreting the results obtained. The system also makes an overall assessment of the entire class in each learning skill. Thus curves are proposed which highlight the performance of each learner and also of the entire class for each learning skill.

4 Results and Interpretations

The system offers several opportunities to improve the level of the learner. It exactly detects the learner's level in each skill, and it proposes solutions to reduce the problematic situations by adopting a progressive method in difficulties. It

Fig. 4 Algorithm for the level i1 skills acquisition

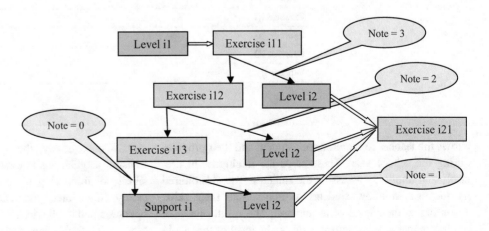

Fig. 5 Algorithm for the level i2
skills acquisition

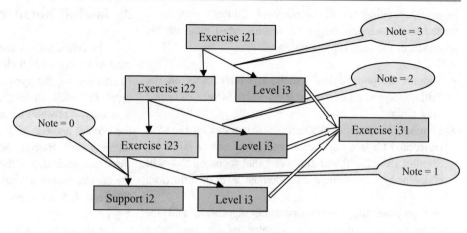

Fig. 6 Algorithm for the level i3
skills acquisition

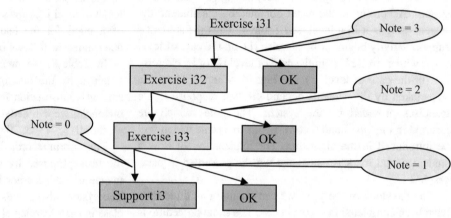

Table 5 Annual skills
assessment of a student

Skills 1	T1	T2	T3	Average
C1	M11	M12	M13	N1
C2	M21	M22	M23	N2
C3	M31	M32	M33	N3
C4	M41	M42	M43	N4
C5	M51	M52	M53	N5

Table 6 Level j scoring
algorithm

Achievement	Notation
Exercise ij1	3
Exercise ij2	2
Exercise ij3	1
Support ij	0

allows the teacher to monitor each student and their progress during the school year. These steps are intelligent in precisely and implicitly assessing the students. The aids offered by the system allow students to appropriate their skills according to their level of acquisition. The result obtained, on the one hand, gives information on the level of the whole class. Consequently, the approach used by the teacher to transmit acquisitions is evaluated. The teacher has the possibility of improving his performance by offering the necessary educational parameters to succeed in his teaching operation and to alleviate the difficulties encountered. On the other hand, the system assigns each student to his category

Table 7 Students' final grade for the assessment of learning skills

Skill	Mean	Classifications	Levels	Results
Ci	Mi	$0 \leq Mi \leq 1$	Level i1	$Ri = n1 + n2 + n3$
		$Mi = 2$	Level i2	$Ri = 4 + n2 + n3$
		$Mi = 3$	Level i3	$Ri = 8 + n3$
		$Mi = 4$	Level i4	$Ri = 12$

Table 8 Assessments and recommendations of the result in learning skills

Skill	Mean	Results	Assessments and recommendations
Ci	Mi	$11 \leq Ri \leq 12$	Perfect, the property is well acquired
		$08 \leq Ri \leq 10$	Good, the property is acquired
		$06 \leq Ri \leq 07$	The acquisition of skill Ci is average
		$03 \leq Ri \leq 05$	Acquisition of skill Ci is insufficient
		$00 \leq Ri \leq 02$	The result is not good enough. You have to work more on Ci skill

of skills acquisition, and it offers a succession of progressive level operations. This allows all students to acquire the skills each one needs for their learning ability. Thus, the intelligent system allows the teacher to make the appropriate decision to continue his course later with the same levels of student skills. Therefore, students can start new stages with the same perquisite skills. The operation provides the student with a homogeneous level of skills which enable them to overcome the various possible insertions into life. The system offers us the progression in skills of each student during all the assessments made by the learner. Thus, the system provides information on the personality of each student and the tendencies to progress in different subsequent learning situations.

The level acquired by the student in the first phase gives him 4 marks while the maximum score that a student who has answered the exercises can reach is 3 marks. The maximum score a student can have is 12 when he has obtained $Mi = 4$ while the maximum score for a student who has validated two levels at the start is 11 ($Ri = 4 + 4 + 3$). The maximum mark for a pupil who has passed all the tests of the three levels is 09 marks ($Ri = 3 + 3 + 3$). The used notation approach motivates the students in their learning, and it offers a progressive scale. Students who have obtained $00 \leq Ri \leq 02$, have enormous difficulties in their Ci learning skill. For this reason, the intelligent system supports them with an aggregate of parallel additional activities. The object is to support students to improve their cognitive aptitudes in this type of skill. For students, who have had $11 \leq Ri \leq 12$, the system encourages them and offers parallel additional activities in which the objective is to improve their acquisition with regard to this type of skill. For other categories, the system also offers supportive parallel activities to improve their skills. The intelligent system, by collecting the results of all the tests, analyzes the progress of each student and presents a typology of that student's personality. The system can also predict and propose the students' orientation through the preponderant natures of their learning skills. Thereby, the used strategy allows permanent follow-up of the different learning skills with an interesting precision.

5 Conclusion

We presented an intelligent strategy for assessing learning skills through a technological system. This takes into account the interaction of the students through well-studied pedagogical situations. It also allows the learning operation to be judicious and meet the requirements of students 'progress in a learning environment. After different operations of learning evaluation of students, each one of them reaches the same level of understanding, which gives students an intrinsic motivation for the next learning sequences. The permanent follow-up of each pupil, by the system, during the school year gives precise information on the learning degree and the improvement for each skill. The teacher, by recovering the results of his students, can also make a self-evaluation of his method for a further teaching performance improvement. We can also present pedagogical situations during the tests to address more than one learning skill. Therefore, we offer activities including the improvement of one or more skills at the same time. We can also adapt the same strategy to other skills that are dedicated to other disciplines through the proposal of well-studied situations and activities. The obtained summative evaluation presents a precise idea of each student's progress and is used to interpret the information collected on the student performance and also of the entire class.

References

Abrahams, L., Pancorbo, G., Primi, R., Santos, D., Kyllonen, P., John, O.P., De Fruyt, F.: Social-Emotional Skill Assessment in Children and Adolescents: Advances and Challenges in Personality, Clinical, and Educational Contexts. Psychological Assessment, Location (2019). http://dx.doi.org/10.1037/pas0000591

Arena, D., Perini, S., Taisch, M., Kiritsis, D.: The training data evaluation tool: towards a unified ontology-based solution for industrial training evaluation. Procedia Manufact. **23**, 219–224 (2018)

Baafi, E., Tolhurst, R., Marston, K.: The work skills development framework, applied to minerals industry employability. In: Proceedings of the international conference on models of engaged learning and teaching, adelaide, pp. 11–13 (2017)

Bandaranaike, S.: From research skill development to work skill development. J. Univ. Teach. Learn. Pract. **15**(4) (2018)

Brundiers, K., Wiek, A.: Beyond interpersonal competence: Teaching and learning professional skills in sustainability. Edu. Sci. **7**(1), 39 (2017)

Ceylan, S., Zeynep, S.A., Seyit, A.K.: STEM skills in the 21st century education. Research Highlights in STEM Education. ISRES Publishing, pp. 81–101 (2018)

Chattopadhyay, S., Shankar, S., Gangadhar, R.B., Kasinathan, K.: Applications of artificial intelligence in assessment for learning in schools. In: Keengwe, J. (ed.), pp. 185–206 (2018)

Diyer, O., Achtaich, N., Najib, K.: Artificial intelligence in learning skills assessment: A pedagogical innovation. NISS2020, March 31-Apr 2, 2020, Marrakech, Morocco. © 2020 Association for Computing Machinery. ACM, Location (2020). ISBN 978-1-4503-7634-1/20/03...$15.00. https://doi.org/10.1145/3386723.3387901

Ganeshan, R., Mary-Ann, S., and Robert, L.: Implementing emerging technologies to support work-integrated learning in allied health education. In: The Journey From Exploration to Adoption Emerging Technologies and Work-Integrated Learning Experiences in Allied Health Education, pp. 266–300, (2018)

Gerardo, H., Lucia, V., Javier, S., Cristina, P., Sergio, C.: On the development of VR and AR learning contents for children on the Autism Spectrum. From real requirements to virtual scenarios. Augment. Real. Enhanc. Learn. Environ. 106–141, (2018)

Gottfredsonk, C., Mosher, B.: Innovative performance support: Strategies and practices for learning in the workflow. McGraw-Hill, (2011)

Harshvardhan, S.: Skill based education system in meeting employer's needs. Indian J. Appl. Res. **4**, 12 (2014)

Jayaprakash, J., Karthik, S.: Need for the skill based curriculum to develop leadership and communication skills. IJEMR. **5**(12) (2015)

Karsenti, T.: Artificial intelligence in education: The urgent need to prepare teachers for tomorrow's schools. Formation et profession **27**(1), 105–111, (2019)

Kennedy, F., Pearson, D., Brett-Taylor, L., Talreja, V.: The life skills assessment scale: measuring life skills of disadvantaged children in the developing world. Soc. Behav. Personal. Int. J. **42**(2), 197–210 (2014)

Mäkiö, J., Mäkiö-Marusik, E., Yablochnikov, E.: Task-centric holistic agile approach on teaching cyber physical systems engineering. In: IECON 2016–42nd annual conference of the IEEE, pp. 6608–6614, (2016)

Nuray, K.F., Bülent, A.: Life skills from the perspectives of classroom and science teachers. Int. J. Progress. Edu. **14**(1) (2018)

Ogunlade, B.O., Bello, L.K.: Pre-Service teachers' perceived relevance of educational technology course, digital performance: Teacher perceived of educational technology. Int. J. Technol. Enabled Stud. Supp. Serv. pp. 41–54 (2019)

Popenici, S.A.D., Kerr, S.: Exploring the Impact of Artificial Intelligence on Teaching and Learning in Higher Education. Popenici and Kerr Res. Pract. Tech. Enhanced Learn. (2017)

Sudana, M., Apriyani, D., Suryanto, A.: Soft Skills evaluation management in Learning processes at Vocational school. J. Phys.: Conf. Ser. **1387**, 012075 (2019)

Sullivan, M., Bandaranaike, S.: Challenges of the new work order: a work skills development approach. In: Proceedings of the 20th World Conference on Cooperative and Work Integrated Learning, Chiang Mai, 5–8 June, (2017)

Tuomi, I.: The Impact of Artificial Intelligence on Learning, Teaching and Education. Publications Office of the European Union, Luxembourg, (2018). ISBN 978-92-79-97257-7

Tvenge, N., Ogorodnyk, D.: Development of evaluation tools for learning factories in manufacturing education. Procedia Manuf. **23**, 33–38, (2018)

Interactivity for Artificial Intelligence Systems: NL2SQL

Karam Ahkouk, Mustapha Machkour, Rachid Mama, and Khadija Majhadi

Abstract

For sustainability purposes, Artificial Intelligence modules that help translating natural language sentences to SQL queries are gaining more and more momentum in the recent decade, since they allow the automatic conversion of questions in English to the SQL without any interference from the user during the process of translation. We show in this paper the utility of the previously proposed models and their added value to the resolution of the complicated problem of inferring complex queries from English sentences. We will also discuss their limits in addition to the necessity of using interactivity between the user and the system in order to enhance the quality of the outputs. In this paper, we show how the majority of models can't independently predict the appropriate queries of SQL without the feedback, the guidance as well as the help of the user. And finally, we present the conclusion together with the future work.

Keywords

Natural language • Database languages • Machine translation • Deep learning

1 Introduction

The main objective of natural language processing (NLP) is to facilitate the interaction between users and the computer without the need of memorizing commands. Translating natural language sentences to relational databases is one of the most critical problems in the field of NLP, which means extracting data from databases using natural language questions; this can be a big step and a huge improvement to the traditional way of using the content of these databases. It offers to the users a simple, uniform, and unlimited access to data without the need to have skills in SQL. We usually obtain data by interacting using the SQL language in the real world rather than the Natural language.

In many of our previous papers like "Comparative study of existing approaches on the Task of Natural Language to Database Language" (Ahkouk et al. 2019) and "Human Language Question To SQL Query Using Deep Learning" (Ahkouk and Machkour 2019) that show how many studies published in the current decade have tried to find solutions using different methods and techniques in order to manage in translating text to SQL without any user intervention. However, the major obstacle lies in how the system will perfectly and automatically understand the natural language sentence (both syntactically and semantically). Until now, this problem is considered as a wide field of research and still an open area of exploration. Another problem is related to the second variable which is more complex related to the schema, even if the semantic analysis is used to understand the meaning of the natural language, the problem of its conversion to a formal query still remains as an issue since this translation requires mapping the understanding of intent into a specific database schema.

Lots of models that are based on the use of shallow and end-to-end approaches have shown very limited and low quality result and the majority of them fail when they are tested with new datasets or new schemas that haven't been seen in the training and the development collections.

From another perspective, we have interactivity that consists of providing more information from the user to help the system understand the question and the links between each component of the schema.

Interactivity in the task of Natural language translation shows promising results since it helps the models to have an idea about how to build the target query. In this context and related to the task of Natural Language to SQL, the inclusion of the users' feedback might be one of the useful approaches

K. Ahkouk (✉) · M. Machkour · R. Mama · K. Majhadi
Ibn Zohr University, Agadir, Morocco
e-mail: k.ahkouk@uiz.ac.ma

© The Editor(s) (if applicable) and The Author(s), under exclusive license to Springer Nature Switzerland AG 2021
M. Ben Ahmed et al. (eds.), *Emerging Trends in ICT for Sustainable Development*,
Advances in Science, Technology & Innovation, https://doi.org/10.1007/978-3-030-53440-0_5

that haven't been studied deeply in the previous research articles published in the recent years. Using relational databases by non-expert-users is creating a big obstacle for these users to benefit from the content stored in this type of systems. In this paper, we discuss the already available research papers, and we will focus on enhancing models by using the interactivity in the query generation process for the aim to generalize for unseen and more complex samples.

2 Interactivity in the NL to SQL

Interactivity has never been taken in consideration while conceiving models. This is due to the aim of getting rid of classical exchanges between the user and the system, or because of the complexity that this interactivity layer adds to the system even if the existing methods have limited capabilities to spot the good elements on the database schema.

Interactivity between the user and the system can play a pivotal role to enhance these models and provide them with better understanding of what to generate and how to build the appropriate queries in the output space. It is a very practical block that should be used in addition to the previous methods like the semantic analysis. We believe that adding interactivity in the already existing models that translate Natural language to SQL is the key solution to success in this task, particularly seeing that the majority of the database systems' users are people who lack knowledge of SQL.

The task of natural language translation to SQL queries can't be resolved using shallow techniques only. There are lots of holdbacks that reduce the relevance of models using end-to-end approaches, especially the semantic parsing one. For example, in lengthy English sentences or even in complex ones, how can the model know the appropriate items to be placed in the SQL clauses? And if there are more than one condition in the WHERE clause, can the model recognize all of them effectively? And what about the values in the user question? How could be treated? Should be used as they are in the input sentence or should be formatted to represent and correspond to the stored values in the tables? How can a model based on semantic parsing deal with a schema that contains more than one table and several columns? Are we modeling for only easy questions and simple databases' schemas or to let the model be able to generalize to unseen questions? What can a model trained on a dataset containing tables about restaurants or movies generate if the question is about employees or another field?

The current state of the art has shown a limited percentage of correctly generated queries even for a small and easy dataset like SPIDER, thing that proves that interactivity is mandatory between the user and the system for the generation process.

In this context, the interactivity between the user and the system can take different forms, initial questions to the user, feedback questions, error corrections, etc. In fact, one of the solutions is asking a sequence of questions to better understand the need and the wanted query. This might be at different stages and it can be mixed with other techniques. For example, the system can ask for the columns to be included in the Select clause or may be the tables to include in the From slot. Feedback messages might be shown as well after a first turn of treatment of the user input, showing the corrections and suggestions in order to guide and limit the space of outputs.

The aim of this work is to construct a model that takes as input, a natural language question and translates it automatically to a structured language query.

3 System Architectures

3.1 The Main Architecture

Our new architecture system is combined with interactivity layers that are included after the semantic generation process in the SQL query.

Since adopting an end-to-end technique is only compounding the complexity of the problem and giving ambiguity in the generate outputs, the interactivity layers come to solve the problem and to enforce the previously understood information from the user input and to catch up the missed things in the generation process. It is considered as an upholding feature that should be integrated together with other linguistic techniques and not to be considered as a substitution approach.

In general, the process starts as follows. The natural language is fed to the interface by the user; it is immediately joined to the schema of the database grabbed from the managing system in order to be presented to the semantic parser layer that is responsible for tokenization of the input, the word embeddings extraction and mapping using the pre-trained module of BERT and then the application of the attention-column mechanism as well as the copying one.

The output of this step is used by several sub-modules; among them, the Select sub-module, Where sub-module, Group by sub-module, etc. Each sub-module is responsible for generating the related part of the final SQL query. For the aim to be validated by the query Validator, the fragments are put together to construct a temporarily non-validated SQL query. The role of the Validator is to check the faulty parts that should be removed or corrected taking into consideration the structure inherent in SQL. If the query is syntactically correct and the items chosen by the modules exceed the threshold, then the SQL query is executed and the result is formatted to be shown to the user.

In case if the query is correct and the probabilities of the candidate columns/tables/values generated by the modules are below the threshold, then the suggestions step is triggered to show to the user the items that have close probabilities' scores and that might be good replacements for the aim to improve the quality of the query. The suggested items are selected from the elements in the pool of candidates. For example, if the Select module computed two columns or more to have nearly the same scores, then all of them will be shown and let the user choose the good ones. The same is done for the rest of the modules. In the same way, the query Validator decides if the query is faulty or not. This is done following the grammar of the SQL language and depending on the schema and the types of columns compared to their operators. When the query is not correct the user is questioned for the goal to better understand his natural language sentence. This includes revisions of the question and the proposition of the possible corrections. Warning also might be triggered if there is ambiguity in the terms, especially the columns and values.

The modifications are applied to the user input directly, and then the process of the translation starts over, and so on. It's particularly important to interact with the user in order to higher the performance of the system and to yield better result. In fact, asking questions like "do you mean…, do you want to group by…, do you want to order the result by…?" will definitely have a positive impact on the output. In the same context, the user might be asked to differently reformulate his or her sentence by substituting the invalid terms and tokens. The Fig. 1 shows the details of the architecture with interactivity.

3.2 The Validator

The validation process starts when the semantic parsing output is ready to be checked. Once the SQL query is predicted, the query reader makes a lexical analysis and a scanning in order to start the syntax check that verifies the compatibilities taking into consideration the structure inherent in SQL.

Once the syntax is correct regarding the SQL compatibilities between the columns and the values, the items of the SQL query are proceeded to check whether the columns names and values included in the user query exist in the database schema and in the rows of the tables.

Afterwards, the query is further exposed to the previously generated queries in the history of the system. This helps the model to give quick response and to learn how to construct the future queries if they are similar to the current one. Earlier queries are stored in the history block to be used if necessary. Fig. 2 shows how the Validator works.

The generated query is checked again against a threshold score. If it fulfills the requirements and its score is higher than the threshold, then it will be executed against the corresponding database and the result will be shown to the user. Otherwise, the suggestion extracted either from the history block if the query is similar to other seen ones or from the already set up ones in the system in order to be displayed as feedback to the user.

4 Related Work

Many improvements have been done in the recent years on the task of Natural language translation to SQL either by using a semantic parsing approach or a syntactic one. In general, those two methods of resolving this task are widely considered as an end-to-end solution to the problem.

The treatment of the user input without taking in consideration its structure and the relation between that natural language sentence and the schema of the database is simply a shallow technique that might work for simple samples but not the advanced ones. It fails in the majority of cases due to many aspects. For example, the work of (Couderc and Ferrero 2015; Yaghmazadeh et al. 2017; Bais et al. 2019) is based on a syntactic approach that split the process of prediction to many other steps including the use of some linguistic techniques like the generation of the syntax tree, the application of free-context rules or even the use of a dictionary of synonyms to proceed with terms substitution in the user question. Fig. 3 shows an example of the system's architecture of (Yaghmazadeh et al. 2017).

In another side, there is the integration of deep learning in works like (Zhong et al. 2017; Wang et al. 2017; Xu and Liu 2017; Yu et al. 2018; Dong and Lapata 2018; Shi et al. 2018; Hwang et al. 2019; He et al. 2019; Liu et al. 2019), as a base approach to perform a better generation of the SQL queries in the output space. This kind of models is trained on a mono-table dataset called WIKISQL (Zhong et al. 2017) that contains more than 8000 natural language questions/SQL queries.

Many datasets have been used either for training or evaluating the models, among them: WIKISQL, YELP, IMDB, SPIDER (Yu et al. 2018), etc. A multitude of studies (Yu et al. 2018; Lee 2019; Lin et al. 2019; Bogin et al. 2019; Wang et al. 2018a; Yao et al. 2019; Min et al. 1909) have worked with the dataset SPIDER which contains more than 10000 natural language sentences/SQL queries distributed on the training, the development and the test set.

The deep learning approach is based on several pillars; among them, the attention mechanism of Bahdanau (Bahdanau et al. 2014) which helps the model to better highlight the important parts of the input question. This is particularly useful for linking the columns that appear in the user's

Fig. 1 Our proposed system architecture using interactivity

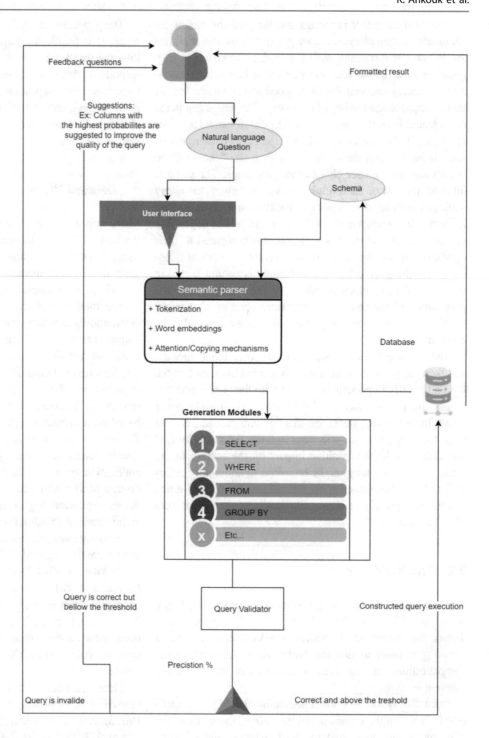

question to the ones on the database schema. Also, the use of the Pointer Network proposed by (Vinyals et al. 2015) is included. It allows pointing to tokens in the input, thing that is necessary for making a copy of the values that should appear in the target SQL query from the user natural language sentence. The majority of the previously cited studies exploit the word embeddings in order to get a meaningful representation for each token. For every word in the user question together with the database schema, a word embedding vector is obtained using one of the existing pre-trained models like Word2Vec, GLoVe, or even the powerful one called BERT (Devlin et al. 2018) which is based on the transformer network by (Vaswani et al. 2017).

Unlike other similar language processing tasks, the one of translating Natural language to SQL has a compounded complexity because the question of the user is always dependent to the schema of the database. The user input is unpredictable and might be random since the user may not

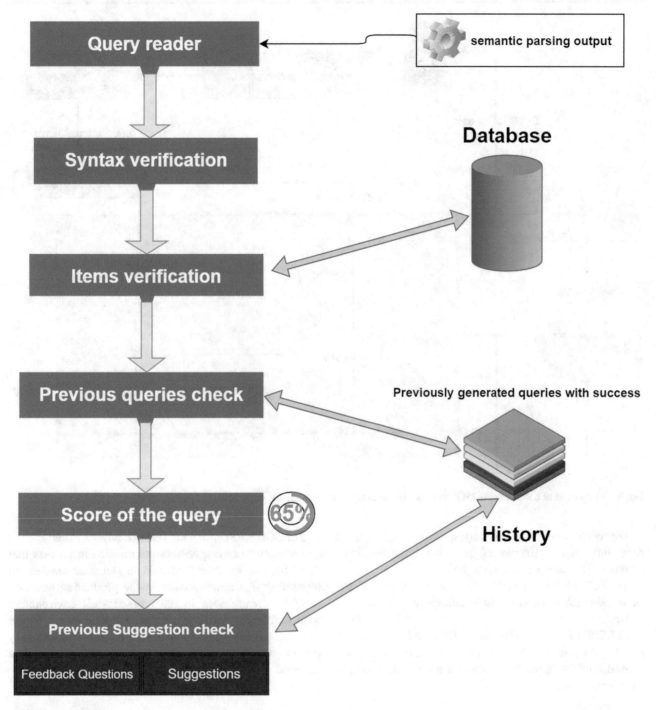

Fig. 2 The validation process in our proposed architecture

follow the proposed grammar of the system. The schema of the database is as important as the natural language question. It is crucial for constructing and predicting the SQL query. Fig. 4 shows the general process of the query generation using semantic parsing.

The role of the schema is to give context to the user question. Since we might have many identical user questions but we can have different SQL queries in the output if the database schema is different. In other words, the output result will not be the same if at least one column is changed or moved from one place to another or from one table structure to the other even in the same database schema. For instance, the English sentence below

Ex1: Get the clients that live in Paris

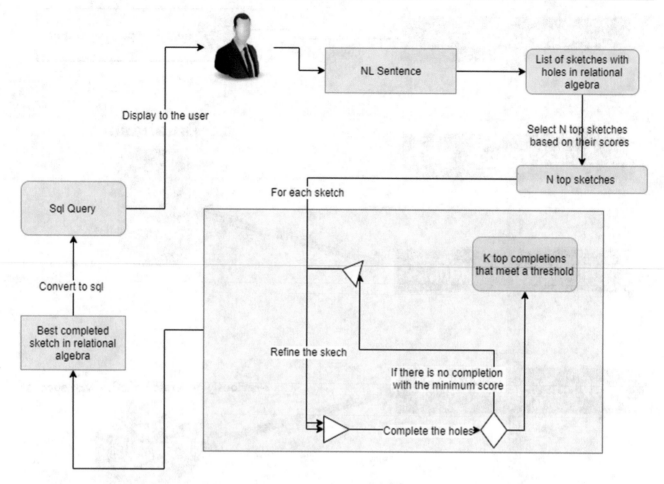

Fig. 3 An example of a previous NL2SQL system (the SQLizer) by (Yaghmazadeh et al. 2017)

We could have different corresponding SQL queries depending on the structure of the schema for that same sentence. The output result might be

SELECT clients.* FROM clients JOIN addresses ON clients.id = addresses.client_id WHERE city = 'Paris'

Or:

SELECT clients.* FROM clients JOIN addresses ON clients.address_code = addresses.code WHERE city = 'Paris'

And it might be the query below if the city column is in the clients' table:

SELECT * FROM clients WHERE city = 'Paris'

Previous work emphasizes the translation process that don't involve the user feedback. In fact, this can lead to overfitting if a new question in the production environment is already seen in the train-dev-test environment with a different database schema. For example, a model that has already seen the sentence in *Ex1* will certainly predict the same query ignoring that the schema has changed.

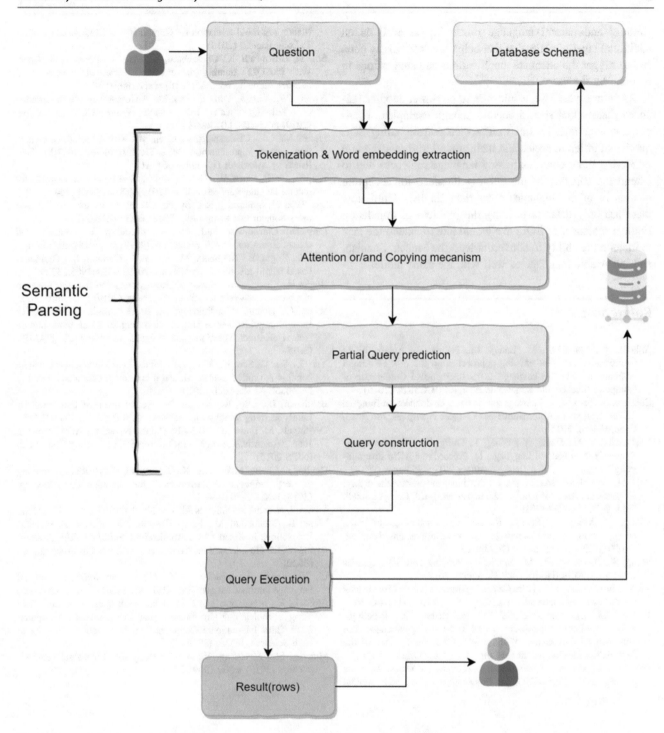

Fig. 4 An example of a previous NL2SQL system (the SQLizer) by (Yaghmazadeh et al. 2017)

5 Conclusion

In this paper, we discussed the utility of using the interactivity in order to enhance the quality of the output queries by the model in the generation process for the goal of generalizing to unseen samples in the production environment.

The main aim of this system is to allow communication between database and human users using interactivity and feedback messages. In order to eliminate the burden of learning new languages each time a user wants to interact with a system for extracting its data, we propose the architecture of a generic natural language query interface for relational databases. This system is based on machine learning and natural language processing as well as an additional layer of validation that helps check for faulty parts and to trigger the adequate modifications and suggestions to be done by the user.

As future work, we would like to construct another balanced dataset that should contain enough examples of natural language questions as well as their corresponding SQL queries of different scales and their ground truth, for the aim of solving more complex natural language sentences and to extend the capacity of the existing interface for accepting queries in other languages other than English. Until now there are only datasets tackling the problem of translation English sentences to SQL. We would like to extend the task to include the ability to convert at least the French, Spanish, and the Arabic language as well with the same model.

References

Ahkouk, K., Machkour, M., Ennaji, M., Erraha, B., Antari, Jilali.: Comparative study of existing approaches on the task of natural language to database language. 2019 International Conference of Computer Science and Renewable Energies (ICCSRE) (2019)

Couderc, B., Ferrero, J.: Interrogation de bases de données en français. 22ème Traitement Automatique des Langues Naturelles, Jun 2015, Caen, France. 2015

Yaghmazadeh, N., Wang, Y., Dillig, I., Dillig, T.: SQLizer: Query Synthesis from Natural Language. In: Proceedings ACM Programming Languages 1, 1, Article 1 (January 2017), 25 pages (2017)

Bais, H., Machkour, M., Koutti, L.: An independent-domain natural language interface for multi-model databases. Int. J. Comput. Intell. Stud. **8**(3), 206–222 (2019)

Zhong, V., Xiong, C., Socher, R.: Seq2sql: Generating structured queries from natural language using reinforcement learning. (2017) arXiv preprint arXiv:1709.00103

Wang, C., Brockschmidt, M., Singh, R.: Pointing Out SQL Queries From Text. MSR-TR-2017-45| November 2017

Xu, X., Liu, C., Song, D.: SQLNet: Generating structured queries from natural language without reinforcement learning. (2017) arXiv:1711.04436v1

Yu, T., Li, Z., Zhang, Z., Zhang, R., Radev, D.: TypeSQL: Knowledge-based type-aware neural text-to-sql generation. The 16th Annual Conference of the North American Chapter of the Association for Computational Linguistics, New Orleans, (2018)

Dong, L., Lapata, M.: Coarse-to-fine decoding for neural semantic parsing. arXiv:1805.04793. In: Proceedings of the 56th Annual Meeting of the Association for Computational Linguistics (Volume 1: Long Papers) (2018)

Shi, T., Tatwawadi, K., Chakrabarti, K., Mao, Y., Polozov, O., Chen, W.: IncSQL: training incremental text-to-sql parsers with non-deterministic oracles. (2018) arxiv:1809.05054

Hwang, W., Yim, J., Park, S., Seo, M.: A Comprehensive Exploration on WikiSQL with Table-Aware Word Contextualization. (2019) arXiv:1902.01069v1

He, P., Mao, Y., Chakrabarti, K., Chen, W.: X-SQL: reinforce context into schema representation. MSR-TR-2019-6| March 2019. Published by Microsoft Dynamics 365 AI

Liu, X., He, P., Chen, W., Gao J.: Multi-task deep neural networks for natural language understanding. (2019) arXiv:1901.11504

Yu, T. et al. SyntaxSQLNet: Syntax tree networks for complex and cross-domain text-to-sql task. (2018) arXiv:1810.05237

Lee, D.: Clause-wise and recursive decoding for complex and cross-domain text-to-sql generation. (2019) arXiv:1904.08835v1

Lin, K., Bogin, B., Neumann, M., Berant, J., Gardner, M.: Grammar-based neural text-to-sql generation. (2019) arXiv:1905.13326

Bogin, B., Gardner, M., Berant, J.: Representing schema structure with graph neural networks for text-to-sql parsing. (2019) arXiv:1905.06241

Wang, C., Huang, P.S., Polozov, A., Brockschmidt, M., Singh, R.: Execution-guided neural program decoding. In ICML workshop on Neural Abstract Machines and Program Induction v2 (NAMPI) (2018a)

Yao, Z., Su, Y., Sun, H., Yih, W.T.: Model-based interactive semantic parsing: A unified framework and a text-to-sql case study. OSU & Facebook AI Research (2019)

Bahdanau, D., Cho, K., Bengio, Y.: Neural machine translation by jointly learning to align and translate. (2014) arXiv:1409.0473v7

O. Vinyals, M. Fortunato, N. Jaitly. Pointer Networks. ArXiv e-prints, June 2015. Advances in Neural Information Processing Systems 28 (NIPS 2015)

Devlin, J., Chang, M.W., Lee, K., Toutanova, K.: BERT: Pre-training of deep bidirectional transformers for language understanding. (2018) arXiv:1810.04805

Vaswani, A. et al.: Attention Is All You Need. (2017) arXiv:1706.03762

Mama R., Machkour, M.: Fuzzy Questions for Relational Systems. Innovations in Smart Cities Applications Edition 3. The Proceedings of the 4th International Conference on Smart City Applications. (2020)

Li, Y., Yang, H., Jagadish, H.V.: NaLIX: an interactive natural language interface for querying XML. SIGMOD Conference (2005)

Ahkouk, K., Machkour, M.: Human language question to sql query using deep learning. In: International Conference of Computer. 2019 Third International Conference on Intelligent Computing in Data Sciences (ICDS) (2019)

Min, Q., Shi Y., Zhang, Y.: A pilot study for chinese sql semantic parsing. (2019) arXiv:1909.13293

Toward an Intelligent Hybrid System Based on Data Analysis and Preprocessing Method

Sara Belattar, Otman Abdoun, and Haimoudi El khatir

Abstract

In recent years, artificial intelligence (AI) represents a crucial domain or technology that can be found everywhere, it can solve many problems facing the researchers. Hence, the powerfulness of Artificial intelligence contributes enormously to the sustainable growth of various domains (e.g., medical, agriculture, data analysis, and so forth). For that, in this paper, we propose the artificial neural network (ANN), and in particular, the paradigm of the Counter propagation artificial neural network. Our objective is to improve the standard Counter propagation artificial neural network in terms of results and classification accuracy by using the principal component analysis (PCA) for making a modified Counter propagation network through the hybridization of PCA and CPN. So, the PCA is a method among data analysis methods. It can reduce the dimensionality space of original data in small data that contained the new elements or objects, and also with this method we can eliminate the dependence and the obstacles between inputs which allowed to obtain the good classification, this hybridization is used in the data analysis field. Thus, the result shows that the proposed approach of modified CPN gave the best performance than the standard CPN in terms of iteration number, mean error, and classification accuracy.

Keywords

Artificial intelligence (AI) • Artificial neural networks (ANN) • Counter propagation artificial neural network (CPN) • Kohonen self-organizing map (SOM) • Data analysis • Principal components analysis (PCA)

S. Belattar (✉) · O. Abdoun · H. El khatir
Computer Science Departement, Laboratory of Advanced Science and Technologies, Polydisciplinary Faculty, University of Abdelmalek Essaadi (UAE), Larache, Morocco
e-mail: sarah.belattar12@gmail.com

1 Introduction

The use of data analysis allowed to discover the new elements that contained the data to solve the research problem in various fields. For that many researchers use data analysis software methods to solve these issues. This method has been widely used in various domains (Martin-Rodilla et al. 2018). The purpose of it is to describe, reduce, classify, and clarify data with a wide variety of perspectives and to study, in other words. The motivation behind the strategies for data analysis is to diminish the huge tables to give a simplified presentation of data, so data analysis techniques are significant for all systems in various fields. During the literature, we found many techniques or methods for data analysis, we mentioned the most used. Principal component analysis (principal component analysis "PCA") permits the specialist to lessen data to a smaller number of object than initial number of variables. These new variables are named "main component", and another technique named correspondence factor analysis (CFA) is utilized to decide and organize all dependencies among lines and columns in the table, so the nature of the customary data analysis techniques is constrained without the utilization of the computer advances instruments that can be prepared and figured with high-power machines.

Advancement in the computer science field is frequently connected with that of todays, computers that are getting all the more effective, yet notwithstanding this specialized and compositional development. The energy of these machines remains restrained within the resolution of particular troubles along with pattern recognition, prevision, and decision-making, especially the resolution of non-linearly separable problems, this kind of issue requires frameworks envisioned by qualities enlivened by the human mind. The terrific researchers in this discipline have come to build up a few models and strategies made of computer systems, among those strategies are machine learning techniques (e.g., support vector machine "SVM", "ANN", etc.). In this study, we

M. Ben Ahmed et al. (eds.), *Emerging Trends in ICT for Sustainable Development*,
Advances in Science, Technology & Innovation, https://doi.org/10.1007/978-3-030-53440-0_6

decided to work with artificial neural networks to develop an intelligent system by the great propriety no longer discovered in traditional systems. The artificial neural network is a non-linear system that represents the abstraction of many simple traits extracted from neurobiological models (Shi et al. 2015; Ahn 2010; Karimi and Gao 2010; Fei and Huang 2008; Valdez et al. 2014). Neural networks are powerful pattern recognizers and classifiers and might clear up several troubles, for example, non-linear problems. For the neural network's architecture, there are numerous architectures, and every model or architecture has the power in a specific problem. In this paper, a model of counter propagation artificial neural network (CPN) is proposed. This model of neural networks has the capability of learning, classification, studying, and easy version complexity to make the new smart systems (e.g., in diagnosis, prediction, discrimination, etc.). The counter propagation artificial neural network (CPN) uses the supervised learning technique algorithm which requires the presence of a learning base that contains the desired outputs. Our objective focuses on improving the counter propagation artificial neural network to give good results in classification by making the coupling of principal component analysis and classical CPN model, to reduce the dimensionally of data and eliminate the dependence between objects or inputs before beginning learning. The modified CPN or coupling of PCA-CPN gave the best results in terms of mean error, iteration numbers, and classification accuracy.

In this paper, we propose to develop a modified counter propagation artificial neural network (CPN) based on principal component analysis (PCA) to increase the classification accuracy and obtained the best results at the level of iteration number, mean error, and classification accuracy. This approach is used in the data analysis field. The outcomes of modified CPN gave the best performance compared to standard CPN.

2 Related Works

After the literature work, various researchers developed their new approach based on principal component analysis. In this work, we will present some papers which (Hu and Cui 2019) developed a coupling of fractional-order-PCA-SVM algorithm in digital medical image recognition in order to improve pattern recognition and for making the diagnosis and treat the diseases. So in their study, they tried to test various techniques of machine learning with principal component analysis for making the comparison at the level of results, and finally, they found that the coupling of fractional-order-PCA-SVM algorithm gave the best outcomes compared to

others methods of machine learning. The average accuracy rate of this approach is 99.2425 % in four experiments.

Salo et al. (2018) employed a new hybrid of dimensionality reduction technique with information gain (IG) and principal component analysis (PCA) based on support vector machine (SVM), multilayer perceptron (MLP), and Instance-based learning algorithms (IBK) for network intrusion detection to develop robust security systems. Their purpose of using the IG and PCA is to reduce and diminish the features and obtain the new datasets of uncorrelated features. Their results showed that the proposed IG-PCA-Ensemble method gave a better performance in terms of classifications accuracy, detection rate and helps the classifiers to take the decision during an attack.

Hoz et al. (2015) proposed the principal component analysis (PCA) mathematical method and fisher Discriminant Ratio (FDR) for feature selection and noise removal based on self-organizing map (SOM) and PCA for network intrusion detection. Their approach gave the 97 % in accuracy, 93 % in sensitivity, and 90 % in specificity.

Haimoudi et al. (2016) developed an improved model of Kohonen self-organizing map to improve the learning process of classification and clustering for dynamic systems based on principal component analysis (PCA) for reducing the data and eliminate the obstacles detected. The results of this proposed work are effective in terms of classification accuracy.

3 The Hybridization of CPN and PCA

After these related works, we can conclude that the researchers got the good outcomes and they found very successful by the coupling of various technologies and methods intelligent with the principal component analysis (PCA) at the level of accuracy, sensitivity, and in specificity (Hu and Cui 2019; Salo et al. (2018); Hoz et al. 2015; Haimoudi et al. 2016) For that purpose, in this work, we propose to make a coupling of artificial neural networks (ANNs) with principal component analysis (PCA) to increase it and got the modified ANN. There are several artificial neural networks models or architecture (e.g., self-organizing map (SOM), counter propagation network (CPN), multilayer perceptron (MLP), etc.). In our study, we propose the counter propagation network thanks to the capacity of classification. In order to make our test we based on this article (Haimoudi et al. 2016), they already proposed the coupling of Kohonen self-organizing map (SOM) with PCA to improve it, because our proposed model of CPN contained two layers, Kohonen layer, and Grossberg layer.

4   The Counter Propagation Artificial Neural Network

In the literature, we discovered many models or architecture that has the power in classification. Among these models, we choose the counter propagation artificial neural network thanks to its simplicity, precision of classification and short learning permit us to create the new systems in various domains that have the problem of classification. CPN model consists of an input layer, hidden layer, and output layer. The hidden layer is called (Kohonen self-organizing map) and the output layer is called (Grossberg layer), these layers connected successively (Buntine et al. 1994) (See Fig. 1), the counter propagation network uses the supervised learning method and it is based on desired outputs for calculating the outputs. The CPN model can be applied in two regimes (e.g., interpolation regime and accreditation regime), but the accuracy rate in the interpolation regime of CPN is low. This model of artificial neural networks is widely used in various works (Zeinali et al. 2016; Chen et al. 2018). Figure 1 depicted the architecture of the counter propagation artificial neural network.

4.1   Learning Procedure of Counter Propagation Network

The CPN architecture was developed and proposed by Robert Hecht-Nielsen in 1987. It contained two layers, Kohonen self-organizing algorithm, and second is Grossberg layer. This model of CPN uses the supervised learning method. Although the Kohonen layer or network uses the unsupervised learning method, the purpose of Kohonen layer is to make the clustering and classification. Its outputs are called wining nodes presented by number 1 and the other nodes neighboring node presented by 0. For obtaining these outputs, we have two methods (e.g., scalar product and euclidean distance) to calculate the distance between inputs objects and all neurons in the grid of Kohonen layer and

after that we must add the algorithm to upload the synaptic weight for the neighboring neuron to take a chance to be a winner. The second layer of Grossberg is based on outputs of the Kohonen layer and the desired outputs in order to give the outputs of CPN.

A.   The Kohonen Neural Network Algorithm

Kohonen neural network also called the competitive filter learning or self-organizing map (SOM) used the unsupervised learning for training. This type of artificial neural network is proposed by Teuvo Kohonen in the 1980 and it works in two modes: training mode and mapping mode, the first one builds the map and second one classifies a new input vector. Kohonen map is a method to dimensionality reduction and for adjusting its weights it uses the competitive learning as opposed to error-correction learning. During the learning, it utilizes the distance euclidean or product scalar for calculating the outputs named wining node based on inputs vectors, weight vector, and all nodes presented on the grid or map in order to give the good classification of dataset. Numerous works suggest the Kohonen self-organizing map to make their approaches and they found many important results (Yu et al. 2019; Wolski and Kruk 2020).

Kohonen neural network contained two layers, input layer and an output layer, the second layer can be two-dimensional and the important difference is that its nodes are connected to each other (Ali et al. 2008) (Fig. 2).

The rule of Kohonen layer is to investigate the input data to present the outputs outcomes represented by winning neurons that associated for each observation or situation to organize them inside the cluster. Its goal is to facilitate the classification process and it could be implemented in numerous fields which have classification issues (e.g., diagnosis systems, prediction systems, discrimination, and so forth).

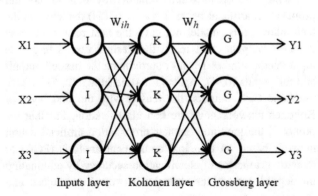

Fig. 1 The architecture of CPN

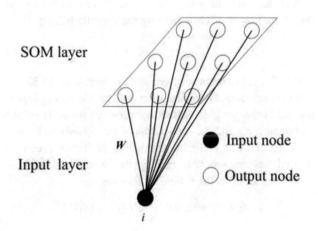

Fig. 2 The self-organizing map (SOM) (Yu et al. 2019)

During the development of the Kohonen network firstly, we add the normalization algorithm for accelerating the learning time in the following formula (Haimoudi et al. 2019).

$$X_i = X_i / \sqrt{\sum_{j=0}^{n=1} X_i^2} \tag{1}$$

where x_i is synaptic weight vector and n is the number of values contained in vector X.

In the next step, we will calculate the distance between the inputs vector and the weight vector of neurons by applying the distance euclidean in order to obtain the outputs of the Kohonen network or map represented by the wining node. So the distance euclidean method is depicted in the following formula:

$$\|xj - x\| = min_n \|w_n - x\| \tag{2}$$

where n is the number of node, w is the synaptic weight vector, and the x is the input object from neuron j.

Finally, we are going to update the synaptic weight of wining node and neighborhood in the following formula [De la Hoz et al. 2015].

$$wj0(t+1) = wj0(t) + \alpha(t)\, hjj0(t)\, (xj - wj0(t)) \tag{3}$$

where x_j is the value of outputs node, $w_{j0}(t+1)$ and $w_{j0}(t)$ are the synaptic weight vector in iteration t and $t+1$, $\alpha(t)$ is learning rate, and h_{jj0} neighborhood function.

B. The Grossberg Algorithm

The second layer named Grossberg is proposed and developed by Stephen Grossberg. It is a self-organizing map, competitive learning. The Grossberg layer was combined with Kohonen network to make a counter propagation artificial neural network (CPN). Hence for beginning the learning procedure of Grossberg layer, we will use the outputs of Kohonen network. These outputs are transmitted to the second layer of Grossberg by the formula below:

$$Gyj = \sum ki.uij \tag{4}$$

where Gy_j is the outputs of Grossberg layer, k_i is the Kohonen network outputs and the same time the Grossberg inputs and u_{ij} is the weights vectors between the ith node of the Kohonen layer and the jth neurons of the Grossberg layer.

In order to approximate the values of inputs vector or desired values, we will modify the synaptic weight of Grossberg layer in the following formula:

$$wij(t+1) = uij(t) + \beta g\,(dj - uij(t))\,ki \tag{5}$$

where βg is learning speed and d_j is the Grossberg desired outputs.

4.2 The Classification Results with CPN

In this study, we collected our data from (Haimoudi et al. 2016) for making our test of the proposed counter propagation network (CPN). In their work, they developed the new approach of Kohonen network to improve it in terms of classification accuracy and clustering. They made two tests, the first one by standard Kohonen self-organizing map and the second one by the modified Kohonen network based on principal component analysis (PCA), because the results obtained by standard Kohonen network are not effective at the level of classification, for that we used their test data at this point. If Kohonen layer did not give the correct classification, then our counter propagation network will give the bad results. These data are structured by a matrix of a dimension of (8X3), the rows of this matrix represent the inputs objects.

In their data, we can notice that inputs 1 and 5 have a similar vector, for inputs 0, 2, 4, and 6 have the linear regularities between its inputs, and finally, inputs 3 and 7 have a normal object (see Table 1). The objective of their work is to eliminate the ambiguities and obstacles detected in order to reach the desired outcomes.

Now, our CPN model is developed. For beginning the learning of this network, we must insert the initial conditions such as iteration rate, learning error (those are the stop condition), and the learning rate represents the learning speed. Those values were entered by user and during the test, we decreased it gradually until stabilization. The values of the initial conditions are:

Iteration rate: 1000, learning rate: 2, learning error: 0.001.

As a result, the CPN finished learning before reaching the initial conditions. For desired outputs of CPN network, we gave randomly desired outputs and we gave the same desired outputs for similar vectors. The results of standard CPN are portrayed in Table 2.

At the base of the results obtained by standard counter propagation artificial neural network (CPN) depicted on the table above, this model not gave the best results, because the CPN outputs can't reach the desired results (e.g., the input vector number 1 can't approximate the desired output) and for inputs (0, 4), inputs (1, 5) and inputs (2, 6) its gave the same CPN outputs, the reason why the first layer of Kohonen network gave the bad classification. For that we proposed the modified counter propagation artificial neural network based on principal component analysis (PCA) to increase it in terms of classification accuracy by eliminating the dependence and the obstacles between input objects and for reaching correct outputs.

Table 1 The normalization of input vectors

Input N°	Initial data	Normalization data
0	8 4 6	0.74 0.37 0.55
1	5 6 7	0.47 0.57 0.66
2	3 5 4	0.42 0.70 0.56
3	11 27 39	0.22 0.55 0.80
4	4 2 3	0.74 0.37 0.55
5	5 6 7	0.47 0.57 0.66
6	9 15 12	0.42 0.70 0.56
7	13 35 42	0.23 0.62 0.74

Table 2 The standard CPN results

Input N°	Desired outputs	CPN outputs	Iteration numbers	Mean error	Learning rate
0	1	[0.009328328512464158]	488	0.0004	1.0181
1	1	[0.991062494485944]			
2	0	[0.00908959986747172]			
3	1	[1.0092921871433367]			
4	0	[0.009328328512464158]			
5	1	[0.9910624944985944]			
6	0	[0.00908959986747172]			
7	1	[0.9907206564757685]			

5 The Modified Counter Propagation Artificial Neural Network

In this section, we suggest the flowchart for explaining the procedure learning of our modified CPN. Firstly, our initial data passes into the block of principal component analysis (PCA) for reducing the dimensionality of data, calculates the correlation matrix and eliminates the dependence between input objects. After that this new data obtained will enter a block of counter propagation neural network for making the learning and classification (Fig. 3).

5.1 The Principal Component Analysis (PCA) Method

PCA is a dimension-reduction tool or method is used in various applications such as face recognition. The purpose of this method is to minimize the space of the dataset. Moreover, it consists to transform a number of correlated objects into a small number of uncorrelated objects, namely principal component. The PCA generates a group of orthogonal base patterns in order that the info is often expressed as a linear combination of that base (De la Hoz et al. 2015).

A. PCA Algorithm

So as to develop the modified counter propagation network based on principal component analysis (PCA), we will develop the various stages of the PCA algorithm for obtaining another matrix that contained the new objects or vectors by eliminating features. The development stages for reducing our data are depicted below.

The originality initial data is presented by a matrix X of $n \times a$ (Fig. 4).

where

The columns represent the variables $x^i = (x_1^j, \ldots, x_n^j)^t$;

The rows represent the individuals $x_j = (x_i^1, \ldots, x_i^a)$;

While x_j^i is the value of the jth variable for the individual i.

First of all, we calculate the vector of the arithmetic means g of each of (a) variable defined in the centered point of individual cloud.

$$g^t = (\bar{x}^1, \ldots, \bar{x}^a) \qquad (6)$$

where

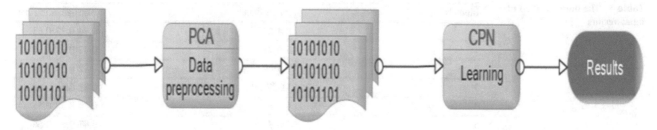

Fig. 3 The proposed modified counter propagation artificial neural network flowchart

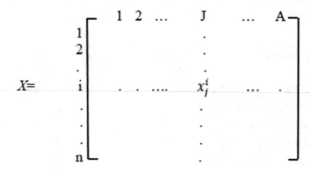

Fig. 4 The presentation of a matrix X

$$\bar{x}^j = \sum_{i=1}^{n} a_i x_i^j$$

In the next step, we must calculate the centered matrix M based on vector g. For calculating this matrix, we need the matrix X. Thus, the centered matrix M is presented by formula below

$$M = X - g^t \qquad (7)$$

g^t is the transposed of the vector g.

In order to calculate the variance–covariance matrix VC, we use the centered matrix M. See formula below:

$$VC = \frac{1}{n} M^t M \qquad (8)$$

M^t is the transposed of the centered matrix M.

The variance–covariance matrix VC is defined by

$$VC = \begin{bmatrix} B_1 & \cdot & \cdot & B_{1a} \\ & B_{21} & \cdot & \cdot \\ & \cdot & \cdot & \cdot \\ B_{a1} & \cdot & \cdot & B_{aa} \end{bmatrix}$$

where B_{kl} is the covariance of the k variable and l, and B_k is the variance of the k variable. The matrix VC also is written according to X and M in the following formula (Decæstecker et al.; Gonzale P.-L; Saporta G.):

$$VC = X^t DX - gg^t = M^t DM \qquad (9)$$

By dividing each column j of the central table M by the standard deviation B_j of the variable j, we construct the table Z of the reduced centered data. The matrix Z is written as a function of X as follows:

$$Z = (X - 1g^t)D_{(1/B)} \qquad (10)$$

where $D_{1/B}$ is diagonal matrix contained $1/B_1,\ldots\ldots,\ 1/B_a$.

Hence, in the last step we determine the correlation matrix C, it's important to calculate it, and it is symmetric. From this matrix, we can remove the linear dependence and influence between (a) variables, the correlation matrix is depicted by the formula below:

$$C = D_{1/B}VC\,D_{1/B} \qquad (11)$$

5.2 The Classification Results with Modified CPN

In order to make the comparison between the standard counter propagation artificial neural network (CPN) and the modified (CPN), we going to use the preprocessed data by principal component analysis (PCA) in our learning of CPN. In this test, we used the same initial conditions of standard CPN. The data obtained by PCA is represented in Table 3.

The principal component analysis (PCA) reduces our original matrix in one vector to simplify it.

As a conclusion, the modified CPN influenced on the maximum number of iterations, minimization error, learning rate, speed and the CPN outputs approximate correctly a desired outputs because the first layer of Kohonen network gave the best classification compared to standard CPN, the Kohonen network gave the bad classification. The results of modified CPN are represented in Table 4.

The results of modified CPN show the best performance, and it reaches correctly of desired outputs and it gave the desired outcomes as we would like. The PCA algorithm could eliminate the dependence between inputs and reduce the original data in small space. This approach was affected on iteration number, means error, learning rate, the speed of application, and it increases the classification accuracy. In

Table 3 The data obtained after PCA method

Input N°	Data obtained by PCA
0	−0.65779
1	−1.02409
2	−1.53019
3	2.30907
4	−1.55335
5	−1.02409
6	0.307247
7	3.17319

Table 4 The modified CPN results

Input N°	Desired outputs	CPN outputs	Iteration numbers	Mean error	Learning rate
0	1	[0.99100806724330255]	280	0.00008	1.3584
1	1	[0.9910455672921619]			
2	0	[0.009466423408190924]			
3	1	[1.0092921871433367]			
4	0	[0.009328328512464158]			
5	1	[1.009580547687483]			
6	0	[0.009100310006078552]			
7	1	[0.9912874816401851]			

this paper (Haimoudi et al. 2016), also they found very important results of classification accuracy for the Kohonen network by PCA.

6 Conclusion

The data analysis has a role important in scientific research, and it is a method widely used in various problems of application in order to improve several approaches. There are many techniques in the data analysis method and each one of these techniques can be used in specific issues, for that we proposed the principal component analysis (PCA) to reduce our initial data to be more practice and more analysis than original data. For making this technique more powerful, we suggested the coupling of principal component analysis (PCA) with counter propagation artificial neural network (CPN) to increase the results and the classification accuracy of standard CPN. The results obtained show that modified CPN gave the best results compared to standard CPN. In future work, we are going to

- Improve the results obtained of the proposed approach,
- Applied this proposed approach in a specific domain to testing its capacity,

- Group this work with this paper (Haimoudi et al. 2016), who have also based on the coupling of principal component analysis (PCA) and Kohonen self-organizing map (SOM).

References

Ahn, C.K.: Passive learning and input-to-state stability of switched Hopfield neural networks with time-delay. Inf. Sci. **180**(23), 4582–4594 (2010)

Ali, A.H.: Member, IEEE self-organization maps for prediction of kidney dysfunction. In: 16th telecommunications forum TELFOR, Serbia, Belgrade, November 25–27 (2008)

Buntine, W.L., Weigend, A.S.: Computing second derivatives in feed-forward networks: a review. IEEE Trans. Neural Netw. **5**(3), 480–488 (1994)

Chen, B.H., Huang, S.C., Yen, J.Y.: Counter-propagation artificial neural network-based motion detection algorithm for static-camera surveillance scenarios. Neurocomputing **273**, 481–493 (2018). https://doi.org/10.1016/j.neucom.2017.08.002

Decæstecker, C., Saerens, M.: Principal component analysis. ULB free University of Bruxelles

Fei, H., Huang, D.-S.: A new constrained learning algorithm for function approximation by encoding a priori information into feed forward neural networks. Neural Comput. Appl. **17**, 433–439 (2008)

Gonzale, P.-L.: Principal component analysis (PCA) School of Industrial Sciences and Information Technology

Haimoudi, E.K., Cherrat, L., Fakhouri, H., Ezziyyani, M.: Towards a new approach to improve the classification accuracy of the kohonen's self organizing map during learning process. Int. J. Adv. Comput. Sci. Appl. **7**(3) (2016)

Haimoudi, E.K., Abdoun, O., Ezziyyani, M.: Towards an intelligent data analysis system for decision making in medical diagnostics. Coast. Res. Lib. 1–13.(2019).https://doi.org/10.1007/978-3-030-11884-6_1

De la Hoz, E., De La Hoz, E., Ortiz, A., Ortega, J., Prieto, B.: PCA filtering and probabilistic SOM for network intrusion detection. Neurocomputing **164**(71), 81 (2015). https://doi.org/10.1016/j.neucom.2014.09.083

Hu, L., Cui, J.: Digital image recognition based on fractional-order-PCA-SVM coupling algorithm measurement. (2019).https://doi.org/10.1016/j.measurement.2019.02.006

Karimi, H.R., Gao, H.: New delay-dependent exponential synchronization for uncertain neural networks with mixed time delays. IEEE Trans. Syst. Man. Cybern. Part B Cybern. **40**(1), 173–185 (2010)

Martin-Rodilla, P., Panach, J.I., Gonzalez-Perez, C., Pastor, O.: Assessing data analysis performance in research contexts: An experiment on accuracy, efficiency, productivity and researchers' satisfaction. Data Knowl. Eng. **116**, 177–204 (2018). https://doi.org/10.1016/j.datak.2018.06.003

Salo, F., Nassif, A.B., Essex, A.: Dimensionality reduction with IG-PCA and ensemble classifier for network intrusion detection. Comput. Netw. (2018). https://doi.org/10.1016/j.comnet.2018.11.010

Saporta, G.: Probability and statistics and data analysis. National Conservatory of Arts and Crafts (2006)

Shi, P., Zhang, Y., Agarwal, R.K.: Stochastic finite-time state estimation for discrete time-delay neural networks with Markovian jumps. Neurocomputing **151**, 168–174 (2015)

Valdez, F., Melin, P., Castillo, O.: Modular neural networks architecture optimization with a new nature inspired method using a fuzzy combination of particle swarm optimization and genetic algorithms. Inf. Sci. **270**(20) 143–153 (2014)

Wolski, G.J., Kruk, A.: Determination of plant communities based on bryophytes: the combined use of Kohonen artificial neural network and indicator species analysis. Ecol. Ind. **113**, 106160 (2020). https://doi.org/10.1016/j.ecolind.2020.106160

Yu, X., Xiao, F., Zhou, Y., Wang, Y., Wang, K.: Application of hierarchical clustering, singularity mapping, and Kohonen neural network to identify Ag-Au-Pb-Zn polymetallic mineralization associated geochemical anomaly in Pangxidong district. J. Geochem. Explor. (2019). https://doi.org/10.1016/j.gexplo.2019.04.007

Zeinali, Y., Story, B.: Structural impairment detection using deep counter propagation neural networks. Procedia Eng. **145**, 868–875 (2016). https://doi.org/10.1016/j.proeng.2016.04.113

Proposed Precautions for Newborn Malware Family Inspired from the COVID19 Epidemic Outbreak

Ikram Ben Abdel Ouahab, Mohammed Bouhorma, Lotfi ElAachak, and Anouar Abdelhakim Boudhir

Abstract

Malicious software attacks cause serious loss to computer users, from personal usage to industrial networks. For this reason, researchers focused more and more on analyzing and detecting malware. Approaches found in literature can well predict a new malware sample belonging to known families, but what about newborn families. In this paper, we perform malware classifiers based on two machine learning algorithms: Random forest and K-Nearest Neighbor. We used the malware visualization technique, so a malware binary is presented as a grayscale image. After that, we calculated the GIST descriptor features for all samples to be ready for the classification. Results obtained are, respectively, 97, 99%, and 98,21% for the two algorithms. Then, we study the behavior of our classifier in the case of new arrival family. Next, we get inspired from the COVID19 disease outbreak. So we proposed precautions to be made as security measures in case of new malware family appearance. With the goal to reduce damage causes by these kind of attacks.

Keywords

Cybersecurity • Malware detection • Machine learning

I. Ben Abdel Ouahab (✉) · M. Bouhorma · L. ElAachak · A. A. Boudhir
Computer Science, Systems and Telecommunication Laboratory (LIST), Faculty of Sciences and Techniques, University Abdelmalek Essaadi Tangier, Tangier, Morocco
e-mail: ikram.benabd@gmail.com

M. Bouhorma
e-mail: mbouhorma@gmail.com

L. ElAachak
e-mail: lotfi1002@gmail.com

A. A. Boudhir
e-mail: boudhir.anouar@gmail.com

1 Introduction

In May 2017, the world has lived a worldwide cyberattack by the WannaCry ransomware cryptoworm. It attacked over 200,000 victims across 150 countries in one day, causing damages in the range of hundreds of millions to billions of dollars. The WannaCry malware was able to spread itself automatically, by making copies of itself and exploiting systems vulnerabilities.

Today in 2020, the COVID19 epidemic attacks humans causing hundreds of death every single day and millions of new affected persons around the globe. To date, we are unable to give exact numbers because they are increasing every hour. The COVID19 epidemic is spreading very quickly and it is hard to stop it (Novel Coronavirus Pneumonia Emergency Response Epidemiology Team 2020). To avoid spreading the virus to more persons, public health officials have taken several precautions such as quarantine and hygiene sanitary practices.

We are not getting into the details; we only got inspired from the way the whole world is taking precautions in order to reduce the risks of infection and the damage of this international epidemic disease.

Otherwise, we perceive many similarities between the WannaCry cyberattack and the COVID-19 epidemic disease. Thus, we had the idea to imitate the COVID-19 precautions in the case of new malware family in the first stage of the malware discovery and analysis. So the proposed solution is given with the goal of reducing risks and damages that could a newborn malware cause in a few days, like what happens with the WannaCry ransomware.

Major works found in literature deal with new malware instances that belong to known families. In this paper, we simulate the appearance of a new family. Then we analyze and see how our system will behave with this new family, and how to know the birth of a new family. In other words, we performed a malware classifier using advanced techniques of malware visualization and machine learning algorithms. Then the malware classifiers are evaluated. After that, we studied a use case of a new malware family using two classifiers based on machine learning.

The rest of this paper is organized as follows: Sect. 2 presents some of the related works to malware classification. Section 3 highlights materials and methods used in the experimentations. Then, Sect. 4 presents the proposed solution in two sub-sections. Next, Sect. 5 displays the obtained results and we discuss it. Finally, Sect. 6 gives conclusion and perspectives for future work.

2 Related Work

A large number of malware analysis and classification techniques have been proposed including static analysis based (Medhat, et al. 2018) (Firdaus 2018), dynamic analysis and behavioral-based techniques (Galal 2016; Pektaş and Acarman 2017), and many other techniques. Among them, the malware visualization technique that we found recently in lots of works like (Akarsh et al. (2019); Ouahab, et al. 2020; Nataraj et al. 2011).

Malware features are classified using machine learning as SVM, KNN, RF, DT, etc. Also, in some works, we found the use of deeper architectures belonging to deep learning models like CNN, RNN, and LSTM.

In (Liu 2017), authors proposed a malware analysis system designed in three modules: data processing, decision-making, and new malware detection. They used grayscale images, Opcode n-gram, import functions, and the shared nearest neighbor (SNN) clustering algorithm. They evaluated their proposed system using 20,000 malware samples from different sources. An accuracy of 98,9% was obtained for unknown malware, and the system was able to successfully detect 86,7% of new malware.

In a previous work (Ouahab, et al. 2020), we perform malware classifier using grayscale malware images from where we extract GIST descriptor features. Then the 320 dimensional features are classified using the K-Nearest Neighbor algorithm. We used Malimg database with all 25 families. Then we got an accuracy of 97%. When testing the model on totally new instances we had 92%.

In (San, et al. 2019), a malware classification system was proposed for 11 families. To classify the malware samples authors used three machine learning algorithms: Random Forest, K-Nearest Neighb,or and Decision Table. Best accuracy was given by RF and KNN with a value of 95,8%.

To sum up, all the related works are present in table 1. We put for each paper technique used to extract features, and the classifier algorithm used each time.

3 Materials and Methods

3.1 Malware Visualization

Malware visualization technique was introduced by Nataraj in (Nataraj, et al. (2011)), where the author convert malware binaries into vector in range of [0,255] that can be visualized as grayscale images (Fig. 1). Malwares of the same family show similarities using many types of image descriptors. For instance, GIST descriptor (Oliva and Torralba xxxx) is used as a feature to classify malware into their families and performed well.

In the experimentation, we use Malimg database which contains 9330 malware samples in the form of grayscale images based on the visualization technique. This database has 25 different malware families.

3.2 Random Forest Classifier

Random Forest (RF) algorithm was formally proposed by Breiman in (Breiman (2001)). Random Forest is a classification method used in machine learning; it is also used in other problems like regression. It constructs a multitude of decision trees in the training phase and output the class that is the mode of the classes of the individual trees.

In our case, data are GIST features, so a decision trees community of 20 trees analyze the given features and each tree makes a prediction. Then majority voting classes give us the final malware family (Fig. 2).

3.3 K-Nearest Neighbor Classifier

The K-Nearest Neighbor (KNN) algorithm is a supervised machine learning algorithm used to deal with many problems including the classification. KNN works by finding the closest points to the unknown input. It used a distance to measure proximity, for instance, the Euclidean Distance is widely used in classification problems.

In our case, we have around twenty classes that represent malware families. For a new instance, we look for the most similar family around (Fig. 3). The value of k is chosen based on previous studies (Ouahab et al. 2020; Ikram et al. 2019).

3.4 Evaluation Metrics

To prove the performance of the previous machine learning models, we use these evaluation metrics: accuracy, recall, precision, f1-score, and hamming loss.

Table 1 Summary of related works

Ref	Features	Classifier	Result	Year
Medhat et al. 2018)	Static analysis based: API functions, file keywords, signatures and files extensions	YARA rules	Précision: 93.3% (Train), 94.14% (Test)	2018
Firdaus 2018)	Android malware features based on static analysis	Naïve Bayes, Functional Trees, J48, Random Forest, and MultiLayer Perceptron	Precision: FT: 95%	2018
Galal 2016)	Dynamic analysis, API calls	Decision Tree, Random Forest, and Support Vector Machine	DT: 97.19% RF: 96.84% SVM: 93.98%	2016
Pektaş and Acarman 2017)	Behavior-based features: File system, network, registry activities observed during the execution traces and n-gram modeling over API-call sequences	Online machine learning algorithms	Accuracy: train: 94% test: 92,5%	2017
Akarsh et al. 2019)	Visualization technique: Malware grayscale images	CNN and LSTM	Accuracy: 94.4%	2019
Nataraj et al. 2011)	Visualization technique: GIST feature	K-Nearest Neighbor (k = 3)	Accuracy: 98%	2011
Ouahab et al. 2020)	Visualization technique: GIST 320-D features	K-Nearest Neighbor (k = 10)	Accuracy: 97%	2019
Liu 2017)	Grayscale images, Opcode n-gram, and import functions	Shared Nearest Neighbor algorithm	Accuracy: 98,9%	2017
San et al. 2019)	API features	Random Forest, K-Nearest Neighbor and Decision Table	RF & KNN: 95.8%	2019

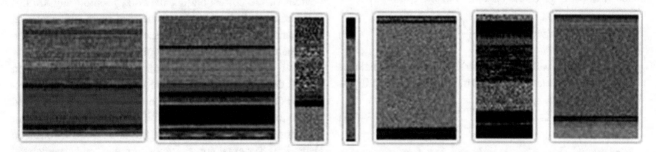

Fig. 1 Examples of malware images: each image belong to a different family

Fig. 2 Random Forest Simplified

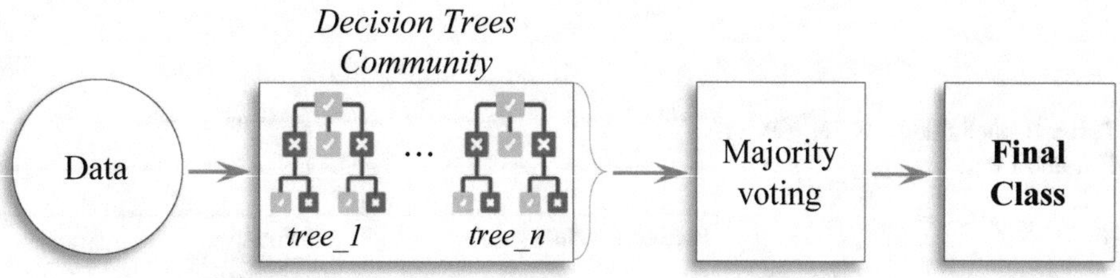

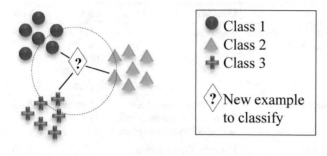

Fig. 3 K-Nearest Neighbor simplified

- $L_H()$ is the Hamming loss between two samples
- $\widehat{y}_j$ is the predicted value for the *j-th* class
- y_j is the true value
- $n_{classes}$ is the number of classes
- $1(x)$ is the indicator function

$$L_H(y, \widehat{y}) = \frac{1}{n_{classes}} \sum_{j=0}^{n_{classes}-1} 1(\widehat{y}_j \neq y_j) \qquad (5)$$

In order to see if the predictions are right or wrong, these metrics are defined using true negative (TN), true positive (TP), false negative (FN), and false positive (FP) as defined in Table 2.

The classification accuracy presents the number of correct predictions divided by all predictions made, and it can be multiplied by 100 to turn it into a percentage (Eq. 1). Since accuracy is not sufficient to evaluate machine learning algorithms we use also others.

$$accuracy = \frac{TP + TN}{TP + FN + TN + FP} \qquad (1)$$

Recall measure also called sensitivity or true positive rate, present the number of positive predictions made divided by the positive class values (Eq. 2).

$$recall = \frac{TP}{TP + FN} \qquad (2)$$

Precision present the number of positive predictions divided by the total number of positive class values predicted (Eq. 3). It gives the exactness of the classifier.

$$precision = \frac{TP}{TP + FP} \qquad (3)$$

F1-score also called f measure passes on a balance between precision and recall. This metric is defined in Eq. 4.

$$f1 = \frac{2 \times precision \times recall}{precision + recall} \qquad (4)$$

Finally, the hamming loss is the fraction of wrongly predicted values. It is mathematically defined in Eq. 5, where

4　Proposed Solution

4.1　Part 1: Inspiration to Make Precautions

In this paper, we got inspired by the actual state that we are living in. The COVID19 virus explosion has affected the entire World causing every day hundreds of deaths and new affected persons. To date, there are no specific drugs to treat this new COVID19 virus. In order to reduce infection and to minimize the propagation of the virus, many precautions have been made all over the world. For instance, quarantine, isolation of possibly affected persons, reduce people movement, and improve daily hygiene and sanitary practices.

For cybersecurity, as example we will talk about a classical computer network. We suggest that a malware attack for a newborn malware family can be a threat for the first time exactly like the COVID19 virus precautions. In other words, in the case of a new malware family attack, precautions to be immediately done are isolate all connected devices of the corresponding network, prohibit any kind of communications by closing connection ports, for example, clean and scan each device to see if the malware has been propagated to all the others or not.

After the trigger of a newborn malware family, reaction can be done using many methodologies. For instance, Network Forensics can be implemented to apply the defined precautions using NIDS and NIPS techniques. Network Intrusion Prevention Systems (NIPS) and Network Intrusion Detection Systems are the most known and specialized methods to find out the malicious attackers Shashidhara and Minavathi: A Survey, 2019.

Table 2 Definitions of TN, TP, FN, and FP

		Actual values	
		Positive	Negative
Predicted values	Positive	TP	FP
	Negative	FN	TN

In Fig. 4, we illustrate the proposed solution, to clarify our inspiration source. Then in the next part, we will develop the part of detecting the new family appearance, based on machine learning algorithms and the visualization technique.

4.2 Part 2: New Family Detection

In this section, we present the new family malware detection process (Fig. 5), in four parts: (1) Split data, (2) Build, train and save the model, (3) Evaluate models, and (4) Test on new family samples.

First, the original Malimg database is composed of 9339 malwares distributed in 25 families. In our experimentations we split it into three parts: TrainDB to train the model, EvaluationDB used to evaluate the model performance, and a whole family is chosen to be tested in a final stage. The new family is "Skintrim.N," and we have 80 malware samples of this family. A detailed distribution of samples per family is given in Table 3.

Second, we made two machine learning classifiers using Random Forest and K-Nearest Neighbor algorithms. We train the two models using the TrainDB part. Then, we save the trained models for later.

Third, we make a performance study of the previously built models. To do so, we use the EvaluationDB part and five evaluation metrics that we defined in detail in Sect. 3.

Fourth, it is time to see how our classifier will behave in the case of a new family. We try to predict the malware family of all the NewFamilyDB part (80 samples). Then we analyze the given results.

5 Results and Discussion

In this section, we present the experimentation results of the machine learning classifiers and the new family tests.

Let us start with testing and evaluating the models: KNN and RF. We use 5 evaluation metrics: accuracy, f1-score, recall, precision and hamming loss. Obtained results are given in Table 4.

In our case, considering all technical details, the best classifier is KNN. It gives an accuracy of 98,218%, which present well-classified malware samples. Then with the evaluation phase, for malware samples that have never been seen by the classifier, we got an accuracy of 95,833%. Also, the loss is very low 1,7% which is a very good result.

Next comes the RF classifier that gives a close result to KNN and it is also considered a good one. For the RF model, the accuracy reached a value of 97,995%, and for new samples we got 91,666%. However, the hamming loss is 2% for testing data and 8,33% for new samples. To visualize clearly obtained results, we put them on a graph in Fig. 6, where higher values represent best results.

In addition to this, the hamming loss is presented separately in Fig. 7. We can see clearly that the losses in testing phase are close in both RF and KNN classifiers. However, in the evaluation phase, we notice an important difference between losses in both models. For the RF the loss is 8,3% and for KNN we got lowest loss of 4,1%.

Apart from that it is time to test the new family samples with the trained KNN and RF models one by one. At this stage, both classifiers don't know this new family, so we are not waiting for a good prediction. However, our goal is to see how the classifier will behave in the case of a new family, and how we can detect the presence of a newborn family. Results are given in Table 5.

Let us start with KNN classifier, in the model predictions using 80 samples we found 2 families: *Instantaccess* and *Alueron.gen!J*. 70% of the new family was predicted to be *Instantaccess*, then 11,25% of the new family was predicted to be *Alueron.gen!J* and the remaining 18,75% got a probability to be both families.

After that, the RF classifier gives us mostly the family: *Alueron.gen!J*. In other words, the highest probability was for the named family, but it shows similarities with 5 other families. The probabilities given using RF are in the range of [0,7–0,45], with the presence of 6 families each time with

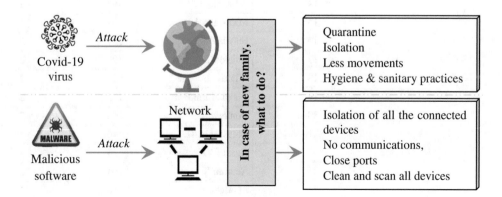

Fig. 4 Analogy description: COVID19 virus and malware attack

Fig. 5 New family detection process

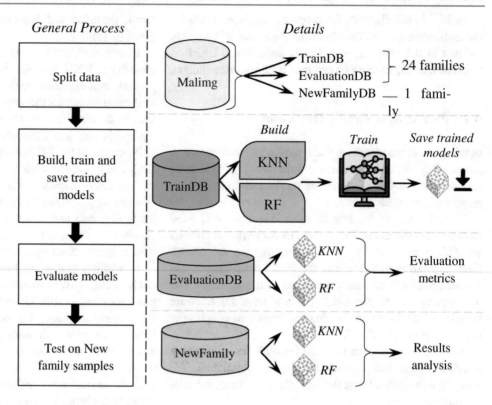

Table 3 Data distribution

ID	Family	Samples per family		
		TrainDB	EvaluationDB	NewFamily
1	Allaple.L	1589	2	-
2	Allaple.A	2947	2	-
3	Yuner.A	798	2	-
4	Lolyda.AA 1	211	2	-
5	Lolyda.AA 2	182	2	-
6	Lolyda.AA 3	121	2	-
7	C2Lop.P	144	2	-
8	C2Lop.gen!g	198	2	-
9	Instantaccess	429	2	-
10	Swizzot.gen!I	130	2	-
11	Swizzor.gen!E	126	2	-
12	VB.AT	406	2	-
13	Fakerean	379	2	-
14	Alueron.gen!J	196	2	-
15	Malex.gen!J	134	2	-
16	Lolyda.AT	157	2	-
17	Adialer.C	120	2	-
18	Wintrim.BX	95	2	-
19	Dialplatform.B	175	2	-
20	Dontovo.A	160	2	-
21	Obfuscator.AD	140	2	-
22	Agent.FYI	114	2	-
23	Autorun.K	104	2	-
24	Rbot!gen	156	2	-
25	Skintrim.N	0	0	80

Table 4 Malware classifier results: Test and evaluation

	RF		KNN	
	Test	Evaluation	Test	Evaluation
Accuracy	0,979955457	0,916666667	0,982182628	0,958333333
F1-score	0,97915544	0,911111111	0,980306256	0,956944444
Recall	0,979955457	0,916660667	0,982182628	0,958333333
Precision	0,979492668	0,916666667	0,979716831	0,965277778
Hamming Loss	0,020044543	0,083333333	0,017817372	0,041666667

Fig. 6 Representation of the malware classifier results: test and evaluation

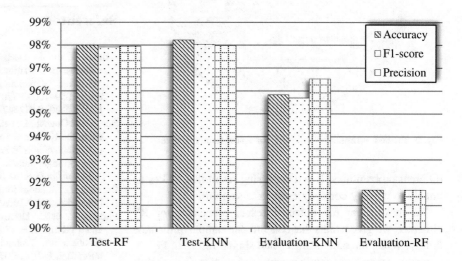

Fig. 7 Hamming Loss results representation

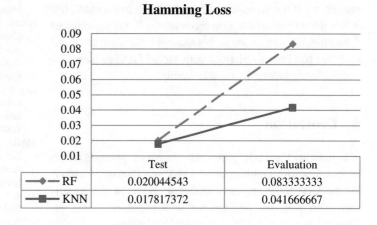

Table 5 New family results summary

KNN Summary			RF Summary		
Samples (%)	Family	Proba	Samples	Family	Proba
70	Instantaccess	1	all	Alueron.gen!J	[0,7–0,45]
11,25	Alueron.gen!J	1			
18,75	BOTH	]0,1[OR (0,1)			

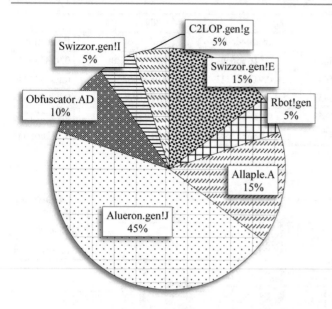

Fig. 8 Prediction results given by RF for a new family sample

different probabilities. It is a confusing result that makes us unable to classify correctly these new samples.

To make clear the obtained result, especially for RF classifier, we give results that concern one sample of the new family. While the other samples had similar result. In Fig. 8, the chart presents all families predicted with their probabilities. Here, the highest value for Alueron.gen!J is 45%, and we can say if the probability of prediction is lower than 50% it is a wrong prediction. One more reason is the presence of 7 families, as we can see in this example. So when the input malware has little similarities with lots of families there is a possibility to belong to another family.

6　Conclusion

We presented in this paper the proposed precautions to reduce newborn malware attacks risk and damage, inspired by the COVID19 epidemic outbreak. Next, in the experimentation part, we perform malware classifiers based on Random Forest and K-Nearest Neighbor algorithms. At this stage, we got an accuracy of 97,99% and 98,21%, respectively. After training and evaluating the models, we test the newborn malware family then we analyzed the obtained results.

Overall, we conclude the necessity to have at least two machine learning classifiers, in order to classify malwares into their corresponding families. This technique benefits in two cases: first for advanced and complicated malware samples belonging to known families, and second for zero-day and newborn families of malware.

As a perspective, we are looking forward to build an efficient system which is able to detect and classify known malware as well as newborn malware. Also, we are working on the classifier itself to make it as efficient as fast, in order to reach the highest accuracy.

Acknowledgements We acknowledge financial support for this research from the "Centre National pour la Recherche Scientifique et technique", CNRST, morocco.

References

Akarsh, S. et al.: Deep learning framework and visualization for malware classification. In: 2019 5th International Conference on Advanced Computing Communication Systems (ICACCS), pp. 1059–1063 (2019). https://doi.org/https://doi.org/10.1109/ICACCS.2019.8728471

Ben Abdel Ouahab, I. et al.: Classification of grayscale malware images using the K-Nearest neighbor algorithm. In: Ben Ahmed, M. et al. (eds.) Innovations in Smart Cities Applications Edition 3. pp. 1038–1050 Springer International Publishing, Cham (2020). https://doi.org/https://doi.org/10.1007/978-3-030-37629-1_75

Breiman, L.: Random forests. Machine Learning. **45**(1), 5–32 (2001). https://doi.org/10.1023/A:1010933404324

Firdaus, A., et al.: Discovering optimal features using static analysis and a genetic search based method for android malware detection. Frontiers Inf. Technol. Electronic Eng. **19**(6), 712–736 (2018). https://doi.org/10.1631/FITEE.1601491

Galal, H.S., et al.: Behavior-based features model for malware detection. J. Comput. Virol. Hack. Tech. **12**(2), 59–67 (2016). https://doi.org/10.1007/s11416-015-0244-0

Ikram, B.A.O. et al.: Machine learning application for malwares classification using visualization technique. In: Proceedings of the 4th International Conference on Smart City Applications. pp. 1–6 Association for Computing Machinery, Casablanca, Morocco (2019). https://doi.org/https://doi.org/10.1145/3368756.3369098

Liu, L., et al.: Automatic malware classification and new malware detection using machine learning. Frontiers Inf. Technol. Electronic Eng. **18**(9), 1336–1347 (2017). https://doi.org/10.1631/FITEE.1601325

Medhat, M. et al.: A New Static-Based Framework for Ransomware Detection. In: 2018 IEEE 16th Intl Conf. on Dependable, Autonomic and Secure Computing, 16th Intl Conf on Pervasive Intelligence and Computing, 4th Intl Conf on Big Data Intelligence and Computing and Cyber Science and Technology Congress (DASC/PiCom/DataCom/CyberSciTech). pp. 710–715 (2018). https://doi.org/https://doi.org/10.1109/DASC/PiCom/DataCom/CyberSciTec.2018.00124

Nataraj, L. et al.: Malware images: visualization and automatic classification. In: Proceedings of the 8th International Symposium on Visualization for Cyber Security - VizSec '11. pp. 1–7 ACM Press, Pittsburgh, Pennsylvania (2011). https://doi.org/10.1145/2016904.2016908

Novel Coronavirus Pneumonia Emergency Response Epidemiology Team: The epidemiological characteristics of an outbreak of 2019 novel coronavirus diseases (COVID-19) in China. Zhonghua Liu Xing Bing Xue Za Zhi. **41**(2), 145–151 (2020). https://doi.org/10.3760/cma.j.issn.0254-6450.2020.02.003

Oliva, A., Torralba, A.: Modeling the Shape of the Scene: A Holistic Representation of the Spatial Envelope. 31

Pektaş, A., Acarman, T.: Classification of malware families based on runtime behaviors. J. Info. Sec. App. **37**, 91–100 (2017). https://doi.org/10.1016/j.jisa.2017.10.005

San, C.C. et al.: Malicious software family classification using machine learning multi-class classifiers. In: Alfred, R. et al. (eds.) Computational Science and Technology. pp. 423–433 Springer, Singapore (2019). https://doi.org/https://doi.org/10.1007/978-981-13-2622-6_41

Shashidhara, D., Minavathi: A survey: on network forensic data acquisition and analysis tools. In: Sridhar, V. et al. (eds.) Emerging Research in Electronics, Computer Science and Technology. pp. 649–659 Springer, Singapore (2019). https://doi.org/https://doi.org/10.1007/978-981-13-5802-9_57

Using Deep Features Extraction and Ensemble Classifiers to Detect Glaucoma from Fundus Images

Stephane Cedric Tekouabou Koumetio,
El Arbi Abdellaoui Alaoui, Imane Chabbar, Walid Cherif,
and Hassan Silkan

Abstract

Early detection of diseases such as glaucoma through regular screening is especially important to prevent vision loss. Glaucoma is one of the leading causes of blindness and visual impairment in adults and the elderly. Several diagnostic techniques are used, ranging from classical expert-centered techniques to modern, sometimes fully computerized diagnostic methods. The implementation of computerized systems based on early detection and classification of clinical signs of glaucoma can greatly improve the diagnosis of this disease. Many authors have proposed models for the automatic classification of clinical signs of glaucoma. However, not only can these models be optimized, but the data used often differ from one experiment to another and would influence their performance in reality. In this article, a classification model is proposed for the detection of early glaucoma from fundus images. The approach consists of three main steps: the first of which is the use of a deep learning model to extract the characteristics of the images, then the PCA and SMOTE techniques, respectively, allow to dimensional reduction and data rebalancing and finally the ensemble classifiers separate glaucomatous and healthy eyes. A test was carried out on a database of 401 fundus images where 301 were healthy and 100 were diseased with very promising results in terms of accuracy, precision, recall, and F_i Score.

Keywords

PCA • Glaucoma • Classification • SMOTE • Fundus images • Ensemble classifiers

1 Introduction

Abnormalities associated with the eye can be divided into two classes: the first includes eye diseases such as cataracts which are the most common, and glaucoma in the second group is classified as a disease related to a lifestyle conditions such as hypertension, arteriosclerosis, and diabetes Hayashi et al. (2001). These pathologies can lead to a reduction in visual acuity, visual impairment, and blindness. The traditional diagnosis and treatment of these abnormalities (pathologies) take a long time. Thus, the automation of diagnosis using new technologies and machine learning will make the diagnosis process faster and more reliable to prevent vision loss. Artificial intelligence is part of the continuity of computer science, whose computing power continues to grow, increased by the availability of large masses of data that the Internet world knows how to aggregate Stacey et al. (2007). Artificial intelligence isn't just about winning the GO World Championships; it now allows cars to run without drivers, robots to become increasingly autonomous, doctors to make more accurate diagnoses, lawyers to make more precise contracts. After information, then knowledge, it is now the turn of expertise to be available everywhere, accessible to everyone. Its scarcity, which up to now has been the source of a legitimate but considerable profit, is on the way to becoming abundance. Only the sharpest experts, who have understood how to take advantage of the novelty that artificial intelligence brings, will

S. C. Tekouabou Koumetio (✉) · H. Silkan
Laboratory LAROSERI, Faculty of Sciences, Department of Computer Science, Chouaib Doukkali University, B.P. 20, 24000 Jadida, El, Morocco
e-mail: ctekouaboukoumetio@gmail.com

E. A. Abdellaoui Alaoui
EIGSI, 282 Route of the Oasis, 20140 Mâarif, Casablanca, Morocco

Faculty of Sciences and Techniques at Errachidia, Department of Computer Science, E3MI Research Team, University of Moulay Ismail, Meknès, Morocco

I. Chabbar
Ophthalmology Department B, Specialty Hospital, CHU IBN SINA, Mohammed V University, Quartier-Souissi, 6220 Rabat, Morocco

W. Cherif
Laboratory SI2M, Department of Computer Science, National Institute of Statistics and Applied Economics, Mohammed V University, 6217 Rabat, Morocco

M. Ben Ahmed et al. (eds.), *Emerging Trends in ICT for Sustainable Development*, Advances in Science, Technology & Innovation, https://doi.org/10.1007/978-3-030-53440-0_8

survive. But this is potentially very good news for the whole of humanity Kononenko (2001).

The implementation of computerized systems based on the early detection and classification of clinical signs of glaucoma has greatly improved the diagnosis of this disease. Several authors have proposed models allowing the automatic classification of clinical signs of glaucoma. However, not only these models are not efficient enough and remain optimizable but also often do not take into account the problem of data instability in their construction and the performance test measures adapted to evaluate them. In this paper, we propose a computerized predictive classification model for glaucoma combining AlexNet for features extraction, PCA for dimensional reduction, SMOTE for imbalance dealing and finally ensemble classifiers for classification. The model was texted using fundus images dataset and performances were evaluated in terms of precision, accuracy, recall, and F_1 score. The rest of this paper is organized as follows: In the Sect. 2, we will make a brief presentation of glaucoma pathology and the challenge of artificial intelligence for their automatic detection. Then in the Sects. 3 and 4, we will present, respectively, the approach and the results obtained, and finally, the Sect. 5 will conclude our paper.

2   Background

2.1   Preliminary About Glaucoma

Glaucoma is an ophthalmologic pathology defined as a neurodegenerative disease resulting in a progressive loss of ganglion nerve cells. Clinically, this damage results in a characteristic degradation of the optic nerve with an enlargement of the papillary excavation on examination of the fundus and a specific alteration of the visual field (Fig. 1).

Glaucoma refers to a group of clinical diseases that share a common characteristic, consisting of a deepening or exca-

vation of the optic nerve head. The main forms of glaucoma are open-angle, angle closure, and congenital glaucoma. The open-angle glaucoma is also called primary or chronic glaucoma because it is the most common type of glaucoma (at least 90% of all glaucoma cases). Although the majority of open-angle glaucoma is age-related and the risk of having glaucoma increases as we get older, this eye disease can affect people of all ages, even newborns. Open-angle glaucoma is caused by the slow clogging of the drainage canals, resulting in increased eye pressure with symptoms and damage that are not noticed Weinreb and Khaw (2004), Weinreb et al. (2014). "Open-angle" means that the angle where the iris meets the cornea is as wide and open as it should be (See Fig. 2).

The angle-closure glaucoma is a less common form of glaucoma that is caused by blocked drainage canals, resulting in a sudden rise in intraocular pressure. It is a type of glaucoma that presents very noticeable symptoms and damage. It is also called acute glaucoma or narrow-angle glaucoma.

The congenital glaucoma occurs in babies when there is incorrect or incomplete development of the eye's drainage canals during the prenatal period. This is a rare condition that may be inherited. When it is uncomplicated, microsurgery can often correct the structural defects. Other cases are treated with medication and surgery.

All these forms make glaucoma the second leading cause of blindness around the globe after cataract. This eye disease could be treated by lowering the intraocular pressure (IOP), accomplished by daily eye drop administration, laser treatment to the eye, or ocular surgery Quigley (2018, 2019).

2.2   Challenge of Machine Learning for Early Diagnosis of Glaucoma Risk

In our literature investigation, the glaucoma risk assessment methods were developed by more than many kinds of machine learning technique, which were meanly based on five kinds

Fig. 1 Image of the right (**a**) and left (**b**) optic nerve of a glaucomatous patient

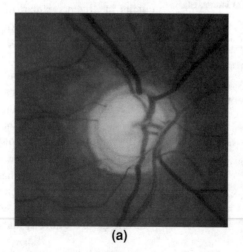
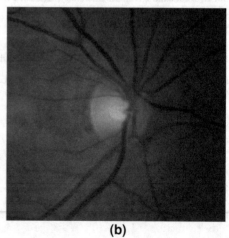

(a)   (b)

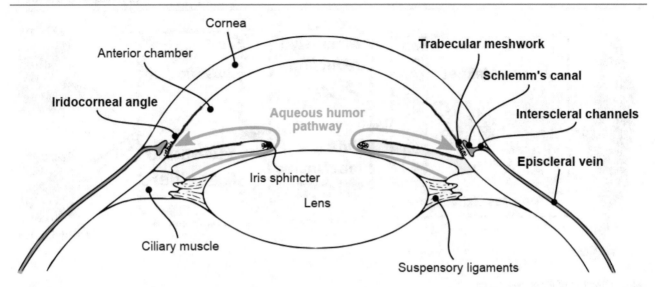

Fig. 2 Schematic representation of the aqueous humor flow. The light-blue lines represent the aqueous humor pathways and the red lines are the iridocorneal angles

of ANN, five kinds of radial basis function (RBF) network, k-nearest neighbor algorithm (kNN), and four kinds of support vector machine (SVM) Hatanaka et al. (2012), Andersson et al. (2013), Ceccon et al. (2013). If to do this the vast majority of authors have used supervised classification methods, the same remains true for a few who have proposed unsupervised classification approaches Sample et al. (2004). Such machine learning methods have been used to improve clinical techniques and some mathematical used measures or censored methods Erler et al. (2014), Bryan et al. (2013). However, such machine learning supervised or unsupervised methods tend to be preferred in classification phase over the deep learning-based approaches which were the most recently used methods for challenging well automatic glaucomatous eyes diagnostic Kucurss et al. (2018), Gulshan et al. (2016). If these algorithms are used for the classification phase, several other more efficient classifiers such as ensemble-based one remain more efficient for the classification tasks. Ensemble classifiers must provide good tools in the modeling process of an approach allowing to predict the automatic classification of patients at high risk of glaucoma from fundus images. In order to do this, it is more than necessary to extract characteristics from the images, which is currently done very well by pre-trained deep learning models such as AlexNet. Preprocessing techniques are also necessary to transform these extracted features by size reduction and resampling.

3   Materials and Methods

Our approach is based essentially on three main steps as summarized by the flowchart of Fig. 3. The first consists of features extraction with pre-trained alexn et al. algorithm based on transfer learning technique. After features extraction, they undergo to preprocessing using principal component analysis for dimension reduction and SMOTE for imbalance dealing. Then, finally, the optimization of the ensemble-based algorithms parameters to discriminate between glaucomatous and healthy fundus images. The rest of this part will present in turn the techniques included in these different steps.

3.1   Fundus Image Data Set

To develop our approach, we used in our study a set of left and right fundus images that were collected on a patient set. Of the 401 fundus images collected, a total of 301 are healthy and disease-free, while 100 are glaucomatous as shown in Fig. 1. Then, the images were stored in a file from which they are read and autocrossed. Preprocessing and scaling to bring all the images back to a 227×227 format suitable for extraction by the AlexNet pre-trained deep neural network through deep transfer learning technique.

3.2   Features Extraction Using AlexNet Network

AlexNet is the result of the very famous ImageNet competition Deng et al. (2009) in 2012, where it obtained an error in the top 5 of only 15.3%, more than 10.8% better than the result of the second one who used the shallow neural network Wang et al. (2019). Since then, the model has become very famous for computer vision problems. The original AlexNet was run on two graphics processing units (GPUs). Nowadays, researchers tend to use only one GPU to implement AlexNet

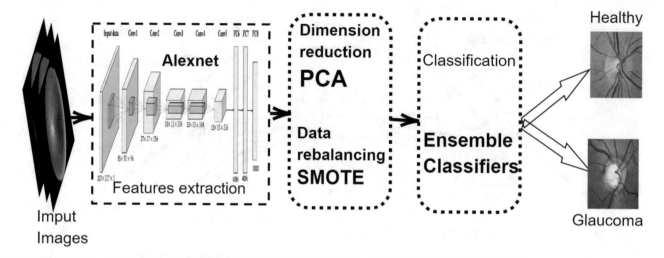

Fig. 3 Flowchart of the proposed approach

mostly from the layers of the model pre-trained using transfer learning techniques. The second block of Fig. 3 illustrates the structure of classical AlexNet. In our study, we count only the layers associated with the learning weights to extract the characteristics of the background images. Thus, AlexNet in its architecture contains five conv layers (CL) and three fully connected layers (FCL), for a total of eight layers. Details of the weights and biases of AlexNet are presented in Table 1. The total AlexNet weights and biases are 60,954,656 + 10,568 = 60,965,224. In Matlab, the variable is stored as a "single-float" type, taking four bytes for each variable. In total, we, therefore, had to allocate 233 MB. The input image format is 227 × 227 × 3 allowing to easily extract the characteristics of the images thanks to transfer learning. Thus, the model returns vectors of extracted characteristics of size 1 × 4096 for each image. These characteristics allow to proceed to the preprocessing and then the classification.

3.3 PCA for Dimension Reduction

Principal component analysis (PCA) Kumar et al. (2010) is a classical dimensional reduction technique that is widely used in dimensional reduction and feature extraction. The main idea of PCA is to first convert the original data into an uncorrelated data called principal components and then select the principal components according to the principle of maximum variance in the transformed space Kang et al. (2020). The calculation of the principal components is relatively simple. It is based on the basic principles of linear algebra such as vectors, matrices, their operations and properties, and in particular vectors and eigenvalues. In this work, PCA is used to reduce the characteristics of images from AlexNet layers. They allow the transition from 1 × 4096 dimensional

data vectors each to 1 × 50 dimensional vectors. PCA is used for dimension reduction for the following reasons. Firstly, the 4096 characteristics of the base layers obtained with AlexNet contain not only additional information but also a large amount of redundant information. Therefore, the PCA is a very good tool for merging the additional information and reduces this redundant information Kherif and Latypova (2020), Kang et al. (2020). Secondly, by merging the base layers with PCA, the influence of noise contained in the original image can be effectively reduced because the detail and noise contained in the original data will be effectively removed. Therefore, PCA can extract the most relevant information from the base layers, and can effectively improve spectral differences among pixels in the image Kang et al. (2020).

3.4 SMOTE Algorithm for Data Resampling

The synthetic minority oversampling technique (SMOTE) is one of the most imbalance methods used for data overampling in minority class. Its strategy consists of creating artificial instances for the minority class according to the following operating process: from an instance x_i of the minority class; create a new artificial instance from xi by first isolating the k closest neighbors to x_i', from the minority class. Then, randomly select a neighbor and finally generate a synthetic example along the imaginary line connecting x_i and the selected neighbor Chawla et al. (2002). This process is clearly described by the algorithm 1. From this strategy, several similar approaches will be derived by SMOTEBoost Chawla et al. (2003), Borderline-SMOTE Han et al. (2005), Majority-Weighted SMOTE Barua et al. (2012), ... However, we haven't used them in this work for reasons of efficiency comparatively to SMOTE for VF data, but they could inspire many other researches.

Table 1 Learnable layers characteristics in AlexNet pre-trained model

Name	Weights	Biases
CL1	11*11*3*96 = 34,848	1*1*96 = 96
CL2	5*5*48*256 = 307,200	1*1*256 = 256
CL3	3*3*256*384 = 884,736	1*1*384 = 384
CL4	3*3*192*384 = 663,552	1*1*384 = 384
CL5	3*3*192*256 = 442,368	1*1*256 = 256
FCL5	4096*9216 = 37,748,736	4096*1 = 4,096
FCL7	4096*4096 = 16,777,216	4096*1 = 4,096
FCL8	1000*4096 = 4,096,000	1000*1 = 1,000
CL Subtotal	2,332,704	1,376
FCL Subtotal	58,621,952	9,192
Total	60,954,656	10,568

Algorithm 1 SMOTE algorithm

Input: • N : number of instances in the minority classes;
  • n amount of SMOTE (in %);
  • k : the number of nearest neighbors;
  • minority data $D = x_i \in X$, where $i = 1, 2, ..., N$.
Output: D': synthetic data from D;
1: n=(int)(n/100);
2: **for** $i = 1 to N$ **do**
3:   Find the k nearest neighbors of x_i;
4:   **while** $n0$ **do**
5:     Select one of the k nearest neighbors of x_i
6:     Select a random number $\alpha \in [0,1]$
7:     $x = x_i + \alpha(x - x_i)$
8:     Append x to D'
9:   **end while**
10: **end for**

3.5 Image Classification

3.5.1 Bagging Classifier (BC)

The bagging approach is to try to mitigate the dependency between the models that are aggregated by building them on bootstrap samples. Due to its simplicity, speed, and efficiency, this algorithm has been widely used for several decades either individually or in combination with preprocessing or dynamic selection strategies.

3.5.2 Random Forest Classifier (RF)

As its name suggests, a random forest consists of aggregating dependent trees on random classification or regression variables. For example, bagging trees (building trees on bootstrap samples) define a random forest. Most often, trees are built with the classification and regression tree (CART) algorithm whose principle is to recursively partition the space generated by the explanatory variables in a dyadic way. More precisely, at each stage of the partitioning, a part of the space is cut into two sub-parts according to a variable X_j Zhang and Haghani (2015). A random forest family stands out among the rest, notably because of the quality of its performance on many data sets. These are the Random Forests introduced by Breiman (2001) in 2005. In many works, the term random forests are often used to designate this family.

3.5.3 AdaBoosting Classifier (ABC)

The adaptive boosting (ABC) ensemble method is based on the fundamental boosting principle. But during the Adaboost process, a new predictor to correct the error of its predecessor simply pays a little more attention to the training instances under which the predecessor has adapted. The result is new predictors that focus more and more on difficult cases. For example, to create an AdaBoost classifier, consider a first classifier that could be for many cases nothing more than a decision tree. This primitive tree formed is used to make predictions on the set of formations Tekouabou et al. (2020). The weight corresponding to the misclassified training instances is then increased. A second classifier is then formed based on these updated weights Mishra et al. (2017). The second classifier again makes predictions about the training game. The weights are then updated, and so on. Once all the predictors have been formed, the set makes predictions very similar to bagging or pasting operations Tekouabou et al. (2020), Mishra et al. (2017). The only difference is that the resulting predictors have different weights based on their overall accuracy over the weighted training set Géron (2017).

3.5.4 Gradient Boosting Classifier (GB)

Gradient boosting (GB) algorithms are very popular boosting algorithm. Gradient Boosting proceeds similarly to AdaBoost by sequentially adding predictors to a set, so that everyone tries to correct the errors of its predecessor. However, instead of adjusting the instance weights at each iteration, as AdaBoost does, this method tries to fit the new predictor to the residual errors committed by the previous one Tekouabou et al. (2020).

4 Experimentation and Results Analysis

4.1 Experimental Protocol and Performance Measures

The implementations have being carried out on a windows platform under python 3.7 using principally sklearn library Géron (2017). The computer used is of the "*Asus*" brand with the following configuration: 8 GB of RAM, intel core i7 processor, and an NVIDIA Geforce 930M for graphic card. For the experiments, the whole dataset has been divided into 5 folds and the tests were performed with cross-validation using 4 folds for training and the one for testing. Most often for classification problems, the predictive accuracy rate, which reports correct predictions on total predictions, is the most commonly used measure. Accuracy score makes it possible to evaluate the quality of the solution produced on the basis of the percentage of correct predictions in relation to the total number of instances. This measure is currently among the most widely used by researchers in the practice of discrimination and selection of the optimal solution and optimization of classifiers. Accuracy is calculated by the formula of Eq. (1).

Regarding our database which is unbalanced Tekouabou et al. (2019), we will evaluate our approach in terms of other performance measures such as precision (Eq. (2)), recall (Eq. (3)), and F_1 score (Eq. (4)) and AUC defined as follows:

$$Accuracy = \frac{a + b}{a + b + c + d} \tag{1}$$

$$Precision = \frac{a}{a + c} \tag{2}$$

$$Recall = \frac{a}{a + d} \tag{3}$$

$$F_1 = \frac{2 * Recall * Precision}{Recall + Precision} \tag{4}$$

a: refers to the number of true positive, b: refers to the number of true negative, c: is the number of false positive, and d: is the number of false negatives.

4.2 Results Analysis and Discussion

Table 4 illustrates the obtained performances according to four metrics used. The accuracy performance from Fig. 4a shows that RF achieves the best score averaging approximately 90.3% followed by GB then BC with, respectively,

88.61% and 87.55%, and finally, ABC with 81.44% gave the lowest accuracy score.

Fig. 4b shows the precision score for the four ensemble algorithms. From this figure, we notice that RF also achieved the best precision score averaging 93.33% which is almost better than the obtained accuracy. It is followed by BC then GB with, respectively, 85.04% and 84.53%, and finally, ABC with 82.90% gave the lowest precision score almost better than it performed accuracy score.

Fig. 4c illustrated the recall score which shows that BC is the most efficient for this measure with 94.90%. It is followed by GB and RF which performed, respectively, 93.63 and 89.82% and finally ABC with 87.71%.

Fig. 4d illustrated the F_1 score which shows that RF is the most efficient for this measure with 90.41%. It is followed by BC and GB which performed, respectively, 88.98 and 88.31% and finally ABC with 81.95%.

Referring to these results, we can globally conclude that ensemble classifiers provide good tools for fundus images classification using AlexNet features. From the experiments, the best ensemble classifier for our task depends on the performance measure chosen. For each of the three measurements used, two of the four experimented classifiers were at least once better depending on the measurement used. By comparing the four performance scores of all these classifiers, we easily deduce that RF classifier is more favorable four discriminating between glaucomatus and healthy eyes from AlexNet features Wang et al. (2019). Note that our experiments were still little limited by the fact that the measurements for each metric were made independently. Knowing that the data split process is random and could affect the performance obtained but without changing the general trend.

5 Conclusion

Artificial intelligence-based techniques and high-performance computing have greatly contributed to advances in various disciplines where practical solutions to their problems have been found, including the automatic diagnosis of signs of glaucoma. In the face of such problems, the optimal use of models based on machine learning algorithms, when used correctly, can lead to a better diagnosis than the expert processes previously used. This article studies an approach to glaucoma diagnosis based on fundus images. The approach uses a pre-trained AlexNet model for feature extraction, PCA for size reduction, and SMOTE for resampling the data with the help of ensemble classifiers. The performances obtained show that this approach is very promising and will be studied further in our future work.

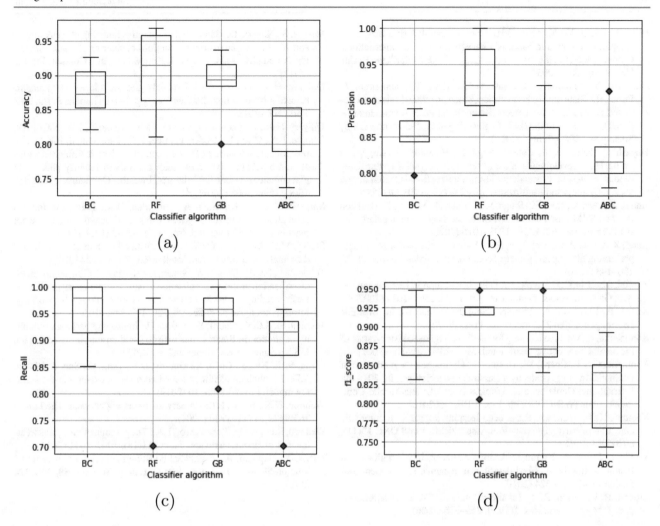

Fig. 4 Obtained results according to ensemble classifiers; **a** accuracy; **b** precision; **c** recall; **d** F_1 score

References

Andersson, S., Heijl, A., Bizios, D., Bengtsson, B.: Comparison of clinicians and an artificial neural network regarding accuracy and certainty in performance of visual field assessment for the diagnosis of glaucoma. Acta Ophthalmol. **91**(5), 413–417 (2013)

Barua, S., Islam, M.M., Yao, X., Murase, K.: MWMOTE-majority weighted minority oversampling technique for imbalanced data set learning. IEEE Trans. Knowl. Data Eng. **26**(2), 405–425 (2012)

Breiman, L.: Random forests. Mach. Learn. **45**(1), 5–32 (2001)

Bryan, S.R., Vermeer, K.A., Eilers, P.H., Lemij, H.G., Lesaffre, E.M.: Robust and censored modeling and prediction of progression in glaucomatous visual fields. Investig. Ophthalmol. Vis. Sci. **54**(10), 6694–6700 (2013)

Ceccon, S., Garway-Heath, D.F., Crabb, D.P., Tucker, A.: Exploring early glaucoma and the visual field test: classification and clustering using bayesian networks. IEEE J. Biomed. Health Inf. **18**(3), 1008–1014 (2013)

Chang, Y.W., Hsieh, C.J., Chang, K.W., Ringgaard, M., Lin, C.J.: Training and testing low-degree polynomial data mappings via linear SVM. J. Mach. Learn. Res. **11**(Apr), 1471–1490 (2010)

Chawla, N.V., Lazarevic, A., Hall, L.O., Bowyer, K.W.: SMOTEBoost: improving prediction of the minority class in boosting. In: European Conference on Principles of Data Mining and Knowledge Discovery, pp. 107–119. Springer, Berlin, Heidelberg, Sept 2003

Chawla, N.V., Bowyer, K.W., Hall, L.O., Kegelmeyer, W.P.: SMOTE: synthetic minority over-sampling technique. J. Artif. Intell. Res. **16**, 321–357 (2002)

Deng, J., Dong, W., Socher, R., Li, L. J., Li, K., Fei-Fei, L.: Imagenet: a large-scale hierarchical image database. In: 2009 IEEE Conference on Computer Vision and Pattern Recognition, pp. 248–255. IEEE, June 2009

Erler, N.S., Bryan, S.R., Eilers, P.H., Lesaffre, E.M., Lemij, H.G., Vermeer, K.A.: Optimizing structure-function relationship by maximizing correspondence between glaucomatous visual fields and mathematical retinal nerve fiber models. Investig. Ophthalmol. Vis. Sci. **55**(4), 2350–2357 (2014)

Géron, A.: Hands-on machine learning with Scikit-Learn and TensorFlow: concepts, tools, and techniques to build intelligent systems. O'Reilly Media, Inc. (2017)

Gulshan, V., Peng, L., Coram, M., Stumpe, M.C., Wu, D., Narayanaswamy, A., Kim, R.: Development and validation of a deep learning algorithm for detection of diabetic retinopathy in retinal fundus photographs. Jama **316**(22), 2402–2410 (2016)

Han, H., Wang, W. Y., Mao, B.H.: Borderline-SMOTE: a new over-sampling method in imbalanced data sets learning. In: International Conference on Intelligent Computing, pp. 878–887. Springer, Berlin, Heidelberg, Aug 2005

Hatanaka, Y., Muramatsu, C., Sawada, A., Hara, T., Yamamoto, T., Fujita, H.: Glaucoma risk assessment based on clinical data and automated nerve fiber layer defects detection. In: 2012 Annual International Conference of the IEEE Engineering in Medicine and Biology Society, pp. 5963–5966. IEEE (2012)

Hayashi, J., Kunieda, T., Cole, J., Soga, R., Hatanaka, Y., Lu, M., ... Fujita, H.: A development of computer-aided diagnosis system using fundus images. In: Proceedings Seventh International Conference on Virtual Systems and Multimedia, pp. 429–438. IEEE, Oct 2001

Jindal, A., Dua, A., Kaur, K., Singh, M., Kumar, N., Mishra, S.: Decision tree and SVM-based data analytics for theft detection in smart grid. IEEE Trans. Ind. Inf. **12**(3), 1005–1016 (2016)

Kang, X., Duan, P., Li, S.: Hyperspectral image visualization with edge-preserving filtering and principal component analysis. Inf. Fusion **57**, 130–143 (2020)

Keerthi, S.S., Lin, C.J.: Asymptotic behaviors of support vector machines with Gaussian kernel. Neural Comput. **15**(7), 1667–1689 (2003)

Kherif, F., Latypova, A.: Principal component analysis. In: Machine Learning, pp. 209–225. Academic Press (2020)

Kononenko, I.: Machine learning for medical diagnosis: history, state of the art and perspective. Artif. Intell. Med. **23**(1), 89–109 (2001)

Koumétio, C.S.T., Cherif, W., Hassan, S.: Optimizing the prediction of telemarketing target calls by a classification technique. In: 2018 6th International Conference on Wireless Networks and Mobile Communications (WINCOM), pp. 1–6. IEEE, Oct 2018

Kucurss, H.G., Sznitman, R.: A deep learning approach to automatic detection of early glaucoma from visual fields. PLoS ONE **13**(11), e0206081 (2018)

Kumar, P., Mittal, A., Kumar, P.: Addressing uncertainty in multi-modal fusion for improved object detection in dynamic environment. Inf. Fusion **11**(4), 311–324 (2010)

Lippert, R.A., Rifkin, R.M.: Infinite-σ limits for Tikhonov regularization. J. Mach. Learn. Res. **7**(May), 855–876 (2006)

Mishra, S., Mishra, D., Santra, G.H.: Adaptive boosting of weak regressors for forecasting of crop production considering climatic variability: An empirical assessment. J. King Saud Univ.-Comput. Inf, Sci (2017)

Quigley, H.A.: Use of animal models and techniques in glaucoma research: introduction. In: Glaucom, pp. 1–10. Humana Press, New York, NY (2018)

Quigley, H.A.: 21st century glaucoma care. Eye **33**(2), 254–260 (2019)

Sample, P.A., Chan, K., Boden, C., Lee, T.W., Blumenthal, E.Z., Weinreb, R.N., Goldbaum, M.H.: Using unsupervised learning with variational bayesian mixture of factor analysis to identify patterns of glaucomatous visual field defects. Investig. Ophthalmol. Vis. Sci. **45**(8), 2596–2605 (2004)

Shrivastava, N.A., Khosravi, A., Panigrahi, B.K.: Prediction interval estimation of electricity prices using PSO-tuned support vector machines. IEEE Trans. Ind. Inf. **11**(2), 322–331 (2015)

Stacey, M.E., Mcgregor, C.: Temporal abstraction in intelligent clinical data analysis: a survey. Art. Intell. Med. **39**(1), 1–24 (2007)

Tekouabou, S.C.K., Cherif, W., Silkan, H.: A data modeling approach for classification problems: application to bank telemarketing prediction. In: Proceedings of the 2nd International Conference on Networking, Information Systems & Security, pp. 1–7, Mar 2019

Tekouabou, S.C.K., Cherif, W., Silkan, H.: Improving parking availability prediction in Smart Cities with IoT and ensemble-based model. J. King Saud Univ.-Comput. Inf, Sci (2020)

Wang, S.H., Xie, S., Chen, X., Guttery, D.S., Tang, C., Sun, J., Zhang, Y.D.: Alcoholism identification based on an AlexNet transfer learning model. Front. Psych. **10** (2019)

Weinreb, R.N., Khaw, P.T.: Primary open-angle glaucoma. The Lancet **363**(9422), 1711–1720 (2004)

Weinreb, R.N., Aung, T., Medeiros, F.A.: The pathophysiology and treatment of glaucoma: a review. Jama **311**(18), 1901–1911 (2014)

Zhang, Y., Haghani, A.: A gradient boosting method to improve travel time prediction. Trans. Res. Part C: Emerg. Technol. **58**, 308–324 (2015)

The Role of Applications Deep Learning in Achieving Sustainable Development Goals

Redouane Lhiadi, Abdelali Kaaouachi, and Abdessamad Jaddar

Abstract

Deep Learning spanning from artificial intelligence (AI), is one of technology which has tremendous potential to revolutionize daily processes in various fields, leading humanity to an era of self-sufficiency and productivity. In particular, there is the need to extend research into Deep Learning and its broader application to many sectors, and to assess its impact on achieving the Sustainable Development Goals (SDGs). Using consensus-based expert opinions, we find that the use of deep learning is widespread across all sectors. However, current research studies overlook some important aspects. The rapid development of Deep Learning needs to be supported by the organizational insight and oversight necessary for AI-based technologies in general, here we present and discuss implications of how Deep Learning enables the delivery Agenda for Sustainable Development.

Keywords

Deep learning • Sustainable • SDGs • Artificial intelligence • Supervised learning • Unsupervised learning

R. Lhiadi (✉) · A. Kaaouachi
High School of Technology, University of Mohammed 1st Oujda, Oujda, Morocco
e-mail: lhiadi.redouane@gmail.com

A. Kaaouachi
e-mail: akaaouachi@hotmail.com

A. Jaddar
National School of Business and Management, University of Mohammed 1st Oujda, Oujda, Morocco
e-mail: ajaddar@gmail.com

1 Introduction

Today, sustainable development goals are closely linked with Deep Learning (DL). The central goals of DL research include reasoning, planning, learning, natural language processing (communication), perception and objects recognition. Current popular DL approaches include statistical methods and computational intelligence. There are a large number of tools used in DL, including versions of search and methods based on probability, bioinformatics, neural networks, etc.

Many ancient machine learning and signal process techniques exploit shallow architectures, that contain one layer of nonlinear feature transformation. Samples of shallow architectures square measure typical hidden Andre Mark off models (HMMs), linear or nonlinear resurgent systems, conditional random fields (CRFs), most entropy (MaxEnt) models, support vector machines (SVMs), kernel regression, and multilayer perceptron (MLP) with one hidden layer. A property common to those shallow learning models is that the easy design that consists of just one chargeable layer for remodeling the raw input signals or options into a problem-specific feature area, which can be unperceivable. Taking the instance of a support vector machine, we can say that it's a shallow linear separation model with one feature transformation layer once kernel trick is employed, and with zero feature transformation layer once kernel trick isn't used. Human scientific discipline mechanisms (e.g., vision and speech), however, recommend the necessity of deep architectures for extracting com-plicated structure and building cognitive content made from sensory inputs (e.g., natural image and its motion, speech, and music). As an example, human vocalization and perception systems square measure each, equipped with clearly bedded ranked structures in remodeling info from the wave shape level to the linguistic level and contrariwise. It is natural to believe that the state of the art may be advanced in process-ing these kinds of media signals if economical and effective deep learning algorithms square measure are developed. Signal process systems with deep architectures square measure composed of the many layers of nonlinear process stages, wherever every lower layer's outputs square measure fed to its immediate higher layer because of the input. The booming deep learning techniques developed thus far, share two further key properties: the generative nature of the model, which generally needs a further prime layer to perform the discriminative task, associate degreed an unsupervised pre-training step that produces effective use of huge amounts of unlabeled coaching knowledge for extracting structures and regularities within the input options.

In this paper, we present and discuss implications of how Deep Learning enables the delivery Agenda for Sustainable Development. The paper is constituted by nine sections. After this section which is devoted to the introduction, the second section defines learning. The third section gives a brief history about deep learning. The fourth section describes an overview of deep learning and its applications. The fifth section details deep learning applications to SDG. The sixth section explores the relationship between learning and data analysis (learning bases). The seventh section presents the two main modes of supervised and unsupervised learning, as well their combination. The paper is closed by a conclusion and a set of main references.

2   Define Learning

The concept of learning, which Hunt, Marin and Stone (1966) described compactly as "[the] capability to develop classification rules from experience" has long been a principal space of machine learning analysis. Supervised idea learning systems are furnished with data concerning many entities whose category membership is thought and turn out from this to get a characterization of every category. One major dimension to differentiate the idea of learning systems concerns the quality of the input and output languages that they use. At one extreme are learning systems that use a propositional attribute-value language for describing entities and classification rules. The simplicity of this formalism permits such systems to influence massive volumes of information and consequently to take advantage of applied math properties of collections of examples and counterexamples of an idea. At the opposite end of the spectrum, logical illation systems settle for descriptions of complicated, structured entities and generate classification rules expressed in first-order logic. These usually have access to information that is pertinent to the domain, so they need fewer entity descriptions. FOIL, the system delineate during this paper, builds on ideas from each one of the teams.

Objects are delineating mistreatment relations and from these FOIL generates classification rules expressed in an exceedingly restricted style of first-order logic, mistreatment strategies custom-made from people who have evolved in attribute-value learning systems (Quinlan 1990).

2.1   A Brief History About Deep Learning

The thought over deep learning was born beside synthetic neural network research Multilayer Perceptron with dense stolen layer which is a proper example concerning the fashions along flagrant architectures. Backpropagation, invented in the 1980s, has been a commonly used algorithm because it studied the weights concerning these networks. Unfortunately, single Backpropagation single does no longer nicely do the employment of action because of discipline networks including greater thana with younger numbers of secret layers (see a decrial then interesting analysis between durability, Bengio 2009). Stability, is the pervasive appearance over regional optima between the nonconvex objective feature over the sound networks and the primary supply about the concern of learning. Backpropagation is based on local gradient class then starts generally at incomplete random initial points. It repeatedly gets trapped in poor regional optima or the rapidity will increase extensively as the deepness regarding the networks increases. This subject is partially accountable for steerage away concerning the computer study or sign technology lookup from neural networks after shallow fashions, bearing many convex deprivation applications (e.g. SVMs, CRFs, MaxEnt models) because it is able to keep efficiently what has been obtained at the charge, regarding less muscular models. The optimization problem associated together with the deep fashions used to be empirically soothed as a moderately efficient, unsupervised learning algorithm was brought into 2006 by way of Hinton durability (Hinton et al. 2006), for a category regarding extreme creative models to those known as awful belief networks (DBNs). A core issue of the DBN is a greedy, layer-by-layer learning algorithm which optimizes DBN weights at time linear complexity, linear in accordance with the quantity over the networks. Separately or with partial surprise, initializing the weights of an MLP including a correspondingly configured DBN, often produces a great deal having higher effects with the random weights (Quinlan 1990). As such, deep networks that are learned with unsupervised DBN pretraining followed by the Backpropagation fine-tuning are also called DBNs in the literature (e.g. Mohamed et al. 2010; Mohamed et al. 2009). A DBN comes with additional attractive properties:

- The learning algorithm makes effective use of unlabeled data;

- It can be interpreted as Bayesian probabilistic generative models;

- The values of the hidden variables in the deepest layer are efficient to compute;

The over fitting problem that is often observed in the models with millions of parameters such as DBNs, and the under fitting problem that occurs often in deep networks is effectively addressed by the generative pre-training step.

The DBN training procedure is not the only one that makes deep learning possible. Since the publication of the seminal work of (Hinton et al. 2006) numerous researchers have been improving and applying the deep learning techniques with success. Another popular technique is to pre-train the deep networks layer by layer by considering each pair of layers as a denoising auto-encoder (Hinton et al. 2006).

We will provide a brief overview of the original DBN work and the subsequent progresses in the remainder of this article.

3   An Overviews of Deep Learning and Its Applications

The primary idea in deep learning algorithm is computerizing the extraction of portrayals (reflections) from the information (Bengio et al. 2013; Bengio 2009, 2013). Profound learning calculations utilize a tremendous measure of solo information to naturally extricate complex portrayal. These calculations are to a great extent persuaded by the field of man-made consciousness, which has the general objective of imitating the human cerebrum's capacity to watch, break down, learn, and decide, particularly for very perplexing issues. Work relating to these intricate difficulties has been a key inspiration driving Deep Learning calculations which endeavor to imitate the various deep learning approach of the human cerebrum. Models dependent on shallow learning designs, for example, choice trees, bolster vector machines, and case-based thinking may miss the mark when endeavoring to extricate valuable data from complex structures and connections in the info corpus. Interestingly, Deep Learning structures have the capacity to sum up in non-nearby and worldwide ways, producing learning examples and connections past prompt neighbors in the information (Bengio 2013).

4   Deep Learning Applications to Sdg

Substantive research and application of deep learning technologies to SDGs is concerned with the development of better data-mining for the prediction of certain events.

This area depicts the establishment of this survey by talking about some fields that have been applied with Deep Learning calculation to work towards sustainability solutions.

5   Automatic Speech Recognition (ASR)

Radio communication has been around since the late 1800s, and in many remote and poor parts of the world, it remains a critical communications tool. But now in Africa (Rosenthal 2019), the United Nations is partnering with policymakers, academics, and local staffers to apply cutting-edge artificial intelligence (AI) technology to radio broadcasts to gain greater insight into how to better serve citizens, especially those who might not have access to more modern technology.

UN and government programs gather qualitative data, it is often difficult to collect it from remote areas, especially where local dialects are spoken. Anecdotes and opinions shared on local radio stations can be turned into useful data that can inform development projects and legislation. This information can then be used to develop early warning systems – not just for natural disasters, but also for the spread of diseases, outbreaks of violence, and even corruption.

Google has reported that Google voice search had gone in a different direction by receiving Deep Neural Networks (DNN) as the center innovation used to show the hints of a language in 2012 (Sak et al. 2015). DNN supplanted Gaussian Mixture Model which has been in the business for 30 a long time. DNN additionally has demonstrated that it is better ready to gauge which sound a client is creating at each moment in time and with this they conveyed conspicuously expanded discourse acknowledgment exactness.

In 2013, DL has increased full energy in both ASR and ML (Deng and Li 2013) DL is fundamentally connected to the utilization of different layers of nonlinear changes to determine discourse highlights, while learning with shallow layers includes the utilization of model-based portrayals for discourse highlights which have high dimensionality however normally empty sections.

6   Image Recognition

Deep max-pooling convolutional neural systems are utilized to identify mitosis in bosom histology pictures was introduced in (Ciresan et al. 2013). Mitosis location is extremely hard. Truth be told, mitosis is a complex procedure during which a phone core experiences different changes. In this approach, DNN as amazing pixel classifier which works on crude pixel esteems and no human info is required.

Subsequently, DNN consequently learns a lot of visual highlights from the preparing information. DNN is tried on a freely accessible dataset and essentially outflanks all contending strategies, with reasonable computational exertion: preparing a 4MPixel picture requires couple of minutes on a standard workstation. Enormous and profound convolutional neural system is prepared to characterize the 1.2 million high resolution pictures in the ImageNet LSVRC-2010 challenge into 1000 distinct classes (Krizhevsky et al. 2012). On the test information, they accomplished top-1 and top-5 mistake paces of 37.5 and 17.0% which is extensively superior to the past cutting edge. From every one of the tests, the outcomes can be improved just by hanging tight for quicker GPUs and greater datasets to wind up accessible.

This technology gives governments the potential for unprecedented control and monitoring over populations. We see the scope of ID systems grow and change, as systems originally used for one purpose begin to be used for others.

7 Natural Language Processing

As of late, profound learning techniques have been effectively applied to an assortment of language what's more, data recovery applications. By misusing profound structures, profound learning procedures can find from preparing information the concealed structures and highlights at various degrees of deliberations valuable for the any undertakings. In this study (Requejo-Castroa et al. 2019), proposed a data-driven Bayesian network (BN) approach to identify and interpret SDGs interlinkages. The Sustainable Development Goals (SDGs) are presented as integrated and indivisible. Therefore, for monitoring purposes, conventional indicator-based frameworks need to be combined with approaches that capture and describe the links and interdependencies between the Goals and their targets.

8 Customer Relationship Management

Within the multiplicity of scientific publications exploring sustainability in customer-related areas, this study (Müller 2014) provides an overview on the current state of research on sustainability-oriented customer relationship management. A structure for independent control of a client relationship in the board's framework has been diagrammed (Tkachenko 2015). First, an altered form of the generally acknowledged Recency-Frequency Monetary Value arrangement of measurements can be utilized to characterize the state space of customers or benefactors is investigated. Second, a technique to decide the ideal direct showcasing

activity in discrete, persistent activity space for the given individual, in view of his situation in the state space, is portrayed. The strategy includes the utilization of the "model Q", figuring out how to prepare a profound neural arrange that relates a customer's situation in the state space to prizes related with conceivable showcasing exercises. The evaluated worth capacity over the customer state space can be deciphered as client lifetime esteem (CLV), and in this manner takes into consideration a fast module estimation of CLV for a given customer. Test results are introduced, in view of Knowledge Disclosure and Data Mining Tools Competition, mailing dataset of gift sales.

9 Bioinformatics

Bioinformatics develops algorithms and suitable data analysis tools to infer the in-formation and make discoveries. Application of various bioinformatics tools in bio-logical research enables storage, retrieval, analysis, annotation and visualization of results and promotes better understanding of biological system in fullness. The com-ment of genomic data is a noteworthy test in science and bioinformatics. Existing databases of realized quality capacities are inadequate and inclined to mistakes, and the bimolecular trials expected to improve these databases are moderate and exorbi-tant. While computational strategies are not a substitute for test confirmation, they can help in two different ways: calculations can help in the curation of quality explana-tions via naturally proposing mistakes, and they can foresee beforehand unidentified quality capacities, quickening the pace of quality capacity revelation. In this work (Wang et al. 2020), They compared the performance of them model with that of two skilled radiologists, and them model has shown much higher accuracy and sensitivity. These findings have demonstrated the proof-of-principle that deep learning can extract CT image features of COVID-19 for diagnostic purposes. Using the supercomputer system, the time for each case takes only about 10 s, and it can be performed remotely via a shared public platform. Therefore, further developing this system can significantly shorten the diagnosis time for disease control. This study represents the first study to apply artificial intelligence technologies to CT images for effectively screening for COVID-19.

In this work (Chicco et al. 2014), a calculation that accomplishes the two objectives utilizing profound auto encoder neural systems is created. With investigates quality comment information from the Gene Ontology venture, it demonstrates that profound auto encoder systems accomplish preferable execution over other standard AI techniques, including the famous truncated particular worth deterioration.

10 The Relationship Between Learning and Data Analysis (Learning Bases)

Information examination changes over information into data and learning, and investigates the connection between factors. Information Analysis is the procedure of efficiently applying measurable as well as legitimate methods to depict and outline, consolidate and recap, and assess information. As indicated by Shamoo and Resnik1 different logical methods "give a method for drawing inductive derivations from information and recognizing the sign (the wonder of enthusiasm) from the commotion (factual changes) present in the information." Comprehension of the information examination techniques will empower you to welcome the significance of the logical technique, which incorporates testing the speculations and measurable criticalness in connection to research questions. There are various issues that analysts ought to be discerning of as for information examination. A portion of the key contemplations in examination and determination of the correct trial of importance are as per the following:

- Having the fundamental aptitudes to investigate;
- Distinguishing information types;
- Distinguishing various sorts of Statistical tests;
- Identifying the determination of a correct test;
- Determining factual essentialness;
- Distinguishing among Parametric and Non-Parametric test with their applying criteria;
- Distinguishing among Correlation and Regression;
- Investigating Inappropriate subgroup;
- Estimating Lack of unmistakably characterized and target result;
- Partitioning 'content' when investigating subjective information;
- Reliability and Validity;
- Extent of investigation (Sharma 2018).

Now, we can get the relation between the learning bases and the data analysis, to develop classification rules from experience that has long been a principal space of machine learning analysis.

11 The Two Main Modes of Supervised and Unsupervised Learning, and the Possibility of Combining Them to Learn Better

11.1 Introduction

Presentation of intellectual thinking into a traditional PC can take care of issues by model mapping like example acknowledgment, grouping and anticipating. Fake Neural Networks (ANN) gives these sorts of models. These are basically numerical models depicting a capacity; however, they are related with a specific learning calculation or a standard to imitate human activities. ANN is described by three kinds of parameters; (an) in light of its interconnection property (as feed forward system and repetitive system); (b) on its application work (as Classification model, Association model, Optimization model and Self-sorting out model) and (c) in view of the learning rule (managed/solo/support and so forth.,) (Fu 2003).

All these ANN models are remarkable in nature and every offer points of an interest of its own. The significant hypothetical and down to earth ramifications of ANN have differing applications. Among these, a great part of the exploration exertion on ANN has concentrated on example order. ANN performs grouping errands clearly and productively due to its auxiliary structure and learning strategies. There is no novel calculation to structure and prepare ANN models since, taking in calculation varies from one another in their learning capacity and level of induction. Consequently, in this paper, we attempt to assess the managed and unaided learning rules and their grouping productivity utilizing explicit model (Sharma 2018).

The general association of the paper is as per the following. After the presentation, we present the different learning calculations utilized in ANN for example arrangement issues and all the more explicitly the learning procedures of directed and unaided calculations in segment II. Area III presents characterization and its necessities in applications and talks about the nature qualification among managed and unaided learning on the example class data. Likewise, we establish framework for the development of characterization arrange for instruction issue of our advantage. Exploratory arrangement and its result of the present investigation are introduced in Section IV. In Section V, we examine the final products of these two calculations of the investigation from alternate point of view. Area VI, closes with some last musings on regulated and unaided learning calculation for instructive arrangement issue.

12 ANN Learning Paradigms

Learning can allude to either gaining or upgrading information. As Herbert Simon says, Machine Learning signifies changes in the framework that are versatile as it empowers the framework to do a similar undertaking or assignments drawn from a similar populace all the more productively and all the more viably whenever. ANN learning ideal models can be delegated administered, unaided and fortification learning. Managed learning model expect the accessibility of an instructor or administrator who characterizes the preparation models into classes and uses the data on the class participation of each preparation occurrence, while, Unsupervised learning model recognizes the example class data heuristically and Reinforcement learning, through experimentation associations with its condition (remunerate/punishment task). In spite of the fact that these models address learning in various ways, learning relies upon the space of interconnection neurons. That is, administered learning learns by changing its entomb association weight mixes with the assistance of blunder signals where as unaided learning utilizes data related with a gathering of neurons and fortification learning utilizes support capacity to adjust nearby weight parameters. Subsequently, learning happens in an ANN by changing the free parameters of the system that are adjusted where the ANN is inserted. This parameter alteration assumes key job in separating the learning calculation as administered or unaided models or different models. Likewise, these learning calculations are encouraged by different learning rules as appeared in the Fig. 2 (Fu 2003).

13 Supervised Learning.

Supervised learning depends on preparing an information test from information source with right order previously doled out. Such strategies are used in feedforward or Multi-Layer Perceptron (MLP) models. These MLP have three unmistakable qualities: 1. At least one layers of shrouded neurons that are not part of the info or yield layers of the system that empower the system to learn and tackle any mind-boggling issues 2. The nonlinearity reflected in the neuronal action is differentiable and, 3. The interconnection model of the system shows a high level of network, in addition to these attributes, alongside learning through preparing tackle troublesome and assorted issues. Learning through preparing in an administered ANN model additionally called as mistake Backpropagation calculation. The blunder remedy learning calculation prepares the system depending on the info yield tests and discovering mistake signal, which is the distinction of the yield determined and the ideal yield and alters the synaptic loads of the neurons that is relative to the result of the mistake signal and the information occurrence of the synaptic weight. In view of this rule, mistake back proliferation learning happens in two passes:

The "Forward Pass": Here, the input vector is exhibited to the system. This info sign spreads forward, neuron by neuron through the system and rises at the yield end of the system as yield signal: $y(n) = \varphi(v(n))$ where $v(n)$ is the prompted neighborhood field of a neuron characterized by $v(n) = \sigma\ w(n)y(n)$. The yield that is determined at the yield layer $o(n)$ is contrasted and the ideal reaction $d(n)$ and finds the mistake $e(n)$ for that neuron. The synaptic loads of the system during this pass are stays same.

In the "Reverse Pass": The mistake signal that is begun at the yield neuron of that layer is proliferated in reverse through system. This figures the nearby slope for every neuron in each layer and permits the synaptic loads of the system to experience changes as per the delta rule as: $\Delta w(n) = \eta * \delta(n) * y(n)$. This recursive calculation is proceeded, with the forward pass pursued by the retrogressive go for each information example, till the system is merged (Rojas 1996; Pei et al. 2006). Regulated learning worldview of an ANN is effective and discovers answers for a few straight and non-direct issues, for example, order, plant control, anticipating, forecast, mechanical technology and so forth (Awodele and Jegede 2009; Rao and Alvarruiz 2007).

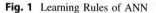

Fig. 1 Learning Rules of ANN

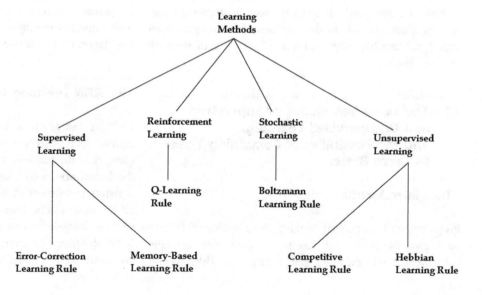

Table 1 Deep learning Algorithm Performance

Performance metric	Internal	External
AUC (95%CI)	0.93(0.86 to 0.94)	0.81(0.71 to 0.84)
Accuracy, %	89.5	79.3
Sensitivity	0.38	0.83
Specificity	0.37	0.87
PPV	0.71	0.55
NPV	0.95	0.90
Kappa*	0.69	0.48
Yoden index	0.75	0.50
F1 score*	0.77	0.63

* Measures the agreement between the CNN model prediction and the clinical

Finally, all of two main modes of supervised and unsupervised learning, have their advantages and there is possibility of combining them to learn better.

14 The Relationship Between Learning and Classification.

Classification is one of the most experienced basic leadership undertakings of human movement. An order issue happens when an item should be doled out into a predefined gathering or class dependent on various watched credits identified with that object. There are numerous modern issues recognized as grouping issues. For models, Stock market expectation, Weather anticipating, Bankruptcy forecast, Medical finding, Speech acknowledgment, Character acknowledgments to give some examples (Moghadassi et al. 2009; Khan et al. 2009). These characterization issues can be understood both scientifically and in a non-direct design. The trouble of taking care of such issue numerically lies in the precision and conveyance of information properties and model abilities (Ali and Smith 2006).

The ongoing exploration exercises in ANN demonstrate, ANN as best characterization model due to the non-direct, versatile and useful guess standards. A Neural Network groups a given article as per the yield enactment. In a MLP, when a lot of information examples are exhibited to the system. The hubs in the concealed layers of the system extricate the highlights of the example introduced. For instance, in a two concealed layers ANN model, the shrouded hubs in the main shrouded layer structures limits between the example classes and the concealed hubs in the subsequent layer shapes a choice locale of the hyper planes that was framed in the past layer. Presently, the hubs in the yield layer coherently joins the choice locale made by the hubs in the concealed layer and arranges them into class 1 or class 2 as indicated by the quantity of classes portrayed in the preparation with least blunders all things considered. Likewise, in SOM, arrangement occurs by separating

highlights by changing of m-dimensional perception info design into q-dimensional element yield space and therefore gathering of articles as per the likeness of the information design.

15 Conclusion

The research findings in this study revealed the importance of laying a sustainable foundation for advanced technologies based on Deep Learning and enable more effective and efficient public policy for sustainability, and more specifically improve access, connectivity and efficiency in sectors ranging from healthcare, sanitation, and education, to farming and transportation. To be ready for this deep learning-powered future, the academic community has an important role to play in preparing the future generations of business leaders and national and international policymakers in addressing the opportunities and the challenges presented by DL and the imperative to advance the Global Goals. Without astute management education, individuals, corporations, and governments may falter and fail in their strife for economic growth with sustainability development.

References

Ali, S., Smith, K.A.: On learning algorithm selection for classification. Applied Soft Computing **6**, 119–138 (2006)

Awodele, O., Jegede, O.: "Neural Networks and Its Application in Engineering", Proceedings of Informing Science & IT Education Conference (InSITE) 2009, pp. 83–95.

Bengio, Y.: Learning Deep Architectures for AI. Now Publishers Inc., Hanover, MA, USA (2009a)

Bengio, Y.: Learning Deep Architectures for AI, vol. 2. 1st edn, pp. 1–127. Foundations and Trends in Machine Learning (2009b)

Bengio, Y.: (2013) Deep learning of representations: Looking forward. In: Proceedings of the 1st International Conference on Statistical Language and Speech Processing. SLSP'13. Springer, Tarragona, Spain. pp 1–37. http://dx.doi.org/https://doi.org/10.1007/978-3-642-39593-2_1.

Bengio, Y., Courville, A., Vincent, P.: Representation learning: A review and new perspectives. Pattern Analysis and Machine Intelligence, IEEE Transactions on **35**(8), 1798–1828 (2013). https://doi.org/10.1109/TPAMI.2013.50

Chicco, D., Sadowski, P., Baldi, P.: Deep autoencoder neural networks for gene ontology annotation predictions. In Proceedings of the 5th ACM Conference on Bioinformatics, Computational Biology, and Health Informatics (pp. 533–540). ACM (2014, September).

Ciresan, D.C., Giusti, A., Gambardella, L.M., & Schmidhuber, J.: Mitosis detection in breast cancer histology images with deep neural networks. In International Conference on Medical Image Computing and Computer-assisted Intervention (pp. 411–418). Springer Berlin Heidelberg (2013, September).

Deng, L., Li, X.: Machine learning paradigms for speech recognition: An overview. IEEE Trans. Audio Speech Lang. Process. **21**(5), 1060–1089 (2013)

Deng, L., Seltzer, M., Yu, D., Acero, A., Mohamed, A., Hinton, G.: "Binary coding of speech spectrograms using a deep auto-encoder," In: Proc. Interspeech, 2010.

Fu., L.: Neural Networks in Computer Intelligence, Tata McGraw-Hill, 2003.

Hinton, G., Osindero, S., Teh, Y.: A fast learning algorithm for deep belief nets. Neural Comput. **18**, 1527–1554 (2006)

Khan, U., Bandopadhyaya, T.K., Sharma, S.: Classification of Stocks Using Self Organizing Map. International Journal of Soft Computing Applications, Issue **4**, 19–24 (2009)

Krizhevsky, A., Sutskever, I., & Hinton, G.E.: Imagenet classification with deep convolutional neural networks. In Advances in neural information processing systems (pp. 1097- 1105) (2012).

Moghadassi, A., Parvizian, F., Hosseini, S.: A New Approach Based on Artificial Neural Networks for Prediction of High Pressure Vapor-liquid Equilibrium. Aust. J. Basic Appl. Sci. **3**(3), 1851–1862 (2009)

Mohamed, A., Dahl, G., Hinton, G.: "Deep belief networks for phone recognition," in Proc. NIPS Workshop Deep Learning for Speech Recognition, 2009.

Mohamed, A., Yu, D., Deng, L.: "Investigation of full-sequence training of deep belief networks for speech recognition", in Proc. Interspeech, Sept (2010)

Müller, A.L.: Sustainability and customer relationship management: current state of research and future research opportunities. Management Review Quarterly **64**(4), 201–224 (2014)

Pei, J.S., Mai, E., Piyawat, K.: "Multilayer Feedforward Neural Network Initialization Methodology for Modeling Nonlinear Restoring Forces and Beyond", 4th World Conference on Structural Control and Monitoring, 2006, pp. 1–8.

Quinlan, J.R.: Learning Logical Definitions from Relations. Machine learning 5(3) 1990

Rao, Z., Alvarruiz, F.: "Use of an Artificial Neural Network to Capture the Domain Knowledge of a Conventional Hydraulic Simulation Model", Journal of HydroInformatics, 2007, pg.no 15–24.

Requejo-Castroa, D., Giné-Garriga, R., Pérez-Foguet, A.: Data-driven Bayesian network modelling to explore the relationships between SDG 6 and the 2030 Agenda. Sci. Total Environ. **710**, 136014 (2019)

Rojas, R.: The Backpropagation Algorithm, Chapter 7: Neural Networks, pp. 151–184. Springer-Verlag, Berlin (1996)

Rosenthal, A.: "When Old Technology Meets New: How UN Global Pulse is Using Radio and AI to Leave No Voice Behind" https://www.unglobalpulse.org/2019/04/when-old-technology-meets-new-how-un-global-pulse-is-using-radio-and-ai-to-leave-no-voice-behind/ , April 18, 2019.

Sak, H., Senior, A., Rao, K., Beaufays, F., Schalkwyk, J.: Google voice search: faster and more accurate (September 2015).

Sharma, B.: "Processing of data and analysis" Sri Aurobindo Medical College & PG Institute, India Correspondence: Dr Balkishan Sharma, PhD, Associate Professor (Biostatistics), Department of Community Medicine, Sri Aurobindo, Received: February 06, 2018, Published: February 20, 2018 Copyright© 2018 Sharma.

Tkachenko, Y.: Autonomous CRM Control via CLV Approximation with Deep Reinforcement Learning in Discrete and Continuous Action Space. (2015). arXiv preprint arXiv:1504.01840.

Wang, Shuai & Kang, Bo & Ma, Jinlu & Zeng, Xianjun & Xiao, Mingming & Guo, Jia & Cai, Mengjiao & Yang, Jingyi & Li, Yaodong & Meng, Xiangfei & Xu, Bo. "A deep learning algorithm using CT images to screen for Corona Virus Disease (COVID-19)", 2020.

Self-Attention Mechanism for Diabetic Retinopathy Detection

Othmane Daanouni, Bouchaib Cherradi, and Amal Tmiri

Abstract

Diabetic Retinopathy (DR) is a high blood sugar level that causes damage to blood vessels and is one of the common causes of blindness in the developed world. Convolutional Neural Networks (CNNs) are one of the fundamental and successful applications of computer vision application, especially in prediction of DR. Recently self-attention has become a recent advance part of models that must capture long-range interactions. In this study, we propose a self-attention mechanism combined with pre-trained convolutions models in this case MobileNet. Self-attention is used to determine how much attention should pay to across Optical Coherence Tomography (OCT) image regions will modeling long-range, multi-level dependencies in order to accurately predict patient with DR. In the interest of demonstrating the feasibility of our method, we used Gradient-weighted Class Activation Mapping (Grad-CAM) to highlight important regions in the OCT used for prediction. Different metrics are used to evaluate the proposed model performance such as Accuracy, Recall, Precision, and Roc Area. The proposed architecture achieved 98% accuracy, 98% precision, and 98% recall.

Keywords

Self-attention • Diabetic Retinopathy (DR) • Convolutional Neural Network (CNN) • Transfer learning • Clinical diagnosis

1 Introduction

Diabetic Retinopathy (DR) distorts vision due to the leakage of fluid in the retinal blood vessels and forms lesions in the retina. It is the most common cause of blindness in adults with diabetes (Faust et al. 2012]. DR is considered asymptomatic or causes noticeable mild vision problems, but eventually, it can cause blindness especially for individuals that have type 1 or type 2 diabetes. Early clinical diagnosis and detection of diabetic retinopathy are vital in order to protect patients from losing their vision (Daanouni et al. 2019, 2020).

Machine Learning (ML) techniques are largely used in many problematic. In medical field, the ML techniques are important results in terms of prediction and classification of patients with or without diseases (Terrada et al. 2018; Laghmati et al. 2019).

Deep Learning (DL) (LeCun et al. 2015) is one of the powerful techniques and research areas of machine learning. It was proposed by Hinton et al. in 2006 and has been successfully applied in the fields of image identification and especially medical image processing.

Recently, Convolution Neural Network (CNN) shows tremendous success in many computer vision applications, especially in image classification (Krizhevsky et al. 2012) and is widely used in the field of Diabetic Retinopathy detection (Mateen et al. 2019; Wang and Yang 2017). This technique largely outperforms previous image recognition methodologies. However, implementing a Convolution Neural Network model from scratch is time-consuming and

O. Daanouni (✉) · B. Cherradi · A. Tmiri
LaROSERI Laboratory, Chouaib Doukkali University, El Jadida, Morocco
e-mail: daanouni34@gmail.com

B. Cherradi
e-mail: bouchaib.cherradi@gmail.com

A. Tmiri
e-mail: b_tmiri@yahoo.fr

B. Cherradi
STIE Team, CRMEF Casablanca-Settat, Provincial section of El Jadida, El Jadida, Morocco

M. Ben Ahmed et al. (eds.), *Emerging Trends in ICT for Sustainable Development*,
Advances in Science, Technology & Innovation, https://doi.org/10.1007/978-3-030-53440-0_10

requires a large amount of data for training and obtaining robust prediction model.

The self-attention mechanism applied in computer vision models yields to boost the performance of traditional CNNs (Hu et al. 2018; Tan et al. 2019) in different tasks like object detection and image classification (Bello et al. 2019).

The rest of this paper is organized as follows: In Sect. 2, we review some recent work in relation with our method. Section 3 presents an overview of the used materials and methods, in particular we introduce the OCT dataset and some rudiments about the DR techniques. In Sect. 4, we present a description of our proposed self-attention CNN for accurately detecting DR diseases. Implementation setup, some results, and discussion are given in Sect. 5. Section 6 concludes this paper and presents some perspectives of this work.

2 Some Related Works

Many researches in the field of DR detection have been proposed using a machine learning technique. Below, we present and discuss some related works that we will consider later for performance evaluation and comparison.

In (Liang et al. 2019), the authors propose a multi-self-attention network structure for accurate detection of diabetic retinopathy. In this study, the authors used a CNN architecture based on inception V3 model to generate a feature map and feed it into a multi-self-attention network to calculate different self-attention features. The results show that the classification result is the best when the self-attention network is used twice (n = 2) and achieved 87.7%.

The authors in (Li et al. 2019) introduce a novel attention network (CANet) composed of two attention model: one for learning useful features and the other for disease internal dependence between DR and Diabetic Macular Edema (DME). These two attention models are then integrating into a deep network in order to leverage attentions models to maximize the overall grading performance. The study achieved promising result on Messidor dataset[1] with 92.6% and 92.4% accuracy for DR and DME, respectively.

Based on these works, we propose in this paper a method that enhances the detection accuracy.

3 Materials and Methods

This section aims to highlight different techniques and datasets used to perform classification of diabetic retinopathy.

[1]https://www.adcis.net/fr/logiciels-tiers/messidor2-fr/.

3.1 OCT Dataset

The repository dataset (Kermany et al. 2018) holds 207.130 OCT images. These images are split into two sets:

- 108.312 training images (37.206 images with choroidal neovascularization, 11.349 images with diabetic macular edema, 8.617 images with drusen, and 51.140 normal images) taken from 4.686 patients.
- 1.000 testing images (250 images from each category) taken from 633 patients.

The two sets of images are split into four directories: CNV, DME, DRUSEN, and NORMAL and identified as follows: (Disease type)-(randomized patient ID)-(image number for this patient). Figure 1 presents an illustration of these different DRs captured in OCT images.

3.2 Convolutional Neural Network (CNN)

A Convolutional Neural Network (CNN) consists of different convolution layers for automatic feature extraction and neural networks for classification. A convolution layer utilizes specific kernel sizes (small neighborhoods) that are convolved with images to encourage the network to learn and detect different patterns such as colors and shapes. CNN is constructed with different layers such as conv2D, pooling layers, and feature maps. These are successfully applied in recognition tasks such as handwriting (Cireşan et al. 2012) and classification (Sayamov 2019). Various CNNs are already available namely MobileNet, AlexNet, VggNet, and others. Figure 2 presents an example of convolution operation.

3.3 Network in Network

One of common model design pattern problems with deep convolutional neural networks is the increasing number of feature maps and number of parameters and computation requirements with the depth of the network. This problem results in a performance penalty.

To address this problem, a Network in Network approach proposed by (Lin et al. 2013) as 1×1 convolutional layer can be used that offers a channel-wise pooling, i.e., (feature map pooling) in order to reduce dimensionality and decreasing the number of features maps this technique used in Inception network architecture (Szegedy et al. 2015).

We use this approach in our architecture as dimensionality reduction modules to remove computational bottlenecks that would have a significant performance penalty (Fig. 3).

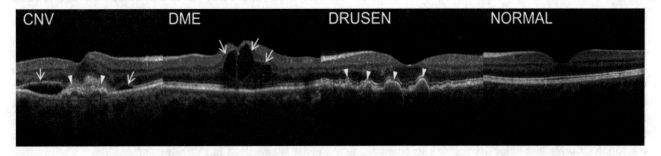

Fig. 1 Illustration of the 4 DR types using OCT images

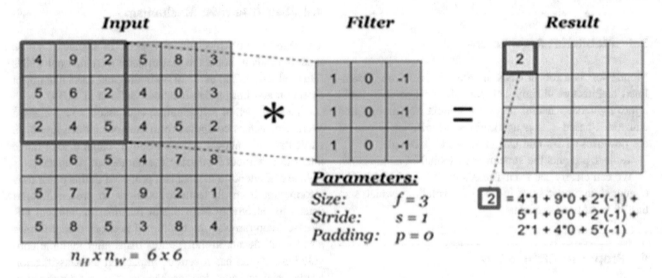

Fig. 2 Example of convolution operation

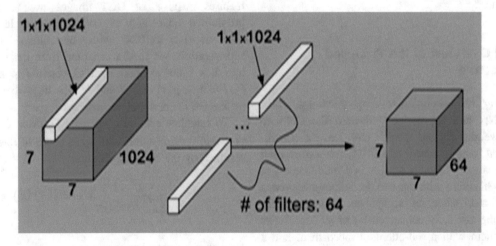

Fig. 3 Example of network in network with $1 \times 1 \times 1024$ convolution filter size

3.4 Bottleneck Layer

In a CNN, bottleneck features are generated from deep neural network in which one of the internal layers has small number of hidden units compared to other layers. The bottleneck layer is added to reduce the number of feature maps, i.e., (channels) in the network which otherwise tends to increase in each layer. This is achieved by using 1×1 convolutions layer that helps the network to compress feature representations and dimension reduction (Yu and Seltzer 2011) (Fig. 4).

3.5 MobileNet Architecture

MobileNet[2] is a robust CNN architecture based on streamlined architecture mainly 1*1 convolution kernel to build lightweight deep neural networks suitable for mobile and embedded-based vision applications. MobileNet is one of the powerful CNNs that was introduced by Google.

Table 1 presents the structure of MobileNet architecture.

We can clearly see from Table 2 that 95% of MobileNet computation time is spent in 1×1 convolutions, which also has 75 % of the parameters.

4 Proposed Architecture

In this section, we present our proposed architecture for DR prediction using CNN Model based on Transfer Learning (TL)[3] technique.

4.1 Global Overview of the Proposed Architecture

TL is a training method aiming to adopt the weights of a pre-trained CNN and appropriately re-train the CNN to optimize the weights for our specific task, i.e., AI classification of retinal image. Indeed, this CNN architecture used for the classification of fundus images is a modification of MobileNet Architecture with the transfer learning approach combined with a self-attention mechanism. To achieve this, we truncated the last fully connected layers of MobileNet and replaced them with a self-attention mechanism and a new SoftMax layer that is relevant to our problem (classification of OCT images) to four classes (DRUSEN, DME, CNV, and Normal) in order to compute the probability of each class. In this manner, we achieved 128 non-trainable parameters in the CNN architecture from y parameters. Figure 5 shows an overview of the proposed architecture with different components and their workflow.

The details of self-attention mechanism proposed in this paper is explained in following sub-section.

4.2 Self-Attention Mechanism

The self-attention mechanism applied in computer vision models yields to boost the performance of traditional CNNs (Hu et al. 2018; Tan et al. 2019) in different tasks like object detection and image classification (Bello et al. 2019).

The core idea of self-attention is to associate a weighted average of values computed from previous hidden layers. In particular, this allows self-attention to capture long-range interaction without increasing the number of parameters.

The self-attention mechanism proposed in this paper uses information from all features locations to process images details to selectively learn useful features for different DR diseases, contrariwise the traditional network model throws away small details in the features maps after multiple convolutions, which has a serious impact on the classification results. The self-attention mechanism steps are described in Fig. 6.

In order to reduce the number of channels and extract the features maps x of OCT images, we used MobileNet pre-trained model with network in network layer with filter size of $1 \times 1 \times 1024$ with 64 filters. To calculate self-attention, we feed x obtained from the previous layer into 1×1 convolution filter to obtain f, g, and h, where $f(x) = w_f x$, $g(x) = w_g x$ and $h(x) = w_h x$ where w_f, w_g, and w_h are the learned weight matrices.

To calculate attention, we used SoftMax function $\beta_{j,i}$ as probability distribution with s_{ij} as score of relevant regions obtained by matrix multiplication

$$\beta_{j,i} = \frac{\exp(s_{ij})}{\sum_{i=1}^{N} \exp(s_{ij})} \, where \, s_{ij} = f(x_i)^T g(x_j) \qquad (1)$$

j, i indicate the extent to which the model attends to the ith location when synthesizing the jth region.

The output of the attention layer is $o = (o_1, o_2, \ldots, o_j, \ldots, o_N) \in R^{C \times N}$ where

$$o_j = v\left(\sum_{i=1}^{N} \beta_{j,i} h(x_i)\right) \qquad (2)$$

---

[2]https://arxiv.org/abs/1704.04861.
[3]https://machinelearningmastery.com/transfer-learning-for-deep-learning/.

Table 1 MobileNet architecture

Type/stride	Filter shape	Input size
Conv/s2 Conv dw/s1	$3 \times 3 \times 3 \times 32$ $3 \times 3 \times 32$ dw	$221 \times 221 \times 3$ $112 \times 112 \times 32$
Conv/s1	$1 \times 1 \times 32 \times 64$	$112 \times 112 \times 32$
Conv dw/s2	$3 \times 3 \times 64$ dw	$112 \times 112 \times 64$
Conv/s1	$1 \times 1 \times 64 \times 128$	$56 \times 56 \times 64$
Conv dw/s1	$3 \times 3 \times 128$ dw	$56 \times 56 \times 128$
Conv/s1	$1 \times 1 \times 128 \times 128$	$56 \times 56 \times 128$
Conv dw/s2	$3 \times 3 \times 128$ dw	$56 \times 56 \times 128$
Conv/s1	$1 \times 1 \times 128 \times 256$	$28 \times 28 \times 128$
Conv dw/s1 Conv/s1	$3x\ 3 \times 256$ dw $1 \times 1 \times 256 \times 256$	$28 \times 28 \times 256$ $28 \times 28 \times 256$
Conv dw/s2 Conv/s1	$3 \times 3 \times 256$ dw $1 \times 1 \times 256 \times 512$	$28 \times 28 \times 256$ $14 \times 14 \times 256$
$5 \times$ Conv dw/s1 $5 \times$ Conv/s1 Conv dw/s2	$3 \times 3 \times 512$ dw $1 \times 1 \times 512 \times 512$ $3 \times 3 \times 512$ dw	$14 \times 14 \times 512$ $14 \times 14 \times 512$ $14 \times 14 \times 512$
Conv/s1 Conv dw/s2	$1 \times 1 \times 512 \times 1024$ $3 \times 3 \times 1024$ dw	$7 \times 7 \times 512$ $7 \times 7 \times 1024$
Conv/s1	$1 \times 1 \times 1024 \times 1024$	$7 \times 7 \times 1024$
Avg Pool/s1	Pool 7×7	$7 \times 7 \times 1024$
FC/s1	1024×1000	$1 \times 1 \times 1024$
Softmax/s1	Classifier	$1 \times 1 \times 1000$

Table 2 Parameters distribution in different layer of MobileNet architecture

Type	Mult-adds (%)	Parameters (%)
Conv 1×1	94.86	74.59
Conv dw 3×3	3.06	1.06
Conv 3×3	1.19	0.02
Fully connected	0.18	24.33

Fig. 4 Example of bottleneck layer

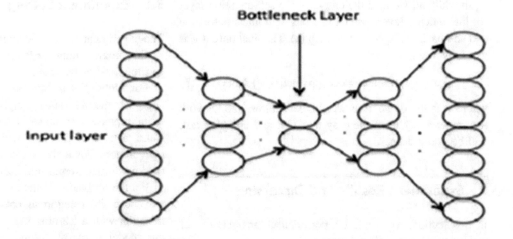

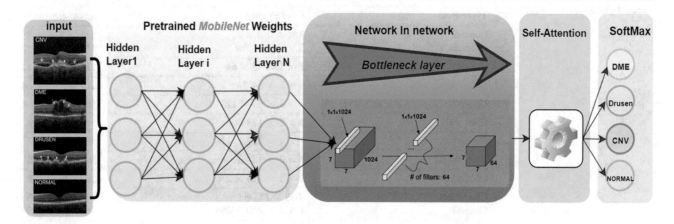

Fig. 5 Different components of the proposed architecture

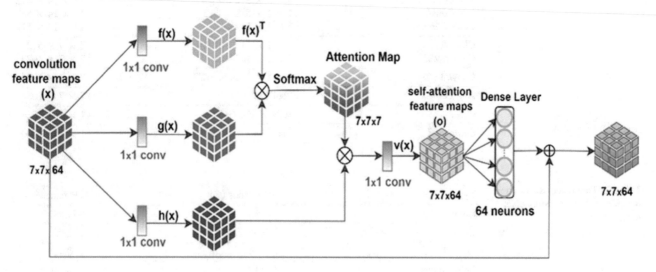

Fig. 6 Self-attention mechanism architecture

In addition, we feed the output of the self-attention layer (o) into a dense layer with 64 neurons with linear activation and add back the input feature map (x). The final output y_i is given by

$$y_i = \alpha o_i + x_i \quad \text{where} \quad \alpha = \text{dot\_product}(o, w) + \text{bias} \quad (3)$$

where w is the 64 neurons weight. This allows the network to progressively learn the easy task first and complex task will be going deeper.

5 Experiment Results and Discussion

In this section, we conduct a performance analysis of our architecture using different metrics.

5.1 Experiment Setting

The experiment set conducted in this research used Python 3 as the programming language with Keras 2.3.1 and TensorFlow 2.0[4] as backend.

For MobileNet architecture, we used a transfer learning approach, the pre-trained weights are not updated during the training phase. For the self-attention mechanism, we used a ReLu activation function with the same padding in 2D convolution. For a dense layer, we used 64 neurons with random normal kernel initialization.

For the Activation Function, we used ReLu activation and we chose the categorical cross-entropy with Adam optimization with a learning rate of 0.001. In order to prevent our model from overfitting, we defined an early stop to monitor validation loss with min delta equal to zero and

---

[4]https://www.tensorflow.org/guide/effective_tf2.

Fig. 7 Training and validation accuracy after 16 epochs

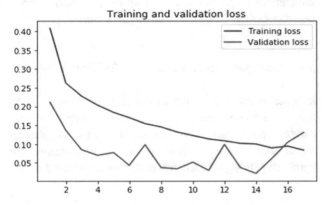

Fig. 8 Training and validation loss after 16 epochs

number of patience = 3 as 3 number of epochs with no improvement after which training would be stopped and we used a batch normalization and dropout regularization technique with 50%.

For resolving the problem of local minimums, we used to reduce on a plateau to monitor validation loss with a factor equal to 0.2 and minimum delta = 0.00001 with three patience. Note that, in order to make the model more robust, we used real-time data augmentation on images on CPU in parallel to training the model on GPU. For the computational task, we used an eight virtual CPUs Intel(R) Xeon(R) CPU @ 2.20 GHz with 2 × NVIDIA Tesla K80 GPUs.

5.2 Performance Evaluation Results

In this section, we define the performance metrics that are commonly used to assess DR prediction and detection. Therefore, the accuracy, precision, and recall are mean indexes considered in the evaluation.

$$\text{Recall} = \frac{\text{TP}}{\text{TP + FN}} \quad (4)$$

$$\text{Precision} = \frac{\text{TP}}{\text{FP + TP}} \quad (5)$$

$$\text{Accuracy} = \frac{\text{TP + TN}}{\text{TP + TN + FP + FN}} \quad (6)$$

The confusion matrix and ROC curve are also used in order to give a valuable evaluation of the classification performance of the model.

After training our self-attention convolution neural network architecture as shown in Figs. 7 and 8, the number of epochs is 16.

After the training stage, we evaluated the network model on the testing set with 1000 OCT images divided into 250 images per class. We used also the matrix confusion and Receiver Operating Characteristic (ROC) to present more the efficiency and performance of our proposed self-attention CNN. The testing scores are shown in Table 3.

The Accuracy achieved by self-attention convolution architecture is promising and equal to 98%, and the highest area under curve among the four classes as shown in fig y is achieved by the CNV, DRUSEN, and Normal classes (area = 99%), however, the lowest area is achieved by the DME class (area = 0.98%).

Figure 9 presents other performance evaluation results in terms of Receiver Operating Characteristic (ROC) variation.

Figure 10 presents the four class prediction results on a confusion matrix for the testing DR dataset.

5.3 Illustration of Neural Network Decision-Making

Gradient-weighted Class Activation Mapping (Grad-CAM) is a class-discriminative localization technique that generates visual explanations (i.e., drawing a bounding around the

Table 3 Evaluation metrics of self-attention CNN

	Accuracy	Precision	Recall	Support
CNV	0.996	0.96	1	250
DME	0.952	1	0.95	250
DRUSEN	0.988	0.98	0.99	250
NORMAL	0.984	0.98	0.98	250

Table 4 Comparison of our results with some related work. '–' indicates no reported result

	Support	Accuracy (%)
Liang et al. [13]	–	87.7
Li et al. [14]	DME	91.2
	DRUSEN	92.6
This work	CNV	**0.996**
	DME	**0.952**
	DRUSEN	**0.988**
	NORMAL	**0.984**

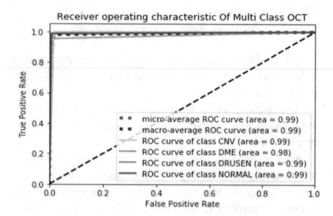

Fig. 9 Receiver operating characteristic performance

object) based on the last convolution layer (Selvaraju et al. 2017). In order to determine important regions in OCT images, Fig. 11 presents three examples showing some illustrations of prediction results by the proposed architecture and their correspondent Grad-CAM mapping.

5.4 Comparison Results with Related Work

In order to show the enhancement in terms of state of the arts DR detection accuracy, we present in Table 4 a comparative performance on the latest works on diabetes retinopathy disease. According to the experiments, our approach gets higher accuracy among the considered works.

Fig. 10 Confusion matrix of 1.000 OCT images classified in four classes

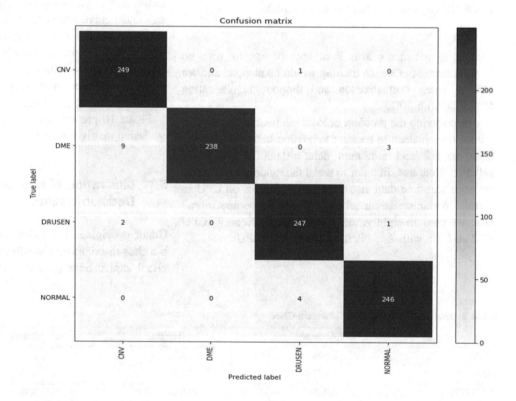

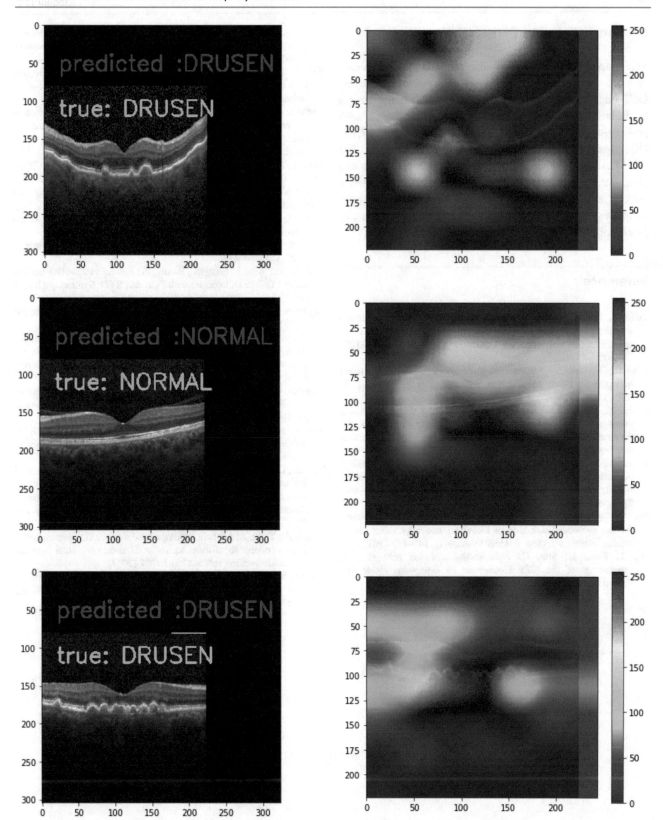

Fig. 11 Visualization of Grad-CAM tested on OCT images

6 Conclusion and Perspectives

This paper proposed a self-attention convolution neural network architecture to detect diabetic retinopathy using OCT images. This approach is based on associating weight to different regions of features maps and combined with pre-trained MobileNet model. The model trained on 207.130 OCT images to classify them into four classes. The Grad-CAM technique is used to show a localization and generates visual explanations of the region of interest directly responsible for DR disease. The self-attention convolution neural network model shows results that are more promising by achieving 98% accuracy with 0.99 Roc Area.

Reference

Bello, I., Zoph, B., Vaswani, A., Shlens, J., Le, Q.V.: Attention augmented convolutional networks. In: Proceedings of the IEEE International Conference on Computer Vision, pp. 3286–3295 (2019)

Cireşan, D., Meier, U., Schmidhuber, J.: Multi-column deep neural networks for image classification. arXiv preprint arXiv:1202.2745 (2012)

Daanouni, O., Cherradi, B., Tmiri, A.: Predicting diabetes diseases using mixed data and supervised machine learning algorithms. In: Proceedings of the 4th International Conference on Smart City Applications, p. 85 (2019)

Daanouni, O., Cherradi, B., Tmiri, A.: Type 2 diabetes mellitus prediction model based on machine learning approach, 3rd edn. In: Innovations in Smart Cities Applications Edition 3, pp. 454–469. Springer (2020)

Faust, O., Acharya, R., Ng, E. Y.-K., Ng, K.-H., Suri, J.S.: Algorithms for the automated detection of diabetic retinopathy using digital fundus images: a review. J. Med. Syst. 36(1), 145–157 (2012)

Hu, J., Shen, L., Sun, G.: Squeeze-and-excitation networks. In: Proceedings of the IEEE Conference on Computer Vision and Pattern Recognition, pp. 7132–7141 (2018)

Kermany, D., Zhang, K., Goldbaum, M.: Labeled optical coherence tomography (oct) and chest X-ray images for classification. Mendeley Data 2 (2018)

Krizhevsky, A., Sutskever, I., Hinton, G.E.: Imagenet classification with deep convolutional neural networks. In: Advances in Neural Information Processing Systems, pp. 1097–1105 (2012)

Laghmati, S., Tmiri, A., Cherradi, B.: Machine learning based system for prediction of breast cancer severity. In: Presented at the 7th International Conference on Wireless Networks and Mobile Communications (WINCOM), Fez, Morocco (2019)

LeCun, Y., Bengio, Y., Hinton, G.: Deep learning. Nature 521 (2015)

Li, X., Hu, X., Yu, L., Zhu, L., Fu, C.-W., Heng, P.-A.: CANet: cross-disease attention network for joint diabetic retinopathy and diabetic macular edema grading. IEEE Trans. Med. Imaging (2019)

Liang, Q., Li, X., Deng, Y.: Diabetic retinopathy detection based on deep learning (2019)

Lin, M., Chen, Q., Yan, S.: Network in network. arXiv preprint arXiv:1312.4400 (2013)

Mateen, M., Wen, J., Song, S., Huang, Z.: Fundus image classification using VGG-19 architecture with PCA and SVD. Symmetry 11(1), 1 (2019)

Sayamov, S.: Weakly supervised learning for retinal lesion detection (2019)

Selvaraju, R.R., Cogswell, M., Das, A., Vedantam, R., Parikh, D., Batra, D.: Grad-cam: visual explanations from deep networks via gradient-based localization. In: Proceedings of the IEEE International Conference on Computer Vision, pp. 618–626 (2017)

Szegedy, C. et al.: Going deeper with convolutions. In: Proceedings of the IEEE Conference on Computer Vision and Pattern Recognition, pp. 1–9 (2015)

Tan, M. et al.: Mnasnet: platform-aware neural architecture search for mobile. In: Proceedings of the IEEE Conference on Computer Vision and Pattern Recognition, pp. 2820–2828 (2019)

Terrada, O., Cherradi, B., Raihani, A., Bouattane, O.: A fuzzy medical diagnostic support system for cardiovascular diseases diagnosis using risk factors. In: 2018 International Conference on Electronics, Control, Optimization and Computer Science (ICECOCS), pp. 1–6 (2018)

Wang, Z., Yang, J.: Diabetic retinopathy detection via deep convolutional networks for discriminative localization and visual explanation. arXiv preprint arXiv:1703.10757 (2017)

Yu, D., Seltzer, M.L.: Improved bottleneck features using pretrained deep neural networks. In: Twelfth Annual Conference of the International Speech Communication Association (2011)

Comparative Study of Supervised Machine Learning Color-Based Segmentation for Object Detection in X-Ray Baggage Images for Intelligent Transportation Systems

Mohamed Chouai, Mostefa Merah, José-Luis Sancho-GÓmez, and Malika MIMI

Abstract

Aviation security screening systems have been considerably strengthened since the attacks of September 11, 2001. To date, almost all used baggage control systems are managed manually by human operators. The latter remains essential in this type of technology; however, it presents a number of risk factors such as distraction or errors of inattention, which constitutes a major risk. Therefore, the need for intelligent transportation systems is convenient. This paper presents an automatic color-based segmentation for object detection in X-ray baggage images at airports. It is based on the use of machine learning to perform the color-based segmentation of X-ray images of scanned baggage. A comparative study between eight different machine learning methods is carried out with the optimization of the hyperparameters for a better design of the different models. The results obtained over our private database (High Tech Detection Systems HDTS) composed by dual-energy X-ray luggage scan images showed a perfect object detection without practically any false detection.

Keywords

Air transport • Security • Baggage control • Segmentation • Machine learning • Fusion • Object detection

M. Chouai (✉) · M. Merah · M. MIMI
Signals and systems laboratory, Department of Electrical Engineering, Mostaganem University, Site 1 Route Belahcel, 27000 Mostaganem, Algeria
e-mail: mohamed.chouai@univ-mosta.dz

J.-L. Sancho-GÓmez
Data Processing and Machine Learning), Information and Communications Technologies Department, Universidad Politécnica de Cartagena, Plaza del Hospital 1, 30202 Murcia, Cartagena, Spain

1 Introduction

In baggage screening, the X-rays pass through a suitcase and are then attenuated by the different densities of objects inside the bag that they encounter. An enlarged "shadow" of the bag and its contents is then displayed on the computer screen of the X-ray machine. As a result, all objects along an X-ray path attenuate the electromagnetic radiation and contribute to the measured final intensity. This is the transparency property which allows X-ray imaging to see through objects. More importantly, unlike reflection images in which each pixel corresponds to a single object, the pixels of the transmission images represent the attenuation of several objects. For regular images, artificial vision approaches decompose an image into disjunct regions (segments) that must roughly correspond to the objects. For radiological images, however, pixels should not be assigned to a single region, but to all overlapping objects that contribute to their attenuation.

This paper proposes a comparative analysis of a machine learning color-based segmentation for object detection within dual X-ray baggage images. First, it used color-based pixel segmentation of images to separate organic, inorganic, mixed, and opaque objects from the background. Second, those five types of images are reduced in the so-called fusion phase and classified into only two: organic and inorganic. Finally, object detection is carried out.

This paper is organized as follows. Section 2 presents a review of the segmentation process in the domain of baggage inspection. Section 3 sets the objective and shows the originality of this work. Section 4 gives a description of the different phases involved in the proposed system. The background information of the evaluation metrics are included in Sect. 5. Section 6 shows the experimental, in which the results are discussed. Finally, Sect. 7 closes the article with conclusions and future work.

2 Related Works

Thresholding is a simple but effective technique for image segmentation Strouthopoulos and Papamarkos (2000). A variety of thresholding methods have been developed. The limitation of thresholding techniques is that they only apply to a single-band image, such as a grayscale image or a single band of a multi-band image. For most multi-thresholding methods, the appropriate number of thresholds should be estimated first. For this purpose, Abidi et al. (2006) improved a weapon detection system in single energy X-ray images through pseudo coloring using multilevel thresholding segmentation in their study. In addition, Muthukkumarasamy et al. (2004) aimed to develop an intelligent banned object detection system to enhance aviation security by converting each image to grayscale after segmenting them using a multiple-thresholding algorithm.

Data clustering is another effective technique for image segmentation Coleman and Andrews (1979). The advantage of the data clustering technique is that it can be applied to a multi-band image. The main disadvantage of the data clustering method is again that the number of clusters must be determined first. Initially, Mery et al. (2017) studied the evolution of modern computer vision techniques for the X-ray test using adaptative K-means clustering segmentation for grayscale images. In addition, In Liang et al. (2003), Liang et al. proposed an automatic X-ray image segmentation method for threat detection based on the Radon transform to determine the optimal number of clusters and to evaluate the segmented images. They applied the clustering with multi-threshold segmentation

The state of the art shows that machine learning techniques have not been almost used in segmentation for X-ray baggage control. This can be explained by the lack of sufficiently large databases.

3 Objective

The objective of this work is to obtain a very effective image segmentation method from the color of pixels using a representative set of machine learning algorithms. The best image segmentation allows a better detection of objects and thus a better classification. The idea is to apply machine learning classification methods to the segmentation process, exploiting the color information present in the images. As we suppose to have a priori knowledge of the nature of the baggage (each material has a specific color), it is possible to classify pixels by means of supervised machine learning.

4 Methodology

Figure 1 shows a synoptic diagram describing the steps of our detection system in X-ray baggage images. Next, there is a description of each block.

4.1 Enhancement

Image enhancement plays a key role in image processing. On the one hand, in an automatic system, this pre-processing aims to highlight regions of interest attenuating others and to reduce the noise. On the other hand, in the case of a manual system, the objective is to propose a visual aspect underlining the invisible structures by balancing the distribution of shades of grey and accentuating the transitions to reduce the blur in order to facilitate the diagnosis by human experts Kase (2002); Besma Roui-Abidi Jimin Liang and Abidi (2004).

In this work, we used the Median Filter (MF), which is a non-linear digital filter often used for noise reduction. The main idea of MF is to replace each pixel value by the median value of its neighborhood. This filter eliminates the outliers without the limitations of averaging, which contaminates the neighboring values. This filter respects the contrast of the image, leaves it unchanged in areas of monotonous intensity, respects contours, and eliminates extreme values Arce (2005). MF can be used with different neighborhoods of $n \times n$ pixels, where $n = 7$ is generally used because large neighborhoods produce severe smoothing Bonnet (2009).

4.2 Segmentation

The contribution of this paper is the segmentation of the X-ray dual-energy color images using machine learning classification algorithms. The color provides information about the material (orange, blue, green, and black for organic, inorganic,

Fig. 1 Phases of the system

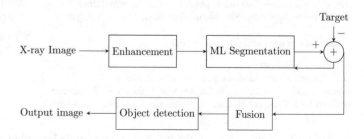

Fig. 2 The segmentation process

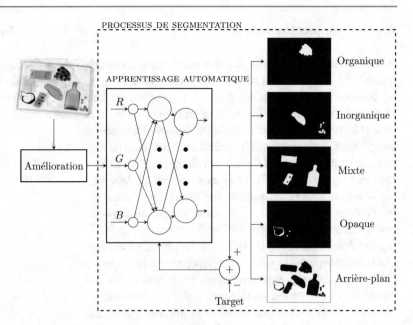

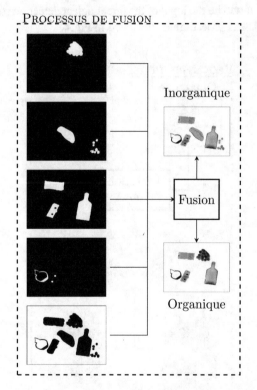

mixed, and opaque objects, respectively). The background of the image is considered as another class, so the segmentation problem becomes a classification problem with five classes, see Fig. 2.

The following algorithms are used in this study: Multilayer Perceptron (MLP), Support Vector Machine (SVM), Random Forest (RF), Naive Bayes (NB), K-Nearest Neighbors (KNN), and Linear Discriminant Analysis (LDA).

4.3 Fusion and Evaluation

Better performance is achieved by reducing the segmentation output of five classes into two classes: organic and inorganic, as it is shown in Fig. 3. This is because the overlap between classes is reduced. Nevertheless, most of the explosives are organic objects and most of the firearms and bladed weapons are inorganic. For this reason, it is convenient to get two separate images containing organic and inorganic objects, respectively.

To obtain these two images, a fusion phase is introduced following the next criteria:

1. Mixed objects are considered organic and inorganic objects.
2. Opaque objects can be metals or very thick organic or inorganic materials. Therefore, particular processing is applied to those types of objects. The main idea is to consider an opaque object as organic or inorganic as long as it contains or touches an organic or inorganic object; otherwise, it will act both as an organic and inorganic object.
3. The background is ignored.

Fig. 3 Fusion process to aggregate organic and inorganic objects

The selection of the best segmentation algorithm is accomplished not only by the confusion matrix, but also by four clustering evaluation metrics, which are the Davies–Bouldin index (DB), the Calinski-Harabasz index (CH), the Dunn Index (DI), and the Hartigan Index (HI), all of them are described in Sect. 5.

4.4   Object Detection

It is important to avoid the detection of small objects, through a threshold parameter that controls the surface of the object. To achieve this objective, we propose a detection method based on binarization and thresholding using the Otsu method Otsu (1979), followed by the labeling technique of the related components. This technique is applied to obtain disjoint sets that can be easily isolated Sezgin and Sankur (2004); di Stefano and Bulgarelli (1999). The model returns the labels of the detected objects, as well as the image coordinates of the corresponding objects. Using these values, the application generates a new image filled with rectangles surrounding the detected objects, in which the red, blue, green, and black boxes represent organic, inorganic, mixed, and opaque objects, respectively. Separate images are generated for the categories of the organic and inorganic objects, allowing the human operator to discriminate between the selected objects on the one hand then facilitate the diagnosis, and on the other hand, it will be the input of the threat object detection (future work). The procedure is illustrated in Fig. 4.

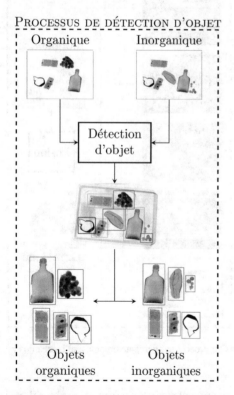

Fig. 4 Synoptic diagram showing the final task of our system (object detection). The output is composed of two images corresponding to organic and inorganic objects

5   Segmentation Evaluating Metrics

As it was already commented in Sect. 4.3, the best segmentation algorithm is selected not only by the confusion matrix but also by four evaluation clustering metrics which are described next.

- The Davies–Bouldin (DB) criterion relies on a relation of inter-cluster and intra-cluster measures. In particular, the DB index is given by

$$DB = \frac{1}{k} \sum_{i=1}^{k} R_i \tag{1}$$

where

$$R_i = \max_{i=1,\dots,k, i \neq j} D_{ij} \tag{2}$$

and

$$D_{ij} = \frac{\overline{d_i} + \overline{d_j}}{d_{ij}} \tag{3}$$

where k is the number of clusters, $\overline{d_i}$ and $\overline{d_j}$ are the average distances between each sample in the i-th and j-th cluster to their respective centroids $\overline{C_i}$ and $\overline{C_j}$, and d_{ij} is the Euclidean distance between these centroids.
The smallest value of DB corresponds to the best segmentation method Davies and Bouldin (1979).

- The Calinski-Harabasz (CH) criterion is described by the following variance ratio

$$CH = \frac{A}{B} \frac{N-k}{k-1} \tag{4}$$

where A is the overall between-cluster variance, B is the overall within-cluster variance, and N is the number of observations Calinski and JA (1974).
The overall between-cluster variance A is defined as

$$A = \sum_{i=1}^{k} n_i \parallel C_i - \mathbf{m} \parallel^2 \tag{5}$$

where $\mathbf{m}$ is the overall mean of the sample data and n_i are the samples in the i-th cluster, and C_i is the centroid of the i-th cluster.
The overall within-cluster variance B is defined as

$$B = \sum_{i=1}^{k} \sum_{i \in c_i} \parallel \mathbf{x} - m_i \parallel^2 \tag{6}$$

Fig. 5 Enhancement process. Reduction of noise and outliers

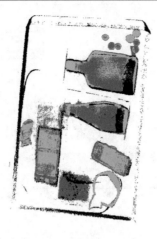 

(a) Without Enhancement   (b) With Enhancement

where $\mathbf{x}$ is a sample and c_i is the i-th cluster. The best segmentation is the solution with the highest Calinski-Harabasz value Calinski and JA (1974).

- The Dunn Index (DI) is the ratio of the minimum separation distance between two elements of different classes and the maximum distance between two elements of the same class. It is an index that does not rely on a particular distance and can, therefore, be used in a wide variety of situations. It is expressed as follows

$$DI = \frac{d_{\min}}{d_{\max}} \qquad (7)$$

where $d_{\min}$ is the minimal distance between samples of different clusters and d_{max} is the largest within-cluster distance. These two distance are defined by

$$d_{\min} = \min_{k \neq k'} d_{kk'} \qquad (8)$$

$$d_{max} = \max_{1 \leq k \leq K} D_k \qquad (9)$$

where $d_{kk'}$ is the distance between clusters c_k and $c_{k'}$ defined by the distance between their closest samples and D_k is the largest distance separating two different samples in the same cluster.

The best method of segmentation is the one that gives the highest value of DI Dunn (2008).

- The last metric used for evaluation is the Hartigan Index (HI) which is calculated as follows

$$HI(k) = (n - k - 1)\frac{E(k) - E(k+1)}{E(k+1)} \qquad (10)$$

where n is the number of entities (in our case the pixels to classify), k is the number of clusters, $k = 1, 2, \ldots, n-2$, and $E(k)$ is given by

$$E(k) = \sum_{j=1}^{k} \sum_{i \in C_i} (\mathbf{x}_i - C_j)(\mathbf{x}_i - C_j)^T \qquad (11)$$

being $\mathbf{x}_i$ the i-th p-dimensional vector of samples in cluster c_i and C_j is the centroid of j-th cluster. The optimal segmentation method is the one that minimizes HI Hartigan (1975).

6 Experiments

The images used in this work come from the HDTS database with 1000 dual-energy X-ray luggage scan images.

First of all, the Median Filter (MF) is applied for noise reduction. MF of 7×7 pixels has been selected since this size provides a good trade-off between noise reduction and smoothing. Figure 5 shows the effect of applying this filter on one of the images after the segmentation task. It can be seen that the filtered images show an improvement in quality, which increases the segmentation performance and, consequently, the object detection and classification performance.

The segmentation methods are tested on the HDTS dataset. Optimization of the hyperparameters using hold-out validation is required to ensure the best segmentation. The dataset is divided into three sets (training, validation, and test). The input samples are the pixels and their 8-adjacent neighboring pixels, containing the calorimetric values (red (R), green

Fig. 6 Hyper-parameter optimization process via the hold-out validation method. Figure 6a–d show the optimal values (red points) of the parameters N, K, T and δ, respectively, N being the number of neurons of MLP, K the number of neighborhoods of KNN, T the number of trees for RF, and δ the polynomial order of the kernel in SVM

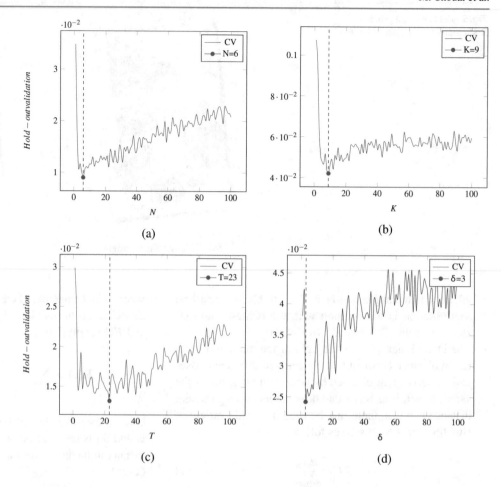

(a)

(b)

(c)

(d)

(G), and blue (B)) and targets indicating the corresponding value of the material (organic, inorganic, mixed, opaque, and the background). Once the datasets are prepared, the search for the optimal hyper-parameter values is performed. These parameters are the number of neurons of MLP, the number of trees of RF, the neighborhood number K of KNN, and those concerning the different kernels used in the SVM network. In particular, the Linear Support Vector Machine (LSVM) (12), Radial Basis Function Support Vector Machine (RBFSVM) (13), and Polynomial Support Vector Machine (PSVM) (14) kernels (KR) have been explored. These kernels are described by the following expressions

$$KR(\mathbf{x}, \mathbf{x}') = \langle \mathbf{x}, \mathbf{x}' \rangle \tag{12}$$

$$KR(\mathbf{x}, \mathbf{x}') = \exp\left(-\left\|\mathbf{x} - \mathbf{x}'\right\|^2\right) \tag{13}$$

$$KR(\mathbf{x}, \mathbf{x}') = \left(1 + \langle \mathbf{x}, \mathbf{x}' \rangle\right)^{\delta} \tag{14}$$

where $\mathbf{x}$ and $\mathbf{x}'$ are vectors in the input space, δ is the parameter to be optimized for the SVM with the polynomial kernel. Note that there are no hyperparameters to be optimized for NB and LDA.

Figure 6a shows that overfitting occurs when the number of nodes of the MLP network exceeds 6 and underfitting occurs for a number below this value, which implies that this number is the optimal one. Similarly, Fig. 6b–d show that 9, 23, and 3 are the optimal values for the hyperparameters of the KNN, RF, and polynomial SVM, respectively.

Once all the images are segmented, the best average results of each evaluation metric are obtained, see Fig. 7. The average segmentation time is also shown in Fig. 8. Table 1 shows the confusion matrices.

In Fig. 7, the Davies-Bouldin index indicates that the Naive Bayes method is the best, rated by 69.66% with an average calculation time of 0.22 s per image in segmentation. In addition, the Calinski-harabasz and Hartigan indexes also confirmed that Naive Bayes is the best algorithm, with rates of 60.23 and 55.97%, respectively. The Dunn index indicates that MLP is the best method followed by Naive Bayes. RF is the worst of the methods, besides the computation time exceeds the average value, which is 20 s for each image, as illustrated in Fig. 8.

Before the object detection process, composed by a binarization and thresholding using the Otsu method, an evaluation of our methods has been carried out by means of a set of experiments. Figure 9 illustrates qualitative examples taken

Fig. 7 Average percentage of the best segmentation obtained by the clustering validation indexes: Davies–Bouldin index (*DB*), Calinski-Harabasz index (*CH*), Dunn Index (*DI*), and Hartigan Index (*HI*). Testing on 1000 images

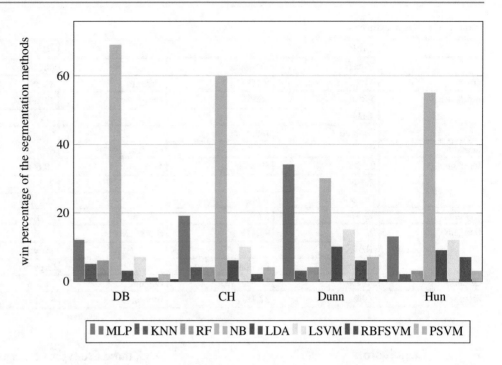

Fig. 8 The average calculation time of segmentation obtained for each method after testing 1000 different images

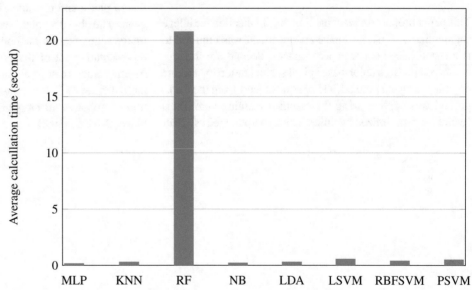

from our dataset. In this figure, the output images are showed with rectangles surrounding the detected objects, in which the red, blue, green, and black boxes represent organic, inorganic, mixed, and opaque objects, respectively. Notice that the obtained results are very satisfactory with a very small number of false detections, which is very encouraging for future works (objects classification).

Despite its "naive" design model and its simple assumptions, the conclusion is Naive Bayes is the best segmentation method used in this application. This is supported by other studies, such as Zhang (2004), which showed that there were theoretical reasons for the unexpected effectiveness of the Naive Bayes classifiers, especially because of conditional independence.

Table 1 Confusion matrices

	MLP			RF	
	Organic	Inorganic		Organic	Inorganic
Organic	94.2%	4.1%	Organic	86.3%	22.9%
Inorganic	5.8%	95.9%	Inorganic	13.%7	77.1%
	LDA			LSVM	
	Organic	Inorganic		Organic	Inorganic
Organic	91.1%	10.2%	Organic	89.4%	8.1%
Inorganic	8.9%	89.8%	Inorganic	10.6%	91.9%
	KNN			NB	
	Organic	Inorganic		Organic	Inorganic
Organic	90.5%	8.6%	Organic	97.5%	0.8%
Inorganic	9.5%	91.4%	Inorganic	2.5%	99.2%
	RBFSVM			PolySVM	
	Organic	Inorganic		Organic	Inorganic
Organic	90.1%	12.9%	Organic	91.5%	14.3%
Inorganic	9.9%	87.1%	Inorganic	8.5%	85.7%

7 Conclusion

This paper proposed a machine learning application for image segmentation. First, an image enhancement was required to increase the segmentation performance. Second, for the segmentation phase, a color-based pixel segmentation to separate organic, inorganic, mixed, and opaque objects from the background was applied using the machine learning algorithms (which were optimized by fitting their hyperparameters). Second, those five types of images are reduced in the so-called fusion phase and classified into only two: organic and inorganic. The objective of this work was to obtain a very effective image segmentation method from the color of pixels using a representative set of machine learning algorithms. A comparative study of several machine learning algorithms over a large dataset of X-ray images is presented for the classification of organic and inorganic objects for future threatening object detection work.

Fig. 9 The figure shows some object detection results extracted from our dataset. The red, blue, green, and black boxes represent organic, inorganic, mixed, and opaque objects, respectively. The results show the perfect object detection with practically no false detection

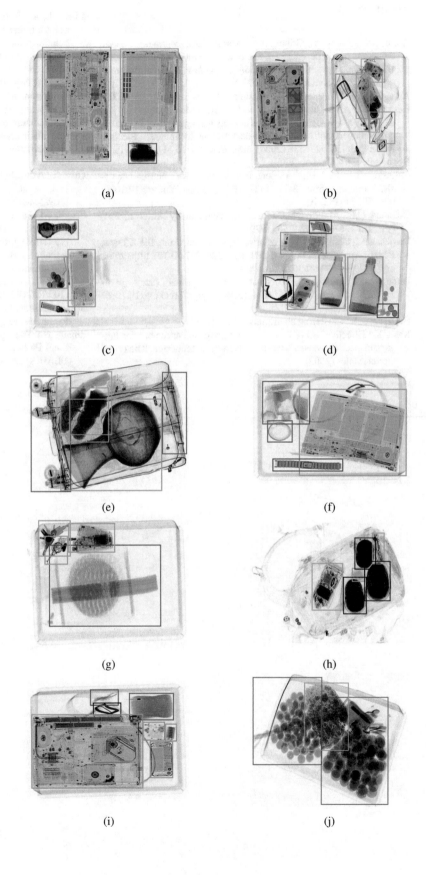

References

Abidi, B.R., Zheng, Y., Gribok, A.V., Abidi, M.A.: Improving weapon detection in single energy x-ray images through pseudocoloring. IEEE Trans. Syst. Man Cybern. **36**(6), 784–796 (2006). http://orcid.org/10.1109/TSMCC.2005.855523

Arce, G.R.: Nonlinear Signal Processing: A Statistical Approach (2005)

Besma Roui-Abidi Jimin Liang, M.M., Abidi, M.A.: Improving the detection of low-density weapons in x-ray luggage scans using image enhancement and novel scene-decluttering techniques. Electron Imaging **3**(13), 523–538 (2004). https://doi.org/10.1117/1.1760571

Bonnet, P.: Filtr. Median (2009)

Calinski, T., JA, H.: A dendrite method for cluster analysis. Commun. Stat. **3**(1), 1–27 (1974). https://doi.org/10.1080/03610927408827101

Coleman, G.B., Andrews, H.C.: Image segmentation by clustering. Process. IEEE **67**(5), 773–785 (1979)

Davies, D.L., Bouldin, D.W.: A cluster separation measure. IEEE Trans. Pattern Anal. Mach. Intell. **PAMI-1**(2), 224–227 (1979). https://doi.org/10.1109/TPAMI.1979.4766909

Dunn, J.C.: Well-separated clusters and optimal fuzzy partitions. J. Cybern. **4**(1), 95–104 (2008). http://orcid.org/10.1080/01969727408546059

Hartigan, J.A.: Clustering algorithms. Wiley (1975)

Kase, K.: Effective use of color in x-ray image enhancement for luggage inspection. Master's thesis, University of Tennessee, Knoxville. United States (2002)

Liang, J., Abidi, B.R., Abidi, M.A.: Automatic x-ray image segmentation for threat detection. In: Fifth International Conference on Computational Intelligence and Multimedia Applications (ICCIMA), pp. 396–401 (2003). https://doi.org/10.1109/ICCIMA.2003.1238158

Mery, D., Svec, E., Arias, M., Riffo, V., Saavedra, J.M., Banerjee, S.: Modern computer vision techniques for x-ray testing in baggage inspection. IEEE Trans. Syst. Man Cybern.: Syst. **47**(4), 682–692 (2017). http://orcid.org/10.1109/TSMC.2016.2628381

Muthukkumarasamy, V., Blumenstein, M., Jo, J., Green, S.: Intelligent illicit object detection system for enhanced aviation security. In: International Conference on Simulated Evolution and Learning (2004)

Otsu, N.: A threshold selection method from gray-level histograms. IEEE Trans. Syst. Man Cybern. **9**(1), 62–66 (1979). http://orcid.org/10.1109/TSMC.1979.4310076

Sezgin, M., Sankur, B.: Survey over image thresholding techniques and quantitative performance evaluation. Electron. Imaging **13**(1), 159–179 (2004). http://orcid.org/10.1117/1.1631315

di Stefano, L., Bulgarelli, A.: A simple and efficient connected components labeling algorithm. In: International Conference on Image Analysis and Processing. Venice, Italy (1999). https://doi.org/10.1109/ICIAP.1999.797615

Strouthopoulos, C., Papamarkos, N.: Multithresholding of mixed-type documents. Eng. Appl. Artif. Intell. **13**(3), 323–343 (2000)

Zhang, H.: The optimality of Naive Bayes. In: Seventeenth International Florida Artificial Intelligence Research Society Conference (FLAIRS) (2004)

A Survey of Artificial Intelligence-Based E-Commerce Recommendation System

Mohamed Khoali, Abdelhak Tali, and Yassin Laaziz

Abstract

Recommendation systems have led to the growth of many e-commerce and content providing companies because they perform well in modelling consumer's habit and providing personalization to users. Recommendation system used in E-commerce has been extensively researched and a number of algorithms/methods have been proposed. These methods can be a linear or a non-linear approach. Neural networks are proven non-linear techniques that outperform the linear techniques and have produced satisfactory results on various available recommendation datasets. Deep learning neural networks have also proven their efficiency in domains such as computer vision and natural language processing and of recent is been applied to recommender systems because of high performance. In this paper, we reviewed deep learning methods that have been applied to recommender systems in E-commerce both in academics and industry.

Keywords

Recommender systems • Deep learning • Neural building blocks

M. Khoali (✉) · A. Tali · Y. Laaziz
Laboratory of Information and Communication Technologies, National School of Applied Sciences of Tangier, Tangier, Morocco
e-mail: khoali.med@gmail.com

A. Tali
e-mail: sakom@hotmail.com

Y. Laaziz
e-mail: ylaaziz@gmail.com

1 Introduction

Recommendation engines have taken over e-commerce platforms because of its proven ability to increase consumer's engagements and purchases. Among the various applications and real-life scenario that recommender systems have been applied to can be suggestion of articles of the reader based on their preferences and interests, movie recommendation based on what the users already watched on the platform. Recommender engines are built on an algorithm that suggests the most relevant data to users.

Recommender systems have been deployed in various consumer product market to provide customers with better user experience such as personalization. Domains in which recommendation systems have been successfully deployed include movie recommendation as reported by Netflix that 80 % of movies watched on the platform by users was recommended by their recommendation engine (Gomez-Uribe and Hunt 2016), e-commerce, restaurant booking, video recommendation on YouTube brought about 60 % video clicks, book recommendation, etc.

Variables used in creating a recommendation engine are mostly user preferences, item features, user-item past interactions, temporal and spatial data (Zhang et al. 2018). Categorized recommendation techniques are based on the type of input data into a content-based recommender system, collaborative filtering and hybrid recommender system. The content-based recommendations provide users with recommendation similar to their previously preferred choice. Collaborative filtering recommendation unlike the content-based recommends products which other customers with similar preferences to the user bought while the hybrid model is a combination of the previously mentioned models (Adomavicius and Tuzhilin 2005).

With the recent improvement in the deep learning field and computer vision, research in the application of these fields in recommendation systems has seen tremendous improvements with major e-commerce companies employing

© The Editor(s) (if applicable) and The Author(s), under exclusive license to Springer Nature Switzerland AG 2021
M. Ben Ahmed et al. (eds.), *Emerging Trends in ICT for Sustainable Development*,
Advances in Science, Technology & Innovation, https://doi.org/10.1007/978-3-030-53440-0_12

recommendation systems on their platform as it provides a significant improvement over traditional methods.

In this paper, we provided the state-of-the-art on deep learning approaches that have been used to achieve recommender systems is reviewed.

2 Deep Learning Techniques

Deep learning, a machine learning technique or a machine learning sub-field, uses an objective function as an optimization scheme employing any variant of Stochastic Gradient Descent (SGD) algorithm. Otherwise, it has the characteristics of any machine learning algorithm. Deep learning techniques that have been used in the recommendation system in literature can be enumerated as the following (Zhang et al. 2018).

Multilayer Perceptron—it is a feed-forward neural network with architecture comprising more than one hidden layers injected between the input and the output layer.

Autoencoder—uses an unsupervised technique to regenerate the input at the output layer with the middle layer used as a feature representation of the input data.

Convolutional Neural Network (CNN)—uses convolutional layers for feature representation by moving the filter in blocks and pooling layers. It is a feed-forward neural network.

Recurrent Neural Network (RNN)—employs loops and memories for computations and mostly used for sequential data such as time-series data.

Restricted Boltzmann Machine—is generative neural networks which use two layers, the visible layer and a hidden layer to learn the internal representation of a problem.

Attentional Models—it operates based on soft content addressing over an input sequence.

Neural Autoregressive Distribution Estimation—is an unsupervised method inspired from the Restricted Boltzmann Machine coupled with a tractable distribution estimator by converting the RBM into a Bayesian network.

Deep Reinforcement Learning—it involves a combination of artificial neural networks with reinforcement learning to provide a trial-and-error architecture for an agent to learn and relearn (optimization) actions to achieve a set of goals.

Adversarial Networks—two neural networks, a discriminator and a generator are trained together in a competitive manner where the generator learn to create fake output which will serve as input for the discriminator network. The objective of the discriminator network is to determine which input is fake and which is real (Zhang et al. 2018).

Identified the various benefits of developing a recommendation system using deep learning techniques (Zhang et al. 2018)

An important benefit DNN has provided for recommendation systems is the ability to model non-linearity in data using the non-linear activation functions (relu, tanh and sigmoid). Linear models cannot properly map out the complex pattern in user-item interaction the way non-linear models of neural networks do.

Deep neural networks perform as automatic feature extractor from varieties of input data (image, video, text, etc.) and can extract descriptive information from these input data. This ability proves another importance to recommendation systems by facilitating the usage of different types of input data to a single network where descriptors can be extracted independently. Also, it saves time spent on some manual labour-intensive tasks involving data preparation.

Also, CNN and RNN models used in sequence modelling have also proven effective in the application to recommendation systems. Frameworks and libraries developed to make deep learning techniques easy to develop have made the techniques flexible and development of models easier.

Not to overplay the benefits of deep learning approaches, the limitations were also identified (Zhang et al. 2018). Limitations observed include very large data requirements of deep learning models to adequately learn, extensive, expensive and iterative hyperparameter tuning to increase model accuracy, and interpretability of the models due to their hidden weights and activation functions.

3 Related Works and Literature

Recommendation systems using deep learning techniques in literature are categorized based on neural network architecture used (previously identified deep learning architectures).

Two main categories have been identified (Zhang et al. 2018). If the recommendation system is developed with neural blocks (neural blocks based) or when two or more neural networks were combined together for the recommendation system (hybrid).

Subsequently, literature that used deep learning models to develop recommendation systems accordingly to the stated categories is reviewed.

3.1 Multilayer Perceptron-Based Recommendation Models

Multilayer perceptrons (MLPs), otherwise called feed-forward neural networks, are the building blocks of most deep learning models. With no feedback connections as the name implies (feed-forward), information only flows in the forward direction to achieve the approximate function from the input to the output (Goodfellow et al. 2016).

Linear models have been extended using this basic neural network blocks (Dziugaite and Roy 2015; He et al. 2017; Guy et al. 2017). Collaborative filtering is a recommendation model method that mostly uses matrix factorization linear model and has been extended using Neural Network Matrix Factorization (NNMF) (Dziugaite and Roy 2015), Neural Collaborative Filtering (NCF) (He et al. 2017) and Deep Factorization Machine (DeepFM) (Guy et al. 2017). We review these neural network extensions of collaborative filtering as a deep learning technique.

Feature vectors are used by matrix factorization for relational data modelling. The function X is approximated by the product of two low-rank matrices $U^T V$, where $U \in R^{D \times N}$ and $V \in R^{D \times M}$, and given D, the rank, is not greater than N and M. U_n is the feature vector that provides the description of the user N, and V_m is the feature vector that provides the description of the item M. An inner product $U_n^T V_m$ approximates the ranking $X_{n,m}$. Using this technique expressed in Eq. 1, an attempt is made to recover missing rankings.

$$\hat{y}_{ui} = f\left(P^T v_u^U, Q^T v_i^I \vee P, Q, \theta_f\right) \quad (1)$$

Equation 1 gives the function where $P \epsilon R^{M \times K}$ and $Q \epsilon R^{N \times K}$, with P and Q, are the feature vectors for users and items, respectively, and k is the dimension of vector. The interaction function f which is a feed-forward neural network has θ model parameters.

The architecture used in (He et al. 2017) is used on both the recommendation system MovieLens dataset and Pinterest dataset and compared the performance with other methods already implemented in literature.

Several follow up research (Niu and J, Cavarlee, and H Lu 2018; Song et al. 2018; Lian et al. 2017; Wang et al. 2017; Xue et al. 2017; Zhang et al. 2018) have attempted to improve the performance and provide an extension to many domains with different variations to the algorithm such as using a pairwise ranking loss function, replacement of one-hot identifier (embeddings) with columns or rows in the matrix.

DeepFM (Guy et al. 2017) is a Factorization Machine-based neural network that combines the power of deep learning model with factorization machines in the recommendation system and provided a new architecture for Click-Through Rate (CTR) recommendation technique. CTR method is used to model the interaction between user clicks in the e-commerce application pages and the application itself. It estimated the probability of a user clicking a specific app.

This model used a factorization machine to model both the low-level feature vectors and high-level feature vectors using the feed-forward neural network.

DeepFM has two components, the factorization machine and deep neural network blocks. The components share the same inputs which are the latent vector as shown in Fig. 2 that describes the architecture of the model.

The efficiency of the developed model was tested on two datasets, Criteo dataset and Company dataset with evaluation metrics set as Area under ROC (AUC) and logloss (cross-entropy). The model was compared with about nine other models and better performance was reported.

To further investigate the ability of multilayer perceptron in recommendation systems, the use of multilayer perceptrons in feature representation is investigated. Deep and wide learning (Cheng et al. 2016) used for Google play app recommendation, and (Covington et al. 2016) for Youtube recommendation are reviewed.

Wide linear models and deep neural networks are combined and trained together in other to tackle memorization and generalization challenges (Cheng et al. 2016).

The wide generalized linear model is defined as a prediction function $y = w^T x + b$, where x is the feature vectors and w are the model parameters with bias b. The deep memorization feed-forward neural network architecture accepts input of feature vectors, which are transformed into low-dimensional space, dense real-valued vector (embedding vector). The output (embedding vectors) is fed into the network of hidden layers that uses the Rectified Linear Units (ReLU) activation function.

After this phase, the components of the model are merged. The output weighted sum log-odds are used as the target in a supervised approach and the network using logistic loss function is trained on these outputs.

Using two networks called candidate generation network and ranking network, (Covington et al. 2016) provides feature representation. The networks have a different set of tasks. The Candidate generation network retrieves video data from the database (Youtube activity history) and trains it on the collaborative filtering network. The ranking network provides ranking by the score to the user. This is done by assigning scores to video according to the desired objective function from a set of features describing the video and user.

3.2 Autoencoder-Based Recommendation Models

A number of autoencoder variations (denoising, variational, contactive and marginalized) have been used as a deep neural networks approach to developing a recommendation system. (Zhang et al. 2018) identified two ways of applying autoencoder to the recommender system. It can either be used to learn low-level feature vectors of the input or generating new values for empty space matrix directly at the reconstruction layer of the network.

The linear model of collaborative filtering has seen tremendous success when enhanced with the non-linear

Fig. 1 The architecture of the
Neural collaborative filtering
model

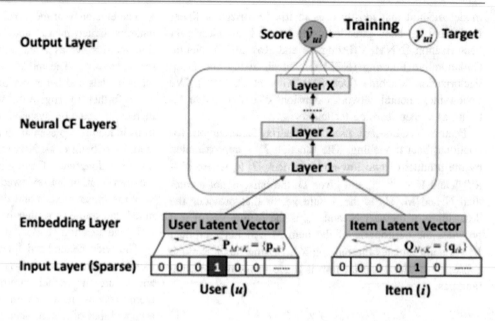

Fig. 1 The architecture of the Neural collaborative filtering model

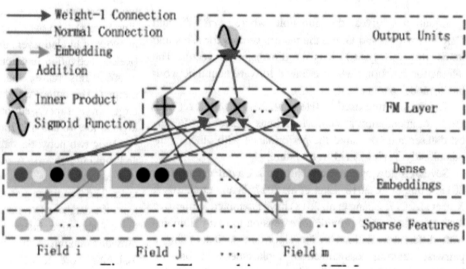

Fig. 2 The architecture of DeepFM (Guy et al. 2017)

autoencoder network as seen in (Sedan et al. 2015; Zhang et al. 2017; Li et al. 2015).

AutoRec (Sedan et al. 2015), a denoising autoencoder network, is built ontop a collaborative filtering model and designed on both an item-based autoencoder network and user-based autoencoder network. The networks accept the latent vector of each item or user as the input to the network, maps it to the subsequent hidden layers of the network and then try to reconstruct the item or user in the output layer of the network. The autoencoder has a single k-dimensional hidden layer with parameters learned using the backpropagation algorithm. Item-based autoencoder variation applies the network to the input vectors by first updating weights parameter of associated inputs during backpropagation and regularizing the parameters to prevent overfitting.

Four points from the AutoRec for deployment have been deduced (Zhang et al. 2018): (1) Item-AutoRec provides better performance when compared to User-AutoRec to generate missing ratings, (2) performance can be improved by combining different activation functions in the network, (3) larger hidden layer sizes can prove effective to improve performance and (4) the deeper the network, the better the performance but the performance is not significantly increased.

Variational Autoencoder has also been applied to recommender system with a considerable performance recorded. (Chen and Rijke 2018) proposed the collective Variational Autoencoder (cVAE) which can learn both user ratings and side information concurrently by using a generation and an inference network. The input of the variational autoencoder network is the item rating given by each user along with the dimension of the side information of all items. They are then fed into the inference network and the generation network. The method combines both collective Sparse Linear Method (cSLIM) and autoencoders to regenerate the side information and the user rating.

AutoSVD++ (Zhang et al. 2017) tackles some challenges faced when using autoencoders for recommender systems and so proposed a hybrid collaborative filtering model that is based on Contractive autoencoder architecture and therefore produces two models, the AutoSVD and the AutoSVD++. The AutoSVD extracts feature vectors from the items and their respective side information. The AutoSVD++ model takes the implicit feedback to tackle the sparsity problem faced when items have limited ratings.

Another autoencoder type used in recommender system is the Marginalized autoencoder (Li et al. 2015). The authors proposed a tightly coupled model that combines matrix factorization-based collaborative filtering and marginalized Denoising Autoencoders (mDAs) to produce a robust architecture. The denoising autoencoder attempts to regenerate input from a corrupted version of the data. The marginalized denoising autoencoder provides marginalization of random feature corruption in the model and can be represented as seen in Fig. 3.

$$L(W) = \frac{1}{2ck} \sum_{j=1}^{c} \sum_{i=1}^{k} \|x_i - Wx \sim_{ij}\|^2 \qquad (2)$$

To obtain $X\sim$, the marginalized denoising autoencoder iterates through the random corrupted set over X (sample set) multiple times with W representing the mapping of the objective function expected to minimize the loss function.

The approach established a hybrid model with a combination of collaborative filtering based on matrix factorization. Figure 3 summarizes the architecture with the input of the user-item rating matrix, the user feature set and the item feature set. The model extracts the latent factors from the ratings and side information, which are extracted at the connected hidden layers of the autoencoder neural network.

Fig. 3 The architecture of Autoencoder with input user-item rating matrix

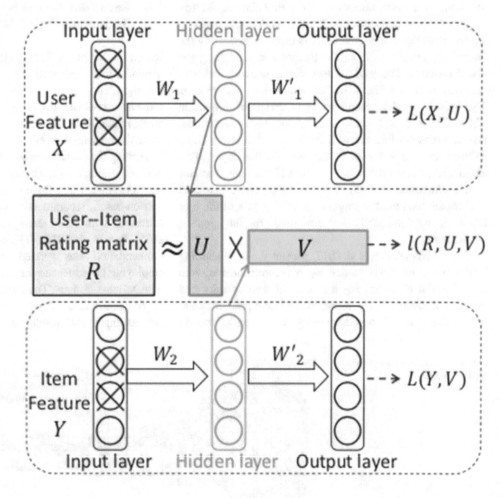

3.3 Convolution Neural Networks (CNN)-Based Recommendation Models

Convolutional Neural network is well known for its success in learning feature representation from both structured and unstructured multimedia data such as images, videos and texts. Its ability to work with relatively all data types has made it popular over the year. These features are learned through some convolution operations and pooling operations that continuously pass over the data.

Extracting feature vectors from images has been a particular and famous niche for CNN with application ranging from image classification, object detection and semantic analysis. Recommender systems have seen a fair usage of CNN too for feature representation. (Wang et al. 2017) improved the Point-of-Interest (POI) recommendation by considering images as a source of improvement in performance. It proposed the Visual Content Enhanced POI recommendation (VPOI) for recommender system. Three different types of objects were identified and served as input to the model. They are the users, locations and the images of the items the user interacted with. Probabilistic Matrix Factorization, a collaborative filtering model, was used as the frame of the model, and CNN is used for extracting the feature representation from the images and modelling the visual contents. The VGG16 deep neural network architecture was used, see Fig. 4. for the summarized architecture. The VGG16 model accepts an input image of dimension $224 \times 224 \times 3$ and consists of about 13 convolutional layers, 5 max-pooling layers, 3 fully connected layers and a softmax layer that serves as the classification layer. This architecture makes the output of the CNN a feature vector with a dimension (length) of about 4096. These extracted features are then used to support the training task of the user latent vector and POI features used in the general architecture.

DeepCoNN (Zheng et al. 2017) presented a deep learning model based on CNN to develop a recommender system with the aim of addressing the issue of data sparsity and quality improvement. The model learned item properties and user behaviour all together using two parallel neural networks that are coupled together in the last layers. The user behaviours from reviews are learned by the first network and the second network learned the item properties from the reviews. This is achieved by using a matrix factorization technique. Word-embedding technique is used on the review texts to embed them and is mapped into a lower-dimensional vector space. The general architecture of the CNN consists of convolutional layers with different kernels, a max-pooling layer and fully connected layer. The output of both networks is finally joined together and serves as the input of the prediction layer of the general model where the factorization machine is applied to map their interactions for rating prediction.

In music recommendation, CNN model has been implemented (Oord et al. 2013). Latent factors were extracted from music audios. The architecture of the CNN used ReLU as the activation function and the output of the network, the feature vector is used to train the prediction model in a collaborative filtering approach.

3.4 Recurrent Neural Network (RNN)-Based Recommendation Models

Recurrent Neural Networks (RNNs) are among the neural networks that work well with processing sequential data. It has the ability to work with data of the varied length of sequences (Goodfellow et al. 2016). Just as CNN is good with images, RNN is great to a sequence of information, and mostly used with a serial data structure.

Among the problem cases of sequential patterns that have not seen enough research interest due to sparse available of data and limited technique to model the problem case is the session-based recommendation. Recently, this has been addressed with RNN deep learning models (Hidasi et al. 2015; Tan et al. 2016). The unpopular session-based recommendation was tackled in (Hidasi et al. 2015) by employing Gated Recurrent Unit (GRU)-based RNN to deal with sequential data. The network accepts the state of the session (an item of actual event or events in the session so far) as input and produced an output (a score) of the

Fig. 4 The architecture of the VGG16 model

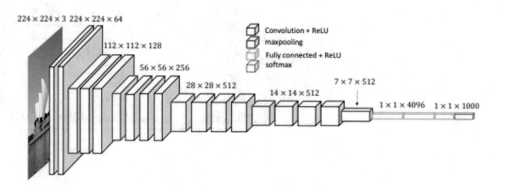

predicted preference of the items as seen in the architecture, as seen in Fig. 5, consists of a combination of typical GRU layers and feed-forward layers between the last layer and the output. Each GRU layer has the hidden state of the previous layer as the input.

For the ranking loss, Bayesian Personalized Ranking and TOP1 ranking were introduced to the GRU layers to have a ranking of items according to relevance.

Tan et al. 2016) proposed improvement to RNN-based recommender system with the application of data augmentation technique to enhance the input data distribution. The strategies the authors took in order to have these improvements can be summarized (1) Input data augmentation based on sequence pre-processing and implementation of drop-out regularization in the deep architecture; (2) Network pre-training and fine-tuning to facilitate the adaptation of learned weight on new sequence data; (3) "privileged information" that cannot be used in training is provided in a supervised matter to "distilled" the network; (4) embedding the output of the network for faster computation.

Session-based recommender systems usually do not consider user identity. RNN, (Wu et al. 2017), a

recommender system for a sequential recommendation based on user identification is discussed. Recurrent Recommender Networks (RRNs) are based on Long Short-Term Memory (LSTM), Recurrent Neural Network are used to model the seasonal evolution of movies and changes of user preferences over time. Two different LSTM networks each modelling dynamic user state and the movie states were used.

RNN has also been successful as a side information feature extraction tool in sequential data. (Bansal et al. 2016) proposed the use of GRUs to encode the text sequences into a latent factor model and used to solve the cold and warm start problems in recommendation systems. In addition, overfitting and sparsity of training data were prevented by using multi-task regularizer. The main task is rating prediction while the auxiliary task is item metadata prediction.

3.5 Restricted Boltzmann Machine-Based Recommendation Model

Salakhutdinov et al. 2007) employed Restricted Boltzmann Machine (RBM) to model relational data, one of the earliest implementation of neural networks to the recommender system. By employing Contrastive Divergence in learning, the authors showed that the model could perform effectively. The rating score of the user on the movie review is represented in a one-hot vector encoding with the visible unit of the RBM restricted to binary values.

Each user is assigned an RBM with a shared parameter and using the conditional multinomial distribution for modelling each column of the visible unit matrix V, conditional Bernoulli distribution for modelling the hidden user features h, the model can be given as follows:

$$p\big(h_j = 1 \vee V\big) = \sigma\Big(b_j + \sum\nolimits_{i=1}^{m}\sum\nolimits_{k=1}^{K} v_i^k W_{ij}^k\Big) \quad (3)$$

where $\sigma = \frac{1}{(1+e^{-x})}$ is the logistic function, W_{ij}^k is a symmetric interaction parameter between feature j and rating k of the movie I, b_i^k is the bias of rating k for movie I and b_j is the bias of feature j.

3.6 Neural Attention-Based Recommendation Model

The attention mechanism is mostly combined with other neural network models such as CNN, RNN and MLP. Attention models can be classified into standard vanilla attention and co-attention (Zhang et al. 2018). The vanilla attention (Chen et al. 2017) utilizes parameters of the context vector in other to learn while the co-attention (Zhang et al. 2018) uses the learning attention weights from

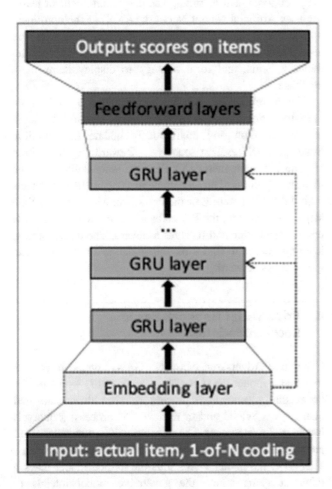

Fig. 5 The architecture of the RNN model

two-sequences. Another special model of attention model identified is the self-attention that can work independently of any other neural networks.

Attentive Collaborative Filtering (ACF) (Chen et al. 2017) addressed the issues of item and component level implicit feedback using a neural network by designing two attention-based networks to model each of item and component feedbacks. The network was able to learn the important informative properties of the item and component in their respective attention network. The model is then integrated into a collaborative filtering technique and trained with Stochastic Gradient Descent algorithm (SDG).

AttRec (Zhang et al. 2018) proposed self-attention model as a sequential model-to-model user's short-term intent and also long-term preference mapping. The self-attention models the short-term intent while collaborative filtering learning is used to model the long-term preference. Doing this, the user's temporal preferences can be deducted from their recent interaction footprints.

The self-attention input consists of the query, key and value, and the output is a weighted sum of the value, which is determined by the query and the key. To model the long-term preference of user, latent factor of the item and the user are evaluated. Euclidean distance is used to measure the distance between the item and the user.

3.7 Neural Autoregressive-Based Recommendation

Collaborative filtering-based Neural Autoregressive Distribution Estimator (CF-NADE) (Zheng et al. 2016) is a neural autoregressive architecture that encloses collaborative filtering tasks used for implicit rating recommendation. User's implicit feedback (such as watch, search, browse, or purchase behaviours of users) are converted into feature vectors and a confidence vector which are then used to model the probability of the "like" vector. The architecture that is modelled as a user U's explicit ratings is given as $r^u = \left(r^u_{m_1}, r^u_{m_2}, \ldots, r^u_{m_D} \right)$, with D as the number of items the user rated and m is the index of the ith rated items. The CF-NADE models the joint probability of the rating vector r as follows:

$$p(r) = \prod_{i=1}^{D} p\left(r_{m_{o_i}} \vee r_{m_{o_i}} \right) \qquad (4)$$

CF-NADE is adapted to implicit feedback. The rating is defined as the number of times a user interacts with a particular item using the implicit feedbacks. The implicit CF-NADE is trained by minimizing the negative log-likelihood directly with incorporation of confidence

levels. This model could predict user's preferences even on unobserved items.

3.8 Deep Reinforcement Learning for Recommendation Model

Mostly recommender does not consider two key real-life issues: real-time feedback to users and generation of a page of items with proper display according to the feedback. Deep reinforcement learning is used to provide a solution to these challenges.

Zhao et al. 2018) modelled a recommender system by implementing deep reinforcement learning to provide solutions to the issues. The model could effectively update recommendation strategy in real-time on new pages. The authors considered page-wise recommendation for the research and developed a model called DeepPage (Zhao et al. 2018) that could optimize a page of items in real-time with proper display based on real-time feedback from users.

The recommendation task was considered as sequential interactions between users and the recommender agent using deep reinforcement learning. The deep reinforcement learning uses Artificial Neural Network (ANN) as the non-linear approximator. In addition, page-wise recommendation was also introduced to generate a set of diverse and complementary items, and form strategy to display items on a webpage.

The recommended solution is modelled as a Markov Decision Process using reinforcement learning to learn optimal solution and continuously update the solution. State-space (S), Action space (A), Reward (R), Transition (P) and Discount factor (Y) elements were defined for the identified process. The Recommender Agent (RA) interacts with the environment E or the users over a sequence of time steps. At each step, the RA takes an action according to the environment state and receives a reward. The reinforcement learning goal is to find a recommendation strategy that can give the best reward.

4 Adversarial Network-Based Recommendation

Generative Adversarial Network (GAN) applied to recommendation model IRGAN (Wang et al. 2017) works on the belief that a generative process between documents and queries is provided and to retrieve this process, a query is issued to the document. The discriminative process learns to predict the score from the input of the query-document pairs. Therefore, the model works thus the unified model plays a minimax game where the generative model intends to produce relevant documents that are similar to labelled documents given to it and then tries to fool the discriminative retrieval model while it tries to differentiate between the labelled documents and the generated ones from the generative retrieval model. The generative model and the discriminative retrieval model are estimated as the probability of a document being relevant to query and are given by activation function (Sigmoid used) of the discriminator score.

The discriminator objective is to maximize the log-likelihood to generate relevant documents and correctly distinguished true values. The generative retrieval model fits the relevance distribution over documents and randomly samples from the entire documents in order to fool the disriminative retrieval model.

5 Conclusion

Existing linear models have established a strong foundation for recommender systems, deep learning techniques have further enhanced the performance and efficiency of the systems to outperform existing popular linear methods such as collaborative filtering. Since the advent of neural networks, research has attempted to apply it to various computation tasks. The recent feat achieved by deep neural networks on tasks such as computer vision and natural language processing has made its application also generalized to recommender systems with proven results that outperform linear methods.

This paper provides a comprehensive review of deep learning techniques that have been applied to the problem of recommendation covering techniques and to research and industrial problems.

It is also established that most deep learning techniques have been used to extend the linear model in other to provide non-linear modelling. Collaborative filtering linear model has been mostly extended in this context. By iterating through deep learning architectures of the model, we identified and reviewed the state-of-the-art methodologies that are used to develop the model and provide architectural strategies to developing light-weight deep network models.

With these strategies, our architectures effectively perform adequately accurate recommendations on the selected dataset. Although our network is trained on a specific dataset, the application is not limited to this dataset and can be scaled accordingly to other platforms.

Deep learning techniques have generally proved to be better at recommendation tasks as a non-linear model than the linear models.

There are many areas of improvement. The model can be scaled to accommodate structure of other types of e-commerce platform that has not seen enough recommender

system research such as B2C and C2C business models. The cold-start problem is a known issue of recommender systems which have been moderately resolved when deep networks are employed but still not very totally eliminated. This is an area of improvement that can be tackled.

In future work, we will focus on deep learning recommendation systems mainly in E-commerce platforms with its different business models. We will identify the types of the E-commerce platform in theory and categorized the recommendation systems discussed based on these types. We will emphasize on developing a model that can scale across the identified business models in E-commerce.

References

Adomavicius, G., Tuzhilin, A.: Toward the next generation of recommender systems: a survey of the state-of-the-art and possible extensions. IEEE Trans. Knowl. Data Eng. **17**(6), 734–749 (2005)

Bansal, T., Belanger, D., McCallum, A.: Ask the GRU: multi-task learning for deep text recommendations. In: Proceedings of the 10th ACM Conference on Recommender Systems, pp. 107–114 (2016)

Chen, Y., De Rijke, M.: A collective variational autoencoder for top-N recommendation with side information. In: DLRS (2018)

Chen, J., Zhang, H., He, X., Nie, L., Liu, W., Chua, T.: Attentive collaborative filtering multimedia recommendation with item-and component-level attention (2017)

Cheng, H., Koc, L., Harmsen, J., Shaked, T., Chandra, T., Aradhye, H., Anderson, G., Corrado, G., Chai, W., Ispir, M., et al.: Wide and deep learning for recommender systems. In: Recsys, pp. 7–10 (2016)

Covington, P., Adams, J., Sargin, E.: Deep neural networks for Youtube recommendations. In: Recsys, pp. 191–198 (2016)

Dziugaite, G.K., Roy, D.M.: Neural network matrix factorization (2015). arXiv:1511.06443

Gomez-Uribe, C.A., Hunt, N.: The Netflix recommender system: algorithms, business value, and innovation. TMIS **6**(4), 13 (2016)

Goodfellow, I., Bengio, Y., Courville, A.: Deep Learning. Chapter 6 (2016)

Guy, H., Tang, R., Ye, Y., Li, Z., He, X.: DeepFM: a factorization-machine based neural network for CTR prediction. In: Proceedings of the Twenty-Sixth International Joint Conference on Artificial Intelligence, pp. 1725–1731 (2017)

He, X., Liao, L., Zhang, H., Nie, L., Hu, X., Chua, T.: Neural collaborative filtering. In: WWW, pp. 173–182 (2017)

Hidasi, B., Karatzoglou, A., Baltrunas, L., Tikk, D.: Session-based recommendations with recurrent neural networks. In: International Conference on Learning Representations (2015)

Li, S., Kawale, J., Fu, Y.: Deep collaborative filtering via marginalized denoising auto-encoder. In: ACM, pp. 811–820 (2015)

Lian, J., Zhang, F., Xie, X., Sun, G.: CCCFNet: a content-boosted collaborative filtering neural network for cross domain recommender systems. In: WWW, pp. 817–818 (2017)

Niu, W., Cavarlee, J., Lu, H.: Neural personalized ranking for image recommendation. In: Proceedings of the 11th ACM International Conference on Web Search and Data Minings, pp. 423–431 (2018)

Salakhutdinov, R., Mnih, A., Hinton, G.: Restricted Boltzmann machines for collaborative filtering. In: Proceedings of the 24th International Conference on Machine Learning (2007)

Sedan, S., Menon, A.K., Sanner, S., Xie, L.: Autorec: autoencoders meet collaborative filtering. In: WWW, pp. 111–112 (2015)

Song, B., Yang, X., Cao, Y., Xu, C.: Neural Collaborative Ranking (2018). arXiv:1808.04957

Tan, Y.K., Xu, X., Liu, Y.: Improved recurrent neural networks for session-based recommendations. In: Recys, pp. 17–22 (2016)

Van den Oord, A., Dieleman, S., Schrauwen, B.: Deep content-based music recommendation. In: NIPS, pp. 2643–2651 (2013)

Wang, X., He, X., Nie, L., Chua, T.: Item silk road: recommending items from information domains to social users (2017)

Wang, J., Yu, L., Zhang, W., Gong, Y., Xu, Y., Wang, B., Zhang, P., Zhang, D.: IRGAN: a minimax game for unifying generative and discruminative information retrieval models (2017)

Wang, S., Wang, Y., Tang, J., Shu, K., Ranganath, S., Liu, H.: What your images reveal: exploiting visual contents for point-of-interest recommendation. In: WWW (2017)

Wu, C., Ahmed, A., Beutel, A., Smola, A.J., Jing, H.: Recurrent recommender networks. In: WSDM, pp. 495–503 (2017)

Xue, H., Dai, X., Zhang, J., Huang, S., Chen, J.: Deep matrix factorization models for recommender systems. In: IJCAL, pp. 3203–3209 (2017)

Zhang, S., Yao, L., Xu, X.: AutoSVD++: an efficient hybrid collaborative filtering model via contractive autoencoders (2017)

Zhang, S., Yao, L., Sun, A., Wang, S., Long, G., Dong, M.: NeuRec: on nonlinear transformation for personalized ranking (2018). arXiv:1805.03002

Zhang, S., Yao, L., Sun, A., Tay, Y.: Deep learning based recommender system: a survey and new perspectives. ACM Comput. Surv. 1(1) (2018)

Zhang, S., Tay, Y., Yao, L., Sun, A.: Next item recommendation with self-attention (2018). arXiv:1808.06414

Zhao, X., Xia, L., Zhang, L., Ding, Z., Yin, D., Tang, J.: Deep reinforcement learning for page-wise recommendations. In: RecSys (2018)

Zheng, Y., Liu, C., Tang, B.: Neural autoregressive collaborative filtering for implicit feedback. In: DLRS (2016)

Zheng, L., Noroozi, V., Yu, P.S.: Joint deep modeling of users and items using reviews for recommendation. In: WSDM (2017)

Green Networks and Intelligent Transportation Systems

Customer-Oriented Dial-A-Ride Problems: A Survey on Relevant Variants, Solution Approaches and Applications

Sonia Nasri, Hend Bouziri, and Wassila Aggoune-Mtalaa

Abstract

The interest of on demand transport systems is constantly growing. This trend is accompanied by increased customer's expectations in the quality of the service provided. In this regard, the quality of service must meet the needs for more customer-oriented features. This article summarizes recent developments in Dial-A-Ride Problems which integrate customer service quality. We provide a classification of the variants examined, as well as the available resolution methodologies and references on the instances used. A discussion on the sate of the art is given with perspectives for future researches.

Keywords

Dial-A-Ride Problem • Customer-dependent • Service quality

1   Introduction

Nowadays, one of the major classes of transportation is that of Transport on Demand (TOD) (Cordeau et al. 2007). This class of public and private transportation cope with the rise of demands with specific needs of people with reduced mobility or within areas not covered by the global public transport network. The research community has thus introduced

S. Nasri (✉)
Higher Business School of Tunis, Manouba University, Manouba, Tunisia
e-mail: sonia15.nasri@gmail.com

H. Bouziri
Higher School of Economic and Commercial Sciences Tunis, Tunis University, Tunis, Tunisia
e-mail: hend.bouziri@gmail.com

W. Aggoune-Mtalaa
Luxembourg Institute of Science and Technology, L-4362 Esch-sur-Alzette, Luxembourg
e-mail: wassila.mtalaa@list.lu

the Dial-A-Ride Problem (DARP) of (Cordeau and Laporte 2003a), as part of the more general TOD problem (Rekiek et al. 2006), and especially as a door-to-door transportation problem (Cordeau and Laporte 2003a; Rekiek et al. 2006; Melachrinoudis and Min 2011). The DARP consists of satisfying TOD requests under a set of problem constraints. This problem is NP-Hard as it is shown by Healy and Moll (1995). In this class of problems, several users are transported by the same vehicle according to the maximal vehicle capacity. The users specify when they wish to be picked up and when they have to reach their destination. This demand is called a request. The users do not specify an exact time of the day, but rather time windows are chosen for the request. The users have to specify either the pickup or the delivery time window and an operator has to assign the other time window. A maximal ride time is also proposed as a bound for satisfying all the requests. Thus, the users have inconvenience times (time spent aboard the vehicle from the pickup to the destination) which must respect the supposed maximal ride time for all transportation requests. The main DARP objective consists of minimizing vehicles tours' costs in a complete transportation network of nodes and arcs. The nodes are the pickup and delivery locations of customers with a single depot which represents the starting and final arrival point of all the vehicles of the fleet. All vehicles' tours have to respect a maximal travel total duration. Each arc has a transit time and cost. A request of transport on demand is related to a set of persons who want to travel from a node of pickup to a node of delivery. These two nodes are successively the origin and the destination of a request. Each couple must be visited by the same vehicle. A service time expresses the time needed for loading and unloading a customer.

1.1   Quality of Service in DARPs

The quality of service is expressed as the degree of conformity to customers' requirements in a general framework. In this review, we focus on the quality of service in Dial-A-Ride problems. Paquette et al. (2009) reviewed in a set of definitions from two classes of quality of service. The customer-based approach is based on customer perceptions while the technical approach relies on technical specifications.

From the customer perceptions viewpoint, Pagano and McKnight (1983), stated that the service quality relies on comfort, reliability, convenience of making reservations, extent of service, vehicle access, and safety. This later is the main motivation for customers with disability or reduced mobility. In this regard, there is also a strong need for assistance in moving from the vehicle to the destination, as well as short distances from the origin or destination to the means of transport. Besides, travel time is also seen from the customer perception as a major indicator of service quality. It relies on waiting time, total tour duration, and pickup time. This attribute is generally considered as operational costs by transportation providers.

On the technical side, large and various specifications for service quality have been defined in Paquette et al. (2009), since 1976. Most of these specifications are included as part of DARPs model as constraints. Some of these are expressed as time windows, see Melachrinoudis et al. (2007), Parragh et al. (2009), Cordeau (2006), Rekiek et al. (2006).

Other constraints define deviations from the desired or promised time at origin or destination like in the works of Jorgensen et al. (2007), Melachrinoudis et al. (2007).

The quality of service is sometimes specified by a maximum ride time and total travel time like in the work of Diana (2004), a deviation between the ride time and the direct travel time as in the contribution of Wolfler Calvo and Colorni (2007), the total waiting time (during the ride or before the pickup) as it is stated in Diana (2004), Jorgensen et al. (2007). The maximum or excess ride time (Jorgensen et al. 2007) and the delay between the call and the delivery time (Wilson et al. 1976), can also help to measure the level of quality of service. The maximum number of stops during the customer route has also been identified by Armstrong and Garfinkel (1982), as a parameter to measure the quality of service. Whatever the dimension of quality by perception or specification, Paquette et al. (2009), outpointed the favor of combining them into a single DARP model. This is intended to find a trade-off between costs and service quality.

1.2 Motivation, Objectives, and Methodology Used for This Survey

Few research works are addressing the specification of services offered to customers according to their preferences. This topic has started to emerge recently (since 2015). The recent work that emphasized this scope is that of Molenbruch et al. (2017a), which highlights the operational effects of the variations in the level of quality of service for the Dial-a-Ride Problem. As compared with this review and the one of Ho et al. (2018b), our focus and research directions are not similar.

We specify our research on advanced services in DARPs like real-life characteristics, new customers preferences, innovative services quality, and decision- making systems. Then, we zoom on customer-based approaches in terms of customers' expectations of nowadays and their effects on TOD systems. We also focus more on resolution methods improving the service quality aspect in dial-a-ride services and diverse trade-offs between costs and service quality. We emphasize on applications where balancing costs and service quality should be considered as insights. For our research review, we accessed Elsevier and Google Scholar databases using the keywords dial-a-ride, on demand transport, on demand mobility, demand responsive transport, service quality, and dial-a-ride services. A total of sixty publications since 2003, were considered with the biggest contribution published in Computers and Operations Research, Transportation Research Part B, Transportation Science, European Journal of Operational Research, Operations Research Letters, Public Transport, and Transportation Research Part C.

Our survey is organized as follows. Section 2 reviews the variants of DARPs, giving the main applications found in the literature. The next section reviews the different ways through which service quality is expressed in customer-oriented DARPs models either in static or dynamic frameworks. A survey on the existing resolution methods and benchmark instances is provided in Sect. 4. A discussion on the literature review is given in the next section, as well as perspectives for future researches. Finally, the conclusion is set.

2 Classical Dial-A-Ride Problem

2.1 Relevant Variants

DARPs have been extensively studied in the literature. Many variants are provided in Cordeau and Laporte (2007), to deal with various routing transportation situations depending on static or dynamic environments involving a single or multiple vehicles. If all data necessary for the decision-making are provided before the start of the operations, the DARP is considered as static. But, if some decisional information arises during the operations, the problem is considered dynamic. A recently reviewed classification can also be found in Ho et al. (2018b), where it was argued that either in the static or dynamic class, a DARP may be treated as deterministic or stochastic. At the decision time, information is known with certainty in a deterministic DARP, while it is unknown or uncertain in a stochastic DARP. Numerous works are found in the literature where DARP was differently modeled and solved. But the largest set of DARP studies is related to static frameworks. In this survey, we propose to highlight the customer service quality while respecting the general DARP classification of Ho et al. (2018b). Quality of service in dial-a-ride operations is studied in Paquette et al. (2009). This latter work outpoints the need for aligning the customer requirements with the specifications of DARP models. The first model which established the relationship between operational costs and service quality was proposed in Paquette et al. (2007). The common and the general quality specifications used by operations researchers are related to time windows and maximum user ride time. Thus, the optimization of the quality of service can be one of the multiple DARP objectives.

Several trade-offs between conflicting criteria are expressed through mono-objective, bi-objective, and multi-objective models. But the largest number of studies was concerned by mono-objective models with multi-criteria decision variables like in Jorgensen et al. (2007), Melachrinoudis and Min (2011), Parragh (2011), Paquette et al. (2013), Lehu et al. (2014). The aim in Jorgensen et al. (2007) was to minimize the objective function through two measures: low costs and a high level for the quality of service for the customers which is proportional to the total excess ride time and total waiting time. In Paquette et al. (2013), the objective consisted in minimizing the total routing costs and the penalties on waiting times when users are aboard. Authors in Melachrinoudis and Min (2011), proposed a multi-depot, multi-vehicle, double request DARP in a healthcare organization. The objective consists of minimizing a weighted sum of the total costs, total client inconvenience times, and the extent of earliness due to a client delivery before service and a late pickup after service. To define the interaction between operational costs and quality of service for users, Lehu et al. (2014), proposed high weights for elements which are related to either vehicle operator preferences or users preferences.

In this paper, the aim is to review the different ways through which service quality is expressed in customer-oriented DARPs models either in static or dynamic frameworks.

2.2   Applications and Benchmarks of DARPs

Numerous applications of DARP are found in the literature either in artificial or real-life studies. Major earliest DARP applications used artificial benchmark instances like the ones of Cordeau et al. (2006), having as main objective the minimization of cost. However, recent DARP applications are motivated by real-life cases. To better address such real problems, real-life benchmark instances have been set up by Chassaing et al. 2016 in Instances of Chassaing (2016). The main and traditional applications focus on elderly and disabled persons like in the works of Faria et al. (2010), Masson et al. (2014), transportation, health care transportation services, transport complementary to public transport, and private on demand mobility.

The first integrated DARP is proposed by Hll et al. (2009), where some part of each request may be performed by an existing fixed route service and the design of integrated routes is the key for this kind of on demand responsive service. Recent related works are proposed by Posada et al. (2017), Ma et al. (2019). Health care transportation services have a focus on various vehicles capacities and prioritized requests in the works of Beaudry et al. (2010), Tlili et al. (2018), Jlassi et al. (2012), Melachrinoudis and Min (2011).

The problem of vehicle reconfiguration to ensure the transportation to various persons is addressed by Cappart et al. (2018), Detti et al. (2017), Molenbruch et al. (2017b), Qu and Bard (2015). Demand responsive transportation which may replace unavailable public transportation services for low demand periods are treated by Mulley and Nelson (2009).

DARPs are tackled for providing services to specific facilities such as airports, markets, public places or to specific areas such as employment areas and service areas such as in the case provided by Enrique Fernindez et al. (2008). Then private on demand mobility is an application of DARPs which allows connecting private drivers and passengers through an integrated platform for mobility, see the papers of Çetin (2017) and Li et al. (2019). This kind of on demand responsive application provides an alternative to private cars in urban mobility.

3   Customer Service Quality in DARPs

The relationship between the service level criteria on operational costs is seriously highlighted in Molenbruch et al. (2017c). The authors stated that the variations in the level of service provided, impact the operational DARP results. To show several levels of trade-offs, they investigate 78 different scenarios for service quality. They are based on maximum deviation from the user's preference time and relative maximum exceedance of the direct ride time. However, the analysis is limited to artificial instances of Cordeau and Laporte (2003b), and cannot be assumed in real-life applications of TOD. The main focus of the service level scenarios is the establishment of a decision system. It will be able to select the appropriate scenario for a specific operating context according to customer preferences. Based on users' preferences, Molenbruch et al. (2017c), outpointed two relevant specifications in DARP services which directly influence the service level experience. Firstly, the time windows width at the pickup and delivery locations is a parameter which helps to measure the allowed deviation from the user's preferred departure or arrival time. Consequently, it influences the waiting time for a user. Secondly, the maximum user ride time is a maximal bound of the direct ride time related to a vehicle's tour. This bound corresponds to the time spent by a user aboard a vehicle and can be used for defining successive time windows.

Furthermore, in the classical DARP of Cordeau and Laporte (2003b), the maximum ride time is defined as a common bound for all users. However, Molenbruch et al. (2017c), antagonize such an assumption by explaining that the ride time is highly customer related and contributes in neglecting a possible vehicle's detour. Thus, the authors define a direct ride time for each user. In what follows, we review the contribution of the literature in the static environment.

3.1 Customer Service Quality in Static DARPs

User inconvenience is commonly expressed by the time spent aboard a vehicle. To model such inconvenience time, authors are attributing customer related maximal ride times as restrictions from a customer-oriented quality of service viewpoint. The maximum ride time is restricted for each customer see Parragh et al. (2009), Molenbruch et al. (2017b), Chassaing et al. (2016) and used to provide time windows. In the problem of Parragh et al. (2009), a request is split into a set of persons' groups and has a revenue. The objective is to maximize the total profit which is equal to the total revenue minus the total cost. Time windows are redefined for only the delivery location using a service time (the time needed to enter and leave a vehicle), a transit time, and a maximal ride time. A cooperative dial-a-ride service was studied in Molenbruch et al. (2017b), as a multi-depot DARP where a request can be exchanged through transportation service providers. The aim of this cooperation is the minimization of the overall routing costs for a system involving multiple service providers. The customized maximal ride time is also bounded by a maximal common ride time for all requests. Moreover, time windows are imposed for each request at the pickup. The results indicated that the benefits of this horizontal cooperation are enhanced by a decrease in the width of the customers time windows. A DARP variant of Chassaing et al. (2016), consists of minimizing travel costs. Constraints on maximal ride time and time windows are aligned to those of Cordeau and Laporte (2003b), with a more restriction on the maximal ride time. Three measures are suggested for the solution quality which are the total duration, the total riding time, and the total waiting time.

Time windows width is a factor which helps in decision-making systems. Different scenarios of variations in the quality of service are defined as multi-criteria models (Markovi et al. 2015). Markovi et al. (2015) proposed a mobile resource management system which manages Dial-A-Ride services. Time windows and route duration are factors that considerably influence operator costs and service quality through diverse demand distributions. Thus, their constraints relaxation has led to an operating cost decrease and fewer vehicles needed for

transportation. The objective was to minimize fixed and variable trip outsourcing costs. The obtained results explain the trade-offs between quality of service and various system characteristics helping in decision-making. As it is outpointed, a larger width of time windows provides more flexibility to system provider, but a lower level of service for customers. The need for redefining time windows was outpointed in the work of Hu and Chang (2015), explaining the correlation between ride time and time windows. Indeed, the time windows width significantly influences the users' ride time: its increase negatively impacts the service quality in terms of the ride and a decrease of the width increases the level of customer service quality. Ho et al. (2018a) addressed a static dial-a-ride problem. This is performed by a time window adjustment using a method of relaxation. The objective consists of minimizing the travel costs under constraints such as vehicle capacity, time window, ride time, and route duration. These constraints are penalized in the objective function. A maximal ride time is offered for all requests and it is used to bound the related ride time of each request. Moreover, the adjustment of the time windows is made by the use of some parameters such as the maximal ride time of all users, the beginning of service, and old bounds of time windows. As stated by Molenbruch et al. (2017b), to enhance a flexible strategy while deciding in the cooperative requests' exchange, all information on customer requests should be shared. The authors outlined that other operational characteristics should also contribute to the decision-making strategy such as the heterogeneity of customers, as well as their expectations.

To the best of our knowledge, the only one model of DARP which addresses customer service quality deeper is that of Nasri and Bouziri (2017). The aim was to minimize the travel costs through more customized vehicles routing plans in a static framework. Their model includes customers-dependent constraints which give more specification and adequacy to passengers. Two terms are suggested in the objective function: a total travel time for all vehicles routes and a penalty for time windows violations. Both time windows on pickup and delivery are redefined and adjusted taking into consideration minimal values for upper bounds and maximal values for lower bounds. For each bound computation, two values are suggested: the initial supposed bound and the new customized-bound. The customers' expectations include for each request related parameters such as the customer-dependent maximal ride time, the related customer service time, the travel time between origin, and destination of a request.

3.2 Customer Service Quality in Dynamic DARPs

Given the complex nature of dynamic DARPs as compared with static DARPs, and considering the customer-oriented quality of service, few contributions were found. There are only two dynamic DARPs which address the customer service quality, the ones of Wong et al. (2014), Carotenuto and Martis (2017) through the definition of a maximal ride time. In the dynamic and stochastic DARP of Wong et al. (2014), the requests' arrivals are defined by a Poisson distribution function without considering any external condition such as the weather, congestion or blocked routes. The objective was to minimize the operational costs in terms of the number of vehicles used and the total distance traveled satisfying service quality-related constraints. They include a customer-dependent maximal ride time and desired time windows. The maximal ride time depends on the urgency of a request for transportation and a maximum acceptable time needed by a passenger for waiting at the pickup location.

The service quality level is enhanced in the work of Carotenuto and Martis (2017), through a good-quality requests' schedule and provides a fast users' satisfaction by a dynamic insertion of the requests. The objective consists of minimizing the variance between the desired and effective time of the requests for pickup and delivery. To improve the user convenience, a redefinition of the maximal ride time is based on the redefinition of the delivery time taking into account the minimal travel time between the pickup and the delivery plus a constant function of time.

4   Resolution Methods for Customer-Oriented DARPs

The choice of a resolution method depends strongly on the nature of the problem treated. It can be an exact method like the branch and bound (Serrano et al. 2013; Bennekrouf et al. 2013), or other branching methods (Amroun et al. 2016). For larger sized problems metaheuristics like local search methods, evolutionary techniques (Aggoune-Mtalaa and Aggoune 2014), or hybridized ones (Rezgui et al. 2017, 2018), are more appropriate. In this section, we will follow the classification given in Ho et al. (2018b), for solving customer-oriented DARPs with exact, heuristics, and metaheuristic methods.

4.1   Exact Methods

In this part of the study, we focus on exploring a set of related works proposed for DARPs with the emphasis on measuring the quality of service for the customer. Given the high complexity of dynamic and real-life DARPs, there are no exact resolutions proposed for such problems. The dynamic nature of the problem increases the complexity at such an extent that it makes its exact resolution timely impossible. Moreover, dynamic instances do not exist in the literature, and

those which are found are modified according to the problem framework.

There is a work which addresses a stochastic and deterministic DARP in Heilporn et al. (2011), using a Branch and Cut algorithm. The arrival of customers is distributed according to probabilistic delays. For each customer, a maximal ride time is defined according to his travel time and an adjusted time windows. The stochastic nature of the problem increases the complexity, which avoids its exact resolution.

The majority of exact methods are dedicated to static and deterministic DARPs. This fact is due to the complexity of realistic DARP instances, especially for high sized instances. Therefore, the exact methods are generally used for artificial instances since it is too complicated to solve real-life cases exactly. Besides, an exact resolution is limited to problems of small size with a fixed computational time. There are few studies where a customer service quality is introduced, especially for the maximal ride time, see for instance Braekers et al. (2014), Parragh et al. (2015). In Parragh et al. (2015), authors introduced more real-life features to the instances of Cordeau and Laporte (2006), and more restrictions to the time windows computation. Their experimentation have shown that the exact resolution is optimal within 4 h on large artificial benchmarks, which is not the case for instances with some real-life and customized distances measurements.

4.2   Heuristics

As discussed previously, finding optimal solutions to complex DARPs is computationally intractable or difficult. Heuristics are methods designed for specific problems which produce a good solution to the problem without guaranteeing an optimum. Moreover, heuristics may be used to construct initial solutions for DARPs or to investigate the problem. Moreover, several heuristics may be combined in a single algorithm in order to produce efficient solutions for DARP such as in Luo and Schonfeld (2007). The authors proposed an insertion heuristic for solving a static multi-vehicle Dial-A-Ride Problem. The heuristic was based on a rejection of infeasible requests' insertion and their reinsertion into the best position. All related request operations are evaluated and improved with respect to all DARP constraints and the optimization of the global objective function. The aim was the minimization of the number of vehicles used for satisfying all requests with service quality-related constraints. Thus time windows constraints are designed considering the desired pickup and delivery time for each customer and a customized maximal ride time. The efficiency of the problem is shown through tests on randomly generated problems. An insertion-based heuristic was provided for solving a DARP with customized constraints in Markovi et al. (2015). The heuristic was based on the parallel insertion heuristic of Jaw et al. (1986). Some customer-oriented constraints are improved to fulfill various users requirements. The heuristic was implemented and compared to the exact method of Cordeau and Laporte (2006), on their benchmark instances (Instances of Cordeau 2006). Results were compared to known best solutions and were competitive on only small instances. To cope with customer-oriented and other real-life features of the problem, a mobile resource management system was implemented for a broader analysis. It provides a decision-making tool using various operational scenarios allowing the analysis of trade-offs between service quality level and some provider system parameters (time windows width, fleet needed for fulfilling the requests, distribution of demand over time, vehicle capacity).

More adapted algorithms for DARPs with customer service quality highlighted the need to design specific methods, see Wong et al. (2014), Carotenuto and Martis (2017). A double dynamic fast algorithm to solve a multi-vehicle Dial-A-Ride Problem in a dynamic framework is proposed in Carotenuto and Martis (2017), to cope with future demand through several degrees of dynamism.

4.3 Metaheuristic Methods

This section presents metaheuristic based approaches dedicated to customer-oriented transport on demand problems. We emphasize on the papers which proposed metaheuristics for problems including customer-oriented quality of service aspects. Cordeau and Laporte (2003b) and Molenbruch et al. (2017b) investigated local search approaches. More precisely, a Tabu Search (TS) was proposed by Cordeau and Laporte (2003b), and was applied with success on static DARP instances.

Several works are based on the Tabu Search of Cordeau and Laporte (2003b), which was adapted to new real-life DARP applications. That of Ho et al. (2018a), was designed for a customer-oriented DARP framework. The new TS was enhanced and compared to that of Cordeau and Laporte (2003b), and tested on their instances (Instance of Cordeau 2003). Results showed that with the redefinition of time windows through a constructive heuristic, a high quality of solutions in term of costs and run time is provided.

A large neighborhood search algorithm of Molenbruch et al. (2017b), outperformed best known solutions of DARP tests on modified instances of Cordeauand Laporte (2006) (Instances of Cordeau 2006). More real-life customers requirements are introduced to cope with unpredictable cases in such TOD problems. The adding computational efforts are guaranteed by the local search technique including additional specific operators to address specific customers' needs. To enhance the search in promising regions, a diversification mechanism has been periodically repeated.

To solve large instances of the multi-depot DARP with heterogeneous users and vehicles, a deterministic simulated annealing metaheuristic was firstly proposed providing an upper bound for the problem, see Braekers et al. (2014). A set of local search operators were investigated. Three modified sets of benchmark instances of Parragh (2011), were used and efficiently solved.

Another work is that of Chassaing et al. (2016), which is related to an evolutionary local search-based approach. Six operators are managed through dynamic probabilities in the local search. This method was proposed for solving the classic DARP of Cordeau and Laporte (2003b), with a customer-oriented maximal ride time and used as a resolution method for recent real-life instances of Instances of Chassaing (2016). Results are obtained using 20 instances introduced by Cordeau and Laporte (Instance of Cordeau 2003), where the Tabu Search is the resolution method. To better analyze the provider's cooperation, more real-life characteristics are added to the artificial instances.

5 Discussion and Prospects

In most of the works, a common objective of the DARP is considered from the service provider perspective. The aim is to minimize costs involved by the total transportation time, total distance traveled by the vehicles, total route duration, and the number of vehicles required. These costs are enriched by operational costs for measuring the quality of service in the objective function. These costs are related to the inconvenience felt by the passengers mainly those related to the total ride time and the waiting time. Since the objective functions are diverse and depend on the problem addressed and its field of application, the ways of seeking a compromise between operational costs and the level of quality offered to customers are also multiple. In some works the terms of quality of service translated in time windows and maximal ride time are expressed in the form of constraints like in the works of Wong et al. (2014) and Chassaing et al. (2016). In this kind of problem, the service level is implicit since it translates the satisfaction of these constraints impacting the costs.

As it is stated by Paquette et al. (2012), the improvement of the service quality in dial-a-ride problems relies on its good integration and design within optimization models. They also underline the fact that a reliable measurement scale of the quality must first be set and then integrated through a multi-criteria optimization model. In this way, a trade-off between costs and service quality is feasible and would help provide optimal routing schedules maintaining a certain level of quality. However, improving quality through such decision-making systems is not enough. In our personal view, the customization of the quality of service deals with a higher level of satisfaction of the customer. This can be possible by using service quality parameters modeled in accordance with customers expectations. Therefore, artificial instances need modifications to be used for such customer-dependent modeling problems. Few real data exist to test customer-dependent modeling except that of Chassaing et al. 2016 (Instances of Chassaing 2016). Moreover, the maximal ride time is hugely customer-dependent, see Molenbruch et al. (2017c), and should be designed for each customer in a problem. There are a few and recent related works which consider this assumption. The maximal ride time was used for the time windows redefinition for all users, see the works of Ho et al. (2018a). The investigation of a customized maximal ride time in the time windows redefinition is a new trend over the recent years followed by Parragh et al. (2015), Markovi et al. (2015), and Nasri and Bouziri (2017). In Parragh et al. (2015), service time and maximal ride time are related to each customer and used for redefining time windows bounds on destination with the consideration of the bounds on the pickup time windows. However, only time windows on the delivery locations are redefined according to some customers expectations. With this concern, time windows on the pick up will not be adjusted to the customer's desired time of pickup. Besides, more customized features should be addressed and considered through time windows setting since they are considered as the main factors influencing service quality in Markovi et al. (2015), and user inconvenience Parragh et al. (2015).

Time windows width is a factor which also helps in decision maker system (Molenbruch et al. 2017b; Markovi et al. 2015). Therefore, all information on customer requests should be incorporated through a decision-making system, as well as the heterogeneity of customers and their expectations. Therefore, the redefinition of both time windows on origin and time window on destination should be considered in any customer-oriented model. This attempt of taking into account the customers' preferences while satisfying their requests enhances the DARP costs minimization. For instance, the trade-off between service provider costs and customers service quality level could be handled through a multi-objective optimization framework to properly examine the different objective function terms and their importance depending on the problem specificity and real-world experimentation for further decision-making.

Regarding the customized DARP resolution, it was outpointed that more customized features of the problem do not lead to high cost. However, the real-life character of the applications or instances tested increases the complexity of an exact search for the optimal solution mainly for large instances.

The role of heuristics in the resolution of DARP is often to build a first initial solution for a more complex method or seek a better final solution to the problem. They contribute in obtaining good results in an acceptable computational time. Given their problem dependency, they are adapted for addressing customized problems.

Despite the great importance of dynamic realistic applications of DARP, no metaheuristic method was provided for DARPs integrating the customer-oriented quality of service (Agatz et al. 2012). To cope with new frameworks for dynamic DARPs, enhanced method should be considered as for instance evolutionary ones and hybridized metaheuristics.

6   Conclusion

To enhance the practical applicability of DARPs within the Transport On Demand domain, researchers are more and more focusing on problem variants that adopt additional real-life characteristics. In this survey, we emphasize the need for assessing new service quality levels defined from a customer-oriented approach coping with customer expectations of nowadays. Dial-a-ride services involve many interactions with customers and their perceptions of the service. Ideally, a model of the dial-a-ride problem would take into account all customers expectations, as well as adding stochastic factors. However, for practical reasons, uncertainty factors should be integrated through a dynamic model. As described in the survey, some researches have already been done on customized and dynamic models for dial-a-ride services. There are other possible avenues of research like dedicated constraints absorbing new real-life features and new solving methods.

References

Agatz, N., Erera, A., Savelsbergh, M., Wang, X.: Optimization for dynamic ride-sharing: a review. Eur. J. Oper. Res. **223**(2), 295–303 (2012)

Aggoune-Mtalaa, W., Aggoune, R.: An optimization algorithm to schedule care for the elderly at home. Int. J. Inform. Sci. Intell. Syst. **3**(3), 41–50 (2014)

Amroun, K., Habbas, Z., Aggoune-Mtalaa, W.: A compressed generalized hypertree decomposition-based solving technique for non-binary constraint satisfaction problems. AI Commun. **29**(2), 371–392 (2016)

Armstrong, G.R., Garfinkel, R.S.: Dynamic programming solution of the single and multiple vehicle pickup and delivery problem with application to dial-a-ride. Working paper 162, University of Tennessee (1982)

Beaudry, A., Laporte, G., Melo, T., Nickel, S.: Dynamic transportation of patients in hospitals. OR Spectr. **32**(1), 77–107 (2010)

Bennekrouf, M., Aggoune-Mtalaa, W., Sari, Z.: A generic model for network design including remanufacturing activities. Supply Chain Forum **14**(2), 4–17 (2013)

Braekers, K., Caris, A., Janssens, G.K.: Exact and meta-heuristic approach for a general heterogeneous dial-a-ride problem with multiple depots. Transp. Res. Part B: Methodol. **67**, 166–186 (2014)

Cappart, Q., Thomas, C., Schaus, P., Rousseau, L.M.: A constraint programming approach for solving patient transportation problems. In: International Conference on Principles and Practice of Constraint Programming, pp. 490–506. Springer, Cham (2018)

Carotenuto, P., Martis, F.: A double dynamic fast algorithm to solve multi-vehicle dial-a-ride problem. Transp. Res. Proced. **27**, 632–639 (2017)

Çetin, T.: The rise of ride sharing in urban transport: threat or opportunity? In: Urban Transport Systems, vol. 191 (2017)

Chassaing, M., Duhamel, C., Lacomme, P.: An ELS-based approach with dynamic probabilities management in local search for the dial-a-ride problem. Eng. Appl. Artif. Intell. **48**, 119–133 (2016)

Cordeau, J.-F.: A branch-and-cut algorithm for the dial-a-ride problem. Oper. Res. **54**, 573–586 (2006)

Cordeau, J.F., Laporte, G.: The dial-a-ride problem (DARP) variants, modeling issues and algorithms. Q. J. Belg. Fr. Ital. Oper. Res. Soc. 89–101, (2003a)

Cordeau, J.-F., Laporte, G.: A tabu search heuristic for the static multi-vehicle dial-a-ride problem. Transp. Res. Part B: Methodol. **37**(6), 579–594 (2003b)

Cordeau, J.F., Laporte, G.: The dial-a-ride problem: models and algorithms. Ann. Oper. Res. **153**(1), 29–46 (2007)

Cordeau, J.F., Laporte, G., Potvin, J.Y., Savelsbergh, M.W.: Transportation on demand. Handbooks in Operations Research and Management Science **14**, 429–466 (2007)

Detti, P., Papalini, F., de Lara, G.Z.M.: A multi-depot dial-a-ride problem with heterogeneous vehicles and compatibility constraints in healthcare. Omega **70**, 1–14 (2017)

Diana, M.: Innovative systems for the transportation disadvantaged: toward more efficient and operationally usable planning tools. Transp. Plann. Technol. **27**, 315–331 (2004)

Enrique Fernindez, L.J., de Cea Ch, J., Malbran, R.H.: Demand responsive urban public transport system design: methodology and application. Transp. Res. Part A: Policy Pract. **42**(7), 951–972 (2008)

Faria, A., Yamashita, M., Tozi, L.A., Souza, V.J., Brito Jr, I.: Dial-a-ride routing system: the study of mathematical approaches used in public transport of people with physical disabilities. In: Proceedings of 12th World Conference on Transport Research Society (2010)

Healy, P., Moll, R.: A new extension of local search applied to the dial-a-ride problem. Eur. J. Oper. Res. **83**, 83–104 (1995)

Heilporn, G., Cordeau, J.F., Laporte, G.: An integer L-shaped algorithm for the dial-a-ride problem with stochastic customer delays. Discret. Appl. Math. **159**(9), 883–895 (2011)

Hll, C.H., Andersson, H., Lundgren, J.T., Vrbrand, P.: The integrated dial-a-ride problem. Public Transp. **1**(1), 39–54 (2009)

Ho, S., Nagavarapu, S.C., Pandi, R.R., Dauwels, J.: An improved tabu search heuristic for static dial-a-ride problem (2018a)

Ho, S.C., Szeto, W.Y., Kuo, Y.H., Leung, J.M., Petering, M., Tou, T.W.: A survey of dial-a-ride problems: literature review and recent developments. Transp. Res. Part B: Methodol. **111**, 395–421 (2018b)

Hu, T.-Y., Chang, C.-P.: A revised branch-and-price algorithm for dial-a-ride problems with the consideration of time-dependent travel cost. J. Adv. Transp. **49**(6), 700–723 (2015)

Instance of Cordeau (2003). http://neumann.hec.ca/chairedistributique/data/darp/

Instances of Cordeau (2006). http://neumann.hec.ca/chairedistributique/data/darp/. Last accessed 2020

Instances of Chassaing (2016). http://fc.isima.fr/~lacomme/Maxime/. Last Accessed 2020

Jaw, J.J., Odoni, A.R., Psaraftis, H.N., Wilson, N.H.: A heuristic algorithm for the multi-vehicle advance request dial-a-ride problem with time windows. Transp. Res. Part B: Methodol. **20**(3), 243–257 (1986)

Jlassi, J., Euchi, J., Chabchoub, H.: Dial-a-ride and emergency transportation problems in ambulance services. Comput. Sci. Eng. **2**(3), 17–23 (2012)

Jorgensen, R.M., Larsen, J., Bergvinsdottir, K.B.: Solving the dial-a-ride problem using genetic algorithms. J. Oper. Soc. **58**, 1321–1331 (2007)

Lehu, F., Masson, R., Parragh, S.N., Pton, O., Tricoire, F.: A multi-criteria large neighbourhood search for the transportation of disabled people. J. Oper. Res. Soc. **65**(7), 983–1000 (2014)

Li, D., Antoniou, C., Jiang, H., Xie, Q., Shen, W., Han, W.: The value of prepositioning in smartphone-based vanpool services under stochastic requests and time-dependent travel times. Transp. Res. Rec. **2673**(2), 26–37 (2019)

Luo, Y., Schonfeld, P.: A rejected-reinsertion heuristic for the static dial-a-ride problem. Transp. Res. Part B: Methodol. **41**(7), 736–755 (2007)

Ma, T.Y., Rasulkhani, S., Chow, J.Y., Klein, S.: A dynamic ridesharing dispatch and idle vehicle repositioning strategy with integrated transit transfers. Transp. Res. Part E: Logist. Transp. Rev. **128**, 417–442 (2019)

Markovi, N., Nair, R., Schonfeld, P., Miller-Hooks, E., Mohebbi, M.: Optimizing dial-a-ride services in Maryland: benefits of computerized routing and scheduling. Transp. Res. Part C: Emerg. Technol. **55**, 156–165 (2015)

Masson, R., Lehu, F., Pton, O.: The dial-a-ride problem with transfers. Comput. Oper. Res. **41**, 12–23 (2014)

Melachrinoudis, E., Min, H.: A tabu search heuristic for solving the multi-depot, multi-vehicle, double request dial-a-ride problem faced by a healthcare organisation. Int. J. Oper. Res. **10**(2), 214–239 (2011)

Melachrinoudis, E., Ilhan, A.B., Min, H.: A dial-a-ride problem for client transportation in a health-care organization. Comput. Oper. Res. **34**, 742–759 (2007)

Molenbruch, Y., Braekers, K., Caris, A.: Typology and literature review for dial-a-ride problems. Ann. Oper. Res. **259**(1–2), 295–325 (2017a)

Molenbruch, Y., Braekers, K., Caris, A.: Benefits of horizontal cooperation in dial-a-ride services. Transp. Res. Part E: Logist. Transp. Rev. **107**, 97–119 (2017b)

Molenbruch, Y., Braekers, K., Caris, A.: Operational effects of service level variations for the dial-a-ride problem. CEJOR **25**(1), 71–90 (2017c)

Mulley, C., Nelson, J.D.: Flexible transport services: a new market opportunity for public transport. Res. Transp. Econ. **25**(1), 39–45 (2009)

Nasri, S., Bouziri, H.: Improving total transit time in dial-a-ride problem with customers-dependent criteria. In: 14th International Conference on Computer Systems and Applications (AICCSA), pp. 1141–1148. IEEE, Tunisia (2017)

Pagano, A.M., McKnight, C.E.: Quality of service in special service paratransit: the users perspective. Transp. Res. Rec. **934**, 14–23 (1983)

Paquette, J., Cordeau, J.-F., Laporte, G.: Une tude comparative de divers modles pour le problme de transport la demande. INFOR **45**, 95–110 (2007)

Paquette, J., Cordeau, J.-F., Laporte, G.: Quality of service in dial-a-ride operations. Comput. Ind. Eng. **56**, 1721–1734 (2009)

Paquette, J., Bellavance, F., Cordeau, J.F., Laporte, G.: Measuring quality of service in dial-a-ride operations: the case of a Canadian city. Transportation **39**(3), 539–564 (2012)

Paquette, J., Cordeau, J.F., Laporte, G., Pascoal, M.M.: Combining multicriteria analysis and tabu search for dial-a-ride problems. Trans. Res. Part B: Methodol. **52**, 1–16 (2013)

Parragh, S.N.: Introducing heterogeneous users and vehicles into models and algorithms for the dial-a-ride problem. Transp. Res. Part C **19**, 912930 (2011)

Parragh, S.N., Doerner, K.F., Hartl, R.F., Gandibleux, X.: A heuristic two phase solution approach for the multi objective dial a ride problem. Netw. Int. J. **54**(4), 227–242 (2009)

Parragh, S.N. Jorge Pinho, S., Bernardo, A.: The dial-a-ride problem with split requests and profits. Transp. Sci. **49**(2), 311–334 (2015)

Posada, M., Andersson, H., Hll, C.H.: The integrated dial-a-ride problem with timetabled fixed route service. Public Transp. **9**(1–2), 217–241 (2017)

Qu, Y., Bard, J.F.: A branch-and-price-and-cut algorithm for heterogeneous pickup and delivery problems with configurable vehicle capacity. Transp. Sci. **49**(2), 254–270 (2015)

Rekiek, B., Delchambre, A., Saleh, H.A.: Handicapped person transportation: an application of the grouping genetic algorithm. Eng. Appl. Artif. Intell. **19**, 511–520 (2006)

Rezgui, D., Chaouachi-Siala, J., Aggoune-Mtalaa, W., Bouziri, H.: Application of a memetic algorithm to the fleet size and mix vehicle routing problem with electric modular vehicles. In: GECCO 2017—Proceedings of the Genetic and Evolutionary Computation Conference Companion, pp. 301–302 (2017)

Rezgui, D., Siala, J.C., Aggoune-Mtalaa, W., Bouziri, H.: Towards Smart Urban Freight Distribution Using Fleets of Modular Electric Vehicles. Lecture Notes in Networks and Systems **37**, 602–612 (2018)

Serrano, C., Aggoune-Mtalaa, W., Sauer, N.: Dynamic models for green logistic networks design. IFAC Proc. Vol. (IFAC-Papers Online) **46**(9), 736–741 (2013)

Tlili, T., Abidi, S., Krichen, S.: A mathematical model for efficient emergency transportation in a disaster situation. Am. J. Emerg. Med. **36**(9), 1585–1590 (2018)

Wilson, N.H.M., Weissberg, R.W., Hauser, J.: Advanced dial-a-ride algorithms research project: final report. Working paper R-76-20 (1976)

Wolfler Calvo, R., Colorni, A.: An effective and fast heuristic for the dial-a-ride problem. Q. J. Oper. Res. (4OR) **5**, 61–73 (2007)

Wong, K.I., Han, A.F., Yuen, C.W.: On dynamic demand responsive transport services with degree of dynamism. Transp. A Transp. Sci. **10**(1), 55–73 (2014)

Big Data Accident Prediction System in Green Networks and Intelligent Transportation Systems

Mouad Tantaoui, My Driss Laanaoui, and Mustapha Kabil

Abstract

Given the mass of information collected during communication in VANETs networks, the analysis of this information is important in order to improve road traffic and remedy various problems, namely accidents in green networks and Intelligent Transportation Systems (ITSs). Intelligent transportation system is a domain of scientific research that attracts researchers' community. These anticipation systems aim to prevent road accidents by developing a safety margin assistant to help drivers avoid risky driving situations. In this paper, we propose a new prediction system in real time with the help of Big Data to improve the VANET network. In the first hand, The Traffic density and average speed are computed in each section of road, and after that the risk of vehicle accident is predicted in instantaneous manner with parallel data processing, which makes execution faster.

Keywords

Intelligent transportation systems • Traffic density • Big data • VANET • Lambda architecture

1 Introduction

Intelligent Transportation System (ITS) is research subject of computer which interests scientific community keeping up with either social and economic challenges or the safety of

M. Tantaoui (✉) · M. Kabil
University Hassan II, Mohammedia, Morocco
e-mail: tantaoui.mouad@gmail.com

M. Kabil
e-mail: kabilfstm@gmail.com

M. D. Laanaoui
University Cadi Ayyad, Marrakech, Morocco
e-mail: d.laanaoui@uca.ma

the citizens, therefore, we need to find new ways to estimate density of the traffic in different roads and cities. This would help us, subsequently to identify, for example, the places where the risk of accidents is higher. To attain this goal, we have to define a road density prediction method, with the help of massive data collection using vehicle-to-vehicle and vehicle-to-infrastructure communication (Cunha and Domingos 2014).

Since their appearance, Vehicular Ad hoc Networks (VANETs) have done a great development. Lots of standards, applications, and mechanisms of data processing have been proposed to satisfy the requirements of this new trend of networks. The design challenges increase mainly from the high mobility of vehicles and the spatio-temporal diversity of traffic density. Given the mass of information collected during communication in VANETs networks, the analysis of this information is important in order to improve road traffic and remedy various problems, namely accidents.

2 VANET Architecture

2.1 On-Board Unit (OBU)

Each mobile node is equipped with this unit to facilitate communication with other mobile nodes (vehicles); Fixed stations (Road-side Units) via DSRC and the ability to communicate using cellular radio networks such as GSM, 4G, WiFi, and WiMAX.

2.2 Road-Side Units (RSUs)

These are base stations that support VANET applications and coordinate actions to share and process information as well as propagate data, provide traffic directories, act as location servers, and connect to the Internet and external centralized or distributed servers.

© The Editor(s) (if applicable) and The Author(s), under exclusive license to Springer Nature Switzerland AG 2021
M. Ben Ahmed et al. (eds.), *Emerging Trends in ICT for Sustainable Development*,
Advances in Science, Technology & Innovation, https://doi.org/10.1007/978-3-030-53440-0_14

2.3 Centralized Cloud

Is a type of computer architecture where all or most of the processing/calculation is done on a central server. Centralized computing allows the deployment of all IT resources, administration and management of the central server.

3 State of Art

The problem of VANET becomes a big data problem. This would help us to manage the large amount of collected data during vehicle-to-vehicle and vehicle-to-infrastructure communication. Thus, different Big Data techniques can be applied in VANET communication to improve the traffic management.

Among works which we find interesting was to implement VANET routing algorithms, like Dijkstra, which is a routing algorithm which uses the shortest path as a metric for determining the best route to destination, this algorithm provides a large amount of information that will be processed by Big Data techniques such as Hadoop Map-Reduce. In article Bedi and Jindal (2014), the authors implemented Dijkstra algorithm in Hadoop Map-Reduce environment and they compared it with Dijkstra on just a simple program in NetBeans IDE. The experiment proved that there is a large difference in the execution time between the Netbeans environment and Hadoop environment. Indeed Hadoop Map-Reduce is faster than Netbeans and also the execution time decreases significantly in multi-node cluster with 2, 3, 4, and 5 nodes.

Another work which interests us is to find the vehicles which are able to act as a relay whose role is gathering some information from the network, sharing them, and distributing them to vehicles which located in its range. They established Ranking algorithms in this context such as InfoRank (Khan et al. 2015) which aims to help vehicles to measure their importance or their centrality in the network. InfoRank utilizes an Information-Centric Networking (ICN) approach that evaluates the centrality of a vehicle based on the pertinence of the information gathered from different localizations visited by the vehicle. The results confirm InfoRank's ability to find important vehicles within the network, to evolve and comply with the principles of ICN.

Nowadays, researchers deepen their research on speed prediction algorithms because it plays a very important role in managing traffic in the field of intelligent transport system; BDDL-SP (Cheng et al. 2017) is an algorithm that predicts the speed of a vehicle in highways and urban environment. Many factors linked to the speed of the vehicle are taken into account for the prediction model development like driver behavior, route information, and weather and traffic conditions.

Today, there are new systems which have emerged to manage Big Data, sets powered in real time by vehicles and Road-Side Units (RSUs) ensuring a rapid processing of data with the intention of making a good road management decisions. These systems can be used as application to compute the estimated time of arrival vehicles and predict accidents and congestions using Distributed Random Forest (DRF) and Naive Bayes (Al Najada and Mahgoub 2016). We quote, for example, the Lambda architecture (Marz and Warren 2015) among these systems that is processing architecture, allows to process big data whether in batch or in real time, it equilibrates throughput, latency, and error tolerance. The system contains three layers: Batch layer, speed layer, and serving layer (Fig. 1).

In article Mahajan and Kaur (2016), an experiment on ITS environment was implemented to assess the traffic density estimation about different cities on different roads and carry out a comparative evaluation relying on that parameter. The scientist is based on the physical properties of vehicles such as vehicle shape and position. And that, in the purpose of increasing the accuracy of the ITS density prediction technique. The proposed method for density estimation, which is based on the interaction of information between vehicles, is more adequate for the scalability of ITS.

In article Lin and Wang (2016), they suggest a system to send with dynamic manner reports against selfish and malicious vehicles. The proposed system utilizes an encryption mechanism to exchange messages, a certification authority for verifying vehicles authenticity for increasing sender reliability. Because of causing congestion by real-time reports in the wireless network, the authors presented a Persistent P-Report (PPR) for dynamically adjusting the probability of sending reports. The authors compared between PPR and Real-Time Reporting (RTR) and Periodic Reporting (PR) in terms of reporting and timeout costs. The experimental results show that the PPR mechanism outperforms RTR and PR because PPR can dynamically modify the probability of sending the adequate reports to the environment.

In article Daniel et al. (2015), the authors propose an architecture for large on-vehicle datasets that allows centralized access to massive data. The proposed method includes centralized data storage, processing mechanism, and a distributed data storage mechanism for real-time processing and analysis. In addition to that two more algorithms have been proposed for pseudo-real-time data processing which is used to organize and analyze a vehicle data flow in a place. Therefore, the result and analysis showed that the proposed system model is for optimal use of massive datasets, intended to transmit data in a pseudo-real-time process in an Intelligent Transportation System (ITS) context in a vehicular environment.

In article Alinani and Alinani (2018), the authors have solved the problem of innumerable applications including

Fig. 1 Overview of Lambda architecture

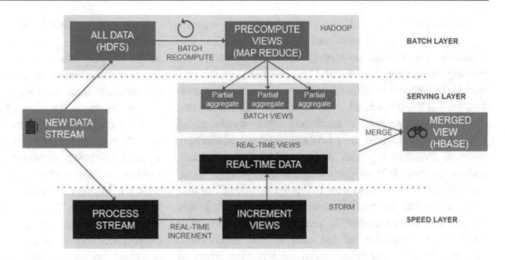

Fig. 1 Overview of Lambda architecture

object tracking, tracking a moving vehicle, and managing an effective communication channel in the VANET, by establishing a new Internet architecture called Named Data Networking (NDN) to find a solution of the weaknesses of the current VANET networks. However, it is compulsory in the vehicle tracking system to collect data continuously for updating the targeted vehicle location without anticipating the interest packet several times. But the Vehicle Name Data Network (VNDN) currently uses a system (pull-based mechanism) which creates an undesirable delay which is not acceptable specifically in critical situations. Thus, the gap must be bridged to take advantage of this technology in real time. The concept of this article provides a push data schema for the VNDN to track a targeted vehicle in real time in a specific area.

In article Gawas et al. (2018), the authors established a new routing protocol that utilizes (link guarantee) and (forwarding movement distance) a node for selecting the next-hop node. The weighted function is utilized in the purpose of normalizing all quality of service metrics. The node which has the maximum value of the weighted function has the highest priority and, thus, they choose the optimal node for the selection of the next hop. The experimental results show that the proposed method can significantly improve the packet delivery rate and minimize the link failure rate during routing in VANET networks.

For the article Al Najada and Mahgoub (2016), Autonomous Vehicle (AV) technology conducts too many economic and social advantages and effects. The idea of trajectory planning is one of the important and compulsory tasks to the autonomous vehicle driving. In this article, they solve the trajectory planning problem for wholly autonomous vehicles. The practical cases of autonomous vehicles are implemented in a cloud environment with connected vehicle. The method proposes an optimal trajectory selection in autonomous vehicles environment. The safety trajectory selection in this

method is relying on the support of Big Data technologies and the real-life accident data analysis and real-time connected vehicle data. The selection of the trajectory is processed with automatic manner and without any human touch. The human intervention in this scenario could be only for defining and classifying driving preferences and concerns at the beginning of the planned trip. In this article, security has always the highest priority of the list of user preferences order. The output of this method is a safety path described by position, Estimated Time of Arrival (ETA), distance traveled, and estimated fuel consumption for the entire trip.

For the article Zhang (2014), authors proposed a method to use Architecture Analysis and Design Language (AADL), ModelicaML and Hybrid Relation Calculus (HRC) for the development of physical systems used by Big Data technologies. The proposed method is showed by a practical case on the specification and modeling of vehicular ad hoc networks (VANETs). The main benefits of the proposed method are its idea to take into account big data properties and physical system properties via specific concepts and models in a precise, simple, and expressive way.

For the article Malik and Pandey (2017), the authors propose an adequate and secure data collection technique which assures data security and confidentiality interacted between vehicles, RSUs, and different equipment of VANET architecture. It is relying on asymmetric encryption that assures the security of communication between vehicles and other agents. In this method, secure authentication is assured between the vehicle and the RSU before the collection of the vehicle data is begun by RSU. Also, the proposed data collection method is compared with currently in existence of data collection methods. The experimental results show that the proposed method is more sure than the currently in existence of data collection techniques. This method serves the beginners for improving the efficiency of data collection in VANETs architecture.

For the article Xu et al. (2017), by significantly expanding the scale of the network and performing real-time and long-term information processing, vehicular VANETs are moving toward the Internet of Vehicles (IoV), which promises effective and intelligent prospects for a future transport system. On the other hand, vehicles are not just consumers; they also generate huge amounts and types of data, called Big Data. In this article, they first examined the relationship between IoV and big data in the vehicular environment, mainly on how IoV supports transmission, storage, computing using big data and how IoV pulls benefit of big data for characterization, performance evaluation, and big data support for a communication protocol design. They then studied the application of big data IoV in autonomous vehicles. Finally, they talked about the emerging issues of IoV with Big Data.

For the article Al Najada and Mahgoub (2016), they used the H2O and WEKA extraction tools to evaluate five classifiers on two large sets of workshop data. The classifiers used are Naive Bayes, C4.5, Random Forest (RF), AdaboostM1 (with the C4.5 basic classifier), and Bagging (with the C4.5 basic classifier). They applied the selection of attributes and they also addressed the problem of class imbalance. From their experiences, Naive Bayes (NB) gave the optimal results, with the shortest calculation time and a practical AUC and ACC. C4.5 and RF gave better results than NB with respect to AUC and ACC but with longer computation time. The analysis and extracted models and results can help decision-makers and practitioners intelligently improve the transportation system and develop new rules. This study revealed some misconceptions about road accidents. The analysis showed that human behavior has a significant impact on traffic and safety decisions. The results revealed that driver characteristics such as age and gender could be predicted correctly by providing other attributes for an accident or victim.

For the article Zhao et al. (2018), first, they used the Synthetic Minority Oversampling Technique (SMOTE) to reconstruct the experimental dataset, the minority samples in the study dataset were oversampled and new samples were synthesized to complete the missing data and to balance the number of samples from each category in the original dataset. The AdaBoost algorithm of trichotomy was used to form a series of weak classifiers of the experimental data set, and then combine them into a powerful classifier to obtain the prediction model. The performance of this prediction model has been tested in the field. Finally, the performances of the system model with different maximum iteration values n are compared with the decision tree model under four aspects: the ROC curve, the AUC, the learning time, and the real classification. The result of the test shows that the system model with the maximum iteration value n of 400 guarantees the maximum accuracy of the accident prediction in the ordinary road condition, and the system model with the maximum value of lower iterations n in a particular situation can improve the speed of prediction.

For the article Yu et al. (2013), they proposed a routing protocol that is based on road vehicle density in real time to provide fast and reliable communications, so it adapts to the dynamic environment from the city. In the proposed routing mechanism, each vehicle calculates the density of the road to which it belongs by using beacon messages and the road information table. Based on real-time road vehicle density information, each vehicle establishes a reliable route for package delivery. In order to evaluate the performance of the proposed mechanism, they compared the proposed mechanism with the GPSR via the NS 2-based simulations and they showed that the mechanism outperformed the GPSR in terms of delivery success rate and system time required for routing.

4 Real-Time Prediction

Our concern is to predict accidents, the proposed method consists of designing a Big Data system that allows receiving real-time data from sensors and vehicles on the road. The Big Data system consisting of lambda architecture for managing the big data gathered relying on Hadoop Map-Reduce that will make sure that the data processing is completed.

The mass of processing is done by the Batch layer, this layer is designed to manage a very big data which is characterized by data diversity: it processes two principal tasks.

Firstly, the layer receives a city map. Then we partition the roads according to sections, this layer computes the number of vehicles which crossed each section of each road every ten minutes. After that the batch computes the average speed per section of each road every ten minutes. Figure 2 summarizes the steps of the batch layer.

Speed Layer is the layer which receives information's vehicles in real time like the road where they are and the section, and then it will compute a probability relying on the pre-calculated table previously established by the batch layer. This layer has to be fast because it has to respond to real-time vehicle requests. As soon as a vehicle reaches a section of a road, Speed Layer looks the table for the average speed and traffic status in that section of road, and following the current time, the layer will compute

$$P = [(V + Vm)/2 * Vmax] * [N/Nmax] \qquad (1)$$

Fig. 2 Batch layer steps

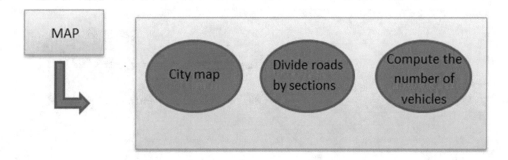

Fig. 3 Speed layer

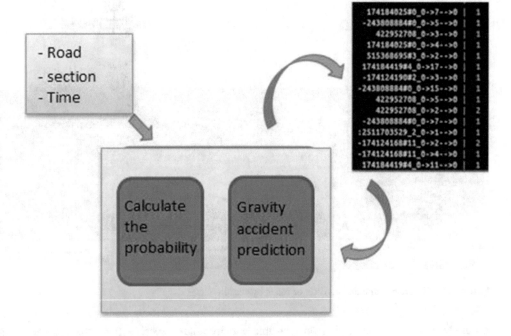

where

- V: the speed of the vehicle
- Vm: the average speed in the section concerned
- N: the number of vehicles that crossed the current section
- Vmax: the maximum average speed
- Nmax: the maximum number of vehicles.

Relying on this probability, Speed layer can predict the gravity of having an accident which will be directly sent to the vehicle concerned: If the probability is less than or equal to 0.3, then the severity of having an accident is low; if it is between 0.3 and 0.7, then the gravity is average; if it is greater than or equal to 0.7, then the accident gravity is high. Figure 3 summarizes the stages of the speed layer.

5 Analysis

Relying on the storage and processing system in parallel and distributed way helps our system to be executed in a fast manner and access to light data with very low latency. Batch processing was realized on all the roads in the city of Mohammedia and that by carving it by sections as is shown in Fig. 4 and computing road density and the average speed every ten minutes in every road section. Our system communicates with the vehicles crossing the sections and that by sending them the accident gravity relying on the current vehicle speed, the current density of the section, and also the current average speed of the section. Accident gravity prediction is a function of instantaneous factors like current and real-time parameters, which assigns to our method good

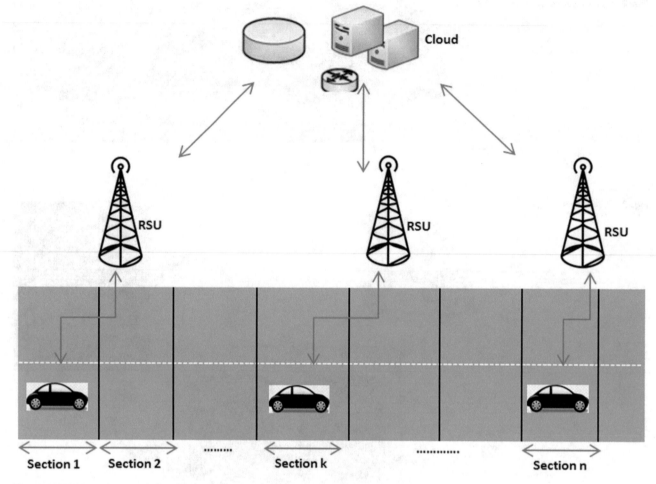

Fig. 4 VANET architecture and division of roads by sections

accuracy and prediction relying on real-time and instantaneous data.

We tested the execution time of our system, we measured
the time between the sending of the request of the vehicle
and the response of the gravity to have an accident predicted
by our system, and it is between 100 and 200 ms.

In the simulation below, Table 1 shows the trips of the
vehicles before and after receiving the notification from
the server, the system calculates the severity of having an
accident according to the table for the average speed and
traffic status in that section of road. In the table, we
mentioned the vehicles at different times which they will

receive from the system the gravity of the accident. For
vehicle which has id number 1, it receives a high gravity
from the system, we suppose in this experiment that
vehicles are cooperative, so the vehicle number 1 will
decrease its speed, not only that vehicle but all vehicle
will decrease their speed, so that after ten minutes, the
average speed will decrease, and so the severity will
decrease, therefore, with this method, we reduce the
probability of having an accident. We notice on the table
that the severity on anytime and anywhere is reduced after
gravity accident prediction on the ten minutes following
the prediction.

Table 1 Trips of vehicle before notification and after

Time	Vehicle ID	Road ID	Severity (t)	Severity (t + 10 min)
12:00	1	1	High	Medium
12:10	2	1	Medium	Low
13:00	3	2	High	Low

6 Conclusion

Human life is very dear to us. Road accidents seriously threaten the lives of human beings, so the community of researchers has to find a solution to solve predicting the risk of vehicle accidents problem. So this paper utilizes the methods of big data for improving traffic management. We developed a real-time vehicle accident risk prediction system with instantaneous manner in parallel with data processing environment that makes execution faster.

References

Alinani, A., Alinani, K.: Real-time push-based data forwarding for target tracking in vehicular named data networking. In: 2018 IEEE SmartWorld, Ubiquitous Intelligence and Computing, Advanced and Trusted Computing, Scalable Computing and Communications, Cloud and Big Data Computing, Internet of People and Smart City Innovation (SmartWorld/SCALCOM/UIC/ATC/CBDCom/IOP/SCI), pp. 1587–1592. IEEE (2018). https://doi.org/10.1109/SmartWorld.2018.00272

Al Najada, H., Mahgoub, I.: Anticipation and alert system of congestion and accidents in VANET using Big Data analysis for intelligent transportation systems. In: 2016 IEEE Symposium Series on Computational Intelligence (SSCI). IEEE (2016)

Al Najada, H., Mahgoub, I.: Autonomous vehicles safe-optimal trajectory selection based on big data analysis and predefined user preferences. In: 2016 IEEE 7th Annual Ubiquitous Computing, Electronics and Mobile Communication Conference (UEMCON), pp. 1–6. IEEE (2016). https://doi.org/10.1109/UEMCON.2016.7777922

Al Najada, H., Mahgoub, I.: Big vehicular traffic data mining: towards accident and congestion prevention. In: International Wireless Communications and Mobile Computing Conference (IWCMC), pp. 256–261. IEEE (2016)

Bedi, P., Jindal, V.: Use of big data technology in vehicular ad-hoc networks. In: 2014 International Conference on Advances in Computing, Communications and Informatics (ICACCI). IEEE (2014)

Cheng, Z., et al.: A big data based deep learning approach for vehicle speed prediction. In: 2017 IEEE 26th International Symposium on Industrial Electronics (ISIE). IEEE (2017)

Cunha, D., Domingos, F., et al.: Data Communication in VANETs: A Survey, Challenges and Applications. Diss. INRIA Saclay, INRIA (2014)

Daniel, A., Paul, A., Ahmad, A.: Near real-time big data analysis on vehicular networks. In: 2015 International Conference on Soft-Computing and Networks Security (ICSNS). IEEE (2015)

Gawas, M.A., Mulay, M., Bhatia, V.: Cross layer approach for neighbor node selection in VANET routing. In: 2018 11th International Symposium on Communication Systems, Networks and Digital Signal Processing (CSNDSP), pp. 1–6. IEEE (2018). https://doi.org/10.1109/CSNDSP.2018.847

Khan, J., Ghamri-Doudane, Y., El Masri, A.: Vers une approche centrée information (ICN) pour identifier les véhicules importants dans les VANETs. In: Proceedings of the Conference CFIP NOTERE 2015 (2015).

Lin, B.-R., Wang, T.-P.: Dynamic Reporting Mechanisms for Trust Management in Vehicular Ad-Hoc Networks (2016)

Mahajan, A., Kaur, A.: Predictive urban traffic flow model using vehicular big data. Indian J. Sci. Technol. **9**, 42 (2016)

Malik, A., Pandey, B.: Asymmetric encryption based secure and efficient data gathering technique in VANET. In: 2017 7th International Conference on Cloud Computing, Data Science and Engineering-Confluence, pp. 369–372. IEEE (2017). https://doi.org/10.1109/CONFLUENCE.2017.7943177

Marz, N., Warren, J.: Big Data: Principles and Best Practices of Scalable Real-Time Data Systems. Manning Publications Co., New York (2015)

Xu, W., Zhou, H., Cheng, N., et al.: Internet of vehicles in big data era. IEEE/CAA J. Autom. Sinica **5**(1), 19–35 (2017)

Yu, H., Yoo, J., Ahn, S.: A VANET routing based on the real-time road vehicle density in the city environment. In: 2013 Fifth International Conference on Ubiquitous and Future Networks (ICUFN), pp. 333–337. IEEE (2013)

Zhang, L.: A framework to specify big data driven complex cyber physical control systems. In: 2014 IEEE International Conference on Information and Automation (ICIA), pp. 548–553. IEEE (2014). https://doi.org/10.1109/ICInfA.2014.6932715

Zhao, H., Yu, H., Mao, T., et al.: Vehicle accident risk prediction over AdaBoost from VANETs. In: 2018 10th International Conference on Intelligent Human-Machine Systems and Cybernetics (IHMSC), pp. 39–43. IEEE (2018)

A Survey of Optimization Techniques for Routing Protocols in Mobile Ad Hoc Networks

Younes Ben Chigra, Abderrahim Ghadi, and Mohamed Bouhorma

Abstract

Mobile ad hoc network (MANET) forms a collection of wireless mobile nodes that communicate in a multi-hop manner without relying on any pre-established infrastructure. Various potential applications of MANETs in civil and military domains are gaining popularity due to the lower cost and the minimum time needed for deployment. Nowadays, emerging applications require the guarantee of a certain level of QoS such as low latency, guaranteed bandwidth, and low packet loss. Consequently, it becomes very crucial to design efficient routing protocol to deal with multiple constraints, of MANET environment, which affect the performance of the network like mobility, overhead, battery drainage, delay, and interference. In this paper, we study and discuss the applicability of the existing optimization techniques to find out the best and optimal solution.

Keywords

MANETs • Optimization techniques • Routing metrics • QoS

1 Introduction

Mobile Ad Hoc networks are prominent networks that can support ubiquitous applications at a lower cost. This is because MANET doesn't rely on any pre-established infrastructure. Only wireless devices that incorporate intelligent routing protocol are sufficient to establish data communication between a pair of source and destination nodes. However, since the advent of MANET, few implementations

have emerged because there are multiple network environments where mobile devices should behave differently to guarantee the QoS required by the end user or application. Facing these challenges, researchers in Ad Hoc routing field have proposed several routing schemes such as proactive, reactive, hybrid, multicast, and swarm intelligent based protocols without succeeding to develop a universal protocol that satisfies the constraints of almost all expected MANET conditions. In this context, optimization techniques might represent efficient tools that allow us to combine multiple routing metrics to reflect network state such as node mobility, network density, and remaining node power and deliver the best performances. In this paper, we are going to study the existing optimization techniques and discuss how far they could help to design an adaptive intelligent routing protocol capable of calibrating its routing process to the dynamic nature of MANET and provide required QoS by the end user.

2 Overview of Optimization Techniques

Since the emergence of MANET, various routing protocols have been proposed to ensure path setup and route maintenance when needed and guarantee data delivery from a source toward a destination in an effective manner. However, developed protocols use mainly single routing metrics such as hop count and establish best effort path without considering the QoS needed to help applications work properly. With the advent of new applications of mobile ad hoc networks such as portable game platforms, multimedia-related traffic introduced new challenges to guarantee the required quality of service level. The provision of QoS by routing protocols needs at first stage an accurate assessment of QoS-level requirement of such applications and the knowledge of additional routing metrics that must be taken in account while routing algorithm decides which route will be chosen for a specific traffic delivery. Moreover,

Y. Ben Chigra (✉) · A. Ghadi · M. Bouhorma
Computing, Systems and Telecommunications Laboratory,
UAE-FSTT, Tangier, Morocco
e-mail: younesbenchigra@gmail.com

real-world applications often need the guarantee of several metrics simultaneously. Furthermore, very probably these metrics can be conflicting, which impose to adopt a balanced solution. In this case, the use of multi-criteria optimization techniques, which is widely used in other areas, may contribute in proving the QoS.

2.1 Fundamental of Optimization Problem

The main purpose of the optimization process is to share limited available resources in the most effective way to satisfy the requirement of conflicting services. This process has been used in economics field under the name of decision-making and operational research.

In general, the optimization process consists of minimizing or maximization an objective function subject to constraints on its variables (Nocedal & Wright 1999). As illustrated in Fig. 1, this process could be achieved in three main steps: First, start with problem modeling that aims to identify objective functions such as distance, time, and energy. The objective depends on system characteristics, called variables. Variable is generally restricted or constrained, for example, the speed or remaining energy of mobile nodes cannot be negative or greater than certain threshold. Once the model is well defined, an appropriate optimization algorithm has to be selected and applied to it. Note that there are several optimization algorithms and each one fits to a specific type of problem. Thus, it is recommended to pay attention while choosing optimization algorithm to guarantee optimal solution. The third step consists of evaluating the solution and decides whether algorithm output represents the expected solution.

There are two main relevant objective functions; we talk about single-criterion problem when there is only one objective function. In contrast, a problem with more than one objective function is called multi-criteria problem. For instance in routing protocols both approaches may be used, single-metric routing or multiple-metrics routing problem.

Single-criterion optimization. Single-criterion optimization is the simplest and classic form of optimization problems. A decision problem could have a single criterion or single aggregate measures such as cost (Fülöp 2005). This optimization problem may have as input a single variable or multiple variables. In the first case, optimization method determines the best possible value of the unique input variable. However, the second case provides the best feasible combination of multiple input variables.

For illustration, we assume that x represents the combination of input parameters; the single-criterion optimization problem takes the following form:

$$\min\{f(x) = z\} \text{ with } x \in S \qquad (1)$$

Where S is a set of feasible solutions, which is the set of points satisfying all the constraints (Loo et al. 2016). Thus, the problem is reduced to determining the unique minimum of objective function f(x) over an input set S. Figure 2 illustrates graphic representation of single-criterion optimization problem.

Multi-criteria optimization. Various real-world applications, in particular routing protocols, are generally based on multiple conflicting objectives or criteria. Therefore optimization problems are intrinsically multidimensional. Actually, decisions are often made according to multiple and opposite criteria. For instance, time goes against money, work against personal life. Consider, for example, a laptop that we want to buy. Purchase decision can be made according to the price, size, power autonomy, reliability, and various other performances. A decision that is made according to one criterion might be worse according to another. Thus, there is no unique optimal solution but rather a set of incomparable options. The main purpose of multi-criteria optimization is to develop methods that take into account multiple objective functions in their problem modeling, and try to find balanced or representative solution of the trade-offs (Trystram 2011). The multi-criteria problem can be formulated as an extension of the single-criterion problem as follows:

$$min\{f_1(x) = z\}$$
$$min\{f_2(x) = z\}$$
.

Fig. 1 Illustration of optimization process

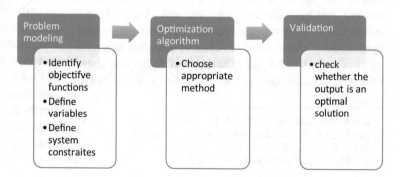

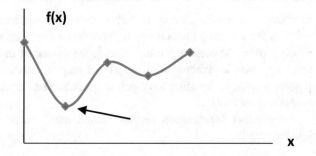

Fig. 2 Single-objective optimization problem

$$min\{f_n(x) = z\}$$
$$with\ x \in S$$

Where f_i represents the ith objective function. The optimal solution consists of finding the vector $x^* = [x1, x2,...., xn]$, which will satisfy (Loo and Mauri 2016):

$$\exists x^* \in S \mid min\{f_i(x^*) = z_i^*\} \ \forall i = 1, 2, ..., n \quad (2)$$

2.2 Optimization Models

As mentioned in the previous section, optimization process must first start with problem modeling or characterization of the criteria to be optimized. In this context, various categories of optimization models are used in practice to represent and solve decision-making problems. Figure 3 illustrates the most successful models like mathematical programming and constraint programming (Talbi 2009).

Linear programming. Linear programming (LP) is the common model used in mathematical programming. It can be defined as the problem of maximizing or minimizing linear objective function subject to linear constraints, which might be equalities or inequalities. Here after we give a simple example for illustration:

The problem consists to find two numbers x and y that maximize the objective function x + y. Variables are subject to following constraints:

$$x, y \geq 0$$
$$x + 2y \leq 4$$
$$4x + 2y \leq 12 \quad (3)$$
$$-x + y \leq 1$$

The present problem has two unknowns and four constraints. These constraints are inequality and linear. The objective function to be maximized is x + y is also linear.

Since there are only two variables, the problem can be optimized by plotting the set of points that satisfy all constraints in a graph and then determine the point of this set that

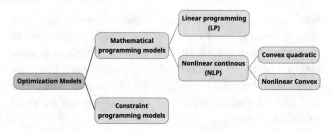

Fig. 3 Optimization models

maximizes objective function, val. The feasible solution space is bordered by the intersection of all the lines that represent each inequality constraint. Figure 4 shows the optimal point of objective function that satisfies all the constraints.

$$4x + 2y = 12$$
$$x + 2y = 4$$
$$-x + y = 1$$

NB:

• When decision variables are discrete we are handling linear integer programming (LIP). In case where the decision variables are both continuous and discrete, we are dealing with mixed-integer programming (MIP);

• The optimal solution is said to be Pareto-optimal point because there is no other point that maximizes the objective function with the respect of constraints.

Nonlinear programming. NLP as well as linear programming is a subset of mathematical programming models. It consists of optimizing a nonlinear problem, which is characterized by nonlinear objective function and/or feasible space is defined by nonlinear constraints.

A continuous nonlinear optimization problem involves minimizing an objective function f: $S \subset \mathbb{R}^n \rightarrow \mathbb{R}$ in a

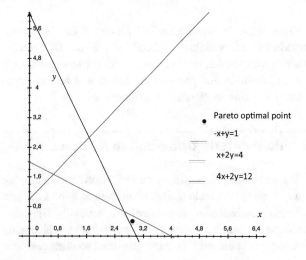

Fig. 4 Illustration of LP model and its solution

continuous space. This model is pretty more difficult to resolve, even if there are many linearization techniques to reduce problem complexity. Of course, linearization techniques due to its approximation approach may introduce extra variables and constraints in the model. Thus, linearization technique should be carefully chosen to avoid unrealistic outputs.

Constraint programming. CP is another common technique for modeling optimization problems. It integrates powerful modeling tools than mixed-integer programming (MIP) models. Though, it does not mean that solving problems within constraint programming is more effective than using mathematical programming. CP has been largely used in the artificial intelligence field (Apt 2003) and successfully applied to numerous optimization problems with tightly constrained search problems such as timetabling and scheduling problems.

In CP, the optimization problem is modeled using a set of variables linked by constraints. The value of variables belongs to a finite domain of integers and the constraints may be expressed mathematically or symbolically. Let us illustrate this concept through an example: Consider that we want to assign n objects $\{o_1, o_2, \ldots o_n\}$ to n different locations $\{l_1, l_2, \ldots l_n\}$ the model will be as follows:

$$\text{all\_different}(y_1, y_2 \ldots y_n) \tag{4}$$

Where y_i represents the ith location in which the ith object is assigned.

In this example, global constraints that refer to variables specify that all variables must be different and represented by $\text{all\_different}(y_1, y_2 \ldots y_n)$.

In case this problem is modeled using integer programming (IP) model, the following variables should be stated:

$$x_{ij} = \begin{cases} 1, & \text{if } 0_i \text{ is assigned to } l_j \\ 0, & \text{otherwise} \end{cases} \tag{5}$$

This model is much more complex than CP, because it introduces n^2 variables instead of n in CP model. Although CP model, in this example, has fewer variables, we cannot conclude that problem resolution will be more efficient than using mathematical programming.

3 Multi-criteria Optimization Methods

After succeeding in modeling an optimization problem, we have to apply a suitable optimization method to find the best solution. Optimization algorithms are generally time and resource consuming because of their heavy calculating process that starts with an initial estimated solution and tries

to reach the optimal solution. In fact, obtaining an optimal solution isn't always guaranteed. It depends on the search strategy of the selected algorithm. Hence, the choice of the most appropriate method is an important step in solving problems without engaging too much resource such as time, computing memory…

An efficient optimization method should meet the following criteria:

- Robustness: achieving high performance for a large variety of similar problems;
- Efficiency: engaging fewer computational resources;
- Accuracy: the output must be as close as possible to the desired solution, without high sensitivity to data error.

In fact, the above criteria are often conflicting, which involves paying special attention to the trade-off between opposite criteria. For instance, rapid convergence may need extra memory allocation or a robust algorithm may be time consuming.

Various approaches have been proposed to deal with multi-criteria optimization problem, such as in MANETs, starting with exact methods to the well-known metaheuristic methods. Other hybrid approaches tried to combine exact and metaheuristic methods (Seghir and Khababa 2018; Wang et al. 2018; Jaafari et al. 2019). In the following, we present the most relevant multi-objective optimization techniques that might be applied in MANETs.

Weighted sum method. Weighed sum method also known as linear aggregation method, commonly applied in multi-objective optimization problems, consists of aggregating vector problem into a scalar problem. As a result, all objectives could be formulated in one expression where weight is attributed to each objective.

Although this method is easy to implement in solving real problems, weighted sum method presents the following limitations:

- Lack of accuracy, which means that small changes in weight may cause big changes in objective function while big variation in weight may result in insignificant changes in result;
- Weight assignment is subjective since the decision-maker provides it;
- Uniform assignment of the weights to the objective functions does not mean that resulting Pareto points will be uniformly distributed;
- Weighted sum method does not solve the problem of nonconvex zones of the Pareto front. This means that the method does not accept a negative value of weight;
- Lack of efficiency in complex systems.

This optimization method is quite complex because the weights must be adjusted to accurately represent the problem under optimization. Some solutions have been proposed to overcome certain limitations of this method. For instance, adaptive weighted sum method proposed by Kim and Wek in (Kim & De Weck 2006). This approach succeeded to output a uniform Pareto-optimal solution and established a solution to nonconvex regions. However, the method could not solve problems with more than two objective functions. Another approach is physical programming. It does not require specifying the weight related to the objective function in modeling phase. This method was proposed to eliminate the need for fine-tuning of weights. Consequently, system efficiency is improved due to reducing the computational rate of large problems.

Various works have applied weighted sum method to develop routing protocol for MANETs. In (Craveirinha et al. 2003), the authors have proposed multiple objective dynamic routing method (MODR) using weighted sum method. It is based on multiple objective shortest path models with QoS constraints and uses implicit costs as metric. Likewise, the authors in (Donoso Meisel 2005) proposed an approach that aggregates multiple QoS criteria, hop count, bandwidth consumption, end-to-end delay, and maximum link utilization into single metric using weighted sum method.

Lexicographic Method. Lexicographic method is another way to handle multi-objective optimization problems. In this method, conflicting objectives are ordered based on their priority level and then optimized one at a time. The decision-maker must define priority levels, which represent the main complex task in lexicographic method. It's pretty useful when optimization problem has to be treated hierarchically and objective functions cannot be scalarized to permit the application of weighted sum method.

Regarding the difficulty of setting preferences, the authors in (Castro-Gutiérrez et al. 2009) proposed a dynamic approach based on changing relative preferences during the search to bypass the need for ordering objectives.

Lexicographic method was applied in various fields such as optimization of network flows for load balancing purpose (Basagni et al. 1998) and in (Ahmed et al. 2015) and the authors used lexicographic max-order objective to solve vehicle routing problem (VRP) in VANET.

Fuzzy logic method. Dr. Lotfi Zadeh of the University of California at Berkeley introduced fuzzy logic technique in the 1960s. He was working on understanding the problem of computer processing of natural language and noticed that the most natural phenomena cannot be represented just based on Boolean logic (0 or 1) but in practice most data might be assigned a value between 0 and 1 while treated by a computer to accurately reflect the phenomena under study. Hence, fuzzy logic can be defined as a new computing approach that is based on the degree of truth rather than classical true and false logic.

Fuzzy logic appears nearer to the reasoning of human brains. It was exploited in various computing fields such as neural networks, expert systems, and other artificial intelligence applications. It provides an effective means for conflict resolution of multiple criteria problems (Singh et al. 2013). Regarding its application in MANET, dynamic routing based on fuzzy logic was introduced by (Chaythanya 2014) to manage routing policies and optimize routing performance. In addition, dynamic source routing (DSR) was extended in [85] using fuzzy logic to select the suitable QoS routing. To determine the most qualified nodes to participate in QoS routing, fuzzy logic schemes consider throughput, latency, and cost of path to integrate required QoS requirements. Moreover, Gomathy and Shanmugavel proposed in (Gomathy & Shanmugavel 2005; Gomathy & Shanmugavel 2005) a scheduler, based on fuzzy logic, for setting data the packets' priority according to multiple criteria. This approach provides enhanced QoS level, which improves network performance such as packet delivery ratio and average end-to-end delay.

Heuristic and Metaheuristic Methods. In real life, problems optimization in science, engineering, economics, and business are complex and difficult to solve using exact methods because they need huge computational resources and too much time to be achieved. In this context, approximate algorithms have emerged as the main alternative to solve this category of problems.

There are two main classes of approximate algorithms: heuristics and metaheuristics. Heuristics algorithms are designed and applicable to solve specific problems and cannot be generalized to other problems. However, metaheuristics algorithms are more general and are applicable to solve a large variety of optimization problems. They serve three main purposes: solving problems faster, solving large problems, and obtaining robust algorithms. Moreover, they are simple, flexible to design and implement (Talbi 2009).

The origin of heuristic term refers to the old Greek word "heuriskein", which means any new approach or strategy to solve problems (Talbi 2009). The suffix meta is also a Greek word that means upper layer methodology. In (Glover 1986), F. Glover introduced the metaheuristic concept.

Metaheuristics belong to optimization branch in computer science and applied mathematics, which are associated with algorithms and computational complexity theory. During the past decades, numerous metaheuristics methods have been developed in various fields, including artificial intelligence, computational intelligence, soft computing, mathematical programming, and operations research (Sörensen et al. 2018).

Most of the metaheuristics imitate natural phenomena in physics and biology to solve complex optimization problems. Hereafter we present some methaeuristic methods that are applicable in MANET problems. For instance, simulated annealing process, tabu search, evolutionary algorithms, and swarm intelligence algorithms.

Simulated Annealing. Simulated Annealing (SA) is one of the oldest metaheuristics methods that have been used to solve several multi-criteria optimization problems. SA is stochastic local search algorithm; it starts from some initial solution and iteratively explores the neighborhood of the current solution (Franzin and Stützle 2019).

This method has been inspired by metallurgy and materials science, which is based on consecutive heating and cooling of materials. The process of heating a material above its recrystallization temperature, maintaining an appropriate temperature for a suitable time, and then cooling alters its physical properties to increase its ductility and reduce its hardness, making it more workable. In other words, this process increases the size of the crystals and decreases the number of imperfections. Increasing the temperatures above certain level pushes the particles to move, which cause substantial changes in its structure. Then decreasing the temperature until the system reaches its fundamental state. This idea was turned into a heuristic method for solving combinatorial optimization problems as follows (Kirkpatrick et al. 1983):

- Physical temperature is mapped into a "temperature" parameter;
- The state of the physical system represents a candidate solution;
- The fundamental state corresponds to the globally optimal solution;
- Change of state corresponds to a move to a neighboring candidate solution.

Various works have applied SA method in MANET routing. For instance, in (Cui et al. 2003) the authors proposed the simulated annealing multi-constrained Path (SA_MCP) as a QoS routing algorithm that apply simulated annealing to Dijkstra's algorithm after converting multiple QoS metrics into a single one. In addition, simulated annealing representation analysis (SA_RA) proposed in (Liu & Feng 2007) searches the optimal path using energy function to translate multiple QoS weights into a single metric. An adaptive MANET multipath routing protocol based on SA was proposed by (Kim 2014). This algorithm achieves greater performances in dynamic network condition.

Tabu Search. Tabu search (TS) is another solution for optimization problems within metaheuristics field. It guides the local heuristic search procedure to explore solution space beyond local optimality. One of the main distinguishing features of TS is the use of adaptive memory, which creates more flexible search behavior. Memory-based strategies of TS enable the implementation of procedures that are capable of searching effectively and produce solutions with suitable quality within reasonable computational effort (Laguna 2018).

A large number of optimization problems, essentially of combinatorial nature, with continuous and discrete parameters have been addressed by TS technique. For example, optimization of home energy management system based on TS technique (Shafiq et al. 2018). Moreover, the authors in (Belfares et al. 2007) proposed TS application to multi-objective resource allocation. In addition, TS was used in reliable multi-level routing protocol based on clustering in VANETs to solve the problem of developing a trap in the local optimum (Moridi & Barati 2017).

Evolutionary algorithms. The word evolutionary algorithm (EA) refers to a class of stochastic optimization methods that mimic the process of natural evolution including all heuristic methods originated from biological evolution, such as natural selection and genetic inheritance. The EAs emerged in the late 1950s, and since the 1970s numerous evolutionary methods have been proposed, such as genetic algorithms, evolutionary programming, and evolution strategies (Zitzler 1999).

The aim of EAs is to find an optimal solution set that minimizes the distance to the Pareto front while maximizing the diversity of candidates to obtain relevant set. Evolutionary algorithms perform such operations to achieve the targeted goal. The process starts by introducing an initial population of candidate solution, which is recombined by means of reproduction technique to generate new solutions. Then, the natural selection process is used to separate individuals or solutions that are closer to the ideal solution, getting rid of the rest (Abraham & Jain 2005).

Figure 5 illustrates the main components of EAs process. These include solution representation, selection strategy, and genetic variation operators. Such components have to be carefully defined in order to efficiently apply EAs. The following subsections provide some detailed information on these components (Dorronsoro 2014).

Selection: represents the first genetic operator engaged in the EA process. It stochastically chooses individuals to be included in the mating pool.

Crossover: also called recombination, produces an offspring by mixing the genetic material of two or more parent individuals with some probability. The goal is to combine the genes of both parents in order to produce better descendants.

Mutation: used to introduce some small changes in individuals to avoid premature convergence to local optima.

Fitness function: represents the objective function as well as in deterministic optimization. It evaluates how good is the feasible solution of the problem. In maximization

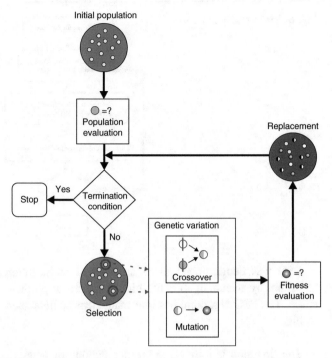

Fig. 5 EA functioning

problem, a feasible solution with higher objective value is better than feasible solution with lower objective value.

Replacement: After genetic variation and fitness evaluation, the resulting offspring is inserted in the fixed size population using some replacement strategy. We note two main strategies: the first one named generational replacement in which all parents are replaced by the offspring in every generation. The second one is the steady-state replacement, where only one parent is replaced by one offspring in every generation.

Termination condition: The EAs algorithms are iteration based. Thus, stopping criterion is defined to avoid infinite computing. Some criterion is based on fixed number of iterations. Another criterion is time based. There are also more complex criteria that stop based on the given population diversity threshold or if there is no improvement of best individual after a predefined number of iterations.

There are different techniques included in evolutionary algorithms field. The first one is evolution strategy (ES) that emulates mutation process over possible solutions to explore the feasible space and avoid holding up in local minimum (Hansen et al. 2015). The second one is evolutionary programming (EP), which is a stochastic optimization technique centered on behavioral link between ascendants and descendants instead of imitating genetic operators observed in nature(Bäck & Fogel 2018). The third one is genetic algorithm (GA) that reproduces the natural selection based on genetics field. It encodes each parameter or design

criterion in a bit chain or parameter code in the form of chromosomes carrying information about every individual. This parameter code allows genetic operator to progress from actual state to the future state with minimum computational effort. Then GA evaluates the appropriateness of each chain to decide if the next iteration will take place or not. Finally, genetic programming (GP) which is similar to GA. The principal difference between GP and GA is that GP automatically creates a formula or computer program that provides a solution to the established problem while GA tries to find the bit chain that represents problem's solution (Hein et al. 2018).

The efficiency of evolutionary algorithms, in terms of finding optimal solutions in a reasonable time, have encouraged researchers to propose effective algorithms to solve optimization problems related to complex network like MANETs. The authors in (Rajan & Shanthi 2015) proposed Hybrid Genetic Based Optimization for Multicast Routing algorithm. This scheme uses the best features of Genetic Algorithm (GA) and particle swarm optimization (PSO) to improve jitter, latency, and PDR with faster convergence. Another GA-based algorithm has been proposed in (Cui et al. 2003) to support soft QoS without hard guarantees, because in mobile environment there may exist transient time periods when the required QoS is not ensured due to path failure. The main purpose of this algorithm is robustness rather than optimality. Hence, it is better to quick establish available route rather than an optimal one out of time.

Swarm intelligence. Swam intelligence (SI) is a subset of bio-inspired computing field as well as non-SI-based techniques such as bat technology, rat technology, and dolphin technology. It constitutes a popular technique for solving optimization problems and can be useful to design algorithms related to distributed systems. Bio-inspired computing emulates the behaviors of swarms of insects or other animals to solve complex real-world problems. There are several domains where bio-inspired algorithms have been successfully applied such as engineering, logistics, and telecommunications (Soni et al. 2018; Slowik & Kwasnicka 2018). Particular advances have been achieved in MANET throughout SI techniques.

Swarm intelligence techniques are characterized by two main concepts: self-organization and division of labors. Self-organization is described as the ability of a system to evolve its agents into suitable form without any external help. SI is driven by four primary principles (Reddy & Kiran 2017):

- Positive feedback is used to highlight the best solution and make other insects follow it up;

- Negative feedback is used to eliminate the out-of-date or invalid solutions. Thus, the other insect will not consider this solution and look forward for the best ones;
- Randomness: SI system explores randomly all available solutions to increase the probability of selecting the best one;
- Multiple interactions: SI system often includes tens to millions of entities and the information collected by this large population is communicated through multiple interactions.

Ant Colony Optimization. Ant colony optimization (ACO) is the most popular swarm intelligence technique. It was proposed in 1992 by Dorigo in (Dorigo 1992) based on foraging behavior of real ants. This algorithm consists of four main components (ant, pheromone, daemon action, and decentralized control) that contribute to the overall system. The ACO algorithm is based on ant's food searching process where they deposit pheromones on the ground as they travel in their search for food. Once the objective is reached, ants go back to the nest and reinforce previously released pheromones. Then, pheromone trails allow other ants to find the best way to food source or the way back home by following those paths with a greater concentration of pheromone (Dorigo & Stützle 2019).

By analogy, ants or imaginary agents drop pheromones when traveling in search space and the concentration of these pheromones indicate the direction that should be followed by ants. Daemon actions are used to gather global information which cannot be done by a single ant and uses the information to determine whether it is necessary to add extra pheromone in order to help the convergence. The decentralized control is used to make the algorithm robust and flexible within a dynamic environment. The importance of having a decentralized system in ACO resides in its flexibility facing ant lost or ant failure (Ab Wahab et al. 2015).

Particle Swarm Optimization. Particle Swarm Optimization (PSO) was introduced by Dr. Kennedy and Dr. Eberhart in 1995 (Kennedy & Eberhart 1995). It is a robust stochastic optimization technique based on the movement and intelligence of swarms. It imitates swarm behavior in birds flocking or fish schooling to guide particles to find global optimal solutions. The PSO has been described with three modest behaviors of separation, alignment, and cohesion (Del Valle et al. 2008):

Separation is the behavior of preventing the crowded local flock-mates;

Alignment is the act of moving toward the average direction of local flock-mates;

Cohesion is the behavior of moving toward the average position of local flock-mates.

Fig. 6 PSO working process

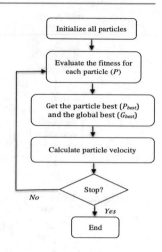

PSO was demonstrated to be an efficient optimization algorithm by searching an entire high-dimensional problem space. The PSO algorithm has four main steps as illustrated in Fig. 6:

- The first step is initializing the population, all particles spread out in order to find the best solution (exploration);
- The second step is calculating the fitness values of each particle;
- The third step is updating individual and global bests;
- The fourth step is getting updated the velocity and the position of the particles.
- The second–fourth steps get repeated until termination condition is satisfied.

Artificial Bee Colony. Artificial Bee Colony (ABC) was proposed by Dervis Karaboga in 2005 (Karaboga 2005). It is a new type of swarm intelligence method that mimics the foraging behavior of honeybees in discovering food sources and relaying this information to other bees in the nest. Due to its simple implementation with very small number of control parameters, various efforts have been done to investigate ABC research in both algorithms and applications. In this approach, artificial agents are defined and classified into three types: the employed bee, the onlooker bee, and the scout bee. In order to achieve algorithm's steps, each of these bees accomplishes its assigned tasks as follows:

- The employed bees memorize the locality of food source. Each employed bee is exclusively associated with one food source, which implies that the number of food sources is always equal to the number of employed bees.
- The onlooker bee collects the information about food source locality from the employed bee in the hive then selects one food source to gather the nectar.

- The scout bees are those unemployed bees that choose their food sources randomly. They are responsible of finding new food sources and the new nectar.

Hereafter, we present the details of each step accomplished during ABC process (Karaboga and Basturk 2007; Li & Yang 2016):

- Step 1: Initialization: Like the other evolutionary algorithms, ABC generates an initial population of food sources randomly and places them uniformly to cover the search space as much as possible. Food sources are represented by (χ_i), where $i = 1, 2 \ldots SN$. (SN: population size);
- Step 2: In this phase each employed bees performs a neighboring search to generate a new vector V_i by updating its vector X_i in order to have more nectar. Once a neighboring food source is found, the probability of selecting a food source or is evaluated. After searching the space, the employed bees return to the hive and share nectar information of their sources with onlooker bees by dancing. Onlooker bees apply roulette wheel selection to select food sources. That is, each onlooker bee prefers a food source depending on nectar information;
- Step 3: Based on calculated probabilities, each onlooker bee that is waiting in the hive, selects a food source using fitness value and information shared by employed bees. Searching procedure of onlooker bees is the same as of employed bees;
- Step 4: It is possible that a food source cannot be improved even if employed bees and onlooker bees have visited it many times. This is known as limit or abandonment criteria, where employed bees become scout bees and all their food sources get abandoned. To maintain necessary population diversity, once an abandoned food source is found, a scout bee is sent to randomly generate a completely new food source to replace abandoned one;
- Step5: The best food source and its position are memorized;
- Step 6: If end condition is met, the program finishes, otherwise the program returns to Step 2 and repeats until stop criteria is met.

Regarding its application in MANET, the BEEDSR routing algorithm proposed in (Tareq et al. 2017) has combined ABC algorithm with DSR routing algorithm to locate possible paths and select the best one using context-aware metrics.

Cuckoo Search Algorithm. Cuckoo Search Algorithm (CSA) is one of the latest metaheuristic methods, which is inspired by the behavior of cuckoo species, like brood parasites, and the characteristics of Lévy flights, such as some birds and fruit flies approaches. Yang and Deb introduced it in 2009 [102]. CSA employs three basic rules or operations in its implementation.

- First, each cuckoo chooses randomly the nest and lays in only one egg;
- Second, eggs and nests with high quality are carried forward to the next generation.
- Third, the number of available host nests is fixed and the egg laid by a cuckoo is discovered by a host bird using probability [0, 1] to decide whether to throw the egg away or abandon the nest and build a completely new nest.

This method was proposed S. Rajalakshmi and R. Maguteeswaran in (Rajalakshmi & Maguteeswaran 2015) to solve multi-constrained Quality of Service Routing (R) problem in MANETs.

4 Applicability of Multi-criteria Optimization

Literature review of multi-criteria optimization method shows a wide variety of techniques that range from deterministic ones like weighted sum algorithms to metaheuristic methods such as evolutionary or swarms intelligence algorithms. The application of advanced multi-criteria optimization algorithms in MANETs routing sounds to be difficult because MANETs devices have limited power, low memory, and computing capabilities. These limitations are conflicting with optimization algorithms, which spend large amount of time to reach the optimal solution. In spite of computing capabilities required by optimization algorithms, multiple routing schemes used these algorithms to solve partial issues in MANETs routing. In this context, artificial intelligence (AI) techniques may reduce required processing capabilities to solve optimization problems. On the other hand, it is possible to address mobility issues, such as frequent link failure using new routing metrics that may reflect the quality of links and intelligently engage an earlier recovery or repairing process to guarantee data delivery under better conditions.

5 Conclusion

Routing protocols in MANET was initially developed to provide best effort service in emergency and military cases in the areas where the deployment of other network infrastructure (fixed or wireless) is not possible. With the emergence of new commercial applications that require some guarantee of QoS levels, an extensive effort was deployed to adapt MANET operations and meet the need of new applications. However, the dynamic nature of MANETs and limited resources of mobile nodes in terms of computation capabilities and power makes it extremely difficult to meet expected QoS levels. Thus, establishing a route that satisfies required service level is a multidimensional problem that involves conflicting parameters. This paper was dedicated to study and discuss the feasibility of using multi-criteria optimization techniques to solve routing problem in MANET environment and establish an efficient distribution of available resources. Nevertheless, these algorithms as well as deterministic or metaheuristic present a high computational cost for real-time MANET communications involving rapidly changing network topologies. In the feature work, we are going to exanimate the use of additional routing metrics to reflect links quality and intelligently engage earlier recovery or repairing process to guarantee data delivery under better conditions.

References

Ab Wahab, M.N., Nefti-Meziani, S., Atyabi, A.: A Comprehensive Review of Swarm Optimization Algorithms. PLoS ONE **10**, e0122827 (2015). https://doi.org/10.1371/journal.pone.0122827

Abraham, A., Jain, L.: Evolutionary multiobjective optimization. In: Evolutionary Multiobjective Optimization. pp. 1–6. Springer (2005)

Apt, K.: Principles of Constraint Programming. Cambridge University Press, Cambridge (2003). https://doi.org/10.1017/CBO9780511615320

Ahmed, A., Hanan, A., Osman, I.: AODV routing protocol working process. Journal of Convergence Information Technology. **10**, 01–08 (2015)

Basagni, S., Chlamtac, I., Syrotiuk, V.R., Woodward, B.A.: A distance routing effect algorithm for mobility (DREAM). In: Proceedings of the 4th annual ACM/IEEE international conference on Mobile computing and networking. pp. 76–84. ACM (1998)

Belfares, L., Klibi, W., Lo, N., Guitouni, A.: Multi-objectives Tabu Search based algorithm for progressive resource allocation. Eur. J. Oper. Res. **177**, 1779–1799 (2007)

Bäck, T., Fogel, D.B., Michalewicz, Z.: Evolutionary Computation 1: Basic Algorithms and Operators. CRC press (2018)

Castro-Gutiérrez, J., Landa-Silva, D., Moreno-Pérez, J.: Dynamic lexicographic approach for heuristic multi-objective optimization. In: Proceedings of the Workshop on Intelligent Metaheuristics for Logistic Planning (CAEPIA-TTIA 2009)(Seville (Spain)). pp. 153–163 (2009)

Chaythanya, B.P.: Fuzzy logic based approach for dynamic routing in MANET. International Journal of Engineering Research. 3, (2014)

Craveirinha, J., Martins, L., Antunes, C.H., Climaco, J.N.: A new multiple objective dynamic routing method using implied costs. J. Telecommun. Inf. Technol. 50–59 (2003)

Cui, X., Lin, C., Wei, Y.: A multiobjective model for QoS multicast routing based on genetic algorithm. In: 2003 International Conference on Computer Networks and Mobile Computing, 2003. ICCNMC 2003. pp. 49–53. IEEE (2003)

Cui, Y., Wu, J.-P., Xu, K.: A QoS routing algorithm by applying simulated annealing. J. Software. **14**, 877–884 (2003)

Donoso Meisel, Y.: Multi-objective optimization scheme for static and dynamic multicast flows. Universitat de Girona (2005)

Del Valle, Y., Venayagamoorthy, G.K., Mohagheghi, S., Hernandez, J.-C., Harley, R.G.: Particle swarm optimization: basic concepts, variants and applications in power systems. IEEE Trans. Evol. Comput. **12**, 171–195 (2008)

Dorigo, M.: Optimization, learning and natural algorithms. PhD Thesis, Politecnico di Milano. (1992)

Dorigo, M., Stützle, T.: Ant colony optimization: overview and recent advances. In: Handbook of metaheuristics. pp. 311–351. Springer (2019)

Dorronsoro, B.: Evolutionary Algorithms for Mobile Ad Hoc Networks. Computer society, IEEE, Wiley, Hoboken, New Jersey (2014)

Fülöp, J.: Introduction to decision making methods. In: BDEI-3 workshop, Washington. pp. 1–15. Citeseer (2005)

Hansen, N., Arnold, D.V., Auger, A.: Evolution strategies. In: Springer handbook of computational intelligence. pp. 871–898. Springer (2015)

Hein, F., Almeder, C., Figueira, G., Almada-Lobo, B.: Designing new heuristics for the capacitated lot sizing problem by genetic programming. Comput. Oper. Res. **96**, 1–14 (2018)

Franzin, A., Stützle, T.: Revisiting simulated annealing: A component-based analysis. Comput. Oper. Res. **104**, 191–206 (2019). https://doi.org/10.1016/j.cor.2018.12.015

Gomathy, C., Shanmugavel, S.: Supporting QoS in MANET by a fuzzy priority scheduler and performance analysis with mixed traffic. In: The 14th IEEE International Conference on Fuzzy Systems, 2005. FUZZ'05. pp. 31–36. IEEE (2005)

Gomathy, C., Shanmugavel, S.: Supporting QoS in MANET by a fuzzy priority scheduler and performance analysis with multicast routing protocols. EURASIP journal on wireless communications and networking. **2005**, 426–436 (2005)

Glover, F.: Future paths for integer programming and links to artificial intelligence. Comput. Oper. Res. **13**, 533–549 (1986). https://doi.org/10.1016/0305-0548(86)90048-1

Jaafari, A., Zenner, E.K., Panahi, M., Shahabi, H.: Hybrid artificial intelligence models based on a neuro-fuzzy system and metaheuristic optimization algorithms for spatial prediction of wildfire probability. Agric. For. Meteorol. **266**, 198–207 (2019)

Karaboga, D.: An idea based on honey bee swarm for numerical optimization. Technical report-tr06, Erciyes university, engineering faculty, computer engineering department (2005)

Karaboga, D., Basturk, B.: A powerful and efficient algorithm for numerical function optimization: artificial bee colony (ABC) algorithm. J. Global Optim. **39**, 459–471 (2007)

Kennedy, J., Eberhart, R.: Particle Swarm Optimization, IEEE International of First Conference on Neural Networks. Perth, Australia, IEEE Press (1995)

Kim, I.Y., De Weck, O.L.: Adaptive weighted sum method for multiobjective optimization: a new method for Pareto front generation. Struct. Multi. Optim. **31**, 105–116 (2006)

Kim, S.: Adaptive MANET multipath routing algorithm based on the simulated annealing approach. Sci. World J. (2014)

Kirkpatrick, S., Gelatt, C.D., Vecchi, M.P.: Optimization by simulated annealing. science. **220**, 671–680 (1983)

Laguna, M.: Tabu Search. In: Martí, R., Pardalos, P.M., and Resende, M.G.C. (eds.) Handbook of Heuristics. pp. 741–758. Springer International Publishing, Cham (2018). https://doi.org/10.1007/978-3-319-07124-4_24

Li, X., Yang, G.: Artificial bee colony algorithm with memory. Appl. Soft Comput. **41**, 362–372 (2016)

Liu, L., Feng, G.: Simulated annealing based multi-constrained QoS routing in mobile ad hoc networks. Wireless Pers. Commun. **41**, 393–405 (2007)

Loo, J., Mauri, J.L., Ortiz, J.H.: Mobile Ad Hoc Networks: Current Status and Future Trends. CRC Press (2016)

Moridi, E., Barati, H.: RMRPTS: a reliable multi-level routing protocol with tabu search in VANET. Telecommun Syst. **65**, 127–137 (2017). https://doi.org/10.1007/s11235-016-0219-6

Nocedal, J., Wright, S.J.: Numerical Optimization. Springer, New York (1999)

Rajalakshmi, S., Maguteeswaran, R.: Quality of Service Routing in Manet Using a Hybrid Intelligent Algorithm Inspired by Cuckoo Search. Sci. World J. **2015**, 1–8 (2015). https://doi.org/10.1155/2015/703480

Reddy, G.R.M., M, Kiran: Mobile ad hoc networks: bio-inspired quality of service aware routing protocols. (2017)

Seghir, F., Khababa, A.: A hybrid approach using genetic and fruit fly optimization algorithms for QoS-aware cloud service composition. J. Intell. Manuf. **29**, 1773–1792 (2018)

Shafiq, S., Fatima, I., Abid, S., Asif, S., Ansar, S., Abideen, Z.U., Javaid, N.: Optimization of Home Energy Management System Through Application of Tabu Search. In: Xhafa, F., Caballé, S., and Barolli, L. (eds.) Advances on P2P, Parallel, Grid, Cloud and Internet Computing. pp. 37–49. Springer International Publishing (2018)

Singh, H., Gupta, M.M., Meitzler, T., Hou, Z.-G., Garg, K.K., Solo, A. M.G., Zadeh, L.A.: Real-Life applications of fuzzy logic. Adv. Fuzzy Syst. 1–3 (2013). https://doi.org/10.1155/2013/581879

Slowik, A., Kwasnicka, H.: Nature inspired methods and their industry applications—Swarm intelligence algorithms. IEEE Trans. Industr. Inf. **14**, 1004–1015 (2018)

Soni, G., Jain, V., Chan, F.T., Niu, B., Prakash, S.: Swarm intelligence approaches in supply chain management: potentials, challenges and future research directions. An International Journal, Supply Chain Management (2018)

Sörensen, K., Sevaux, M., Glover, F.: A History of Metaheuristics. In: Martí, R., Panos, P., and Resende, M.G.C. (eds.) Handbook of Heuristics. pp. 1–18. Springer International Publishing, Cham (2018). https://doi.org/10.1007/978-3-319-07153-4_4-1

Talbi, E.-G.: Metaheuristics: From Design to Implementation. John Wiley & Sons, Hoboken, N.J (2009)

Tareq, M., Alsaqour, R., Abdelhaq, M., Uddin, M.: Mobile Ad Hoc Network Energy Cost Algorithm Based on Artificial Bee Colony, https://www.hindawi.com/journals/wcmc/2017/4519357/, last accessed 2019/03/28. https://doi.org/10.1155/2017/4519357

Trystram, D.: Multi-Criteria Optimization and its Application to Multi-Processor Embedded Systems, (2011)

Wang, H., Wang, W., Cui, L., Sun, H., Zhao, J., Wang, Y., Xue, Y.: A hybrid multi-objective firefly algorithm for big data optimization. Appl. Soft Comput. **69**, 806–815 (2018)

Zitzler, E.: Evolutionary Algorithms for Multiobjective Optimization: Methods and Applications. Citeseer (1999)

Rajan, C., Shanthi, N.: Genetic based optimization for multicast routing algorithm for MANET. Sadhana. **40**, 2341–2352 (2015)

Modeling and Performance Analysis for Transportation Systems of ULA and UCA Massive-MIMO Basing on Spherical Wave

Abdelhamid Riadi, Mohamed Boulouird, and Moha M'Rabet Hassani

Abstract

The fifth-generation (5G) of cellular networks is becoming the most promising technology to meet the needs of Intelligent Transportation Systems (ITS). Including 5G in the ITS can help to progress the manner to send or receive the information, and also to improve the infrastructure and the needs of users. Massive-MIMO (mMIMO) system is one of its pillars used to enhance its capabilities and performance. The mMIMO is based on antenna structures widely used in 5G which manage several hundred channels connecting to thousands of objects. In this paper, Multi-polarized-Uniform-Linear-Array (MULA) and Multi-polarized-Uniform-Circular-Array (MUCA) systems based on Spherical Wave (SW) are established. Different parameters vary depending on the proposed antenna structures or the environment under consideration. These parameters have a remarkable effect on the Channel Orthogonality (CO) and the performance of the multi-polarized system. The one-dimensional structure is more promising for 5G and suggests a suitable choice for the new generation of cellular networks.

Keywords

Multi-polarized • Massive-MIMO • Uniform linear array • Uniform circular array • Spherical wave

A. Riadi (✉) · M. Boulouird · M. M. Hassani
Instrumentation, Signals and Physical Systems (I2SP) Group, Faculty of Sciences Semlalia,
Cadi Ayyad University, Marrakesh, Morocco
e-mail: abdelhamid.riadi@edu.uca.ac.ma

M. Boulouird
e-mail: m.boulouird@uca.ac.ma

M. M. Hassani
e-mail: hassani@ucam.ac.ma

M. Boulouird
National School of Applied Sciences of Marrakesh (ENSA-M), Cadi Ayyad University, Marrakesh, Morocco

1 Introduction

Massive-MIMO technology is a promising system for the next generation of wireless communications 5G (Paulraj et al. 2003; Simon 2002; Telatar 1999). On account of a high throughput, 5G will be implemented in a real communication environment. Although, the communication between transceiver suffers from canal phenomena such as diffusion, reflection, scattering, etc. Therefore, the favorable propagation property or low CO betwixt various users (Cheng et al. 2018; Riadi et al. 2019b, 2020b) is one of the popular perceptions. The CO is a key element in mMIMO system. A higher CO means greater crosstalk between users. This increases the overlap of frequency resources, and makes it impossible to distinguish between users. While low CO means that we are back to good communication. This means that crosstalk between users will be reduced. This facilitates the detection and pre-coding of data. In addition to that, mMIMO system presents many configurations for example rectangular, spherical, linear, and cylindrical (Riadi et al. 2017; Yang and Hanzo 2015; Zheng et al. 2014). These new architectures of antennas are progressed to become implemented in a real environment. In the literature, several scenarios have been proposed to explain the CO under various parameters, three-dimensional MULA-mMIMO systems is proposed in Cheng et al. (2018) basing on Plane Wave (PW), where another scenario of MULA-mMIMO basing on SW is proposed and investigated in Riadi et al. (2019b); similarly MUCA-mMIMO is proposed and analyzed in Riadi et al. (2019a). The results obtained by the new typical antenna array design are efficient to improve system performance in a real situation (Erceg et al. 2002).

The remainder of the paper is organized as follows, in Sect. 2 we present the channel pattern of antenna array configuration, in which MULA-mMIMO based on SW is established with various parameters. While the compared channel pattern of MUCA-mMIMO is defined in the Sect. 3. In Sect. 4, Zero-Forcing (ZF) detector is presented basing on channel matrix of each antenna configuration. The simulation results are presented and discussed in Sect. 5. Section 6 summarizes the results of this work and draws conclusions.

2 Channel Pattern of Multi-polarized ULA

In this study, a new structure of MULA-mMIMO is discussed (Fig. 1). From this configuration, it can be seen that the SW is arrived from signal origin S with randomly AAoA (i.e., α_i where $i \in \{1, 2, \ldots, N_r\}$) and EAoA (i.e., β where $i \in \{1, 2, \ldots, N_r\}$), respectively. The uniform Power Azimuth Spectrum (PAS) model is defined to characterize a rich-scattering environment, it is used also to describe the AAoA and EAoA, which is expressed by Cheng et al. (2018), Schumacher et al. (2002), Cho et al. (2010):

$$p(\Psi) = \frac{1}{2\Delta\Psi}, \quad -\Delta\Psi + \Psi_0 \leqslant \Psi \leqslant \Delta\Psi + \Psi_0 \quad (1)$$

From Eq. 1, $\Delta\Psi$ presents the Azimuth Angle Spread/Elevation Angle Spread (AAS/EAS), which used to establish the AAoA/EAoA distributions; in addition to that Ψ_0 is the mean of AAoA/EAoA.

Furthermore, from Fig. 1, the (x, y, z) marker is considered to be an orthogonal marker, we assume that antennas parallel to the x-axis are horizontal antennas and antennas parallel to the z-axis are vertical antennas. In the same way, noted d as a distance between two adjacent antennas, and h is the elevation of S from the horizontal plane $y - S_1 - x$, noted also S' is the projection of S onto $y - S_1 - x$. Otherwise, we consider the right-angled triangles $S'S_iA$ and SS_iS' with $i \in \{1, 2, \ldots, N_r\}$, as shown in Fig. 1. By the application of the Pythagorean theorem, we find the distances between S and each antenna are

$$d_{SS_1} = \sqrt{d_x^2 + d_y^2 + h^2} \quad (2)$$

$$d_{SS_2} = \sqrt{(d_x + d)^2 + d_y^2 + h^2} \quad (3)$$

$$\vdots$$

$$d_{SS_m} = \sqrt{(d_x + d(m - 1))^2 + d_y^2 + h^2} \quad (4)$$

$$\vdots$$

$$d_{SS_{N_r}} = \sqrt{(d_x + d(N_r - 1))^2 + d_y^2 + h^2} \quad (5)$$

Similarly, for the different angles α_i and β_i, $\forall i \in \{1, \ldots, N_r\}$, d_x and h can be denoted by

$$d_x = \tan(\alpha_i) \times d_y - d(i - 1) \quad (6)$$

$$h = \tan(\beta_i)\sqrt{\tan^2(\alpha_i) \times d_y^2 + d_y^2} \quad (7)$$

we assume that the antenna index one is a reference antenna (i.e., S_1). Hence, one path of multi-path at antennas, for example, S_1 and S_{N_r} can be noted by

$$h_1^{ULA} = \sqrt{P_H}e^{j(\phi + 2\pi\sqrt{d_x^2 + d_y^2 + h^2}/\lambda)} \quad (8)$$

$$\vdots$$

$$h_{N_r}^{ULA} = \sqrt{P_V}e^{j(\phi + 2\pi\sqrt{(d_x + d(N_r - 1))^2 + d_y^2 + h^2}/\lambda)} \quad (9)$$

From these equations, ϕ is an independent and identically distributed random receiving phase on $[\pi; \pi]$, P_V and P_H are the vertically and horizontally powers amplitudes, respectively.

Otherwise, during signal transmission, the power is degraded due to the disruptive phenomena in the channel. The polarization changed both the edges of the transceiver.

Fig. 1 3-D multi-polarized ULA-massive-MIMO system transmission scenario (Cheng et al. 2018)

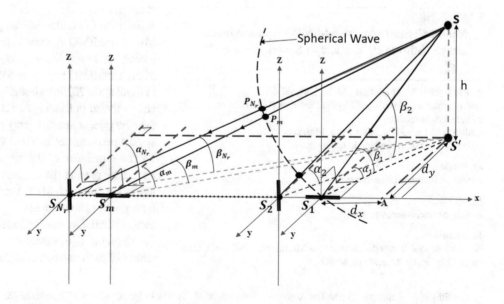

2.1 Cross-Polarization Discrimination

In the same way, in a real environment, for example, the horizontal polarization at users' sides can be changed to a vertical polarization at base station antennas, due to environment phenomena (i.e., diffractions, reflections, etc.). Thus these phenomena can be modeled by a cross-polarization discrimination (XPD) (Cheng et al. 2018; Riadi et al. 2019a; Jiang 2007; Riadi et al. 2020b) as:

$$XPD = \frac{E\{|h_{VV}^{ULA,SW}|^2\}}{E\{|h_{VH}^{ULA,SW}|^2\}} = \frac{E\{|h_{HH}^{ULA,SW}|^2\}}{E\{|h_{HV}^{ULA,SW}|^2\}} = \frac{1-a}{a} \tag{10}$$

From Eq. (10), a is a power escaped during switch of polarization between transceiver. In the case when $a = 0$ the power escaped does not exist (i.e., the polarization remains unchanged). In the case when $a \in]0, 1]$, the power escaped exists. In addition to that, $h_{XX}^{ULA,SW}$ and $h_{XY}^{ULA,SW}$ are the channels where $X, Y \in (V, H)$, and $E\{\}$ is the expectation operator. Therefore, by substitution of (10) in (8) and (9) yields two cases firstly:

$$h_{1,XPD}^{ULA} = \sqrt{P_H(1-a)}e^{j(\phi+2\pi\sqrt{d_x^2+d_y^2+h^2}/\lambda)} \tag{11}$$

$$\vdots$$

$$h_{N_r,XPD}^{ULA} = \sqrt{P_V(1-a)}e^{j(\phi+2\pi\sqrt{(d_x+d(N_r-1))^2+d_y^2+h^2}/\lambda)} \tag{12}$$

where $P_X(1-a)$, $X \in (V, H)$ the power is conserved in the case there is no escape (Cheng et al. 2018; Riadi et al. 2019a; Ma et al. 2009). Secondly,

$$h_{1,XPD}^{ULA} = \sqrt{P_H a}e^{j(\phi+2\pi\sqrt{d_x^2+d_y^2+h^2}/\lambda)} \tag{13}$$

$$\vdots$$

$$h_{N_r,XPD}^{ULA} = \sqrt{P_V a}e^{j(\phi+2\pi\sqrt{(d_x+d(N_r-1))^2+d_y^2+h^2}/\lambda)} \tag{14}$$

where $P_X a$, $X \in (V, H)$ the power escaped between the transceiver (Cheng et al. 2018; Riadi et al. 2019a; Ma et al.

2009). Hence, the channels vectors of all antennas at a location of users e and f are giving by

$$\mathfrak{h}_e^{ULA,XPD} = \begin{bmatrix} h_{1,e,XPD}^{ULA} \\ h_{2,e,XPD}^{ULA} \\ \vdots \\ h_{m,e,XPD}^{ULA} \\ \vdots \\ h_{N_r,e,XPD}^{ULA} \end{bmatrix}$$

$$= \begin{bmatrix} \sqrt{P_H a}e^{j(\phi_e+2\pi\sqrt{d_{x,e}^2+d_{y,e}^2+h_e^2}/\lambda)} \\ \sqrt{P_H(1-a)}e^{j(\phi_e+2\pi\sqrt{(d_{x,e}+d)^2+d_{y,e}^2+h_e^2}/\lambda)} \\ \vdots \\ \sqrt{P_H(1-a)}e^{j(\phi_e+2\pi\sqrt{(d_{x,e}+d(m-1))^2+d_{y,e}^2+h_e^2}/\lambda)} \\ \vdots \\ \sqrt{P_H(1-a)}e^{j(\phi_e+2\pi\sqrt{(d_{x,e}+d(N_r-1))^2+d_{y,e}^2+h_e^2}/\lambda)} \end{bmatrix}, \tag{15}$$

$$\mathfrak{h}_f^{ULA,XPD} = \begin{bmatrix} h_{1,f,XPD}^{ULA} \\ h_{2,f,XPD}^{ULA} \\ \vdots \\ h_{m,f,XPD}^{ULA} \\ \vdots \\ h_{N_r,f,XPD}^{ULA} \end{bmatrix}$$

$$= \begin{bmatrix} \sqrt{P_V(1-a)}e^{j(\phi_f+2\pi\sqrt{d_{x,f}^2+d_{y,f}^2+h_f^2}/\lambda)} \\ \sqrt{P_V a}e^{j(\phi_f+2\pi\sqrt{(d_{x,f}+d)^2+d_{y,f}^2+h_f^2}/\lambda)} \\ \vdots \\ \sqrt{P_V a}e^{j(\phi_f+2\pi\sqrt{(d_{x,f}+d(m-1))^2+d_{y,f}^2+h_f^2}/\lambda)} \\ \vdots \\ \sqrt{P_V a}e^{j(\phi_f+2\pi\sqrt{(d_{x,f}+d(N_r-1))^2+d_{y,f}^2+h_f^2}/\lambda)} \end{bmatrix}. \tag{16}$$

Now we can derive the channel matrix of all users N_t and all BS antenna N_r, according to Eqs. (15) and (16) as expressed below:

$$\mathbb{H}_{ULA} = [\mathfrak{h}_1^{ULA,XPD}, \ldots, \mathfrak{h}_f^{ULA,XPD}, \ldots, \mathfrak{h}_e^{ULA,XPD}, \ldots, \mathfrak{h}_{N_t}^{ULA,XPD}] \tag{17}$$

Fig. 2 3-D multi-polarized
UCA-massive-MIMO system
transmission scenario (Riadi et al.
2019a, 2020b)

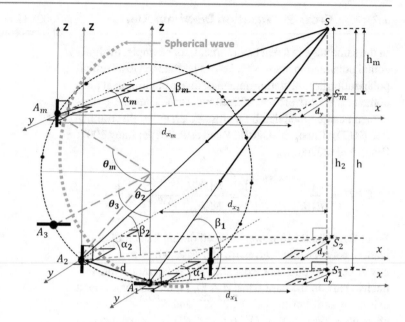

In the next of this work, another structure of MUCA-mMIMO systems is discussed based on SW as shown in Fig. 2.

3   Channel Pattern of Multi-polarized UCA

Furthermore, Fig. 2 presents another architecture of three-dimensional MUCA-mMIMO systems. According to this configuration of antennas, a spherical wave is considered, in which the SW arrives from signal origin S. Moreover, $\{S_1, S_2, \ldots, S_m\}$ are the projection of S on the horizontal plane $\{x - A_1 - y, x - A_2 - y, \ldots, x - A_m - y\}$, respectively. In the same way, the SW arrived from arbitrary AAoA (i.e., α) and EAoA (i.e., β). Similarly, their distributions are modeled by the PAS Eq. (1). Whereas, d_y and d_x are the distance between x and y axis and each projection $\{S_1, S_2, \ldots, S_m\}$, h is the height of the signal origin S and the horizontal plane $(x - A_1 - y)$ (Riadi et al. 2019a). Similarly, from Fig. 2, the distances between the signal origin S and antennas index A_1 and A_m are (Riadi et al. 2019a):

$$d_{SA_1} = \sqrt{d_{x,1}^2 + d_y^2 + h_1^2} \tag{18}$$

$$\vdots$$

$$d_{SA_m} = \sqrt{(d_{x,1} + r\sin(m\theta))^2 + d_y^2 + h_m^2} \tag{19}$$

where

$$d_{x,1} = d_y * \tan(\alpha_1) \tag{20}$$

$$h_m = \tan(\beta_m) * \sqrt{d_y^2 + d_y^2 * \tan(\alpha_m)^2} \tag{21}$$

$$r = \frac{d}{2\sin(\frac{\theta}{2})} \tag{22}$$

From Fig. 2, we can see that the radius expressed by $\theta = \frac{2\pi}{M}$ (M is the number of antenna at the BS), where it has been defined as $\theta_2 = \theta, \theta_3 = 2\theta, \ldots, \theta_m = (m-1)\theta$. We assume that the A_1 antenna is a reference, one path of multi-path at antennas index A_1 and A_m can be noted by

$$h_1^{UCA} = \sqrt{P_H} e^{j\left(\phi + 2\pi\sqrt{d_{x1,g}^2 + d_y^2 + h_1^2}/\lambda\right)} \tag{23}$$

$$\vdots$$

$$h_{N_r}^{ULA} = \sqrt{P_V} e^{j\left(\phi + 2\pi\sqrt{(d_{x1} + r\sin((m-1)\theta))^2 + d_y^2 + h_m^2}/\lambda\right)} \tag{24}$$

In the same way as before, and according to XPD expression (Eq. (10)), Eqs (23) and (24) can be rewritten as In the first case when there's no escape from power,

$$h_1^{UCA} = \sqrt{P_H(1-a)} e^{j\left(\phi + 2\pi\sqrt{d_{x1,g}^2 + d_y^2 + h_1^2}/\lambda\right)} \tag{25}$$

$$\vdots$$

$$h_{N_r}^{ULA} = \sqrt{P_V(1-a)} e^{j\left(\phi + 2\pi\sqrt{(d_{x1} + r\sin((m-1)\theta))^2 + d_y^2 + h_m^2}/\lambda\right)} \tag{26}$$

in the second case when there is an escape from power,

$$h_1^{UCA} = \sqrt{P_H a} e^{j(\phi + 2\pi\sqrt{d_{x1,g}^2 + d_y^2 + h_1^2}/\lambda)} \tag{27}$$

$$\vdots$$

$$h_{N_r}^{ULA} = \sqrt{P_V a} e^{j(\phi + 2\pi\sqrt{(d_{x1} + r\sin((m-1)\theta))^2 + d_y^2 + h_m^2}/\lambda)} \tag{28}$$

Thus, the channels vectors of all antennas at users' position g and w, using spherical wave (Fig. 2) as noted by Riadi et al. (2019a)

$$\mathfrak{h}_g^{UCA} = \begin{bmatrix} h_{1,g,XPD}^{UCA} \\ h_{2,g,XPD}^{UCA} \\ \vdots \\ h_{m,g,XPD}^{UCA} \\ h_{m+1,g,XPD}^{UCA} \\ \vdots \\ h_{N_r,g,XPD}^{UCA} \end{bmatrix}$$

$$= \begin{bmatrix} \sqrt{P_H(1-a)} e^{j(\phi_g + 2\pi\sqrt{d_{x,1,g}^2 + d_{y,g}^2 + h_{1,g}^2}/\lambda)} \\ \sqrt{P_H a} e^{j(\phi_g + 2\pi\sqrt{(d_{x,1,g} + r\sin(\theta))^2 + d_{y,g}^2 + h_{2,g}^2}/\lambda)} \\ \vdots \\ \sqrt{P_H a} e^{j(\phi_g + 2\pi\sqrt{(d_{x,1,g} + r\sin((m-1)\theta))^2 + d_{y,g}^2 + h_{m,g}^2}/\lambda)} \\ \sqrt{P_H(1-a)} e^{j(\phi_g + 2\pi\sqrt{(d_{x,1,g} + r\sin(m\theta))^2 + d_{y,g}^2 + h_{m+1,g}^2}/\lambda)} \\ \vdots \\ \sqrt{P_H a} e^{j(\phi_g + 2\pi\sqrt{(d_{x,1,g} + r\sin((N_r-1)\theta))^2 + d_{y,g}^2 + h_{N_r,g}^2}/\lambda)} \end{bmatrix} \tag{29}$$

$$\mathfrak{h}_w^{UCA} = \begin{bmatrix} h_{1,w,XPD}^{UCA} \\ h_{2,w,XPD}^{UCA} \\ \vdots \\ h_{m,w,XPD}^{UCA} \\ h_{m+1,w,XPD}^{UCA} \\ \vdots \\ h_{N_r,g,XPD}^{UCA} \end{bmatrix}$$

$$= \begin{bmatrix} \sqrt{P_V a} e^{j(\phi_w + 2\pi\sqrt{d_{x,1,w}^2 + d_{y,w}^2 + h_{1,w}^2}/\lambda)} \\ \sqrt{P_V(1-a)} e^{j(\phi_w + 2\pi\sqrt{(d_{x,1,w} + r\sin(\theta))^2 + d_{y,w}^2 + h_{2,w}^2}/\lambda)} \\ \vdots \\ \sqrt{P_V(1-a)} e^{j(\phi_w + 2\pi\sqrt{(d_{x,1,w} + rs\in((m-1)\theta))^2 + d_{y,w}^2 + h_{m,w}^2}/\lambda)} \\ \sqrt{P_V a} e^{j(\phi_w + 2\pi\sqrt{(d_{x,1,w} + r\sin(m\theta))^2 + d_{y,w}^2 + h_{m+1,w}^2}/\lambda)} \\ \vdots \\ \sqrt{P_V(1-a)} e^{j(\phi_w + 2\pi\sqrt{(d_{x,1,w} + r\sin((N_r-1)\theta))^2 + d_{y,w}^2 + h_{N_r,w}^2}/\lambda)} \end{bmatrix} \tag{30}$$

Furthermore, the channel matrix $\mathbb{H}_{UCA}$ of MUCA-mMIMO system with all users N_t and all BS antennas N_r is noted by

$$\mathbb{H}_{UCA} = [\mathfrak{h}_1^{UCA,XPD}, \ldots, \mathfrak{h}_g^{UCA,XPD}, \ldots, \mathfrak{h}_w^{UCA,XPD}, \ldots, \mathfrak{h}_{N_t}^{UCA,XPD}] \tag{31}$$

Otherwise, from the channel vectors defined above, the channel orthogonality of two channel vectors $\mathfrak{h}_g^{UCA}$ and $\mathfrak{h}_w^{UCA}$ (or $\mathfrak{h}_e^{ULA}$ and $\mathfrak{h}_f^{ULA}$) defined by

$$\delta_{x,y} = \frac{|(\mathfrak{h}_x^{UCA})^H \mathfrak{h}_y^{UCA}|}{||\mathfrak{h}_x^{UCA}|| \cdot ||\mathfrak{h}_y^{UCA}||} \text{ or } (\delta_{x,y} = \frac{|(\mathfrak{h}_x^{ULA})^H \mathfrak{h}_y^{ULA}|}{||\mathfrak{h}_x^{ULA}|| \cdot ||\mathfrak{h}_y^{ULA}||}) \tag{32}$$

where $x, y \in \{(e, f), (g, w)\}$ and $||\cdot||$ is the Euclidean norm.

4   Linear Detection

In this section, Zero-Forcing (ZF) detector is presented, basing on the channel vectors above, the channels matrix of three-dimensional MUCA-mMIMO and MULA-mMIMO of N_t users, each with single antenna are defined by Eqs. (17) and (31). In this paper and in related references Cheng et al. (2018), Riadi et al. (2019a), it was adopted that the number of users with horizontally polarized antenna are equal to the users with vertically polarized antenna. In this way, the ZF detector defined by the following expressions:

$$\mathbb{H}_{ULA}^+ = (\mathbb{H}_{ULA}^H \mathbb{H}_{ULA})^{-1} \mathbb{H}_{ULA}^H \tag{33}$$

$$\mathbb{H}_{UCA}^+ = (\mathbb{H}_{UCA}^H \mathbb{H}_{UCA})^{-1} \mathbb{H}_{UCA}^H \tag{34}$$

In addition to that, the ZF is a linear detection scheme which forces the interference to zero. However, it may result in increase in noise level (Yang and Hanzo 2015; Chockalingam and Sundar Rajan 2014; Rajeev et al. 2014; Riadi et al. 2019d, c, 2020a).

5   Simulation Results

In this part, the Monte Carlo simulation technique is used, in which 10,000 channel samples are generated, and the average channel orthogonality is calculated. In the same way, the mean of AAoA and EAoA are considered equal to $0°$, and the power is normalized.

Figure 3 illustrates the CO of MULA/uULA-mMIMO and MUCA/uUCA-mMIMO systems. The AAS and EAS are equal to $3°$ and $30°$, respectively, and the XPD is considered equal to $8\,dB$. While the antenna spacing (i.e., d) is varied. In the case when $d = 0.2\lambda$ and the BS antennas equal to 20 (i.e., $N_r = 20$), the CO is close to 0.526 and 0.513 for uUCA-mMIMO and uULA-mMIMO systems, respectively. In addition to that, the CO is close to 0.3374 and 0.3311 for MUCA-mMIMO and MULA-mMIMO systems, respectively. While the antenna spacing is increased to 5λ. The CO is close to 0.476 and 0.4604 for uUCA-mMIMO and uULA-mMIMO

Fig. 3 Channel orthogonality of MUCA-mMIMO and MULA-mMIMO with various antenna spacing

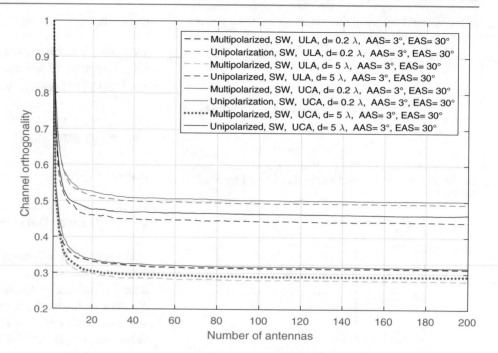

systems, respectively. In the same way, it is close to 0.3036 and 0.2982 for MUCA-mMIMO and MULA-mMIMO systems, respectively. According to these results, it has been found that the CO is affected by the distance between the antennas. In addition to that, a minimum of the CO is achieved by MULA-mMIMO systems, despite there is a few scattering (i.e., AAS = 3°). The whole CO of all systems is decreased when the BS antenna increases.

Figure 4 depicts the CO of MULA-mMIMO and MUCA-mMIMO systems with various AAS. The antenna spacing is considered equal to 0.5λ, mostly used in massive-MIMO systems (Li et al. 2014; Rusek et al. 2012; Gao et al. 2011; Hoydis et al. 2012), and the $XPD = 8$ dB. From this figure, we can see that the CO is declined over the range of BS antennas. In the case where the AAS is equal to 3° and the BS antenna equal to 20. The CO is close to 0.246 and 0.2235 for MUCA-mMIMO and MULA-mMIMO systems, respectively. While it is close to 0.1533 and 0.1462 for MUCA-mMIMO and MULA-mMIMO systems, respectively, in the case where AAS = 20° (i.e., numerous scattering). When we increase the AAS to 30°, the CO of the MUCA-mMIMO system was close to 0.135. Whereas increasing the BS antenna can help to decrease more the CO. Consequently, from these results also, we can see that the CO is affected by a numerous scattering environment.

Furthermore, Fig. 5 presents the CO for various XPD values. The AAS is considered equal to 3° (i.e., few scatter-

ing) and EAS = 30°, and the antenna spacing is equal to 0.5λ. Otherwise, in a BS antennas equal to 20 and XPD of 0 dB. The CO is close to 0.3529 and 0.3178 for MUCA-mMIMO and MULA-mMIMO systems, respectively. While it's close to 0.2287 and 0.2143 for MUCA-mMIMO and MULA-mMIMO systems, respectively, with a XPD of 15 dB. Thus, the CO is affected by a high XPD; that is to say, more power will be conserved between the transceiver.

Figure 6 presents the performance of MULA/uULA-mMIMO and MUCA/uUCA-mMIMO systems. The AAS is equal to 3° and the EAS is equal to 30°. While the antenna spacing is considered equal to 0.5λ, the XPD is set to be 8 dB. In the same way, the ZF detector is used with the transformation matrix defined above (Eqs. (33) and (31)); the number of users and the BS antenna are taken equal to 8 and 200, respectively. From this figure, it can be seen that the Bit Error Rate (BER) decreases over the range of Signal-to-Noise-Ratio (SNR). In the same way, taking an example of an SNR equal to 22 dB. The BER is equal to 26.1×10^{-4}, 17×10^{-4}, 6.98×10^{-4} and 4.5×10^{-4} for uUCA-mMIMO, uULA-mMIMO, MUCA-mMIMO, and MULA-mMIMO systems, respectively. From these results, it can be seen that the MULA-mMIMO system provides better performance, due to a lower CO and the one-dimensional structure of the MULA-mMIMO system against the two-dimensional structure of the MUCA-mMIMO system.

Fig. 4 Channel orthogonality of MUCA-mMIMO and MULA-mMIMO with various azimuth angle spread

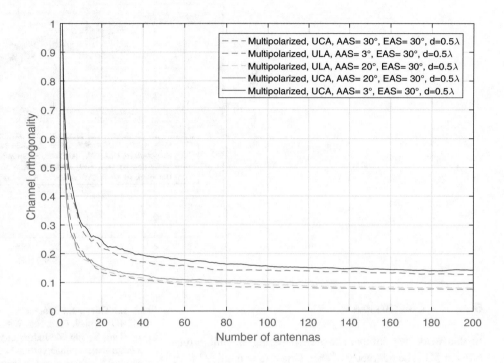

Fig. 5 Channel orthogonality of MUCA-mMIMO and MULA-mMIMO with various cross-polarization discrimination

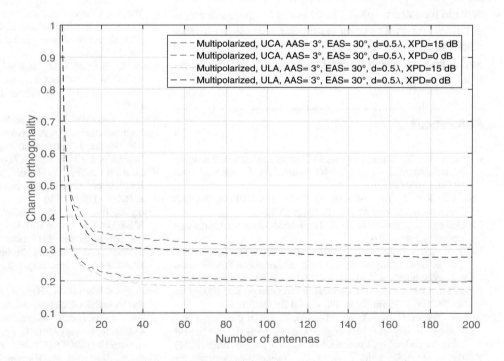

Fig. 6 Bit error rate versus SNR

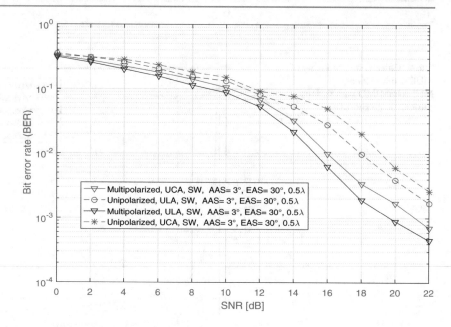

6   Conclusion

In this work, we compare two encouraging antenna structures in 5G. The geometry of each structure is modeled with a spherical wave. It was found that the minimum channel orthogonality is obtained for the linear multi-polarized system in front of the circular multi-polarized system. While single polarity antenna systems have poor channel orthogonality, more power will be retained, so a reduction in the demand for antenna spacing is suggested for the linear multi-polarized system. The linear multi-polarized system provides better performance than other systems. It has been suggested as the most promising system for Massive-MIMO 5G Wireless Communications.

References

Cheng, X., He, Y., Zhang, L., Qiao, J.: Channel modeling and analysis for multipolarized massive MIMO systems. Int. J. Commun. Syst. **31**(12), e3703 (2018)

Cho, Y.S., Kim, J., Yang, W.Y., Kang, C.G.: MIMO-OFDM Wireless Communications with MATLAB. Wiley (2010)

Chockalingam, A., Sundar Rajan, B.: Large MIMO Systems. Cambridge University Press (2014)

Erceg, V., Soma, P., Baum, D.S., Paulraj, A.J.: Capacity obtained from multiple-input multiple-output channel measurements in fixed wireless environments at 2.5 GHz. In: 2002 IEEE International Conference on Communications. Conference Proceedings. ICC 2002 (Cat. No. 02CH37333), vol. 1, pp. 396–400. IEEE (2002)

Gao, X., Edfors, O., Rusek, F., Tufvesson, F.: Linear pre-coding performance in measured very-large MIMO channels. In: 2011 IEEE Vehicular Technology Conference (VTC Fall), pp. 1–5. IEEE (2011)

Hoydis, J., Hoek, C., Wild, T., Brink, S.: Channel measurements for large antenna arrays. In: 2012 International Symposium on Wireless Communication Systems (ISWCS), pp. 811–815. IEEE (2012)

Jiang, L., Thiele, L., Jungnickel, V.: On the modelling of polarized MIMO channel. Proc. Eur. Wirel. **2007**, 1–4 (2007)

Li, J., Zhao, Y., Tan, Z.: Indoor channel measurements and analysis of a large-scale antenna system at 5.6 GHz. In: 2014 IEEE/CIC International Conference on Communications in China (ICCC), pp. 281–285. IEEE (2014)

Ma, Y., Zheng, Z., Zhou, Y.: Characteristics of MIMO channel in consideration of polarization. In: 2009 5th International Conference on Wireless Communications, Networking and Mobile Computing, pp. 1–3. IEEE (2009)

Paulraj, A., Rohit, A.P., Nabar, R., Gore, D.: Introduction to Space-Time Wireless Communications. Cambridge University Press (2003)

Rajeev, P., Prabhat, A., Norsang, L.: Sphere detection technique: an optimum detection scheme for MIMO system. Int. J. Comput. Appl. **100**(2), (2014)

Riadi, A., Boulouird, M., Hassani, M.M.: An overview of massive-MIMO in 5G wireless communications. In: Colloque International TELECOM 2017 and 10èmes JFMMA, pp. 1–4. Rabat, Morocco (2017)

Riadi, A., Boulouird, M., Hassani, M.M.: 3D polarized channel modeling for multipolarized UCA-massive MIMO systems in uplink transmission. Jordan. J. Comput. Inform. Technol. (JJCIT), **5**(3), 231–243. https://doi.org/10.5455/jjcit.71-1563643647 (2019a)

Riadi, A., Boulouird, M., Hassani, M.M.: 3D geometric channel model for multi polarized ULA massive MIMO systems. J. Eng. Technol. **8**(2), 101–116 (2019b)

Riadi, A., Boulouird, M., Hassani, M.M.: Performance of massive-MIMO OFDM system with M-QAM modulation based on LS channel estimation. In: 2019 International Conference on Advanced Systems and Emergent Technologies (IC-ASET), pp. 74–78. Hammamet, Tunisia. https://doi.org/10.1109/ASET.2019.8870991 (2019c)

Riadi, A., Boulouird, M., Hassani, M.M.: ZF/MMSE and OSIC detectors for uplink OFDM massive MIMO systems. In: 2019 IEEE Jordan International Joint Conference on Electrical Engineering and Information Technology (JEEIT), pp. 767–772. Amman, Jordan. https://doi.org/10.1109/JEEIT.2019.8717380 (2019d)

Riadi, A., Boulouird, M., Hassani, M.M.: Least squares channel estimation of an OFDM massive MIMO system for 5G wireless communications. In: Bouhlel, M., Rovetta, S. (eds.) Proceedings of the 8th International Conference on Sciences of Electronics, Technologies of Information and Telecommunications (SETIT'18), vol. 2, pp. 440–450. SETIT: Smart Innovation, Systems and Technologies, vol. 147. Springer, Cham. https://doi.org/10.1007/978-3-030-21009-0_43 (2020a)

Riadi, A., Boulouird, M., Hassani, M.M.: Massive-MIMO configuration of multipolarized ULA and UCA in 5G wireless communications. In: The 3rd International Conference on Networking, Information Systems & Security (NISS2020), pp. 1–5, Marrakech, Morocco, March 31–April 2, ACM, NY, New York, USA. https://doi.org/10.1145/3386723.3387871 (2020b)

Rusek, F., Persson, D., Lau, B.K., Larsson, E.G., Marzetta, T.L., Edfors, O., Tufvesson, F.: Scaling up MIMO: opportunities and challenges with very large arrays. IEEE Signal Process. Mag. **30**(1), 40–60 (2012)

Schumacher, L., Pedersen, K.I., Mogensen, P.E.: From antenna spacings to theoretical capacities-guidelines for simulating MIMO systems. In: The 13th IEEE International Symposium on Personal, Indoor and Mobile Radio Communications, vol. 2, pp. 587–592. IEEE (2002)

Simon, M.K.: Digital communication over generalized fading channels: a unified approach to performance analysis (2002)

Telatar, E.: Capacity of multi-antenna Gaussian channels. Eur. Trans. Telecommun. **10**(6), 585–595 (1999)

Yang, S., Hanzo, L.: Fifty years of MIMO detection: the road to large-scale MIMOs. IEEE Commun. Surv. Tutorials **17**(4), 1941–1988 (2015)

Zheng, K., Suling, O., Yin, X.: Massive MIMO channel models: a survey. Int. J. Antennas Propag. **2014**, (2014)

Enhancing Wireless Transmission Efficiency for Wireless Body Area Sensor Networks Based on Transposing of Sensors

Rahat Ali Khan, Shahzad Memon, and Qin Xin

Abstract

Wireless Body Area Sensor Network (WBASN) has gained much significance in recent times and a lot of researches are being carried in this field because it is related to the betterment of humans. Sensing devices called sensors are used on the human body to observe several physiological parameters. These sensors are easily carried on the human body because of their size being very tiny and able to communicate the observed data wirelessly. Wireless communication has advantages over wired communication being very less complex to be carried. The transmission being wireless has to experience path loss, which makes it weaker and at the destination end the signal is not what needs to be; so in this work we present a new scheme for wearable on-body sensors so that the path loss can be reduced. The proposed scheme is based on transposing of the sensors used on the human body. Path loss is reduced based on its distance parameter using the Euclidean distance algorithm.

Keywords

WBASN • Sensors • Energy • Transposition • Distance • Path loss

List of Abbreviations

BAN	Body Area Networks
BS	Base Station
BSN	Body Sensor Networks
CF	Cost Function
CH	Cluster Head
Co-LAEEBA	Cooperative Link-Aware and Energy Efficient protocol for wire-less Body Area networks
CSI	Channel State Information
dB	Decibels
ECG	ElectroCardioGram
EDA	Euclidean Distance Algorithm
EE	Energy Efficiency
EECBR	Even Energy Consumption and Backside Routing
EEG	ElectroEncephaloGram
E-HARP	Energy-efficient Harvested Aware clustering and cooperative Routing Protocol for WBAN
EH-RCB	Energy Harvested-aware Routing protocol with Clustering ap-proach in Body area networks
ELR-W	Energy aware Link efficient Routing approach for WBANs
EMG	ElectroMyography
EMSMO	Enhanced Multi-objective Spider Monkey Optimization
EOCC	Energy Optimized Congestion Control
ICT	Information and Communication Technology
IoMT	Internet of Medical Things
IoT	Internet of Things
MEMS	Micro Electro Mechanical Systems
MH	Multi Hop
MHz	Mega Hertz

R. A. Khan (✉)
Department of Telecommunication Engineering, Faculty of Engineering and Technology, University of Sindh, Jamshoro, Pakistan
e-mail: rahat.khan@usindh.edu.pk

S. Memon
Department of Electronic Engineering, Faculty of Engineering and Technology, University of Sindh, Jamshoro, Pakistan
e-mail: shahzad.memon@usindh.edu.pk

Q. Xin
Faculty of Science and Technology, University of the Faroe Islands, Tórshavn, Faroe Islands
e-mail: qinx@setur.fo

M. Ben Ahmed et al. (eds.), *Emerging Trends in ICT for Sustainable Development*,
Advances in Science, Technology & Innovation, https://doi.org/10.1007/978-3-030-53440-0_17

MICS	Medical Implant Communication Service
NEMS	Nano Electro Mechanical Systems
OMNeT++	Objective Modular Network Testbed in C ++
pH	potential Hydrogen
SDN	Software Defined Networking
SH	Single Hop
SN	Sensor Node
SNR	Signal-to-Noise Ratio
SO	Square Odd
TARA	Temperature Aware Routing Algorithm
TDMA	Time Domain Multiple Access
WBAN	Wireless Body Area Network
WBASN	Wireless Body Area Sensor Network
WBSN	Wireless Body Sensor Network
WSN	Wireless Sensor Networks

1 Introduction

Developing trends in Information and Communication Technology (ICT) have been so rapid that within a small time researches have emerged to develop small-scale devices (Jafri et al. 2011; Khan et al. 2017). These small-scale devices known as Microelectromechanicalsystems (MEMS) and Nanoelectromechanical Systems (NEMS) are very small in size so that they can be fitted on a piece of cloth. These devices have multiple abilities like they can observe parameters, process the observed data, and to communicate the data wirelessly. These devices can be used to form Wireless Sensor Network (WSN). WSN is a type of network requiring small infrastructure or no infrastructure. WSN comprises of a large number of these devices, which form a network whenever there is a need (2006).

Wireless Body Area Sensor Network (WBASN) introduced by Van Dam in 2001 (Dam et al. 2001) is a subfield of WSN (Dam et al. 2001). WBASN are sometimes alternatively named as Wireless Body Area Network (WBAN), Wireless Body Sensor Network (WBSN), Body Area Networks (BAN), and Body Sensor Networks (BSN) (Khan et al. 2018, 2016). The WBASN is related to those sensors which can be used to monitor physiological changes. These sensors can be categorized as either being invasive or non-invasive (Jung et al. 2014; Yang et al. 2015; Al Rasyid et al. 2015; Muramatsu et al. 2015; Hess et al. 2015; Figueiredo et al. 1504; Sajatovic et al. 2015; Kölbl et al. 2014; Rathbun et al. 2015; Sugiura et al. 2017; Rao and Chiao 2015; Peter et al. 2016; Zheng et al. 2016; Xu and Hua 2016) as shown in Table 1.

The noninvasive sensors are also known as wearable sensors comprising of ECG, EEG, EMG, and other sensors while pacemaker, deep brain stimulator, and others come into the category of invasive sensors. Invasive sensors are also known as implantable sensors means a sensor is inserted into the human body by surgery.

The invasive sensors are those devices, which can be placed inside the human body while non-invasive sensors are those devices which can be placed on the human body. These sensors are required to be placed on a specific point so that they may observe the physiological changes in the human body as shown in Fig. 1. It can be observed in Fig. 1 that there are places where sensors can be placed for monitoring. These sensors have capabilities of sensing, processing, and then communicating the data through wireless medium making them easier for use because initially we had wired sensing devices which were complicated to use and carry around.

WBASN architecture (as shown in Fig. 2) works as: sensors communicate the sensed data to a device called as Base Station (BS) or sink node. The base station collects data from entire sensing nodes and communicates to the entities, which can further handle the data like if data is critical then it can send the signal to ambulance service. Simultaneously the data is sent to the doctor so that any timely action may be performed on arrival and also to the database to store the data.

These sensors operate on batteries which they carry within their hardware. As the size of the sensors is very small so they contain small batteries with them. As these sensors are related to human health and charging of the batteries is somehow a complicated process. So there is a need to design an algorithm in such a way that the sensors may operate for longer span of time. In this chapter, transposing is performed

Table 1 Noninvasive/wearable sensors invasive/implantable sensors

Invasive/implantable sensors	Noninvasive/wearable sensors
Pacemaker	Electrocardiogram (ECG)
Wireless capsule endoscope (electronic pill)	Glucose sensor
Deep brain stimulator	Electromyography (EMG)
Retina implants	Electroencephalogram (EEG)
Cochlear implants	Temperature
Electronic pill for drug delivery	Pulse oximeter
Implantable defibrillators	Blood pressure Oxygen, pH value

using Euclidean Distance Algorithm (EDA) to design and develop an algorithm to make the sensors work efficiently by saving maximum amount of energy.

WBASN can also be considered as the Internet of Medical Things (IoMT). Internet of Medical Things is a technology in which the Internet is the backbone for medical devices. Whatever data the devices record or observe is sent using the Internet to the authorities like in WBASN.

Further, this chapter is organized as: Related work is briefly described in Sect. 2, Mathematical modeling is presented in Sect. 3.

2 Related Work

Researchers in (Liu et al. 2017) proposed Time-Domain Multiple Access (TDMA) based protocol. Time is allocated to data in order to use energy in a better managed way. TDMA adjusts transmission. This minimization optimizes energy consumption.

Researchers in (Sahndhu et al. 2015) propose a routing protocol in which they have utilized intermediate nodes. These nodes have been used to gather to collect data from sensor nodes and after gathering data aggregation is performed. This aggregated data is then sent to the sink node. Threshold value has been set for the sensor nodes. If any of the sensors depletes its energy and its energy is approaching threshold then it will only transmit critical instead of sending normal data.

The authors in (Soontornpipit 2014) have measured data for radio propagation frequencies of 402 megahertz (MHz) to 405 megahertz. They have used Medical Implant Communication Service (MICS) band for evaluating path loss. Their simulation results are based on measurement for two scenarios. One is with furnished and second is with unfurnished room.

In (Kumari et al. 2019), researchers have focused on the cognitive fatigue level monitoring of vehicle drivers. They have named some parameters which they have described as

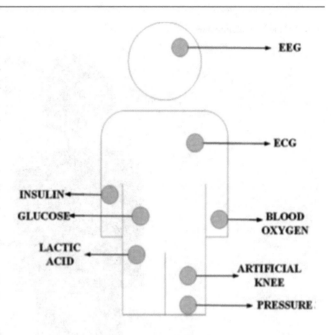

Fig. 1 Placement of sensor on the human body

contributors toward increasing fatigue levels. These parameters are driving comfort, type of vehicle being driven, health of drivers, drowsiness or insufficient sleep of drivers, continuous driving, road conditions, and traffic on road. These parameters affect driving and driver's behavior toward driving. The authors have tried to propose a method to detect a cognitive fatigue level so that the accidents caused by drivers during driving vehicles may be reduced. They have used K-means algorithm in the proposed system. They divided testing into two classes as pre-driving mode and post-driving mode. In the proposed algorithm Square Odd (SO) scanning technique is used for detecting any object. SO saves a significant amount of power as it is efficient and effective in detecting objects. It also switches sensors used between two modes, i.e., sleep mode and awake mode which is very useful is saving sensor power.

A new energy-efficient routing protocol for wireless body area sensor network has been proposed by researchers in (Yang et al. 2020). Their proposed system has involved scheduling of the data transmission also. The proposed system has taken into account several parameters like path loss, type of traffic, node energy, and others. In order to make reduction in traffic flooding, they have proposed to use channel competition. The proposed system first initializes, then there is setup phase in which routes for communication are selected and they are selected on the basis of having abundant energy. In this way, the sensor nodes consume lesser energy hence increasing network lifetime.

Rahat et al. in (Khan et al. 2018) have proposed a novel energy-efficient routing protocol for wireless body area sensor networks in which they have proposed to use eight

Fig. 2 WBASN architecture

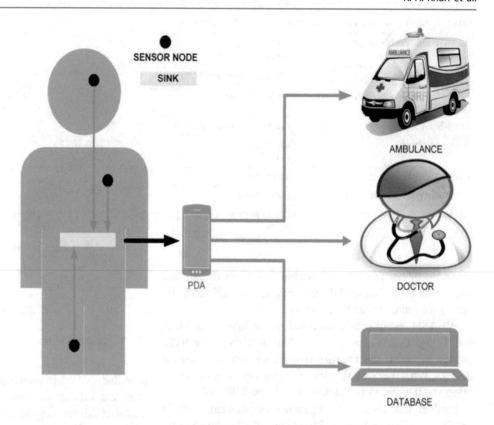

wearable sensor nodes. These sensor nodes are given coordinates in their x and y coordinates and sink node is placed at the center of the body. They have compared their proposed protocol with the one that was an already existing protocol. They have compared path loss, throughput, number of dead nodes, and residual energy. They have compared both protocols on the basis of the number of rounds each protocol needs to perform all of the aforementioned tasks. The comparison was clear that their proposed protocol performed better in terms of all the parameters. They also used forwarder nodes to gather data from other sensor nodes and forward to the outside world.

Authors in (Ahmed et al. 2020) have developed an Energy-Optimized Congestion Control based on Temperature Aware Routing Algorithm (EOCC-TARA). They have used Enhanced Multi-objective Spider Monkey Optimization (EMSMO) for Wireless Body Area Sensor Networks based on Software-Defined Networking (SDN). Their proposed algorithms reduce thermal effects in routing, provides communication which is free of congestion, and is energy efficient. Their proposed algorithm first considers thermal effects into routing and provides route selection to sensor nodes after the temperature parameter is taken into account the proposed algorithm looks for any congestion has occurred. Congestion avoidance scheme, link reliability, energy efficiency, and path loss are added based on EMSMO so that optimized routing is performed.

In paper (Amjad et al. 2020), researchers have tried to optimize energy efficiency (EE) of wireless body area sensor networks. They have formulated a problem. This problem is to optimize EE of the network from sensor nodes acting as transmitters to the aggregator sensor nodes. In this formulation channel state information (CSI) is not considered.

In (Archasantisuk and Aoyagi 2019), the authors have proposed transmission power control. They have done this due to variations in path loss. The proposed algorithm is based on motion-aware temporal correlation model. Their proposed algorithm is the combination of transmission power control and human motion classification. This combination has been performed to provide energy-efficient and reliable communication. Human movement is one of the major factors which has significance on the characteristics of WBASN communication. This movement is classified on the basis on the identification of the motion which is perform using received signal strength.

Zahid et al. in (Ullah et al. 2019) have proposed energy protocol named as "Energy-efficient Harvested-Aware clustering and cooperative Routing Protocol for WBAN (E-HARP)". The proposed algorithm uses a selection of Cluster Head (CH) based on multiple attribute technique. This is the main aim of the proposed algorithm in order to save the energy in the sensor nodes. As the round starts the proposed scheme selects CH. This selection is from all the sensor nodes that form a cluster. The CH selection criteria is based on the Cost Factor (CF). CF is to be calculated by various parameters like communication link signal-to-noise ratio (SNR), cumulative energy loss of the network, transmission power needed, and residual energy of the sensor node. The CF is to be calculated in every round so that the every CH should have equal load distribution and the network may remain stable. They have compared their results with several existing algorithms like (i) Energy Harvested aware Routing protocol with Clustering approach in Body area networks (EH-RCB), (ii) Cooperative Link Aware and Energy Efficient protocol for wireless Body Area networks (Co-LAEEBA), (iii) Even Energy Consumption and Backside Routing, and (iv) Energy-aware Link efficient Routing approach for WBANs (ELR-W).

A Soft computing approach based energy-efficient algorithm has been proposed by (Rakhee 2018). They have performed the balancing of data packets of all sensor nodes based on clustering with Ant Colony algorithm. They have used ant colony probabilistic function for data routing. They have used Objective Modular Network Testbed in C++ (OMNeT++) for simulation and have found their proposed system performing better as compared to conventional schemes.

3 Mathematical Modeling of Energy

In this section, the mathematical formulae is justifying why energy is the most important parameter in WBASN. There are two modes of communication in WBASN as either Single Hop (SH) or Multi-Hop (MH). Energy that is consumed during SH communication is mathematically represented as shown in Eq. 1

$$E_{SH} = E_{TX} \tag{1}$$

E_{TX} is the energy consumed by the transmission so this means that in SH communication there is only one parameter that consumes energy.

The second mode of communication is the MH which is mathematically represented as shown in Eq. 2

$$E_{MH} = S * N * \left[E_{TX} + (E_{DA} + E_{RX}) * \frac{N-1}{N}\right] \tag{2}$$

In Eq. 2, S is the packet size, N being the number of nodes used in multi-hop communication, energy used in data aggregation is represented by E_{DA} and E_{RX} is the energy that is used in receiving of the signal.

Power that is consumed in transmission is mathematically represented as shown in Eq. 3

$$E_{TX} = (E_{AMP} + E_{ELEC}) \times S \times D^2 \tag{3}$$

The cumulative energy that a sensor node consumes is given in Eq. 4

$$E_{NODE} = E_{TX} + E_{RETX} + E_{ACK} + E_{ACC} \tag{4}$$

E_{RETX} is energy consumed when the signal needs to be retransmitted, channel processing energy is represented as E_{ACC} and E_{ACK} in the power consumed when an acknowledgement packet is transmitted.

From Eq. 4, it can be observed that out of 4 parameters 3 are related to transmission. It makes communication more important in terms of energy consumption. Sensors in WBASN have capabilities to perform functions like sensing, processing and to set the observed data wirelessly so these functionalities consume energy. In Fig. 3, it can be observed that energy consumed in communication is more than that for the other two parameters (Khan et al. 2017). So communication needs to be accurate and reliable. If for instance communication is not accurate then the data needs to be retransmitted. This retransmission again consumes energy again. This will drain the battery of the sensor much faster. As we know that the sensors that are used in WBASN are of a tiny size so they need to be efficient enough so that they may operate for a longer span of time. As the data transmission is wireless in nature so they are several parameters which affect it.

In wireless communication, path loss is a parameter which has a significant impact on communication. It is a parameter that makes transmission weaker and is denoted mathematically as

$$PL_{(f,D)} = PL_o + 10 \times n \times log_{10}\frac{D}{d_o} + S \tag{5}$$

$$PL_o = 10 \times log_{10}\left[\frac{4 \times \pi \times D \times f}{c}\right]^2 \tag{6}$$

$$PL_{(f,D)} = 10log_{10}\left[\frac{4\pi Df}{c}\right]^2 + 10nlog_{10}\frac{D}{d_o} + S \tag{7}$$

4 Proposed Algorithm

In this section, the proposed algorithm is defined. In the proposed scheme, eight sensor nodes are to be used on the human body and one sink node that has to be placed on the center of the human body.

The proposed scheme is to be compared with the already existing scheme M-ATTEMPT (Javaid et al. 2013) and then the comparison is performed that which algorithm performs better. Path loss is based on two parameters (i) frequency and (ii) distance of the sensors and base station. In the

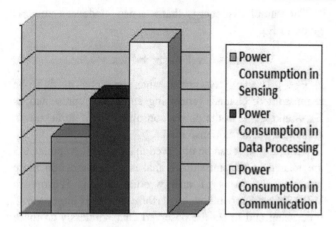

Fig. 3 Comparison of power consumption

proposed scheme, the main focus is on the distance parameter in order to reduce path loss in wireless communication.

$$D_i = \sqrt{(N_{xi} - S_x)^2 + (N_{yi} - S_y)^2} \qquad (8)$$

Equation 8 represents the EDA of the *ith* sensor node. N_{xi} represents the ith node's x coordinate, and N_{yi} is the *ith* node's y coordinate value. X and y coordinates of the base station are given as S_x and S_y.

4.1 Sensor Position Deployment

The proposed scheme uses eight sensor nodes. So we propose the coordinated of the sensors as shown in Table 2. These sensor nodes are given distances described in centimeters (cm) in x and y axes of the human body. Height of the human body is given y coordinate and width as y coordinate. Base station is proposed to be at 0.40 m and 0.87 m as x and y coordinates, respectively.

5 Simulation Results

In this section, the results of both schemes are given and described in detail. The simulation tool used here is MATLAB. All the required parameters are written into it as a code and the results are discussed.

5.1 Distance Comparison

First the distance of both the schemes is computed. In Table 3, individual distances of entire sensor nodes of M-ATTEMPT (Javaid et al. 2013) scheme are presented, d1 is the distance of sensor node 1 and the series goes on to d8.

The distances of the proposed scheme are presented in Table 4.

Distance comparison of individual sensor nodes is presented in Fig. 4. Distance values of sensor node 1, 2, 3, and 5 of M-ATTEMPT (Javaid et al. 2013) scheme are less than the proposed scheme. The proposed scheme has lesser distance values at sensor nodes 4, 6, 7, and 8. In Fig. 5, the cumulative distance comparison is presented in which it can be observed that the proposed scheme has obtained far lesser distance as compared to M-ATTEMPT (Javaid et al. 2013). The total distance of M-ATTEMPT (Javaid et al. 2013) is 341.8895 cm while the total distance of the proposed scheme is 336.3679 cm.

The simulation results of ATTEMPT (Javaid et al. 2013) are presented in Fig. 6. The simulation is distance versus path loss. SN-8 is denoted by black colored dot, SN-7 is represented by yellow color, SN-2 by green color, SN-1 is presented in red, magenta represents SN-6, and blue is for SN-3, SN-4, and SN-5.

The simulation results of the proposed scheme are shown in Fig. 7. SN-8 and SN-4 are in black colored dot, SN-1 is in red color, SN-2 is given in green color, blue color represents

Table 2 Sensor coordinates

Sensor Node (SN) number	X Coordinate (m)	Y Coordinate (m)
SN-1	0.25	0.10
SN-2	0.40	0.15
SN-3	0.27	0.55
SN-4	0.45	0.56
SN-5	0.60	0.56
SN-6	0.17	0.56
SN-7	0.24	0.50
SN-8	0.40	0.80

Table 3 Individual distances of M-ATTEMPT scheme

Distance							
d1	**d2**	d3	d4	d5	d6	d7	d8
36.0555 cm	28.2843 cm	31.6228 cm	31.6228 cm	31.6228 cm	41.2311 cm	60.8276 cm	80.6226 cm

Table 4 Individual distances of proposed scheme

Distance							
d1	**d2**	**d3**	d4	d5	d6	d7	d8
78.4474 cm	72.0000 cm	34.5398 cm	31.4006 cm	36.8917 cm	40.3113 cm	35.7771 cm	7.000 cm

Fig. 4 Comparison of individual distances

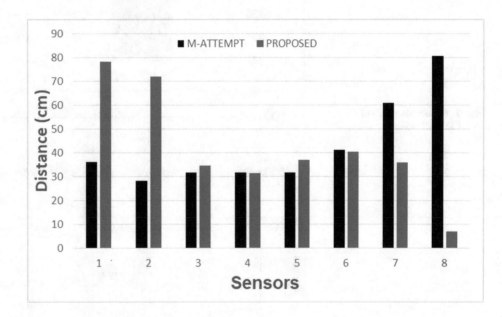

SN-3, yellow is for SN-5, SN-6 is represented by magenta color, and SN-7 is denoted by cyan.

The path loss results of both the schemes are presented in Table 5.

It is observed that path losses PL1, PL2, PL3, and PL5 of ATTEMPT (Javaid et al. 2013) are lesser as compared and path loss values PL4, PL6, PL7, and PL8 of the proposed scheme are lesser than that of the ATTEMPT (Javaid et al. 2013) scheme. The cumulative path loss of ATTEMPT (Javaid et al. 2013) scheme is 5926 decibels (dB) and 5726 dB is the cumulative path loss is of the proposed scheme. There is a difference of 180 dB. The proposed scheme performs better by the transposition suggested making the sensors perform for a longer amount of time and making them consume less energy as the distance between the base station and the sensors has been reduced.

5.1.1 Conclusion

In this work, the transposition scheme has been described that has to be used in Wireless Body Area Sensor Networks is proposed. WBASNs are networks related to physiological parameters and their monitoring on a continuous basis and the sensor need to be wireless so that there is no complexity in carrying them. So these sensors need to perform for a much longer span of time durations so that continuous monitoring is uninterrupted. The main aim of this research is to focus on the sensing points so that if sensors are placed on these positions then there should be two things that need to be achieved (i) No comprise on sensing and (ii) least distance between the sensor and the base station. In the proposed scheme, the selection of the sensors has been proposed to be used in any WBASN architecture so that sensors will perform better. The proposed scheme uses a Euclidean Distance Algorithm for observing the sensor

Fig. 5 Comparison of
cumulative distance

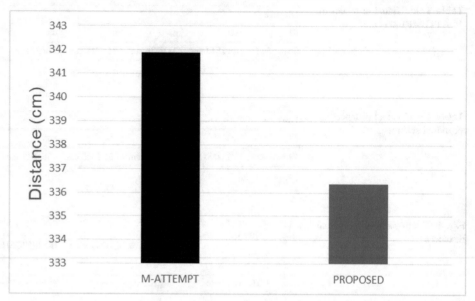

Fig. 6 Simulation result of
M-ATTEMPT scheme

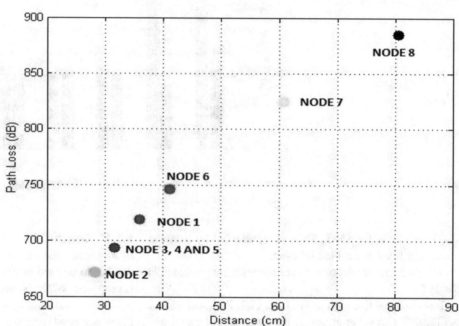

Fig. 7 Simulation result of the
proposed scheme

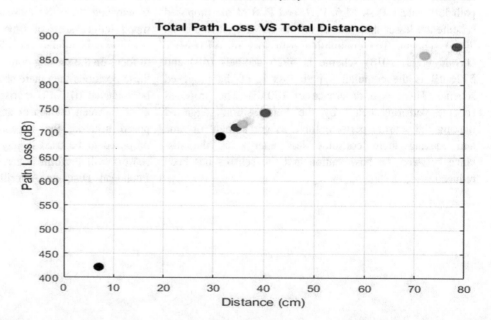

Table 5 Path loss comparison

Scheme	Path Loss Values								
	PL1 (dB)	PL2 (dB)	PL3 (dB)	PL4 (dB)	PL5 (dB)	PL6 (dB)	PL7 (dB)	PL8 (dB)	PLT (dB)
M-ATTEMPT scheme	719.1	671.7	693.3	693.3	693.3	745.8	825.2	884.6	5926
Proposed scheme	878.5	860.6	710.6	691.9	723.6	741.2	717.5	421.9	5746

distances. After the successful implementation of the sensor positions, the simulations were performed. The distance as well as the path loss comparison were performed on MATLAB. The simulation results were obtained and it was observed that the proposed scheme was able to achieve lesser distance when compared to the existing scheme making the proposed scheme better. The decrease in distance did allow the proposed scheme to obtain lesser path loss because path loss is a function of distance.

References

Ahmed, O., Ren, F., Hawbani, A., Al-sharabi, Y.: Energy optimized congestion control-based temperature aware routing algorithm for software defined wireless body area networks. IEEE Access **8**, 41085-41099 (2020)

Al Rasyid, M.U.H., Lee, B.H., Sudarsono, A.: Implementation of body temperature and pulse oximeter sensors for wireless body area network. Sens. Mater. **27**(8), 727–732 (2015)

Amjad, O., Bedeer, E., Ali, N.A. and Ikki, S.: Robust energy efficiency optimization algorithm for health monitoring system with wireless body area networks. IEEE Commun. Lett. (Early Access) (2020)

Archasantisuk, S., Aoyagi, T.: Transmission power control using human motion classification for reliable and energy-efficient communication in WBAN. IEICE Trans. Commun. **102**(6), 1104–1112 (2019)

Figueiredo, I.N., Leal, C., Pinto, L., Figueiredo, P.N., Tsai, R.: An elastic image registration approach for wireless capsule endoscope localization. arXiv preprint arXiv:1504.06206 (2015)

Hess, P.L., Al-Khatib, S.M., Han, J.Y., Edwards, R., Bardy, G.H., Bigger, J.T., Buxton, A., Cappato, R., Dorian, P., Hallstrom, A., Kadish, A.H.: Survival benefit of the primary prevention implantable cardioverter-defibrillator among older patients: does age matter? An analysis of pooled data from 5 clinical trials. Circul. Cardiov. Qual. Outcomes **8**(2), 179–186 (2015)

Jafri, I.H., Soo, M.S., Khan, R.A.: ICT in distance education: improving literacy in the province of Sindh Pakistan. The Sindh Uni. J. Educat.-SUJE **40**, 37–48 (2011)

Javaid, N., Abbas, Z., Fareed, M.S., Khan, Z.A., Alrajeh, N.: M-ATTEMPT: A new energy-efficient routing protocol for wireless body area sensor networks. Proc. Comput. Sci. **19**, 224–231 (2013)

Jung, S.J., Shin, H.S., Chung, W.Y.: Driver fatigue and drowsiness monitoring system with embedded electrocardiogram sensor on steering wheel. IET Intel. Transport Syst. **8**(1), 43–50 (2014)

Khan, R.A., Memon, S., Zardari, S., Dhomeja, L.D., Usman, M.: Transposition technique for minimization of path loss in wireless on-body medical sensors. Sindh Uni. Res. J.-SURJ (Science Series), **48**(4) (2016)

Khan, R.A., Awan, J.H. Memon, S., Zafar, H.: Influence of cell phone in ICT sector of Pakistan towards advancement. Pakistan J. Comput. Inf. Sys. (PJCIS), **2**(1), 17–30 (2017)

Khan, R.A., Memon, S., Awan, J.H., Zafar, H., Mohammadani, K.H.: Enhancement of transmission efficiency in wireless on-body medical sensors. Eng. Sci. Technol. Int. Res. J. **1**(2), 16–21 (2017)

Khan, R.A., Mohammadani, K.H., Soomro, A.A., Hussain, J., Khan, S., Arain, T.H., Zafar, H.: An energy efficient routing protocol for wireless body area sensor networks. Wireless Pers. Commun. **99**(4), 1443–1454 (2018)

Kölbl, F., N'Kaoua, G., Naudet, F., Berthier, F., Faggiani, E., Renaud, S., Benazzouz, A., Lewis, N.: An embedded deep brain stimulator for biphasic chronic experiments in freely moving rodents. IEEE Trans. Biomed. Circuits Syst. **10**(1), 72–84 (2014)

Kumari, R., Nand, P., Astya, R., Chaudhary, S.: Implementation of square-odd scanning technique in WBAN for energy conservation. In: Khanna, A., Gupta, D., Bhattacharyya, S., Snasel, V., Platos, J., Hassanien, A. (eds.) International Conference on Innovative Computing and Communications, vol. 1059, pp. 1–10. Advances in Intelligent Systems and Computing 2019. Springer, Singapore (2019)

Liu, B., Yan, Z., Chen, C.W.: Medium access control for wireless body area networks with QoS provisioning and energy efficient design. IEEE Trans. Mob. Comput. **16**(2), 422–434 (2017)

Muramatsu, D., Koshiji, F., Koshiji, K., Sasaki, K.: Effect of user's posture and device's position on human body communication with multiple devices. In: 2015 International Conference on Electronics Packaging and iMAPS All Asia Conference (ICEP-IAAC), pp. 124–127. IEEE. Japan (2015)

Pathan, A.S.K., Lee, H.W., Hong, C.S.: In: 2006 8th International Conference Advanced Communication Technology 2006, vol. 2, pp. 1043–1048. IEEE, Korea (2006)

Peter, S., Pratap Reddy, B., Momtaz, F., Givargis, T.: Design of secure ECG-based biometric authentication in body area sensor networks. Sensors **16**(4), 570 (2016)

Rakhee, Srinivas M.B.: Energy efficiency in load balancing of nodes using soft computing approach in WBAN. In: Yadav, N., Yadav, A., Bansal, J., Deep, K., Kim, J. (eds.) Harmony Search and Nature Inspired Optimization Algorithms. Advances in Intelligent Systems and Computing 2018, vol 741, pp. 423–430. Springer, Singapore (2018)

Rao, S., Chiao, J.C.: Body electric: wireless power transfer for implant applications. IEEE Microwave Mag. **16**(2), 54–64 (2015)

Rathbun, D.L., Jalligampala, A., Stingl, K., Zrenner, E.: To what extent can retinal prostheses restore vision? In: 2015 7th International IEEE/EMBS Conference on Neural Engineering (NER). pp. 244–247. IEEE. France (2015)

Sahndhu, M.M., Javaid, N., Imran, M., Guizani, M., Khan, Z.A., Qasim, U.: BEC: a novel routing protocol for balanced energy consumption in wireless body area networks. In: International Wireless Communications and Mobile Computing Conference (IWCMC) 2015, pp. 653–658. IEEE, Dubrovnik (2015)

Sajatovic, M., Levin, J.B., Sams, J., Cassidy, K.A., Akagi, K., Aebi, M. E., Ramirez, L.F., Safren, S.A., Tatsuoka, C.: Symptom severity, self-reported adherence, and electronic pill monitoring in poorly adherent patients with bipolar disorder. Bipolar Disord. 17(6), 653–661 (2015)

Soontornpipit, P.: Study of 403.5 MHz path loss models for indoor wireless communications with implanted medical devices on the human body. ECTI Trans. Elec. Eng. Elect. Commun. 10(2), 173–178 (2014)

Sugiura, T., Imai, M., Yu, J., Takeuchi, Y.: A low-energy application specific instruction-set processor towards a low-computational lossless compression method for stimuli position data of artificial vision systems. J. Inf. Proc. 25, 210–219 (2017)

Ullah, Z., Ahmed, I., Khan, F.A., Asif, M., Nawaz, M., Ali, T., Khalid, M., Niaz, F.: Energy-efficient Harvested-Aware clustering and cooperative Routing Protocol for WBAN (E-HARP). IEEE Access 7, 100036–100050 (2019)

Van Dam, K., Pitchers, S., Barnard, M.: Body area networks: Towards a wearable future. In: Proc. WWRF kick off meeting, pp. 6–7. Germany (2001)

Xu, H., Hua, K.: Secured ECG signal transmission for human emotional stress classification in wireless body area networks. EURASIP J. Inf. Sec. 5 (2016)

Yang, Y., Chae, S., Shim, J., et al.: EMG sensor-based two hand smart watch interaction. In: Adjunct proceedings of the 28th annual ACM symposium on user interface software & technology, pp. 73–74. ACM, New York (2015)

Yang, G., Wu, X.W., Li, Y., Ye, Q.: Energy efficient protocol for routing and scheduling in wireless body area networks. Wireless Netw. 26(2), 1265–1273 (2020)

Zheng, G., Fang, G., Shankaran, R., Orgun, M.A., Zhou, J., Qiao, L., Saleem, K.: Multiple ECG fiducial points-based random binary sequence generation for securing wireless body area networks. IEEE J. Biomed. Health Inf. 21(3), 655–663 (2016)

New Method to Detect the Congestion for Green Networking in MANET

Abdellah Nabou, My Driss Laanaoui, Mohammed Ouzzif, and Mohammed-Alamine El Houssaini

Abstract

Mobile Ad hoc Network (MANET) is defined as collection of mobiles devices that use the wireless channel to ensure the connectivity between them. MANET has many constraints in its functions due to infrastructure-less and absence of a central admin. Optimized Link State Routing Protocol (OLSR) is table-driven MANET routing protocol dedicated to the large network density. However, it can affect in its performances by many factors, one of them is the congestion that can increase the energy consumption, lost packets, the End-to-End Delay (EED) and can minimize the throughput of the protocol. In this paper, we propose a new method which applied the concept of Normality Test that can be used in other domains like the statistic in order to detect the congestion in MANET and reduce power consumption which is considered as a step toward green communication. Our method achieves the goal of detection without any modification in the routing protocol or managing other additional control messages. The Shapiro–Wilk (W) is the method that is used to analyze the results of Throughput and End-to-End delay, which can be affected by the congestion. To test the effectiveness of our proposed method, we simulated different number of nodes in two scenarios by changing the area of simulation.

A. Nabou (✉) · M. Ouzzif
RITM Laboratory, CED Engineering Sciences, EST, ENSEM
Hassan II University of Casablanca, Casablanca, Morocco
e-mail: a.nabou@ensem.ac.ma

M. D. Laanaoui
Department of Computer Faculty of Sciences and Technology,
Cadi Ayyad University, Marrakech, Morocco
e-mail: d.laanaoui@uca.ma

M.-A. El Houssaini
ESEF Chouaib Doukkali University, El Jadida, Morocco
e-mail: elhoussaini.m@ucd.ac.ma

Keywords

MANET • OLSR • Congestion • Shapiro–Wilk • Throughput • End-to-end delay

1 Introduction

A Mobile Ad hoc Network (MANET) is a group of wireless equipment like smartphone, computers, tablets, or any other wireless device that communicate between them without any fixed infrastructure using a shared wireless multi-hop channel (Dimitris Kanellopoulos 2019). MANET is very useful in many applications such as disaster recovery, search, and military missions. MANET is characterized by the mobility, limited bandwidth, energy constraints, and lack of security. For these reasons, MANET uses specific ad hoc routing protocols to ensure the communications between all the nodes in the network without any intervention by either central equipment or network admin. MANET routing protocols are classified into three main categories: reactive or on-demand routing protocols, proactive table-driven routing protocols, and hybrid routing protocols that combine both the reactive and proactive protocols.

Optimized Link State Routing Protocol (OLSR) is the most popular MANET proactive routing protocol. The protocol is an optimization of the classical link state algorithm tailored to the requirements of a mobile wireless LAN (Clausen and Jacquet 2003). OLSR uses new concept of Multipoint Relays (MPRs). These selective nodes are responsible for forward broadcast messages during the flooding process. Due to shared medium of the MANET and its characteristics complicate Quality of Service (QoS) provision, and impose various challenges in the design of congestion control. Like all other MANET protocols, OLSR can be affected by the congestion due to the higher number of packets sent in the network compared to the network capacity (number of packets a network that can handle)

M. Ben Ahmed et al. (eds.), *Emerging Trends in ICT for Sustainable Development*,
Advances in Science, Technology & Innovation, https://doi.org/10.1007/978-3-030-53440-0_18

(Maheshwari et al. 2014). The situation of congestion can decrease the performance of any MANET routing protocol by long delay to select an alternate new path, high overhead of the network, and higher number of packets lost.

In this paper, we propose a new method to detect the congestion in MANET without any modification or additional control messages. Our approach count on the concept of Normality Test that is used in other domains like statistics in order to analyze the results of data in normal distribution or not. There are many methods to test the normality, in our case, in order to detect the congestion in MANET, we choose the Shapiro–Wilk (Shapiro and Wilk 1965) method thanks to its effectiveness and powerfulness for small numbers of sample size n ($n \leq 50$). Rest of the paper is organized as follows: Sect. 2 presents related work, and defines the OLSR routing protocol, Sect. 3 presents the effect of congestion in MANET followed by our proposed method to detect the congestion in Sect. 4. The results and analysis are discussed in Sect. 5, and finally, the conclusions and future work in Sect. 6.

2 Related Work

Vadivel (2017) Resolve the congestion and route errors by proposing an Adaptive Reliable and Congestion Control Routing Protocol (ARCCRP) that use bypass route selection in MANETs, after detection of the congestion on a local outgoing link L. The node that detects the congestion calculates the multipath routes to destinations for which the path contains link L by using a traffic splitting function. In addition, when a node cannot resolve the congestion, it uses the congestion indication bit to signal its neighbors.

Sirajuddin et al. (2016) Propose a new TCP congestion control scheme called by TCP-R for detecting the congestion and they propose Ad hoc Distance Vector with Congestion Control (ADV-CC) that combines AODV routing protocol with the proposed congestion mechanism in order to control congestion in MANET.

Holland and Vaidya (2002) Use the Explicit Link Failure Notification (ELFN) in order to improve the performance of TCP that can be affected in its throughput due to link failure and congestion. This technique sends ELFN message to the source node when a route failure occurs, after the reception of this message, the source node freezes its current state in terms of windows size and timers. The source returns to its normal state when it received ACK message for the new request of the source node.

Sundaresan et al. (2005) Present a new reliable transport layer protocol for ad hoc networks called by Ad hoc Transport Protocol (ATP) this new protocol separates the congestion control and reliability at the transport layer by applying the network feedback mechanism to ensure the reliability.

The authors (Ikeda et al. 2012) use congestion control for multi-flow traffic in wireless mobile ad hoc networks, and they consider OLSR routing protocol for evaluating the throughput and congestion window of a MANET in two scenarios. The first one depends on different number of nodes and the second test depends on the area size by sending multiple flows transported over UDP or TCP; the simulation was done by NS3.

Rath et al. (2017) propose Real Time Token Bucket (RTB) method by upgrading traffic shaping mechanisms in TCP/IP protocol for congestion control in real-time traffic, this method checks the incoming traffic flow and streamlines it before inflowing to the network in order to control the rate of sending. The simulation results of RTB method show that the PDR of network increases when the density of network increases too, also reduces the average of End-To-End delay.

Chen et al. (2007) Suggest a Congestion-Aware Routing protocol for Mobile ad hoc networks (CARM) that uses two mechanisms the Weighted Channel Delay (WCD) and Effective Link Data-rate Categories (ELDCs). All these mechanisms assume to improve the routing protocol adaptability to congestion either by selecting high throughput routes with low congestion or by avoiding of mismatched link data-rate routes. The simulation of the proposed mechanism demonstrates adaptability to congestion compared to DSR protocol.

Yang et al. (2009) Propose the F-ECN mechanism in order to discriminate a packet loss whether it is due to wireless link failure or to the congestion. Fuzzy logic controller based on queue length and packet arrival rate has been deployed to measure congestion. The performance of F-ECN takes a tradeoff between the throughput and the delay time.

Soundararajan and Bhuvaneswaran (2012) Present a new approach Multipath Load Balancing and Rate Based Congestion Control (MLBRBCC) for mobile ad hoc Network. In this method, the source node sends the routing packet to the destination through intermediate nodes based on the percentage of channel utilization and queue length estimation. In other to verify the congestion status, when the destination receipts the packet it checks the rate information in the packet's IP header fields and sends an acknowledgement packet as feedback to the sender.

Kopparty et al. (2002) Develop a scheme called by Split TCP that separates the TCP functions of congestion control and reliable packet delivery and they propose to achieve this by introducing proxy agents that split TCP into localized segments. The proxy agents facilitate the separation of the congestion control and the end-to-end reliability semantics of TCP.

The authors (Pillai and Singhai 2020) propose a mechanism to control the congestion in AODV routing protocol by modifying local route establishment process and using the mechanism of fuzzy-based reliable node detection. They proposed different scenarios to simulate the effectiveness of their method by varying the number of nodes, the metrics used for analysis are PDR, Normalized Routing Load (NRL), and the Throughput. To summarize, they observe an improvement in these metrics compared to existing congestion control implemented in on-demand routing protocol.

The researchers (Haveliwala et al. 2020) discuss TCP Westwood DCC in order to perform bandwidth estimation and setting up the congestion window, especially for link failure scenario. They analyzed the performance of Throughput, PDR, and average delay compared to other algorithms. The results of simulation show that when they increase error rate, the TCP Westwood DCC performs well among other compared algorithms.

The authors (Sharma et al. 2020) use two types of safety messages: beacon and event-driven. The first one provides the necessary information about the neighbor of node's status and second one for informing if a danger has been detected. This method is used for the network VANET in order to control the congestion vehicular. The authors analyze the results of PDR, EED, and throughput and transmit frequency to compare the effectiveness of their proper method to DCC_plain mechanism.

2.1 OLSR Routing Protocol

The Optimized Link State Routing Protocol (OLSR) operates as table-driven, proactive routing protocols that means all nodes of network exchange topology information between them in regular way. OLSR is considered as the most suitable MANET routing protocol for the large network density (Clausen and Jacquet 2003) thanks to its new concept that is used for forwarding the control traffic in the network. The new concept is Multipoint relays (MPR) which are selective nodes by their 1-hop neighbors. In OLSR, each node selects its MPRs node from its 1-hop neighbor set by using specific algorithms in order to reach its 2-hop neighbor. MPRs provide an efficient mechanism for flooding control traffic by reducing the number of re-transmissions required (Clausen and Jacquet 2003).

The nodes that use OLSR as MANET routing protocol must periodically exchange two essential control messages. The first control message is HELLO message, it is generated by each node every two seconds and forwarded to the set of 1-hop neighbor. HELLO message contains a list of 1-hop and 2-hop neighbors and its MPR set. The second control message is TC message generated and forwarded only by the MPR nodes. TC messages advert the link states or the topology declaration and they are diffused with the purpose

of providing each node in the network with sufficient link state information to allow route calculation. The routing table in OLSR protocol has been building from both HELLO and TC control messages by selecting the route that has short distance (number of hops) to the destination. The OLSR nodes can send MID-messages to declare the presence of multiple interfaces on a node and HNA messages in order to inject external routing information into an OLSR MANET for casing nodes with non-MANET interfaces.

3 Congestion in Manet

The congestion is defined as a situation in which too many packets are present in a part of subnet in network communication (Maheshwari et al. 2014). The congestion in MANET may fall when the number of sending packets in the networks is greater than the capacity of the wireless channel. The congestion in MANET have negative effects on the performance of any MANET routing protocols, for example, the congestion leads to packet losses, degrades the bandwidth, consumes the energy, and wastes the time. Due to the shared wireless channel of MANET, the congestion will not overload the mobile nodes; however, it has an effect on the entire coverage area (Kopparty et al. 2002).

OLSR routing protocol uses the concept of MPR nodes to optimize the overhead of the network. However, the congestion can be found in some situations, especially when the number of MPR nodes increases due to a higher number of nodes or to small size in area of communication. The effect of the congestion on OLSR protocol can be presented as follows:

Higher Lost Packet: Due to the mechanism of the congestion control, the sender reduces the number of sending packets, or intermediates nodes can drop the packet or both, the techniques can be executed. The result is many packet losses with minimum throughput.

Long Delay: The standard technique used by MANET routing protocol in the congestion is selecting an alternative path, on the other hand, the prevailing on-demand routing protocol can increase the delay of the route searching process.

High overhead: OLSR broadcasts HELLO and TC messages in periodic way, when the number of nodes increases, the overhead of the network increases too. If the multipath routing is utilized, it needs additional effort for upholding the multipaths regardless of the existence of alternate route (Maheshwari et al. 2014).

In the next section, we will present our idea to detect the congestion by using the concept of Normality Test.

4 Proposed Work

In our work, we propose a new method for detecting the congestion in MANET by using the Normality Test that is used in statistics domain, in order to determine if a data set is well modeled by a normal distribution. Lots of work suggest adding or to modify the routing protocol either in physical layer or in TCP or UDP protocols. Our approach of detection proposes to use Shapiro–Wilk formula in order to verify that the empirical distribution of observations is compatible with a specific theoretical distribution.

The Normality Tests are used generally in statistics to check whether real data follow a normal distribution or not and to compute how likely it is for a random variable underlying the data set to be normally distributed. The assumption of normality needs to be checked for many statistical procedures. In our method, we propose to use Shapiro–Wilk test (Shapiro and Wilk 1965). It is very popular and powerful for the small numbers of sample size n ($n \leq 50$). Some researchers recommend the Shapiro–Wilk test as the best choice for testing the normality of data (Kopparty et al. 2002). The Shapiro–Wilk test is based on the W statistic that is calculated as follow:

$$W = \frac{\left[\sum_{i=1}^{\left[\frac{n}{2}\right]} a_i \left(x_{(n-i+1)} - x_{(i)}\right)\right]^2}{\sum_i (x_i - \bar{x})^2} \quad (1)$$

where
$x_{(i)}$: corresponds to the series of sorted data;
$\left[\frac{n}{2}\right]$: is the whole part of the report $\frac{n}{2}$;
$\bar{x}$ is the average of the sampling with

$$\bar{x} = \frac{\sum x_i}{n} \quad (2)$$

a_i are constants generated from the average and the matrix of variance and covariance of quantities for a sampling of size n that followed the Normality Test, (Shapiro and Wilk 1965) present all values of a_i for the W test.

The results W can, therefore, be interpreted as the coefficient of determination between the series of the sampling generated from the Normality Test called by $W_{Critical}$ and the empirical sampling obtained from the data. The higher W is, the more the compatibility with the Normality Test is credible. The values of $W_{Critical}$ that resumes all normal values with different risks α and effectiveness are reading from Shapiro–Wilk table (Shapiro and Wilk 1965). We reject the normality of W calculated when

$$W_{Calculated} < W_{Critical} \quad (3)$$

To detect the congestion in MANET that can affect the performance of Throughput and EED due to many packets

losses, we select to test the normality of Throughput metric that is defined as the total of bites that is received successfully by the destination node in a specified time. In addition, we test Normality for EED delay that is defined as the time taken by the packet to achieve the destination.

We propose two scenarios of simulation related to areas size that are considered as one of MANET congestion reasons.

The first scenario simulates different number of nodes in normal area of network (1000 * 1000 m), and in the second, we are reducing the area size to 500 * 500 m.

The results W calculated of Throughput and EED for different sampling n have two significations in our analysis:

- If calculated W is greater than $W_{Critical}$, normality is accepted and there is no congestion.
- If calculated W is less than $W_{Critical}$, Normality Test is rejected and we detect the congestion in the network.

Our proposed method checks just if there is congestion in the network without any intervention to resolve it. The detection of the congestion in MANET by the Normality Test can be considered as a simple method for any MANET routing protocol without any modification in their algorithms.

5 Simulation and Results

To apply the Normality Test in MANET for detection of the congestion, we simulate different number of nodes in two scenarios by NS3 Simulator that is free software targeted primarily for research and educational use in Internet subject. The first one we select is normal area of simulation (1000 *1000 m). On the other hand, the second scenario the nodes are sending the routing packets in the small area size (500*500 m). For the mobility of nodes, we select the Random Waypoint Mobility Model (RWPM) (Hyytiä and Virtamo 2007) with no pause time for changing the direction or speed of node, the second parameter of RWPM is speed node that is fixed at maximum value of 5 m/s. Table 1 describes all other parameters used in the simulation.

5.1 Shapiro–Wilk of Throughput

The results obtained from the simulation for different number of nodes show that the W of 20 nodes has higher values than the critical W for the sampling of 10, 15, and 20 s of simulation. Thereafter the W of 20 nodes will decrease for the sampling of 25 s. However, this value will be close to the critical value of Shapiro–Wilk, by the end the W of 20 nodes

Table 1 Parameters of simulation in NS3

N°	Parameter	Value
1	PC simulator	Dell Intel Xeon ® CPU ES-2407 0 @ 2.20 GHz 7GiB
2	Simulator	NS3 (3.25)
3	Number of nodes	20,50,80, and 100
5	Simulation time	100
5	Wi-Fi mode	Ad-hoc
6	Wi-Fi Rate	2 MB
7	Mobility model	Random Waypoint mobility model
8	N. of Source/Sink	10
9	Sent data rate	2.048Kbps
10	Packet size	64 Bytes
11	Number of packets send	4
12	Protocols used	OLSR
13	Connection type	TCP
14	Node speed	5
15	Pause time	0
16	Network size	1000 × 1000 m
17	The output per second	Throughput, EED

confirms that there is no congestion in the traffic due to normal value of Throughput and the small number of nodes in big area of the network.

For W of 50 nodes, we remark that for the sampling of 10 and 15 s the W calculated has higher value compared to the Critical W. In this case, the traffic is normal without any congestion. We can explain that by the nature of OLSR protocol which needs approximately 10 s to build the topology of the network. After 20 s of simulation, we show that the value of W becomes more and more decreasing. We explain this decrease by the higher packets send in the network after the building of routing table for each node, also the higher number of nodes can increase the overhead of the network. In the results, we detect the congestion in the network from the 20 s of starting the communication.

For W of 80 and 100 nodes in area size of 1000*1000 m, the results are rather convergent. In the first 10 sampling of simulation, the traffic of communication is normal due to the higher value of W compared to the critical value. The explanation of the no congestion of communication in this period of simulation is due to time taken by the OLSR routing protocol for building the topology of the network, because the TC messages were generated by the MPRs nodes each 5 s, and the first TC message generated and forwarded in the beginning takes 10 s. The congestion of the traffic can be detected for the W of 80 nodes after 15 s of simulation. On the other hand, the detection of congestion for 100 nodes can be detected after 20 s. The congestion of the network for

either 80 or 100 nodes is caused by the higher number of the nodes in the network that mean diminution in the Throughput. In addition, we remark that the W between 50 and 80 has the less value of W for 50 nodes compared to 80 and 100 nodes, is thanks to a new concept of MPR nodes that used in OLSR protocol which minimizes the overhead of the traffic for the large density of the network.

In this second scenario, we reduced the area of the network to 500*500 m in order to detect the congestion by our proposed method. The first remark is the congestion was detected for all number of nodes, with one exception in the first sampling of 10 s for the number of 50 nodes. In addition, when the time of simulation increases the W calculated decreases too. For any number of nodes, these results are very logical due to more packets send in the network by the time. The higher congestion is found for 20 nodes. We explain that by the number of the MPR nodes, for example, when the nodes were far from each other, each node could have more than one MPR to reach its 2-hop neighbors. Moreover, the role of OLSR protocol is when the number of MPR nodes increases that means more TC messages are generated and forwarded in the network. The less value of W is detected for 80 nodes. We explain this result by the less number of MPR nodes in the small network size. Many nodes in the network can have the same MPR node which reduces the amount of the control messages forwarded in the network, by the end we detect less congestion compared to all other number of nodes.

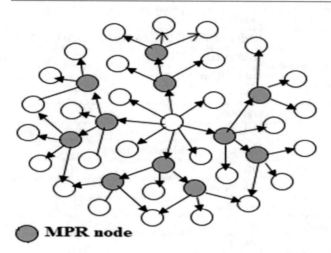

Fig. 1 MPR nodes flooded control message in OLSR

After the 15 s of simulation, we detect the congestion in the network for 50 and 100 nodes with similar values. All results of simulations for any number of nodes show that the small area of the network provokes the congestion in the traffic, due to higher control messages send and received in the same shared canal. For OLSR protocol, the congestion was detected more in the less number of nodes due to the number of MPR possible for each node. However, this congestion can be reduced for the higher number of nodes. For this reason, OLSR protocol is more suitable for the large density of the network.

5.2 Shapiro–Wilk of End-To-End Delay

For the second metric EED that can be affected by the congestion in MANET, Fig. 4 shows that W of EED for 20 nodes is close to critical W during all first 30 s of simulation that means absence of negative effect of the congestion in this normal area.

Another remark is that values of W for all different number of nodes in the first 10 s are similar due to absence of Topology Control messages generated by MPR nodes. For number of nodes 50, the value calculated for EED delay confirms the presence of congestion during all times of simulation. When we compare these values with other number of nodes, we show that EED of 50 nodes has the less W. For 80 nodes, we observe the same result of calculated W in the first 15 s, however, this value becomes increasing during other period of simulation. In the same way, the number of 100 nodes have less W values compared to critical W. By applying the role of Normality Test, we can detect the congestion in the network for all number of nodes in area of 1000 * 1000 m with just an exception of small number of nodes in our case is 20 nodes.

Figure 5 shows the results of calculated W for the metric EED in small area of simulation fixed at 500*500 m. Firstly, we observe that Shapiro–Wilk of EED metric has less values for different number of nodes during all sampling of simulation. The meaning of these results is the presence of the congestion in MANET, for that reason we found the

Fig. 2 W calculated for different number of node in areas 1000*1000 m

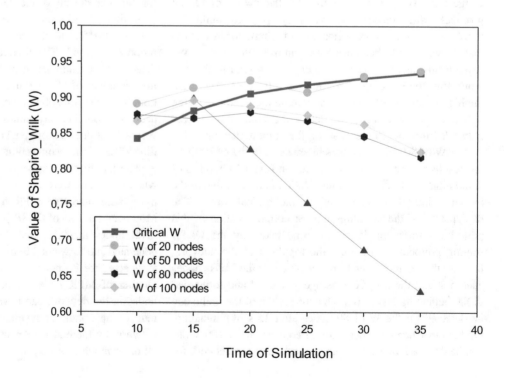

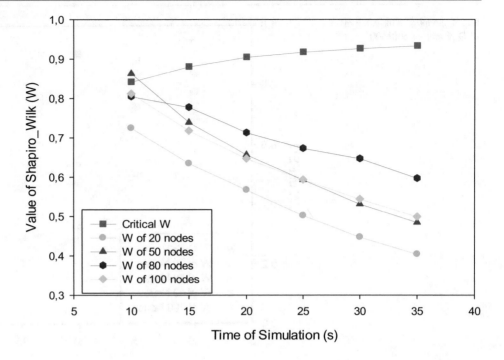

Fig. 3 W calculated for different Number of Node in areas 500* 500 m

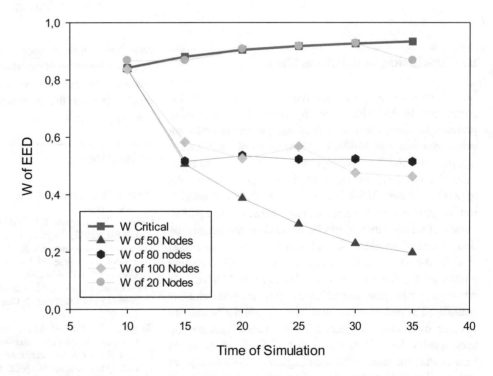

Fig. 4 Shapiro–Wilk value of EED in area of 1000*1000

hypothesis of normality in EED metric is rejected due to negative effect of the congestion. On the other hand, when the network supports 50 nodes, the W of EED has less value compared to all other number of nodes. For 80 nodes, the normality of EED is close to critical during first 15 s of simulation, after that the value of W was decreasing until the end, likewise the result of 100 nodes. The calculated W of 20 nodes confirms the detection of congestion during all time of simulation, in addition, when the time increases we remark that the value of W decreases too for different nodes. These results are close to logical of OLSR protocol that waits for the same period in the beginning before building the topology of the network by using both control messages HELLO and TC.

Fig. 5 Shapiro–Wilk value of
EED in area of 500*500

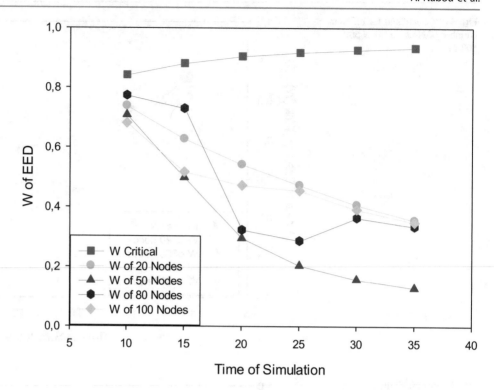

6   Conclusion and Future Work

Lot of research proposes different method to detect the
congestion in MANET either by modifying the routing
protocol or using other additional equipments in order for
safe performance of MANET routing protocol. OLSR routing protocol uses new concept called MPR nodes to minimize the overhead of the control messages send in the
network. For that OLSR is considered as the most suitable
routing protocol for the large density. However, the performance of OLSR protocol can be reduced by the congestion
due to shared wireless channel and the control messages
send in the same region of network or to higher routing
packet send end are received by the nodes in MANET. In
our work, we propose to use Shapiro–Wilk method, which is
considered as one of many other ways to test the Normality
in other to detect the congestion in the network. This method
uses specific formula to calculate the W of two essential
metrics that are affected by the congestion: the Throughput
and the End-to-End delay. The result of simulations shows
that W of 20 nodes for both metrics in normal area gives
values close to the critical W, and we confirm the absence of
the congestion. In the other hand, we can detect the congestion for all other number of nodes after the 15 s of
simulation, and we are explaining the time taken by OLSR
to build the topology of the network. For the second scenario
that simulated all nodes in small area, we observe that the
results of W for different number of nodes in the small area

have less values compared to the critical W which signifi-
cates the detection of the MANET that uses OLSR as routing
protocol. In the new work, we will focus on the new
mechanism for the congestion control in other MANET
routing protocols.

References

Dimitris Kanellopoulos: Congestion control for MANETs: An over-
  view (2019)
Clausen, T., Jacquet, P.: Optimized Link State Routing Protocol
  (OLSR) (2003). https://tools.ietf.org/html/draft-ietf-manet-olsr-11
Lochert, C., Scheuermann, B., Mauve, M.: A survey on congestion
  control for mobile ad hoc networks. Wireless Commun Mobile
  Comput. **7**, 655–676 (2007)
Maheshwari, G., Gour, M., Chourasia, U.K: A survey on congestion
  control in MANET. Int. J. Comput. Sci. Inf. Technol. **5**, 998–1001
  (2014)
Shapiro, S.S., Wilk, M.B.: An analysis of variance test for normality
  (complete samples). Biometrika **52**, 591 (1965)
Vadivel, R.: Murali Bhaskaran: adaptive reliable and congestion control
  routing protocol for MANET. Wireless Netw. **23**, 819–829 (2017)
Sirajuddin, M.D., Rupa, Ch., Prasad, A.: Advanced congestion control
  techniques for MANET. Inf. Sys. Des. Int. Applicat. Int. Sys.
  Comput. **433**, 271–279 (2016)
Holland, G., Vaidya, N.: Analysis of TCP performance over mobile Ad
  Hoc Networks. Wireless Networks, 275–288 (2002)
Sundaresan, K., Anantharaman, V., Hsieh, H.-Y., Sivakumar, R.: ATP:
  a reliable transport protocol for Ad Hoc networks. IEEE Trans
  Mobile Comput. **4** (2005)
Ikeda, M., Kulla, E., Hiyama, M., Barolli, L.: Rozeta Miho. Congestion
  control for Mul i-flow traffic in wireless mobile Ad-hoc networks,
  Makoto Takizawa (2012)

Rath, M., Rout, U.P., Pujari, N., Nanda, S.K., Panda, S.P.: Congestion control mechanism for realtime traffic in mobile Adhoc networks. Comput. Commun. Network. Inter. Sec. (2017). https://doi.org/ https://doi.org/10.1007/978-981-10-3226-4_14

Chen, X., Jones, H.M., Jayalath, A.D.S.: Congestion-aware routing protocol for mobile Ad Hoc Networks, (2007)

Author(s), Yang, X., Ge, L., Wang, Z.: F-ECN: a loss discrimination based on fuzzy logic control (2009)

Soundararajan, S., Bhuvaneswaran, R.S.: Multipath load balancing & rate based congestion control for mobile Ad Hoc Networks (MANET), (2012)

Kopparty, S., Krishnamurthy, S.V., Faloutsos, M., Tripathi, S.K.: Split TCP for mobile ad hoc networks. Global Telecommun Conf. (2002)

Pillai, B., Singhai, R.: Congestion control using fuzzy-based node reliability and rate control. In: Soft Computing for Problem Solving Advances in Intelligent Systems and Computing (2020)

Haveliwala, S., Vyas, D., Shah, H.: Dynamic Congestion control mechanism in mobile Adhoc network: TCP westwood-DCC, (2020)

Sharma, S., Chahal, M., Harit, S.: Fair channel distribution-based congestion control in vehicular Ad Hoc Networks, (2020)

Thode, H.J.: Testing for normality. Marcel Dekker, New York (2002)

NS3 Homepage, https://www.nsnam.org

Hyytiä, E., Virtamo, J.: Random waypoint mobility model in cellular networks. pringer Science Business Media, LLC 2007 (2006). https://doi.org/https://doi.org/10.1007/s11276-006-4600-3

Benchmarking Study of Machine Learning Algorithms Case Study: VANET Network

Sara Ftaimi and Tomader Mazri

Abstract

In the last few years, the expansion of mobile equipment and the deployment of wireless technologies has experienced rapid growth. Today's ongoing advancements in communication systems are opening up new areas of research, such as Intelligent Transport Systems (ITS). The vehicle ad hoc network (VANET) is a promising technological breakthrough in the transport field. Due to its features, namely, the high mobility, the dynamic topology, frequent disconnection, etc., the network became highly susceptible to threats. In accordance with these an amount of research in machine learning has been carried out to secure the VANET network and improve other aspects in VANET such as routing. They have obtained satisfying results. In this article, we will carry out a benchmarking study of the most commonly used machine learning algorithms in the context of the VANET network focusing on five criteria with the objective of selecting the most relevant one to be used for further work. We will also provide an overview of the attacks against VANET Network that may be useful to the average reader to get an idea of the security issues in VANET.

Keywords

VANET • Machine learning • Attacks

S. Ftaimi (✉) · T. Mazri
Laboratory of Electrical and Telecommunication Engineering,
Ibn Tofail Science University Kenitra, Kenitra, Morocco
e-mail: saraftaimi@gmail.com

T. Mazri
e-mail: tomader20@gmail.com

1 Introduction

Technology is changing the way we live and redesigning cities and societies. A smart city's vision is to make life simpler for citizens in all its many dimensions, and intelligent transport systems have become an essential part of this. Intelligent transport can radically change the way passengers travel in dense urban areas and help governments reduce budgets, provide better services to citizens, and better control safety and security. The Intelligent Transport System (ITS) is intended to achieve traffic efficiency by reducing traffic issues.

It empowers users with advance traffic information, etc., which reduces travel time for passengers and makes them safer and more comfortable. In recent years, the vehicle network has gained much attention in the industry (Zhang 2017). Vehicle Ad hoc Networks (VANETs) are an emerging type of mobile ad hoc networks with robust applications in intelligent traffic management systems (Dinesh and Deshmukh 2014).

VANET allows communication between vehicles on the road and communication devices alongside the road. In this type of network, vehicles are considered communication nodes, equipped with a communication unit called OBU, which is used to exchange messages between the nodes and/or with RSUs, that are the fixed access points located around the roads. These messages contain information about road conditions and information from other vehicles. The VANET differs from the MANET by its properties, namely, the network's dynamic topology since the nodes move continuously, the network disconnection constraint, etc. Because of these properties, the network is more vulnerable to attacks. The security in VANETS refers to the security of the information to be shared and the vehicle's confidentiality, i.e., the communication keeps all the aspects related to authentication, confidentiality, integrity, and availability of data. The network's security is a very important aspect; several studies have been carried out to secure communication between vehicles (V2V) and the infrastructure (V2I). In addition to V2X communication, it also presents a challenge since vehicles are connected to smart devices used by the hacker to control the network, since the communication channel between the vehicle and the smartphone is established via WIFI, Bluetooth, and GSM protocols, which are vulnerable. Recent research has shown using advanced learning machine technology in combination with VANET can be useful in detecting intrusions, making predictions, and improving autopilot efficiency.

In this paper, we will focus on machine learning algorithms used in VANET networks, and we will provide an overview of machine learning and then conduct a comparative study of these algorithms.

In the first part of this article, we will present a VANET network, its features, the types of communications in this network, and the equipment required to establish this communication. In the second part, we will explore the type of attacks targeting VANET. The next part is devoted to the machine learning algorithms used in VANET. The paper concludes with a comparative study of the different machine learning algorithms used in VANET and a conclusion.

2 VANET Network

As previously mentioned, the VANET network is a specific type of MANET network. It allows inter-node communication. Several research types have shown that using the VANET network will considerably reduce the number of accidents as well as road congestion, which will allow the driver to travel in safety and comfort. In this section, we will look at the equipment needed in a VANET network as well as the types of communication and the characteristics of this network (Zhang 2017; Hasrouny et al. 2017; Engoulou et al. 2014).

2.1 Equipment Needed

In this type of network, the vehicles are considered communication nodes. Two types of equipment must be installed in the vehicle to make this communication possible, namely: in-vehicle equipment (On Board Unit: OBU) and external equipment to the vehicles (Road Side Unit: RSU).

RSU: The RSU is composed of network devices for dedicated short-range communication (DSRC) on the basis of IEEE 802.11p radio technology. It is installed along the roadside, and it supplies a local connection to the moving vehicle. It can also be used to communicate with other network devices in other infrastructure networks (Al-Sultan et al. 2014).

OBU: Is a GPS-based navigation device that is usually onboard the vehicle. The main function is to connect to the RSU or other OBUs via an IEEE 802.11p (2012) wireless link by exchanging messages. Besides, the OBU gathers information from the sensors installed in the car, namely, the Global Positioning System (GPS), Event Data Recorder (EDR), and obtains the input power from the vehicle battery (Al-Sultan et al. 2014).

2.2 Communication Type

In VANET networks there are three modes of communication as shown in Fig. 1.

V2V: In this communication mode, a vehicle can communicate directly with another vehicle located in the same communication perimeter; the architecture of such mode is decentralized. It is very efficient to transmit information on road safety services, but cannot guarantee a permanent connection between vehicles; due to the nodes' variable speed and mobility of the nodes (Sheikh and Liang 2019; Al-Sultan et al. 2014).

V2I: This mode of communication allows the exchange of information between the vehicle and the infrastructure (RSU, antenna, satellite). It provides better use of the common resources, and the connection's status is more stable in this mode than V2V communication (Sheikh and Liang 2019; Al-Sultan et al. 2014).

V2X: A technology system that allows vehicles to use short-range wireless signals to communicate with the traffic and the surroundings around them. V2X consists of several sub-components, namely, Vehicle-to-Vehicle (V2V), Vehicle-to-Infrastructure (V2I), Vehicle-to-Pedestrian (V2P), and Vehicle-to-Network (V2N) communications. In this diverse ecosystem, vehicles will communicate with other vehicles, with infrastructures such as stoplights, pedestrians using smartphones, and data centers via cellular networks (Sheikh and Liang 2019).

2.3 Characteristics of VANET Network

Vehicular networks have specific features that make them different from mobile ad hoc networks, namely,

Fig. 1 Types of Communication in VANET Network

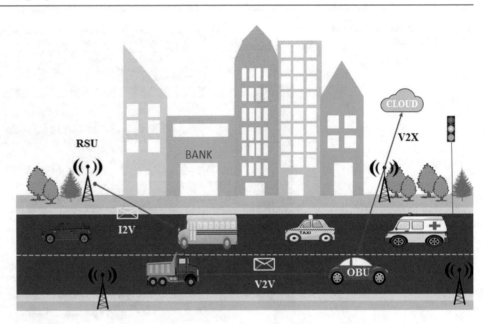

High mobility:	the highly mobile nature of the nodes is one of the most relevant features in the VANET network. Vehicles are constantly moving at variable speed and in many directions (Engoulou et al. 2014; Mejri et al. 2014).
Dynamic topology:	the VANET topology changes rapidly (Engoulou et al. 2014; Mejri et al. 2014) since a vehicle (node) may join or leave a cluster of vehicles in a relatively short time.
Frequent disconnection:	Due to both features mentioned above, combined with weather change and traffic congestion, vehicles can frequently disconnect from the network (Dinesh and Deshmukh 2014).
Time constraint:	Some applications require messages to be received within a specified time (Engoulou et al. 2014; Mejri et al. 2014).
Unlimited power supply and battery storage:	VANET does not suffer from energy or storage problems (Engoulou et al. 2014; Mejri et al. 2014), Unlike MANET networks where the energy constraint is a challenge.
Unbounded network size:	VANET network size has no geographical limits (Engoulou et al. 2014; Mejri et al. 2014).

3 Attacks and Type of Attackers

In this section, we will classify attacks according to security requirements: the availability, confidentiality, authenticity, integrity, and repudiation; also, we will highlight some examples of attacks in Fig. 2. In a further step, we will present the type of attackers that can threaten a network.

3.1 Attack on Availability

The network and applications must remain available even if the network fails, which protects the system, and makes it fault-tolerant at all times, thereby potentially leading to a major security breach (Mejri et al. 2014).

3.2 Attacks on Authenticity and Identification

The main purpose of authentication in a VANET network is to control the authorization levels of the vehicle. Once authenticated, the vehicle has a unique identifier. As a consequence, the vehicle cannot have multiple identifiers and pass as multiple vehicles. Therefore this technique will prevent attacks such as Sybil, and any violation of this service may expose the entire network to serious problems (Mejri et al. 2014).

3.3 Attacks on Confidentiality

Confidentiality is a very important mechanism for reliable communication. By sending encrypted messages, only the

Fig. 2 Taxonomy of attacks in
VANET Network

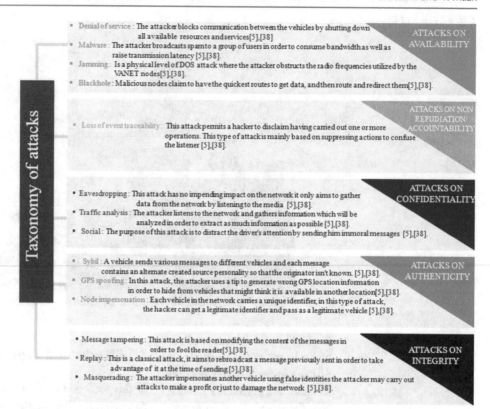

vehicle for which the message is intended can read it. Any breach of this protocol can lead to threats, as described in the table below (Mejri et al. 2014).

3.4 Attacks on Integrity and Data Trust

The integrity mechanism ensures that the data has not been corrupted during the sending. Integrity protects against the destruction and alteration of the message during transmission (Mejri et al. 2014).

3.5 Attacks on Non-Repudiation/accountability

The non-repudiation mechanism checks that the message sender and destination are the identities that respectively claim to have sent and received the message (Mejri et al. 2014). any failures of this mechanism can lead to attacks such as the attack described in Fig. 2.

3.6 Attackers

The attacker's methodology depends on his capabilities, motivations, and knowledge about the targeted machine. The attacker, who has great knowledge about vulnerabilities in a system, can gather enough information about the exploit techniques to craft a specific attack using that vulnerability. The motivations of the attacker can also influence the methodology used to construct the attack. Indeed, if the attacker wants to interrupt the normal production process of a system he can choose attacks that aim the availability of the targeted machine. However, if he desires to cause dysfunction in a specific instance, he can opt for integrity attacks.

Raya et al. classify the attackers according to four dimensions presented in this section.

Insider versus Outsider:	the internal attacker is perceived as a normal member of the network, and he can communicate with the other members; for this reason, he is considered more dangerous since it is difficult to detect him. The external attacker is very restricted in carrying out attacks. An outsider can sneak into the network to collect information; besides, he may also shut down the network with messages (Engoulou et al. 2014).
Malicious versus Rational:	A malicious attacker does not seek any gain from the attacks. He merely enjoys damaging the network with no particular objective (Hao et al. 2008), while a rational attacker can be more

dangerous (Sabahi 2011), as he looks for a gain from the attacks besides being more predictable (Engoulou et al. 2014).

Active versus Passive:	The active attacker is not restricted to carrying out all kinds of attacks on the VANET network, whereas passive attackers stalk the communication between the other nodes of the network to gather data that can be used for further threats (Engoulou et al. 2014).
Local versus extended:	A local attacker controls several entities but has a limited range compared to the extended attacker, who has an unlimited range. This feature makes the extended attacker powerful to conduct attacks against privacy and wormholes.

4 Machine Learning

Machine learning is a form of artificial intelligence that allows computers to learn without programming. It is a modern science for finding patterns in data and making predictions based on statistics, data mining, pattern recognition, and predictive analysis. Machine learning is currently used in many fields, such as autonomous vehicle development and fraud detection. The three most commonly used machine learning methods are supervised learning, unsupervised learning, and reinforcement (Prevention and in VANET 2019; Xiao et al. 2019).

Supervised learning:	In supervised learning, algorithms are trained using labeled data. The algorithm receives inputs and the corresponding correct outputs and learns by comparing the outputs with the expected correct results to detect errors. After being trained, the algorithm performs labeling on the unlabeled data (Ye et al. 2018).
Unsupervised learning:	The input data are not labeled. The algorithm must explore the data and find a structure within it (Ye et al. 2018).
Reinforcement:	Reinforced learning is a technique that teaches the algorithm to learn from its mistakes by interacting with the surrounding environment in order to learn how to make the correct choices. If the choice is erroneous, it is sanctioned. On the other hand, if it makes the right choice, it is rewarded. The algorithm

performs its best to improve its choice to obtain more and more rewards (Ye et al. 2018; Alsheikh et al. 2014; Liang et al. 2019).

We can classify the machine learning algorithms into five classes, namely, classification, regression, clustering, neural network, and policy learning, Fig. 3. It illustrates the algorithms that belong to each class (Ye et al. 2018).

5 Benchmarking Study of Machine Learning Algorithms in VANET

In this section, we will compare different machine learning algorithms, this research will facilitate understanding and learning about ML algorithms and the function of each algorithm in the context of VANET. This section, has consolidated the different results found by researchers using ML in VANET networks to create our review. We have also put forward five criteria to compare these algorithms: the choice of the criteria is based essentially on the impact of these ones on the execution of the algorithm. The first criterion is the outlier detection since such a value can affect the structure of an ML algorithm. Then there is the speed of the algorithm because the processing time is very important in VANET, then the storage capacity, in fact, given that we are in a network that exploits and generates a large amount of data, finding an algorithm that will be able to handle security problems without the need of a large storage capacity will be very useful. Afterword there is the criterion of scalability. Finally the management of voluminous data, since in VANET networks we will be dealing with a large number of vehicles that circulate and share information continuously, we will need a robust algorithm to process a large amount of data.

VANET (Vehicular Ad Hoc Networks) is a new emerging technology for mobile ad hoc networks (MANETs), where mobile nodes are intelligent vehicles, implemented with very high technology equipment. VANET enables inter-vehicular (V2V) and infrastructure (V2I) communications. The different nodes can exchange any alerts or useful information to improve the safety of the road traffic, they can also exchange data (music, advertising, etc.) to make the time spent on the road more enjoyable. Like any other network, VANET also confronts problems and threats. The machine learning technique has been widely applied to VANET networks to solve complex problems such as security, intrusion detection, and prediction. To prevent spoofing, anti-jamming, and malware detection, we can use the Q-learning algorithm, which improves the authentication process (Xiao et al. 2018; Aref et al. 2017). As we can use SVM (Alsheikh et al. 2014) and KNN to detect network

Fig. 3 Machine learning algorithms

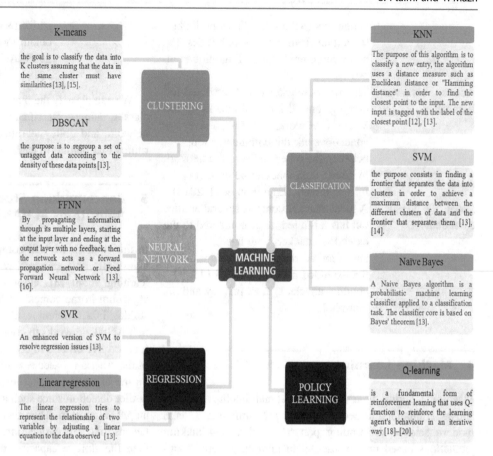

intrusions (Branch et al. 2006) we can also apply SVM against spoofing attacks (Ozay et al. 2016). The vehicles can use Naive Bayes to detect intrusions (Alsheikh et al. 2014) and malicious behaviors (Grover et al. 2012). Besides, K-means can be used to detect traffic congestion (Taherkhani and Pierre 2016).

KNN is an intuitive supervised classification approach it can be used in both classification and regression. KNN is widely used in classification and must store all training data in memory, so it is perfect for small cases. Although it has been demonstrated that KNN algorithms can lead to erroneous results when solving large problems (more than 10 to 15 measurements) (Narudin et al. 2016; Alsheikh et al. 2014). Yi Zhoo, Huanhuan Li, chenhao Shi, Ning Lu, and Nan Cheng (2018) employ the KNN algorithm embedded in the RSU to anticipate the next vehicle movement. The challenge for (Zhou et al. 2018) is to find out the number k of the closest neighbors. Other researchers (So et al. 2018) have explored the possibility of an ML-based model that uses Basic Safety Massage (BSM) data to identify and categorize the type of attacks. Steven So, Prinkle Sharma, and Jonathan Petit (So et al. 2018) extracted feature vectors from BSM and incorporated them into KNN and SVM to identify and classify the type of attack. SVM is a supervised machine learning algorithm. SVM is designed to deal with very large

datasets as well as being memory efficient. Nevertheless, once a solved SVM model is generated, it may not be easy to interpret its parameters. Steven So, Prinkle Sharma, and Jonathan Petit (So et al. 2018) discovered that KNN performs better than SVM in some cases. Some of the cases where SVM has a degraded performance were when the training set of both algorithms included attacks that showed high levels of misconduct and only a few attacks were common errors. Moreover, on the basis of recall accuracy, they observed that both SVM and KNN produce better results in some types of attacks while SVM exceeds KNN in certain attacks and for KNN.

Khattab, Gruebler, and Klaus (2015) implemented SVM and Feed Forward Neural Network (FFNN) to build an intelligent IDS able to recognize and to stop two current types of attacks, namely, grey hole attacks and Rushing Attacks (Alheeti et al. 2015). The key of using these artificial intelligence networks lies in the fact that numerous studies have demonstrated the efficiency of such networks, especially in autonomous vehicles (Alheeti et al. 2015). FFNN is a computational system that attempts to mimic the neural connections of our nervous system in order to develop algorithms that can be used to solve complex prediction problems. Furthermore, the neural network system can train and shape complex relationships, which is incredibly important since, in real life, numerous relationships that exist between I/Os are complex and non-linear. The FFNN is also rapid to handle new data (Mahdavinejad et al. 2018; Kulkarni and Venayagamoorthy 2009).

In addition (Alheeti et al. 2015) they found that FFNN-based IDS are better than SVM-based IDS at detecting malicious vehicles with lower false-negative alarm rates. However, they noticed that the speed of SVM performance is much quicker than FFNN.

The K-means is one of the most well-known algorithms of machine learning; it is fast, upgradeable, and easy to implement. Moreover, the K-means is efficient in handling voluminous dataset, and being suitable for any kind of information; however, occasionally, it can provide different results by adjusting the initial conditions since it is sensitive to the centroid of the largest nodes (Zhou et al. 2018). K-means cannot merge fastly, particularly for large datasets, and as we define the class number at the beginning, this may cause errors in certain cases (Xiao et al. 2019; Sabahi 2011; Ye et al. 2018). Moreover, it is not robust to outliers (Zhou et al. 2018). Wei Kuang Lai, Mei-Tso Lin, and Yu-Hsuan Yang (2015) employed K-means as part of a system conceived to design routing protocols for urban environments to anticipate vehicle motion and identify correct routing paths.

A Naive Bayes classifier is a probabilistic machine learning model applied to classification tasks. The kernel of the classifier is based on Bayes' theorem. It is a classifier that needs a small number of data points to be trained, and it can handle very high amounts of data. It is rapid, with high scalability and easy implementation. However, its main limitation is the obligatory independence of the predictors, and in most real cases, the predictors are dependent, which hinders the performance of the classifier (Mahdavinejad et al. 2018; Grover et al. 2012). Jyoti Grover, Nitesh Kumar Prajapati, Vijay Laxmi, and Manoj Singh Gaur (Grover et al. 2011) have used various forms of misconduct in VANET. They apply machine learning methods to classify the

behaviors as legitimate or malicious. D-48, Naive Bayes, and other algorithms can identify various types of misbehavior during the classification phase. Unfortunately, naive Bayes classifiers are not performing well compared to other classifiers such as J-48.

The Q-learning algorithm is the well-known reinforcement algorithm due to the clear evidence of convergence that accompanied its publication. This algorithm could be applied easily in a distributed architecture such as WSN (Wireless Sensor Networks), in which each node attempts to select actions that will enhance its long term benefits.

It is noteworthy that Q-learning has been widely and effectively used in the WSN routing problem (Alsheikh et al. 2014). Have proved that a continuous Q-learning algorithm achieves faster and increasingly viable learning on many tasks, unlike continuous actor-critical methods, and the simplicity of this methodology will facilitate practice. As described earlier in VANET, rapid change of topology complicates the search and maintenance of the best end-to-end path, but the Q-Learning algorithm can achieve the quickest path from a source node to a destination node by interacting continuously with its environment. Degan Zhang, Ting Zhang, and Xiaohuan Liu (2019) recommend a vigorous, versatile routing algorithm by improving the Q-Learning algorithm. A proficient routing calculation can essentially expand the information transmission rate; the Q-learning calculation utilized for routing has delivered generally excellent outcomes. Besides, the Q-learning calculation has been utilized to tackle numerous issues experienced by the VANET organize. For example, Q-learning was utilized to fortify validation execution against anti-jamming transmissions.

The DBSCAN algorithm can be effective on larger datasets. It is both fast and robust against outliers. The model performance is very sensible to the metric of distance utilized to identify whether an area is dense. In the case of a dataset with large density differences, the resulting clusters are devoid (Mahdavinejad et al. 2018) (Table 1).

Table 1 Comparative study of machine learning algorithms

Algorithms	High Dimensional Data	Fastness	Outliers	Memory	Scalability
KNN	–			–	
Naïve Bayes	+	+			+
SVM	+		+	+	
Linear regression			+		
K-means		+	–		+
DBSCAN	+	+	+		
FFNN	+	+		+	
Q-learning	+	+			

6   Conclusion

This paper highlights two aspects: attacks and vulnerability in the VANET network, and the second one is machine learning algorithms used to detect intrusions or make predictions. An overview of the VANET network and the different attacks that can target it is given in this article. Then, we provide some samples of machine learning algorithms that can be applied in a VANET network. Furthermore, we conducted a comparative study of machine learning algorithms according to five factors. The study carried out in this article can be helpful for the overreader to learn about the safety issues in VANET as well as a survey of ML algorithms. From this benchmarking study, we can use it as a basis for future work to select the right algorithm for a given case study and enhance its efficiency.

References

Alheeti, K.M.A., Gruebler, A., McDonald-Maier, K.D.: On the detection of grey hole and rushing attacks in self- driving vehicular networks. In: 2015 7th Computer Science and Electronic Engineering Conference (CEEC), Colchester, United Kingdom, pp. 231-236 (2015)

Alheeti, K.M.A., Gruebler, A., McDonald-Maier, K.D.: An intrusion detection system against black hole attacks on the communication network of self-driving cars. In: 2015 Sixth International Conference on Emerging Security Technologies (EST), Braunschweig, Germany, pp. 86–91 (2015)

Aref, M.A., Jayaweera, S.K., Machuzak, S.: Multi-agent reinforcement learning based cognitive anti-jamming. In: 2017 IEEE Wireless Communications and Networking Conference (WCNC), pp. 1–6. San Francisco, CA, USA (2017)

Al-Sultan, S., Al-Doori, M.M., Al-Bayatti, A.H., Zedan, H.: A comprehensive survey on vehicular Ad Hoc network. J. Network Comput. Applicat. 37, 380–392 (2014 janv). doi: https://doi.org/10.1016/j.jnca.2013.02.036

Alsheikh, M.A., Lin, S., Niyato, D., Tan, H.-P.: Machine learning in wireless sensor networks: algorithms, strategies, and applications. IEEE Commun. Surv. Tutorials 16(4), 1996–2018 (2014)

Branch, J., Szymanski, B., Giannella, C., Wolff, R., Kargupta, H.: In-network outlier detection in wireless sensor networks. In: 26th IEEE International Conference on Distributed Computing Systems (ICDCS'06), pp. 51–51. Lisboa, Portugal (2006). doi: https://doi.org/10.1109/ICDCS.2006.49

Dinesh, D., Deshmukh, M.: Challenges in vehicle Ad Hoc Network (VANET). Int. J. Eng. Technol. 2(7), 14 (2014)

Engoulou, R.G., Bellaïche, M., Pierre, S., Quintero, A. : Vanet security surveys. Comput. Commun. 44, 1–13 (2014 mai)

Gu, S., Lillicrap, T., Sutskever, I., Levine, S.: Continuous deep Q-learning with model-based acceleration, p. 10

Grover, J., Laxmi, V., Gaur, M.S.: Misbehavior detection based on ensemble learning in VANET. In: Thilagam, P.S., Pais, A.R., Chandrasekaran, K., Balakrishnan, N. (eds.) Advanced Computing, Networking and Security, vol. 7135, pp. 602–611. Berlin, Heidelberg: Springer Berlin Heidelberg (2012)

Grover, J., Prajapati, N.K., Laxmi, V., Gaur, M.S.: Machine learning approach for multiple misbehavior detection in VANET. In: Abraham, A., Mauri, J.L., Buford, J.F., Suzuki, J., Thampi, S.M. (eds.) Advances in Computing and Communication, vol. 192, pp. 644–653. Berlin, Heidelberg: Springer Berlin Heidelberg (2011)

Hao, Y., Cheng, Y., Ren, K.: Distributed key management with protection against RSU compromise in group signature based VANETs. In: IEEE GLOBECOM 2008–2008 IEEE Global Telecommunications Conference, New Orleans, LA, USA, p. 1–5 (2008)

Hasrouny, H., Samhat, A.E., Bassil, C., Laouiti, A.: Vanet security challenges and solutions: a survey. Vehicular Commun 7, 7–20 (2017 janv)

IEEE P1609.0/D5, September 2012: IEEE Draft Guide for Wireless Access in Vehicular Environments (WAVE)—Architecture. Place of publication not identified: IEEE (2012)

Intelligent Accident Prevention in VANETs.: ijrte, 8(2), 2401–2405 (2019 juill). doi: https://doi.org/10.35940/ijrte.B1805.078219

John Cornish Hella Watkins, C.: Learning from delayed rewards. King's College (1989)

Kulkarni, R.V., Venayagamoorthy, G.K.: Neural network based secure media access control protocol for wireless sensor networks. In: 2009 International Joint Conference on Neural Networks. Atlanta, Ga, USA, pp. 1680–1687 (2009)

Lai, W.K., Lin, M.-T., Yang, Y.-H.: A machine learning system for routing decision-making in urban vehicular Ad Hoc networks. Int. J. Distrib. Sensor Networks 11(3), 374391 (2015 mars)

Liang, L., Ye, H., Li, G.Y.: Toward intelligent vehicular networks: a machine learning framework. IEEE Internet Things J 6(1), 124–135 (2019 févr)

Li, Y., Quevedo, D.E., Dey, S., Shi, L.: SINR-based DoS attack on remote state estimation: a game-theoretic approach. In: IEEE Trans. Control Netw. Syst. 4(3), 632–642 (2017 sept)

Lu, X., Wan, X., Xiao, L., Tang, Y., Zhuang, W.: Learning-based rogue edge detection in VANETs with Ambient Radio Signals, p. 6

Mahdavinejad, M.S., Rezvan, M., Barekatain, M., Adibi, P., Barnaghi, P., Sheth, A.P.: Machine learning for internet of things data analysis: a survey. Dig. Commun. Networks 4(3), 161–175 (2018, août)

Mejri, M. N., Ben-Othman, J., Hamdi, M.: Survey on VANET security challenges and possible cryptographic solutions. Vehicul. Commun. 1(2), 53–66, avr (2014)

Narudin, F.A., Feizollah, A., Anuar, N.B., Gani, A.: Evaluation of machine learning classifiers for mobile malware detection. Soft Comput. 20(1), 343–357 (2016 janv)

Ozay, M., Esnaola, I., Yarman Vural, F.T., Kulkarni, S.R., Poor, H.V.: Machine learning methods for attack detection in the smart grid. IEEE Trans. Neural Netw. Learning Syst. 27(8), 1773–1786 (2016 août). doi: https://doi.org/10.1109/TNNLS.2015.2404803

Raya, M., Hubaux, J.-P.: Securing vehicular ad hoc networks, p. 30

Sabahi, F.: The security of vehicular Adhoc networks. In: 2011 Third International Conference on Computational Intelligence, Communication Systems and Networks, Bali, Indonesia, pp. 338–342 (2011)

Sheikh, M.S., and Liang, J.: A comprehensive survey on VANET security services in traffic management system. Wireless Commun. Mobile Comput. 2019, 1–23 (2019 sept). doi: https://doi.org/10.1155/2019/2423915

So, S., Sharma, P., Petit, J.: Integrating plausibility checks and machine learning for misbehavior detection in VANET. In: 2018 17th IEEE International Conference on Machine Learning and Applications (ICMLA), Orlando, FL, pp. 564–571 (2018)

Taherkhani, N., and Pierre, S.: Centralized and localized data congestion control strategy for vehicular Ad Hoc networks using a machine learning clustering algorithm. IEEE Trans. Intell. Transport. Syst. 17(11), 3275–3285 (2016, nov)

Technical, R.: Using artificial intelligence to create a low cost self-driving car

Thandil, R. K.: Security and privacy in Vehicular Ad Hoc Network (VANET): a survey. Int. J. Comput. Applicat, **5**

Tsitsiklis, J.N.: Asynchronous stochastic approximation and Q-learning, p. 18 (1994)

Xiao, L., Lu, X., Xu, D., Tang, Y. Wang, L., Zhuang, W.: UAV relay in VANETs against smart jamming with reinforcement learning. IEEE Trans. Veh. Technol. **67**(5), 4087–4097 (2018 mai)

Xiao, L., Dai, H.: A mobile offloading game against smart attacks, vol. 4, p. 11 (2016)

Xiao, L., Li, Y., Huang, X., Du, X.: Cloud-based Malware Detection Game for Mobile Devices with Offloading, p. 10

Xiao, L., Zhuang, W., Zhou, S., Chen, C.: Learning-based VANET Communication and Security Techniques. Cham: Springer International Publishing (2019)

Ye, H., Liang, L., Li, G.Y., Kim, J., Lu, L., Wu, M.: Machine learning for vehicular networks : recent advances and application examples. IEEE Veh. Technol. Mag. **13**(2), 94–101 (2018 juin)

Zhang, Q.: A pervasive prediction model for vehicular Ad- hoc network (VANET). Nottinhgham Trent University (2017 August)

Zhou, Y., Li, H., Shi, C., Lu, N., Cheng, N.: A fuzzy-rule based data delivery scheme in VANETs with intelligent speed prediction and relay selection. Wireless Commun. Mobile Comput. **2018**, 1–15 (2018 août)

Zhang, D., Zhang, T., Liu, X.: Novel self-adaptive routing service algorithm for application in VANET. Appl. Intell. 49(5), 1866–1879 (2019 mai)

A Comparative Study of Detection Algorithm in VANET Network

Manale Boughanja and Tomader Mazri

Abstract

Due to the current trend of development of new technologies and the development of mobile communications as well as the development of vehicles, the vehicles have become increasingly intelligent and equipped with radio communication interfaces. VANET is one of the new technologies that characterize the development of traffic. This new development faces enormous challenges, particularly in terms of network security. To solve this problem, research was carried out, particularly in terms of detection. In this paper, we will present the architecture of VANET, the different security challenges that face this new technology. At last, we will study different intrusion detection algorithms that could enhance the security of VANET networks

Keywords

VANET • Security • Intrusion detection algorithm

1 Introduction

Recently, mobile technology was a subject of significant development. These remarkable technological advances have surprisingly led to a significant improvement in mobile networks. An ad hoc mobile network is an emerging paradigm in networking (Erritali 2013), it is a distributed system composed of self-sufficient entities capable of communicating with one another without the presence of a centralized infrastructure.

M. Boughanja (✉) · T. Mazri
Electrical System and Telecommunication Engineering, Ibn Tofail Science University, Kenitra, Morocco
e-mail: boughnja.manale@gmail.com

T. Mazri
e-mail: tomader20@gmail.com

Vehicular ad hoc network (VANET) is a new form of mobile ad hoc network (MANET). In the mobile ad hoc network, each cellular is considered as a node, while in vehicular ad hoc network (VANET), each vehicle is taken as a node. A VANET can be a first step toward the realization of an intelligent traffic system (ITS).

VANET is considered as one of the most pertinent technologies developed. With the advent of this technology, several threads emerged (Tanuja et al. 2015). Problems must be identified before they can affect the vehicle network. One of the adviser's methods for preventing and detecting security issues is the intrusion detection system (IDS). An ID is a device monitor's specific network or host activity, detects, and can react to any intrusion attempt. In other words, intrusion detection is the process of monitoring events occurring in a node or network and analyzing signs of intrusion. An intrusion event is defined as an attempt to disrupt the security requirements of a node or network.

In this paper, we aim to present an overview of the VANET architecture as well as the issues that faces this new technology. For that, our paper is organized as follows: in Sect. 2, we will present the VANET network. Section 3 provides the security requirements. Then, we will introduce some challenges that face VANET. In Section 5, we will present some researches made on the intrusion detection algorithm in order to improve security, then in Sect. 6, we will present a comparative study of the intrusion detection algorithm and we will summarize in this Sect. 7.

2 VANET Network

With the development of mobile operating systems in vehicles, there is no doubt that the demand for real-time internet access in vehicles can increase rapidly (Engoulou et al. 2014). Consequently, for a vehicle heterogeneous network formed by a cellular network and VANET, an efficient network selection is essential to guarantee the quality of service (QoS) of the vehicle in order to avoid performance degradation.

2.1 VANET Architecture

The architecture of VANET is composed of three layers:

- The access layer: the vehicles are using the cellular base station (eNB) to connect the vehicle to each other through 3G/LTE or by using the roadside unit (RSU).
- The data aggregation layer: both eNB and RSU are connected to a central controller in order to access to the internet.
- The application layer: this layer provides several services in the VANET network.

The Fig. 1 shows the hierarchical architecture of vehicular networks.

The infrastructure is composed of the base station eNB and RSU, the vehicle and the traffic controller that connect the infrastructure to the internet. Finally, the cloud provides services to the VANET network.

2.2 Communication Types in VANET

Vehicular ad network (VANET) represents a significative evolution. Researchers are increasingly interested in the field. VANET network has three types of communication:

- Vehicle to vehicle (V2V): is designed to permit vehicle to speak to each other.

- Vehicle to infrastructure (V2I): permits vehicles to communicate with the roadside infrastructure (RSU). This type of communication forms a multi-hop vehicle-to-vehicle communication.
- Hybrid communication: in this case, both types are considered V2V and V2I communication; both in a signal hop or multi-hop style, depending on the distance which mean if the vehicle is able to get to its destination or not.

The Fig. 2 shows the communication types in VANET network.

2.3 Standard Communication for VANET

VANET uses several wireless communication technologies to communicate with vehicles, base station (Liang et al. 2018). In general, to improve the traffic control, we can categorize the standard communication type for VANET into three categories:

- Long-range communication: in which we use the different mobile technologies (2G, 3G, and 4G). The WiMAX technology that provides an access to internet with a long distance.
- Medium range communication: consists of several technologies WLAN/Wi-Fi and DSRC (dedicated short-range communication) which is developed to meet the requirement of VANET such as self-configuration and high mobility.
- Short-range communication: consists of several types such as Bluetooth, ZigBee, and UWB. This type of communication does not bear long distances to transfer the signal.

Fig. 1 Hierarchical architecture of vehicular network

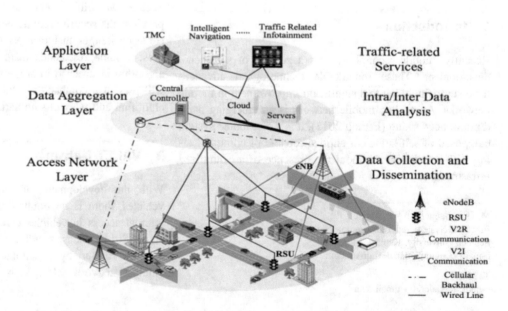

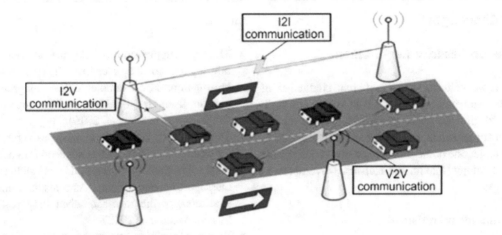

Fig. 2 Communication types in VANET

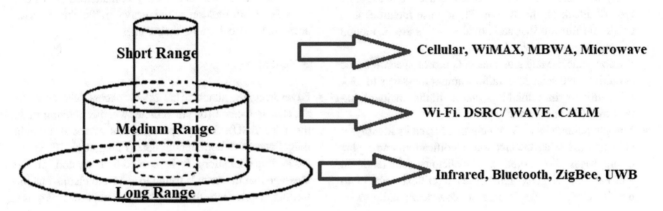

Fig. 3 Standard communication for VANET

The Fig. 3 summarizes the different standard of communication in VANET.

3 Security Requirement

Before addressing the security issues of VANET technology, it is important to understand the security requirements, in order to build a reliable system capable of handling all threads. The main security requirements are as follows (Parul and Deepak 2014):

- Authentication: Provides a guarantee of reliable data/messages generated by the user. The authentication of all users and messages is very important because VANET technology can respond to nodes based on the information received. We can find two types of authentication (Anwer and Guy 2014).
  - **ID authentication**, which allows nodes to uniquely identify the sender of a message.

  - **Attributed authentication**, which allows you to determine the attributes of a node and determine its type. (Vehicle, RSU, etc.).
- Integrity: Ensure that third parties have not modified the information received during transmission. Integrity prevents unauthorized creation or destruction of data.
- Confidentiality: Is a mechanism where unrelated ones cannot understand the data during communication. This mechanism is preserved by various methods, such as encryption.
- Non-repudiation: is defined as the impossibility for one of the entities involved in a communication (vehicles or infrastructure) to deny its participation in one or all of the events. This mechanism protects against false rejection involved in the communication.
- Availability: Since data is managed immediately, availability requirements must be taken into account so that data can be accessed quickly and easily.

4 VANET Challenges

4.1 Attacks on Security Requirement

In this section, we will discuss the different challenges of VANET. Since we cannot conceive all possible attacks in VANET, we will provide a general classification of attacks based on security requirements (Parul and Deepak 2014). Attackers can target one or more of the security requirements of VANET technology (Maxim and Hubaux, November), as shown in Fig. 4:

Authenticity and identification

- Sybil attack: In this attack, a vehicle forges the identity of many vehicles. These identities can be used to launch any type of attack on the system. These false identities also create the illusion that additional vehicles are in motion (Nidhal Mejri et al. 2014).
- Illusion attack: Malicious attackers create specific traffic conditions and send fake traffic warning messages to trick other drivers into thinking that a traffic incident has occurred (Lo and Tsai 2007).
- Masquerading attack: A vehicle that forges its identity. It is performed using the creation, modification, and replay of messages. For example, a malicious vehicle or an attacker can disguise himself as a critical vehicle to deceive other vehicles in order to slow down and give in.

Integrity

- Message suppression attack: the attacker sends false information to the user, which may contain critical information. For example, delete congestion alerts to prevent users from taking another approach to avoid congestion (Rushit et al. 2019).
- Timing attack: Whenever a malicious vehicle receives an emergency message, it does not send it to nearby vehicles at the right time, but it does add time slots to the original message to create a delay. As a result, vehicles near the attackers got the message when they really needed it (Saeed Al-kahtani 2012).
- Bogus Information attack: An attacker sends false information to the network in order to gain personal profit (Rashmi et al. 2016a). For example, malicious nodes can send false information about heavy traffic due to road accidents and determine their routes.

Confidentiality

- Eavesdropping attack: This attack occurs wherever the attacker is located (in the vehicle, whether moving or in the false RSU). The purpose of this attack is to gain illegal access to confidential data (Liang et al. 2018).
- Node impersonation: It is a modified version of the message that the node is trying to send and claims that the message came from the sender for an unknown purpose.

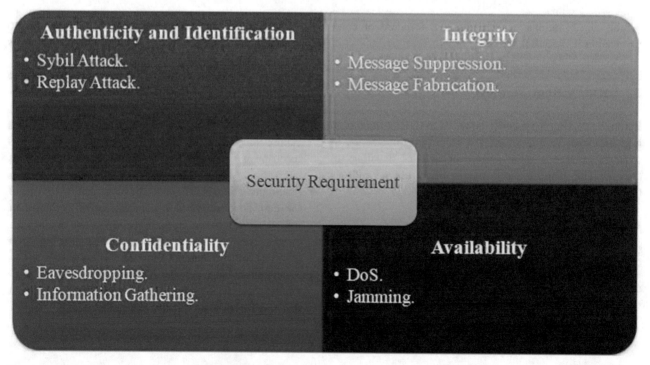

Fig. 4 Example of threats and attacks on VANET security requirement

- Traffic Analysis attack: In a traffic analysis attack, the attacker listens and/or violates certain parts of VANET to match the sender of the message to the recipient (Isaac et al. 2010).
- Men in the Middle attack: In this attack, harmful vehicles listen to the communication between the vehicles and inject false information between the vehicles.

Non-Repudiation

- Loss of event and traceability attack: This attack allows the attacker to deny that one or more operations were carried out on the VANET network (Maxim and Hubaux 2005).
- Black Hole attack: In this attack, the compromised node sends a false route with fewer hops than the source to attract it, and when the source node sends a data packet to this route, the affected node drops the packet(Bibhu et al. 2012).

Availability

- DoS attack: Attackers overload the communication channels. The aim is to send messages to disconnect or prevent the vehicle from accessing the services of the VANET network (Liang et al. 2018).
- Jamming attack: Is the transmission of radio signals which interfere with communication by reducing the signal-to-noise ratio (Punal et al. 2015). The VANET network is vulnerable to this kind of attack because it makes the victim node unavailable to other legatine users.
- Malware and spam: These attacks increase transmission latency, which can be mitigated by using centralized management.

- Gray Hole attack: In terms of removing packets, this attack is similar to a black hole attack, but differs in the context in which these droppings are selected. In other words, only certain types of selected packages are deleted and then selected according to the needs and intention of the attacker (Bibhu et al. 2012).

We summarize in Table 1 other existing attacks for VANETs and we defined the compromised service:

4.2 Attackers

In the previous section, we were able to introduce some attacks against VANET technology. However, these attacks should be classified according to their target. Attackers can be classified into three categories (Ajay et al. 2012). Figure 5 shows the classification of attackers.

We will present the different types of attackers:

- The basis of membership: This type of attacker can be an authorized person or an unauthorized person; it is designed to cover the entire network. We can distinguish two types:
  - **Internal attackers**: are considered to be approved nodes (Nidhal Mejri et al. 2014).
  - **External attackers**: considering intruders trying to enter the network (Parul and Deepak 2014).
- The basis of activity: In this case, an attacker is active and can change the network. There are two types:
  - **The active attacker**: attempts to modify network information and generates malicious packets and signals (Rashmi et al. 2016b).

Table 1 Existing attacks for VANET and compromised services

Attacks	Authentication	Integrity	Confidentiality	Non-repudiation	Availability
Jamming					*
Traffic analysis		*	*		
Bogus information	*	*			
Masquerading	*				
Black hole					*
Message Fabrication/Suppression/Alteration		*		*	*
Men in the middle	*	*	*		
Gray hole	*	*	*	*	*
Malware/Spam	*	*	*	*	*

Fig. 5 Classification of attackers

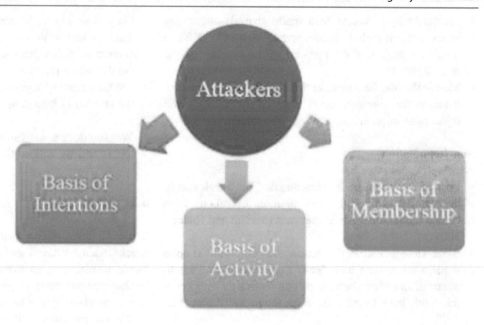

- **The passive attacker**: who cannot change or change anything in the network simply feels it (Parul and Deepak 2014).
- The basis of intension: Any attack is linked to the attacker's target; we can divide the attacker's intention into two types (Ajay et al. 2012):
  - **Relational attacker**: where malicious nodes seek personal advantage from their attacks.
  - **Malicious attackers**: that gains no personal benefit from their attacks. His basic motivation was to create obstacles to the functions of the network.

5   Related Works

New technologies in the world are evolving more and more, making life easier for individuals. The ad hoc vehicular network (VANET) is one of these technologies and is considered as an emerging model of networking.

VANET faces security issues and affects people lives. In the previous section, we introduced some attacks that can affect user authenticity/identity, confidentiality, integrity, and even network availability. Therefore, methods to detect and prevent security problems should be implemented. Detection includes the detection of various network security issues. Although prevention allows us to foresee possible security problems, we can distinguish two types: active prevention, in which we can use encryption methods to protect data or passive prevention; by assuming that all nodes exist in a secure environment. Therefore, nodes are vulnerable to most attacks if the assumption is not verified.

As mentioned above, VANET faces several security issues. Therefore, there must be a robust security audit mechanism that includes recording all or part of the operations performed on the system to provide data. In addition to determining how the security of the system was compromised.

In this study, we focused on intrusion detection systems (IDS) as a security audit mechanism. IDS stands for Intrusion Detection System. It is a device that monitors specific network or host activity, detects, and can respond to any intrusion attempt. In other words, intrusion detection is the process of monitoring events occurring in a node or network and analyzing signs of intrusion. An intrusion event is defined as an attempt to disrupt the security requirements of a node or network.

Many researchers have focused on detecting behavior on the VANET network. In this article, we will explain these methods. In (Golle et al. 2004), they approved a method for detecting and correcting malicious data in VANET. In this system, the vehicle will build a specific model that specifies the regulations and attributes of the VANET environment. The principle is simple: when a node receives a message, it compares it to the VANET version. If the message obtained does not conform to the predefined template, the message is considered to be invalid. In (Ghosh et al. 2009), they proposed a misbehavior detection scheme (MDS), which took into account certain parameters, such as the effect on mobility. The on-board unit (OBU) is responsible for identifying false messages and taking into account the driver's behavior toward an event. In (Raya et al. 2007), they developed a misbehavior detection system (MDS) which excluded the malicious behavior of the vehicle during communication. Its principle is to refuse to attack the nodes. Encryption keys belonging to a malicious node are revoked upon detection (during the process) and executed using a clustering algorithm to determine normal and abnormal behavior. In (Schmidt et al. 2008), they proposed a model based on reputation. Vehicles are based on shipping requirements. In this way, they created a normal behavior model for the nodes in VANET. If the node behaves differently from normal driving, it will be marked as suspect. In (Kim and Bae 2008), they proposed a misbehavior-based reputation management system (MBRMS). MBRMS consists of three main parts: misbehavior detection, reboardcast, and eviction locking algorithm. The system is responsible for detecting false alert messages by observing the actions of the nodes. In (Daeinabi and Rahbar 2013), they proposed a detection algorithm called DMV or misbehavior detection vehicle. The main function of the system is to discover malicious nodes by observation. It isolates malicious nodes from authentic nodes by filtering out nodes that duplicate or drop packets. Another system introduced in (Wahab et al. 2014) is based on the principle of maintaining the quality of service (QoS) during the movement of the nodes. They proposed QoS-OLSR, which reduces the number of malicious nodes and false negative messages. In (Coussement et al. 2013), they proposed an intrusion detection system capable of detecting malicious nodes. This study separated two IDS-based vehicles installed in each vehicle and is responsible for detecting problems during vehicle activation. IDS-based RSU; this system retains more security since detection is also carried out in the road unit, that is to say outside the vehicle. Zone-based intrusion detection system (ZBIDS); Another IDS principle responsible for detecting anomalies consists of dividing the network into zones without overlap. ZBIDS uses local and collaborative detection methods (Sun et al. 2003). Jaydis and Sen proposed a semi-centralized cluster architecture which integrates local intrusion detection (Sen 2010). In this architecture, the network is divided into clusters which can be controlled by the main cluster and the inter-cluster cluster. The head and the cluster start a conversation and communicate with a database which continues to identify the attacks. Table 2 summarizes our research.

6 Comparative Study of Intrusion Detection Algorithm

In the VANET network, the nodes cooperate with each other to manage the establishment of communication, the distribution of path interruption notifications, and the retransmission of data. Because of this feature, attacks become relatively easy. Attacks are designed to modify the network topology, such as declaring a false neighbor, using a false identity (Sybil attack), or reducing its performance. Attacks such as black hole attacks, the declaration of non-optimal routes, or the creation of routing loops in order to make it inoperative, either by internal malicious nodes or by nodes compromised by an attacker during the exploitation of the network. In this paper, we study different intrusion detection algorithm which aims to detect anomalies in the VANET network.

The intrusion detection system (IDS) has two modes: anomaly detection and signature recognition.

- Anomaly detection includes the detection of anomalies related to normal traffic profiles. The implementation always includes a learning phase during which IDS will discover the normal operation of the monitored elements.
- Signature detection is a method which consists of searching for the fingerprint (or signature) of a known attack in the activity of a monitored element. This type of IDS is purely reactive. It can only detect signed attacks.

In this section, a comparative study of different intrusion detection algorithms that has been carried out this study is based on the comparison between three main criteria: Detection method, respond type, and the frequency of use.

- **Detection method**: the way that an attack is detected, we divide the detection method into two types:
  Signature detection where the attack is considered as known.
  Behavior detection by filtering the behavior of vehicles in the VANET network.
- **Respond type**: the way that the vehicle act, we can split the respond type into two types:
  Active detection where the algorithm responds and acts against an attack.
  Passive don't act against an attack.
- **Frequency of use**: in this case, we can divide between two types: periodic or continuous.
  Periodic detection where the algorithm checks periodically on the VANET network.
  Continue checking continuously the traffic in the VANET network.

Table 3 present a comparison of intrusion detection algorithm:

7 Conclusion

The vehicular ad hoc network attracted various researchers. In reality, VANET network is susceptible to attacks because of their shared communication channel, and lack of traditional safety infrastructure. The intrusion detection algorithm

Table 2 Intrusion detection algorithms

Algorithm	Architecture	Methodology	Advantages	Inconvenient	Mechanism
DCMN	Defined by the system	Comparing messages sent in the network and the architecture predefined	Keep privacy	Invalidity of collected data could affect the system	Delete malicious nodes
MD Scheme	Autonomous	Node monitoring	authentication	Issue to access to the certificate authority (CA)	Exclude false negative messages
MD System	Clustered	Determinate malicious nodes	Cooperation between nodes	Fixed parameters permit better detection	Doesn't cooperate with malicious node and decrease false negative messages
VEBAS	Distributed and cooperative	Reputation	Dynamic	The general communication costs involved in the system still low	Defining regular self behavior for each node by observing neighbors
MBRMS	Distributed and cooperative	Cooperative detection	Predefined list of attacks	Possibility to fall in a malicious node that can affect the system	Local detection regardless the other nodes
DVM	Distributed and cooperative	Reputation	Authentication	The attacker can modify the ID of the vehicle and will be considered as sincere node	Self-observation of neighbors and avoiding malicious node in routing search
QoS-OLSR	Autonomous	Monitoring nodes that provide routing	Scalability	Issue to find good path and keeping of QoS	Decrease false negative messages
ZBIDS	Clustered	Cooperative detection	Abstraction of sub-models	Complex	Mobile agent and Markov chain
Jaydis & Sen	Clustered	Signature	Reduce malicious nodes	Detect known attacks	The detection is made by the head cluster

Table 3 Comparaison of intrusion detection algorithm

Algorithm	Detection method		Respond type		Frequency of use	
	Signature	Behavior	Active	Passive	Periodic	Continue
DCMN		*	*		*	
MD scheme		*	*			*
MD system		*	*			*
VEBAS		*		*	*	
MBRMS	*		*			*
DVM		*	*			*
QoS-OLSR		*		*		*
ZBIDS		*	*			*
Jaydis & Sen	*		*			*

can be beneficial to improve and strength the security of vehicular ad hoc network, besides the traditional security methods such as encryption and authentication. in this paper, we have explained the VANET network, its architecture, communication types, and standard communication. The security requirement should be respected in order to enhance the security. Finally, we have presented a comparative study of different intrusion detection algorithm in order to understand the importance of this system. Our perspective is to build a system that will benefit from the advantages of the intrusion detection algorithm in order to detect and prevent security issues in VANET network.

References

Ajay, R., Santosh, S., Rama, S.: VANET: Security attacks and its possible solutions **3**, 301–304 (2012)

Bibhu, V., Roshan, K., Singh, K.B., Singh, D.K.: Performance analysis of black hole attack in VANET. Int. J. Comput. Netw. Inf. Secur. **4**, 47–54 (2012). https://doi.org/10.5815/ijcnis.2012.11.06

Coussement, R., Amar, B., Biskri, I.: Decision support protocol for intrusion detection in VANETs 31–38 (2013)

Daeinabi, A., Rahbar, G.: Detection of malicious vehicle(DVM) through monitoring in vehicle ad hoc networks **66**, 325–338 (2013)

Engoulou, R., Bellaïche, M., Pierre, S., Quintero, A.: VANET security surveys **44**, 13 (2014). https://doi.org/10.1016/j.comcom.2014.02.020

Erritali, M.: Contribution à la sécurisation des réseaux ad hoc véhiculaires. UNIVERSITÉ MOHAMMED V –AGDAL FACULTÉ DES SCIENCES, Rabat (2013)

Ghosh, M., Varghese, A., Kherani, A., Gupta, A.: Distributed misbehavior detection in VANETs. IEEE Wirel. Commun. Netw. Conf. WCNC (2009). https://doi.org/10.1109/WCNC.2009.4917675

Golle, P., Dan, G., Staddon, J.: Detecting and correcting malicious data in VANETs 29–37 (2004)

Isaac, J.T., Zeadally, S., Cámara, J.S.: Security attacks and solutions for vehicular ad hoc networks. IET Commun. **4**, 894 (2010). https://doi.org/10.1049/iet-com.2009.0191

Kim, C., Bae, I.-H.: A misbehavior-based reputation managment system for VANETs. Lect. Notes Electr. Eng. LNEE **181**, 441–450 (2012). https://doi.org/10.1007/978-94-007-5076-0_54

Liang, X., Weihua, Z., Sheng, Z., Cailian, C.: Learning based VANET communication and security techniques **140**, (2018). https://doi.org/10.1007/978-3-030-01731-6

Lo, N.W., Tsai, H.C: Illusion attack on VANET applications - a message plausibility problem. GLOBECOM - IEEE Glob. Telecommun. Conf. (2007). https://doi.org/10.1109/GLOCOMW.2007.4437823

Anwer, M.S., Guy, C.: A survey of VANET technologies **5**(9), (2014)

Maxim, R., Hubaux, J.-P.: The security of vehicular ad hoc networks. In: SASN'05 - Proceedings of the 2005 ACM Workshop on Security of Ad Hoc and Sensor Networks, vol. 2005, pp. 11–21 (2005). https://doi.org/10.1145/1102219.1102223

Nidhal Mejri, M., Jalel, B.-O., Hamdi, M.: Survey on VANET security challenges and possible cryptographic solutions. (2014). https://doi.org/10.1016/j.vehcom.2014.05.001

Parul, T., Deepak, D.: A taxonomy of security attacks and issues in vehicular ad-hoc networks (VANETs) 91 (2014). https://doi.org/10.5120/15893-5040

Punal, O., Pereira, C., Aguiar, A., Gross, J.: Experimental characterization and modeling of RF jamming attacks on VANETs. IEEE Trans. Veh. Technol. **64**, 524–540 (2015). https://doi.org/10.1109/TVT.2014.2325831

Rashmi, M., Akhilesh, S., Rakesh, K.: VANET security: Issues, challenges and solutions 1050–1055 (2016a)

Rashmi, M., Akhilesh, S., Rakesh, K.: VANET security: Issues, challenges and solutions 1050–1055 (2016b)

Raya, M., Papadimitratos, P., Aad, I., Jungels, D., Hubaux, J.: Eviction of misbehavior and faulty nodes in vehicular networks **25**, 1557–1568 (2007)

Rushit, D., Evelyn R, wells B., Kaushik, R.: Efficient data privacy and security in autonomous cars **7**, 32–36 (2019). https://doi.org/10.12691/jcsa-7-1-5

Saeed Al-kahtani, M.: Survey on security attacks in vehicular Ad hoc networks (VANETs) 9 (2012). https://doi.org/978-1-4673-2393-2/12

Schmidt, R.K., Leiinmuller, T., Schoch, E., Held, A., Schaefer, G.: Vehicle behavior analysis to enhance security in VANETs. IEEE V2Vcom (2008)

Sen, J.: An intrusion detection architecture for clustered wireless ad hoc networks 202–207 (2010)

Sun, B., Wu, K., Pooch, W.: Alert aggregation in mobile ad hoc networks. In: Proceedings of the Workshop on Wireless Security, pp. 69–78. (2003). https://doi.org/10.1145/941311.94132310

Tanuja, K., Sushma, T., Bharathi, M., Arun, K.: A survey on VANET technologies **121**, 9 (2015)

Wahab, A., Otrok, H., Mourad, A.: A cooperative watchdog model based on dempster-shafer for detecting misbehaving vehicles **41**, 43–54 (2014)

An Enhanced Energy-Efficient Hierarchical LEACH Protocol to Extend the Lifespan for Wireless Sensor Networks

Fatima Es-sabery[ORCID] and Abdellatif Hair

Abstract

Wireless sensor networks (WSNs) have an essential role in improving the efficiency and the effectiveness of existing intelligent transportation systems. Actually, gathering traffic information for traffic organization and management is carried out mostly via wireless sensors, against in the past the wired sensors are used. Therefore, hierarchical routing is an interesting approach to simplify the operation of a large-scale wireless sensor network into intelligent transportation systems. This type of routing is based on the virtual hierarchy of networks, which is one of the main tools for saving energy in each node of the network. Therefore, the lifetime of the overall system is extended. The hierarchical technique is used to partition the network into sub-sets to facilitate network management. Partitioning is based on cluster building algorithms. These hierarchical routing algorithms have particular advantages related to scaling and efficiency in communication; among these methods, we find the two protocols LEACH (Low Energy Adaptive Clustering Hierarchy) and PEGASIS (Power-Efficient Gathering in Sensor Information Systems). In this paper, we propose a new hybrid protocol that combines the chain communication approach inspired by the PEGASIS protocol with the cluster communication approach proposed by the LEACH protocol. The experimental results provide statistics about the lifetime of the overall network and the energy-consuming. The obtained results from the performed experiments show that our proposed approach (chained approach + clustered approach) is more efficient in terms of network lifetime compared to LEACH and PEGASIS protocols.

F. Es-sabery (✉) · A. Hair
Faculty of Sciences and Technology, Sultan Moulay Slimane University, B.P.523, Beni-Mellal, Morocco
e-mail: fatima.essabery@gmail.com

A. Hair
e-mail: abd_hair@yahoo.com

Keywords

WSNs • Hierarchical routing protocol • LEACH • PEGASIS • Energy consumption

1 Introduction

Advances in WSNs have contributed to the development of the Internet. Therefore the users have full access to all regardless of their geographic location (Wang 2018). Evolution of wireless communications and mobile computing is gaining popularity, and mobile components are becoming more common. Like many technologies, WSNs have emerged for military purposes such as surveillance on the battlefield. Then they found their way to a civilian application (Benkhelifa et al. 2016). Today, WSNs have become a key technology for different types of intelligent environments (Suryadevara et al. 2015); for example, they provide us to have a sound home security system. These networks have particular importance when a large number of sensor nodes have to be deployed, in dangerous situations. For example, for disaster management (Huang et al. 2010), a large number of sensors can be dropped by a helicopter. These sensors carry out the rescue operations by locating survivors, identifying risk areas, or briefing the rescue team. This usage of WSN increases the efficiency of rescue operations and ensures the safety of the rescue team (Harbouche et al. 2017).

WSNs have become a valuable tool to collect, transmit, and receive environmental data of several applications such as Airport Logistics (Wang 2018), Disaster Management (2016), Environmental Monitoring (Suryadevara et al. 2015), Home Security (Huang et al. 2010), Military Surveillance, and Medical Monitoring (Harbouche et al. 2017) by providing efficient, credible and secure communication, reliable control, and carrying out applications. They are characterized by many restrictions: resource limitations, the lifetime of the overall network, energy consumption, and high bandwidth demand. Wireless sensor networks are non-wired networks. Generally, they are composed of numerous units for detecting physical phenomena (humidity, radioactivity, temperature, etc.), which are densely distributed in a geographic area of interest called "capture filed". These units are tiny devices, called sensor nodes, varying from a few tens of elements to several thousand. The wireless sensor nodes are self-organizing to form a credible and active communication network, in order to transmit collected information over a wireless medium, hence the name of Wireless Sensor Network (Song and Yao 2017).

A sensor node is typically composed of the following elements: detection and measurement unit, processing and storage unit, a communication unit, and a power resource unit (Es-sabery and Hair 2020). The detection and measurement unit is made up of electronic sensors capable of measuring some physical quantities: deformations, temperature, brightness, pressure, and humidity in the environment where the sensor nodes are deployed. The processing component is responsible for collecting and processing the captured data. The wireless communication component is responsible for transmitting and receiving data captured from one sensor node to another sensor node, it ensures the transfer of all this measured and received data to another node or to a final destination represented by the base station. And the power resource unit of the sensor node is a battery used to power all of its components (Harbouche et al. 2017; Song and Yao 2017). Therefore, in most applications of the sensor network (Benkhelifa et al. 2016), the nodes are deployed in hard-to-reach areas. As a result, the overall lifetime of the network is entirely dependent on that of the battery. The use of energy recharging systems can increase the lifetime of the network based on the intermittent energy re-covered from the studied environment. Solar cells or Rectennas are some examples of these systems. However, the amount of energy that can be extracted from it is variable and limited. Therefore, to date, WSNs are limited in terms of available energy (Huang et al. 2010; Heinzelman et al. 2000).

Due to the miniaturization constraints (Heinzelman et al. 2000), the sensor nodes have minimal resources in terms of computing capacity, data storage space, transmission, and energy flow. These constraints are part of the research questions in the area of wireless sensor networks. Especially the constraint related to the energy consumption, which is considered as a big problem. Indeed, all sensor nodes need the energy to work; the control of the energy-consuming by the sensor node is the only way for increasing its lifetime (Es-sabery and Hair 2020; Lindsey and Raghavendra 2002). Hierarchical routing is deemed as an effective way of decreasing energy consumption compared to other routing approaches. Many hierarchical routing protocols are proposed. Namely, Protocol LEACH is a self-organizing protocol based on adaptive clustering thus uses the randomized rotation of cluster heads to distribute the loaded energy in an inequitably way between the sensor nodes (Heinzelman et al. 2000), and PEGASIS protocol uses the principle of constructing a chains sensor node (Lindsey and Raghavendra 2002).

In this paper, we propose a new approach that improves the LEACH protocol using PEGASIS protocol. As known, the LEACH routing protocol is based on the clustered approach. It divides the network into clusters in a distributed way, Cluster-Head nodes are elected then used as relays to reach the base station. Therefore, in our improved LEACH, we apply the chained approach of PEGASIS within clusters formed by LEACH, and we relate the cluster heads between them using the chained approach. Generally, our proposed method combines the advantages of both approaches; chained approach and clustered approach.

The rest of this paper is structured as follows: In the next section, we present the literature review. In Sect. 3 we define the energy consumption model, we describe both protocols LEACH, PEGASIS and their energy consumed in WSN. In Sect. 4 we present our proposed approach. In Sect. 5 we evaluate the performance of our proposed method and the other protocols LEACH and PEGASIS. Finally, we summarize our work, and we give perspectives in Sect. 6.

2 Literature Review

Recent papers have been performed on WSNs regarding hierarchical routing protocols. These research papers are mainly centering on auto-configuration, energy-consuming and how to save it, node mobility, and scalability. And they discuss the various techniques applied for enhancing the performance and the effectiveness of WSNs as explained below:

There are several challenges and constraints that wireless routing algorithms should regard itself with. Authors of the paper (Abu Salem and Shudifat 2019) proposed a new improvement in LEACH protocols. This improvement aims to examine the power sensitivity when the member node chooses a suitable cluster head that has a minimum distance to the sink and has reduced the energy consumption, therefore, increasing the lifetime of the overall network. They performed an experiment to compare the proposed approach and LEACH, and the simulation results reveal that the consuming of battery power is reduced, and hence the lifetime of the overall network increased. The comparison between LEACH and their suggested method shows the proposed one performs better than LEACH.

Krishna Kumar et al. suggested new schemas for LEACH protocol called "Multi Power Amplification of Energy-Efficient LEACH Protocol for WSNs". They followed the same stages as of in LEACH algorithm, but the only difference is in the selection of cluster head, where they suggested a new function to choose an efficient cluster head. In their experiments, the energy used by the amplifier, and threshold to elect the cluster head are selected as metrics. And they compared their proposed approach with the existing methods, and the results showed that their approach is more efficient in increasing the lifetime of the network than others.

"Improvement of LEACH Protocol for Wireless Sensor Networks" is proposed by Takele et al. (Takele et al. 2019). In their proposed approach, the clusters have formed as in the traditional LEACH on the first-round, then normal nodes have scheduled for working as a CH on each iteration turn by turn sequentially. The scheduled normal node works as CH until 75% of its total energy is wasted. After 75% of the total energy has consumed, re-clustering will be performed to eliminate the dead sensor nodes and sensor nodes with low power from the schedule. The proposed method has been examined relatively utilizing a simulator called Castalia 3.3, and it outperforms the traditional LEACH protocol.

In (Nguyen et al. 2019), Nguyen et al. proposed a novel method to enhance LEACH protocol using Fuzzy Logic theory. They considered that the collision probability influences energy consumption and affects the transmission process, such as node density or communication traffic. Then they decided to derive the relationship between the number of transmitted packets, the transmission rate, the number of neighbouring nodes, and the total number of nodes in WSN. They used the Fuzzy Logic to represent this relationship among inputs by using some rules to generate the desired output. The experimental results show that their suggested protocol improves the energy efficiency of a WSN compared to the original LEACH.

In this paper (Singh et al. 2019), multiobjective routing techniques are suggested for minimizing the energy consumption, increasing the lifetime of the overall network, raising the network throughput, and reducing network delay. A fitness method has been proposed for homogenous and heterogeneous WSN. This fitness technique is used to elect an efficient CH for energy optimization and load balancing of CHs. Based on this proposed fitness function, a novel routing algorithm is proposed for the clustering process. The experimental results show that the hybrid clustered proposed approach achieves excellent results in terms of increasing the lifetime of the overall network, where the proposed method improves the lifetime by 26%, 63%, and 10% compared with three heterogeneous approaches: EDDEEC, DEEC, and ATEER, respectively.

3 Energy Consumption Model

The sensor node consumes energy to perform three main tasks: detecting, communication, and data processing. The energy utilized for the disclosure of physical phenomenon is negligible. As well as the one used for the treatment is lower than the energy of communication. For example, the prerequisite energy to transfer 1 KB over a distance of 100 m is approximately equal to the consumption energy using to run 3 million instructions with a speed of 100 million instructions per second (Es-sabery et al. 2017). While the necessary energy for processing the data is calculated by applying the following formula:

$$E_{DA} = \frac{5\,nanojoule}{1\,bit} \tag{1}$$

Since communications consume much more energy than other tasks, Heinzelman et al. proposed a power radio consumption model (Heinzelman et al. 2000). Thus, the necessary energies to emit E_{tx} and receive E_{rx} the message are given by:

- To send a message of k bits over a distance of d meters, the transmitter consumes:

$$
\begin{aligned}
E_{tx}(k, d) &= E_{tx-elec}(k) + E_{tx-amp}(k, d)\\
&= k.E_{elec} + k.E_{fs}d^2 \ if\ d < d_0 \\
Or\ &= k.E_{elec} + k.E_{amp}d^4 if\ d \geq d_0
\end{aligned}
\tag{2}
$$

- To receive a message of k bits, the receiver consumes:

$$E_{rx}(k) = k.E_{elec} \tag{3}$$

Where $E_{tx\text{-}elec}$ (k) is the transmission energy, $E_{tx\text{-}amp}$ (k, d) is the amplification energy, E_{elec} is the amount of energy consumed by a bit, and E_{fs} is the signal amplification in a lower distance to the threshold d_0. If the distance transmission is superior to d_0 the amplification energy E_{amp} is used such as

$$d_0 = \sqrt{\frac{E_{fs}}{E_{amp}}} \tag{4}$$

3.1 Energy Consumed by the LEACH Protocol

In (Heinzelman et al. 2000), Heinzelman et al. proposed a distributed clustering algorithm called LEACH, used for routing in the homogeneous sensors networks. LEACH randomly selects the Cluster-Head nodes and attributes this role to the various nodes according to turnstile management policy to ensure equitable surge absorption between nodes. In order to reduce the amount of information transmitted to the base station, the cluster heads aggregate the data captured by the member nodes belonging to their cluster and send an aggregated package to the base station. LEACH is founded on two basic assumptions (Es-sabery et al. 2017):

– The base station is fixed and is placed away from other sensors.
– All nodes in the network are homogeneous and limited in terms of energy.

The energy dissipation of a WSN using the LEACH hierarchical routing protocol is calculated as follows:

4 Initialization Phase Set-Up

According to the radio energy consumption model presented in Sect. 3, the energy consumption by CH nodes and all normal nodes in the cluster formation phase is calculated as follows:

E_{ch1}: is the energy consumed by the CH to broadcast an announcement message p to the entire network, assuming that the broadcast range for the CH is the size of its cluster.

$$E_{ch1} = p.E_{elec} + p.E_{fs}.d^2_{ch,noch} \qquad (5)$$

$E_{no\ ch1}$: is the energy spent by each member node receiving the announcement message broadcasted by the CH of the cluster is as follows:

$$E_{noch1} = p.E_{elec} \qquad (6)$$

$E_{no\ ch2}$: is the energy required for a member node to send its join-request message to its CH is

$$E_{no\_ch2} = p.E_{elec} + p.E_{fs}.d^2_{noch,ch} \qquad (7)$$

E_{ch2}: is the energy consumed by each CH that receives the join-request messages sent by the members of its cluster is

$$E_{ch2} = p.E_{elec}.\left(\frac{n}{k} - 1\right) \qquad (8)$$

E_{ch3}: is the energy expended by the CH in order to create the TDMA transmission schedule for the member nodes of its cluster:

$$E_{ch3} = \frac{n}{k}.p.E_{planningTDMA}. \qquad (9)$$

E_{ch3}: is the energy consumed by the CH to send the TDMA schedule to the members of its cluster:

$$E_{ch3} = (p.E_{elec} + p.E_{fs}.d^2_{ch,no\_ch}).\left(\frac{n}{k} - 1\right) \qquad (10)$$

$E_{no\ ch3}$: is the energy consumed by a member node to receive the TDMA message sent by its CH is

$$E_{noch3} = p.E_{elec} \qquad (11)$$

In summary, the energy spent by the CH node in the cluster construction phase is

$$E_{initialization,ch} = E_{ch1} + E_{ch2} + E_{ch3}$$

$$= 2\left(\frac{n}{k} - 1\right).p.E_{elec} + \frac{n}{k}.p.E_{fs}.d^2_{ch,no\_ch} \qquad (12)$$

Thus, the energy consumed by the member node in the cluster construction phase is

$$E_{initialization,non\ ch} = E_{no\ ch1} + E_{no\ ch2} + E_{noch3}$$

$$= 3.P.E_{elec} + p.E_{fs}.d^2_{noch,ch} \qquad (13)$$

Where **n** is the total number of sensor nodes in the network, **k** is the number of clusters in the network, **p** is the announcement message, and **d** is the distance between a normal node and the CH.

5 Transmission Phase Steady State

The data transmission phase is divided into several frames. Within each frame, the members of each cluster transmit the data to the CH in their time slots. Once the CH receives all the data, it performs the aggregation of data to obtain a common signal, and then transmits it to the BS. In this process, the members of each group dissipate the energy in order to transmit their data to the group CH and the CHs dissipate their energy in order to receive data sent by the nodes members of their group, the aggregation of these data. And their transmission to BS. Therefore, the energy consumed by the CH and the member nodes at a single frame is presented as follows:

E_{ch} is the energy consumed by the CH node during a single frame is

$$E_{ch} = 1.E_{elec}.\left(\frac{n}{k} - 1\right) + \left(\frac{n}{k} - 1.E_{DA}\right) + 1.E_{elec} + 1.E_{amp}.d^4_{ch,sb}$$

$$(14)$$

$E_{no\ ch}$: is the energy dissipated by the member node during a single frame is

$$E_{no\_ch} = 1.E_{elec} + 1.E_{fs}.d_{noch,ch}^2 \qquad (15)$$

Assuming there is an M frame in the data transmission phase. So the energy consumed by a group in the transmission phase for a given cycle is

$$E_{group} = M.E_{ch} + \left(\frac{n}{k} - 1\right).M.E_{no\_ch}$$

$$= M.\left[1.E_{elec}.\left(\frac{n}{k} - 1\right) + \left(\frac{n}{k}.1.E_{DA}\right) + 1.E_{elec} + \right.$$

$$\left. 1.E_{amp}.d_{ch,sb}^4\right] + \left(\frac{n}{k} - 1\right).M$$

$$.(1.E_{elec} + 1.E_{fs}.d_{no\_ch,ch}^2) \qquad (16)$$

From these two steps, it is deduced that the energy dissipation of each group at each cycle is calculated as follows:

$$E_{cycle} = E_{initialization,ch} + \left(\frac{n}{k} - 1\right).E_{initialization,non\_ch} + E_{group}$$

$$= \left(2.\frac{n}{k} - 1\right).p.E_{elec} + \frac{n}{k}.p.E_{fs}.d_{ch,no\_ch}^2$$

$$+ \left(3.p.E_{elec} + p.E_{fs}.d_{no\_ch,ch}^2\right)\left(\frac{n}{k} - 1\right)$$

$$+ \left[1.E_{elec}.\left(\frac{n}{k} - 1\right) + \left(\frac{n}{k} - 1.E_{DA}\right) + \right.$$

$$\left. 1.E_{elec} + 1.E_{amp}.d_{ch,sb}^4\right].M$$

$$+ \left(\frac{n}{k} - 1\right).M.\left(l.E_{elec} + l.E_{fs}.d_{no\_ch,ch}^2\right) \qquad (17)$$

5.1 Energy Consumed by the PEGASIS Protocol

In (Lindsey and Raghavendra 2002), Lindsey and Raghavendra proposed an improved version of LEACH called PEGASIS. The main idea of PEGASIS is to form a chain between nodes. The collected data is transmitted from one node to another one witch aggregates them, until they arrive at a particular node that transmits them to the base station. The nodes that transmit data to the base station are selected alternately according to a round-robin policy to reduce the average energy expended by a node during a period. Unlike LEACH, PEGASIS prevents the formation of clusters and provides a single node in the chain to send data to the base station. Moreover, PEGASIS assumed that the nodes are able to change their power transmission. Therefore, the energy consumption by the sensor nodes in a network using the PEGASIS protocol is calculated as follows:

5.1.1 Case 1: Node Leader Will Be the Head Node or Tail Node of the Chain

E_1: is the energy consumed by the sensor node to broadcast a message to its neighbours.

$$E_1 = 2.1.E_{DA} + 1.E_{elec} + 1.E_{fs}.d_{neighbors}^2 \qquad (18)$$

E_2: is the energy consumed by the head node or the tail node of the chain to send the message l.

$$E_2 = 1.E_{elec} + 1.E_{fs}.d_{neighbors}^2 \qquad (19)$$

E_3: is the energy consumed by the leader node to send its message to the BS.

$$E_3 = 2.1.E_{DA} + 1.E_{elec} + 1.E_{fs}.d_{BS}^2 \qquad (20)$$

In summary, the energy spent by the sensor nodes all along the chain in the data transmission phase is

$$E_{total} = (n - 2).E_1 + E_2 + E_3$$

$$= (n - 2).(2.1.E_{DA} + 1.E_{elec} + 1.E_{fs}.d_{neighbors}^2) + 2.1.E_{DA} + 1.E_{elce}$$

$$+ 1.E_{fs}.d_{BS}^2 + 1.E_{elec} + 1.E_{fs}.d_{neighbors}^2$$

$$= (2n - 2).1.E_{DA} + n.1.E_{elec} + 1.E_{fs}.(d_{BS}^2 + (n - 1)d_{neighbors}^2) \qquad (21)$$

5.1.2 Case 2: Node Leader Will Be a Sensor Node Differ from the Head/tail Node of the Chain

E_1: is the energy consumed by the sensor node to spread its message to its neighbors.

$$E_1 = 2.1.E_{DA} + 1.E_{elec} + 1.E_{fs}.d_{neighbors}^2 \qquad (22)$$

E_2: is the energy consumed by the head/tail node of the chain to spread its message l to its neighbors.

$$E_2 = 1.E_{elec} + 1.E_{fs}.d_{neighbors}^2 \qquad (23)$$

E_3: is the energy consumed by the leader node to send its message to the BS.

$$E_3 = 3.1.E_{DA} + 1.E_{elec} + 1.E_{fs}.d_{BS}^2 \qquad (24)$$

In summary, the energy spent by the sensor nodes all along the chain in the data transmission phase is

$$E_{total} = (n - 3).E_1 + E_3 + 2.E_2$$

$$= (n-3).(2.1.E_{DA} + 1.E_{elec} + 1.E_{fs}.d_{neighbors}^2) + 2.1.E_{DA} + 1.E_{elec}$$

$$+ 1.E_{fs}.d_{BS}^2 + 2.(1.E_{elec} + 1.E_{fs}.d_{neighbors}^2)$$

$$= (2n-4).1.E_{DA} + n.1.E_{elec} + 1.E_{fs}.(d_{BS}^2 + (n-1).d_{neighbors}^2)$$
$$\tag{25}$$

6 Proposed Hybrid Routing Protocol

In this section, we will discuss the different phases of our work and to present the methodology of our hybrid routing protocol. As we have said previously, the objective of our suggested hybrid protocol is to improve the LEACH protocol using the concept of PEGASIS protocol, in order to increase the network lifetime and the scalability. The improvements made on LEACH using the concepts of PEGASIS protocol are an improvement made at the all normal nodes of each cluster and another improvement made at all cluster heads. As we presented earlier, the PEGASIS is based on a chained approach; that to say, it organizes all sensor nodes in WSN as the form of the chain. Therefore in our approach, we applied the chained method as PEGASIS described at normal nodes of each cluster and at nodes CH. We can summarize our work in the two following steps:

- Application of the chained approach, which described by the PEGASIS protocol at all member nodes in each cluster.
- Application of the chained method, which defined by the PEGASIS protocol at all cluster heads in WSN.

6.1 Phase 1: Application of Chained Approach at All Member Nodes in Each Cluster

In this phase, the member nodes of each cluster are organized as a chain. Hence, the communication between them will be as follow: the sensor node SN_0 forwards its collected data to its nearest sensor node SN_1, SN_1 assembles its collected data with the data transmitted from the node SN_0 and sends them to its closest member node until they attain to the leader node that sends them to the CH. Therefore, in this first phase (chained cluster), all member nodes in the group will send their collected information to their CH by passing them by the formed chain, while each leader node transmits all aggregated data to its cluster head. The forming of all member nodes of each cluster in a chain can enhance and

adjust the energy consumption, which minimizes the load of CH. Also, each member node is connected only with its closest neighbors and not immediately with its CH (In LEACH protocol each member node communicates directly with its CH), which reserves more energy and gives an efficient utilization of the bandwidth. Without forgetting, the utilization of the aggregation process at the level of each member node has an essential role in preserving energy. Figure 2 shows how we formed the cluster using the LEACH protocol or our proposed approach.

6.2 Phase 2: Application of Chained Approach at All Cluster Heads

In this phase, the cluster heads in the network are organized as a chain. Hence, the communication between them will be as follow: the cluster-head CH_0 transmits its collected data to its nearest cluster-head CH_1, CH_1 assembles its collected data with the data transmitted from the cluster-head CH_0 and sends them to its closest cluster head until they attain to the leader cluster head that sends them to the Base Station (BS). Therefore, in this first phase (chained cluster heads), all cluster heads in the network will transmit their collected data to the BS by passing them by the formed chain.

As we said previously, this second phase aims to form the nearest neighbor chain between the cluster-head nodes, which increases the lifetime of the furthest cluster-head node from the sink and prohibits them from dying quickly. Also, the applied aggregation process at the level of each cluster head minimizes the number of transmission to the sink (base station) to one transmission performed by the closest cluster-head (leader cluster head) to the base station, which decreases the load on the sink. These enhancements in the way used by cluster heads to communicate with the base station allow saving the energy consumption, hence to maximize the lifetime of the overall network.

The formation of the chained cluster and the chained cluster heads can be carried out in a distributed method by the cluster heads or in a centralized approach by the base station. For better results, in terms of a similar distribution of normal nodes within the clusters, we used the centralized method suggested by LEACH (Takele et al. 1026), where each normal node transmits a data packet to the sink. The transmitted packet consists of the node location in the network, the residual energy of the node, and the identifier of the sensor node. Figure 3 illustrates how the cluster heads send their data to the base station using the LEACH protocol or the proposed approach.

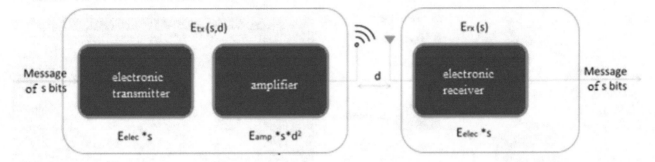

Fig. 1 Energy consumption model

Fig. 2 Formation of the cluster using the LEACH protocol or our proposed approach

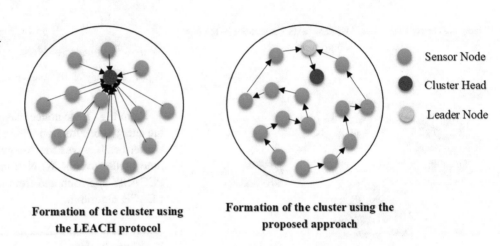

Formation of the cluster using the LEACH protocol

Formation of the cluster using the proposed approach

- Sensor Node
- Cluster Head
- Leader Node

7 Simulation Experiments and Analysis

The first performed experiment in our work is to compare to both protocol LEACH and PEGASIS. In this simulation, we use a network of 100 nodes randomly distributed the following Fig. 4 shows a random network of 100 nodes. The base station is located at (25m,150m) in a field of 50 m x 50 m, and located at (50m, 300m) in a field of 100 m x 100 m. The simulations were performed to determine the number of round's communication when 1%, 20%, 50%, and 100% of nodes die for both protocols as well as the direct transmission of data to the base station. Once a sensor node loses almost all of its energy, it is considered dead in the rest of the simulation.

The simulation shows that the protocol PEGASIS accomplished:

- Approximately two times the number of cycles, compared to the LEACH protocol when 1%, 20%, 50%, and 100% of nodes die for a 50 m x 50 m network.
- Approximately three times the number of cycles, compared to LEACH when 1%, 20%, 50%, and 100% of nodes die for a network of 100 m x100m.
- The sensor nodes consume their energy in a balanced manner so there is full use of the network.

Fig. 5 shows the number of cycles until 1%, 20%, 50%, and 100% of nodes die for a network of 50m × 50 m and Fig. 6 illustrates the number of cycles but for a network of 100m × 100 m. The initial energy value for the nodes is 0.25 J. PEGASIS is approximately 2 times better than LEACH in all cases for a 50m × 50 m network. For a 100 m x 100 m network, PEGASIS extended the network lifetime by 3 times compared to the LEACH protocol.

The results of the simulation have shown that PEGASIS helps to extend the lifetime of the network better than LEACH. Such a the performance gain is achieved by eliminating the energy over-consumption caused by the periodic reelection of CHs, which is automatically followed by a configuration of the cluster in LEACH, reduction in the number of nodes transmitting data to the base station in one node and the optimization of the amount of data sending to the base station.

The second experiment in our work is to compare the lifetime network using the proposed approach with both algorithms LEACH and PEGASIS. We compute the remaining energy of sensor nodes in each round to obtain the number of transmission rounds when 1%, 20%, 50%, and 100% of nodes lose all energy and hence die; the result is illustrated in Fig. 7:

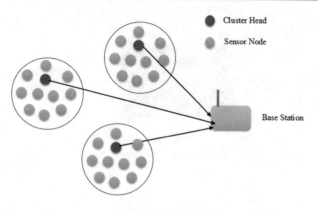

How the cluster heads send their data to the base station using the
LEACH protocol

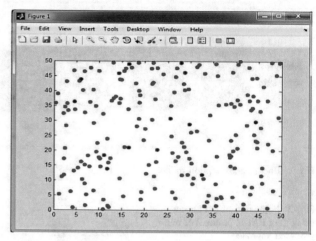

Fig. 4 Topology of 100 random nodes

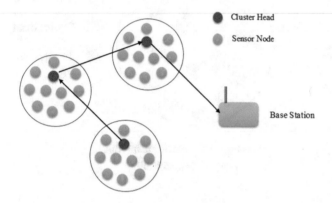

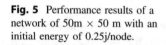

How the cluster heads send their data to the base station using the
proposed approach

Fig. 3 How the cluster heads send their data to the base station using
the LEACH protocol or the proposed approach

From Fig. 7, we notice that our propose approach reduces
the energy consumption within the cluster and at the level of
cluster heads, saves the energy gain, and increases the lifetime of the overall WSN from 12% to 20% compared to
PEGASIS approach and from 49% to 76% compared to the
LEACH algorithm.

8 Conclusion

In this paper, we studied and analyzed both protocols
LEACH and PEGASIS. We implemented both algorithms
using MATLAB software, and the simulation results show
that the PEGASIS protocol outperforms the LEACH protocol in terms of network lifetime and the scalability.

Fig. 5 Performance results of a
network of 50m × 50 m with an
initial energy of 0.25j/node.

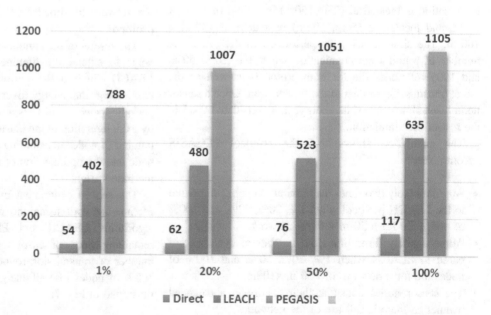

Fig. 6 Performance results of a network of 100 m × 100 m with an initial energy of 0.25 J/node.

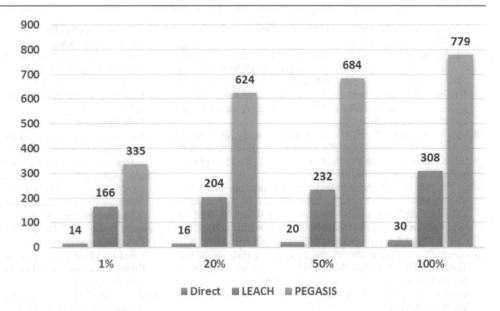

Fig. 7 The performance results of a network of 50 m × 50 m with an initial energy of 0.25j/node

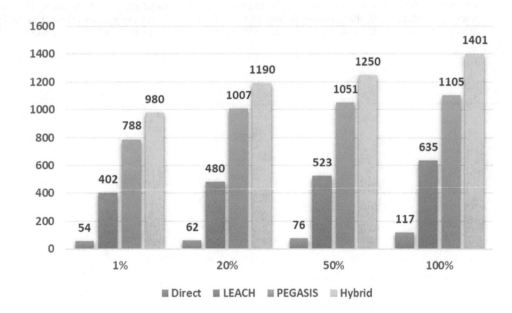

Motivated by the results obtained through the analysis and the study of both protocols; we decided to improve the performance of LEACH protocol using the chained approach of the protocol PEGASIS. The improvements made on LEACH using the concepts of PEGASIS protocol are an improvement made at the all normal nodes of each cluster and another improvement made at all cluster heads. The experimental results illustrate that our hybrid routing protocol performs better than both protocols LEACH and PEGASIS.

As we have known, the chained approach takes into consideration only the distance to form the chain between the nodes. In our future work, we plan to improve the performance of our hybrid routing protocols by also considering the remaining energy of each node to form the chain and to choose an efficient leader cluster head.

References

Wang, L.: Application of wireless sensor network and RFID monitoring system in airport logistics. Int. J. Online Eng. **14**(01), 89–103 (2018)

Benkhelifa, I., Nouali Taboudjemat, N., Moussaoui, S.: Disaster management projects using wireless sensor networks: an overview. Informatique, science de l information et bibliothconomie **21**(2), (2016)

Suryadevara, NK., Mukhopadhyay, SC., Kelly, SDT., Gill, SPS.: WSN-based smart sensors and actuator for power management in intelligent buildings. In: IEEE/ASME Transactions on Mechatronics **20**(2), 564–571 (2015)

Huang, H., Xiao, S., Meng, X., Xiong, Y.: A Remote home security system based on wireless sensor network and GSM technology. 2010 Second International Conference on Networks Security, Wireless Communications and Trusted Computing, vol. 1, pp. 535–538. Wuhan, Hubei (2010). https://doi.org/https://doi.org/10.1109/NSWCTC.2010.132

Harbouche, A., Djedi, N., Erradi, M., Ben-Othman, J., Kobbane, A.: Model driven flexible design of a wireless body sensor network for health monitoring. Comput. Netw. **129**, 48–571 (2017). https://doi.org/10.1016/j.comnet.2017.06.014

Song, Y., Yao, X.: Design of routing protocol and node structure in wireless sensor network based on improved ant colony optimization algorithm. 2017 International Conference on Computer Network, Electronic and Automation (ICCNEA), pp. 236–240. Xi'an, China (2017). https://doi.org/https://doi.org/10.1109/ICCNEA.2017.54

Es-sabery, F., Hair, A.: Evaluation and comparative study of the both algorithm LEACH and PEGASIS based on energy consumption. In: Proceedings of the 3rd International Conference on Networking Information Systems & Security, Marrakech, Morocco, (2020) March 31-April 2,

Heinzelman, WR. , Chandrakasan, A., Balakrishnan, H.: Energy-efficient Communication Protocol for Wireless Micro-sensor Networks. Proceedings of the IEEE Hawaii International Conference on System Sciences, vol. 2, pp. 10. Maui, HI, USA, USA (2000).

Lindsey, S., Raghavendra, C.S.: PEGASIS: Power-efficient gathering in sensor information Systems. Proceedings IEEE Aerospace Conference, vol.3, pp.1125–1130. Big Sky, MT, USA, USA (2002). https://doi.org/https://doi.org/10.1109/AERO.2002.1035242

Abu Salem, A.O., Shudifat, N.: Enhanced LEACH protocol for increasing a lifetime of WSNs. Pers Ubiquit Comput **23**, 901–907 (2019)

Krishna Kumar, A., Anuratha, V.: Energy-efficient LEACH protocol with multipower amplification for wireless sensor networks. In: Bhargava, D., Vyas, S. (eds.) Pervasive computing: a networking perspective and future directions. Springer, Singapore

Takele, A.K., Ali, T.J., Yetayih, K.A.: Improvement of LEACH protocol for wireless sensor networks. In: Proceedings of the information and communication technology for development for Africa conference, vol. 1026, Cham (2019).

Nguyen, T., Hoang, T.M., Pham, V.Q., Nguyen, T.T., Nguyen, N.S.: Enhancing energy efficiency of WSNs through a novel fuzzy logic based on LEACH protocol. In: 19th International Symposium on Communications and Information Technologies (ISCIT), pp. 108–112, Ho Chi Minh City, Vietnam, Vietnam (2019).

Singh, P., Singh, R.: Energy-efficient QoS-Aware intelligent hybrid clustered routing protocol for wireless sensor networks. J. Sens. **2019**, 8691878 (2019)

Es-sabery, F., Ouchitachen, H., Hair, A.: Energy optimization of routing protocols in wireless sensor networks. Int. J. Inf. Commun. Technol. **6**(2), 76–85 (2017)

A Survey of Security and Privacy for 5G Networks

Ahmed Ziani and Abdellatif Medouri

Abstract

In the near future, 5G networks will exploit the high-frequency technology in order to meet the vision of enabling billions of devices and millions of users to share the spectrum, to communicate and provide services. The fifth generation mobile communication technologies (5G) are also intended to respond to new use cases, such as the Internet of Things (IoT), and services. A major challenge to improve security and privacy at all the layers of the network becomes a fundamental aspect. This paper studies and discusses the existing security architecture and its main requirements, procedures, and topics. In addition, the security issues arising from the new business and trust models to be used in the 5G network and the standardization approach are also discussed in the main section of the paper.

Keywords

5G • Security architecture • IoT security • Privacy

1 Introduction

5G systems provide all the enhanced benefits of previous generations such as 2G, 3G, and 4G), and its evolution will move us into an era of ubiquitous, high-capacity radio. By bringing us to ever lower latency levels and extensive Gbps capacity. On the other hand, 5G evolutions will pave the way for highly intelligent Internet of Things (IoT)

A. Ziani (✉)
Computing, Systems and Telecommunications Laboratory,
UAE-FSTT, Tangier, Morocco
e-mail: zianiahmed@gmail.com

A. Medouri
Applied Remote Sensing Lab (RS&AID), UAE-ENSAT,
Tetuan, Morocco
e-mail: abdellatif.medouri@gmail.com

technologies and person-to-person connectivity. As the number of connected devices and applications continue to grow exponentially, the economic value that depends on their integrity will also increase. Intelligent networks will carry an abundance of consumer data and industry, moving the nature of business competition. In the era where everything will be connected over mobile broadband, 5G networks need to meet the requirements of security and privacy connectivity (Chandramouli et al. 2019; Marsch et al. 2018).

5G networks are generally divided into four logical parts: radio access network, core network, transport network, and interconnect network. Each logical part includes three so-called planes, each of which is responsible for transporting a different type of traffic, namely, the control plane which carries the signaling traffic; the user plane which carries the actual traffic; and the management plane which carries the administrative traffic. In terms of network security, the three planes can each be exposed to unique types of threats or to uniform threats simultaneously. For that, the security of telecommunications networks is defined by the components listed below (5G System Design: Architectural and Functional Considerations and Long Term Research; Rodriguez 2015):

- Standardization: The process by which operators, vendors, and other stakeholders set standards for how networks around the world will work together. This also includes how best to protect networks and users against malicious attacks.
- Network design: operators and suppliers develop and implement adequate standards for functional network elements and systems, which play a vital role in making the final product both functional and secure.
- Network configuration: during the deployment phase, networks are constructed for a targeted level of security, which is essential for defining security parameters and increasingly strengthening network security and resilience.
- Network deployment and operation: the operational processes that allow networks to operate and provide

targeted security levels are extremely dependent on the deployment and operations of the network itself.

In this paper, we aim to present security aspects in 3GPP 5G networks by presenting an overview of 5G security architecture and its security domains. Our contributions made in this work are to give an overview of Significance of Security and Privacy in the 5G networks. We mainly focus on the emerging risk prospects and Key Security Challenges in 5G networks. We analyze their Security Characteristics and discuss Internet of Things Security Issues. Finally, we give some 5G Security Topics for these new wireless networks.

The rest of the paper is organized as follows. First, we give an overview of the Security Standardization Approach for the 5G technology. Section 3 explains the Key Security Challenges in 5G. Section 4 introduces 5G security architecture. Sections 5 present the Significance of Security and Privacy. Section 6 describes the Security Characteristics of 5G and Internet of Things Security Issues, in the Sect. 7, the 5G Security Topics are explored. Finally, Sect. 8 presents our conclusions and provides suggestions for future work user.

2 Overview of 5G Security Standardization Approach

As highlighted in Table 1, various key organizations delivered immense contributions toward 5G security standardization development. The major task was to propose 5G security architecture by analyzing requirements and threats. In January 2016, the SA3 group (Holma et al. 2020) of 3GPP worked on standardization of 5G security aspects and make contributions, such as subscription privacy is one of the core security areas focused in 3GPP SA3. In February 2017, 3GPP published "Service Requirements for the 5G System" (TS22.261) that describes performance targets in several scenarios such as indoor, urban, rural, and different applications (Liyanage et al. 2018). In addition, the NGMN P1 WS1 5G security group focused mainly on the security requirements identification for MEC and the corresponding recommendations proposition (5G White Paper 2020). The SA3 group of 3GPP covered all security aspects including RAN security, network slicing, authentication mechanisms, and among others (SA3—Security 2020) 0.3GPP publishes 5G Phase-1specifications in 2018 as Release 15 and it publishes Phase-2 in 2020 as Release 16. Recognizing that 5G should be completely converged with Internet protocols, the standards produced by the Internet Engineering Task Force (IETF) are expected to play an axial role (Liyanage et al. 2018).

The relevant working groups of IETF are, for example, IP Wireless Access in Vehicular Environments (IPWAVE) Working Group (WG) and Host Identity Protocol (HIP) working group on secure mobility protocols. If 5G networks serve safety crucial applications as envisaged, the ISO (International Organization for Standardization) will introduce standards such as Common Criteria (ISO 15408). For instance, for car connectivity, a specific standard is ISO 26262, which covers car safety requirements. The following non-exhaustive list highlights some working groups and areas that are of relevance within IETF when it comes to prominent 5G topics (Liyanage et al. 2018):

- Routing area
- Internet area
- Applications and the Real-time area
- Transport area
- Security area.

ETSI is also a known contributor to the Network Functions Virtualization Industry Specification Group (ISG). The ETSI Industry Specification Group (ISG) for NFV Security (ISG NFV SEC) was responsible for security specifications of NFV platforms. ISG NFV SEC has highlighted the need for a standard interface in ETSI NFV architecture for addition security functions capable of reacting to threats in real time (Liyanage et al. 2018; Work Groups 2020).

3 Key Security Challenges in 5G

The 5G technology requires robust security architectures and solutions because it will connect all aspects of life to communication networks. Consequently, we investigate the important security and privacy challenges in 5G networks and we present the potential solutions that could lead to secure 5G systems. The basic challenges in 5G highlighted by Next Generation Mobile Networks (NGMN) (Liyanage et al. 2018) and much discussed in the literature are as follows (Liyanage et al. 2018; Jayakody et al. 2019):

- **Flash network traffic**: There will be a large number of new things (IoT) and end-user devices.
- **Security of radio interfaces**: Encryption keys on the radio interface are sent over insecure channels
- **User plane integrity**: On the user data plane, no cryptographic integrity protection is presented.
- **Mandated security in the network**: On the security architecture, Service-driven constraints lead to the optional use of security measures.

Table 1 Standardization bodies contributions

Standardization bodies	Work groups	Major security areas in focus	Milestones
3GPP	Service and System Aspects Security Group (SA3)	Security architecture, RAN security, authentication mechanism, subscriber privacy, network slicing	TR 33.899 Study on the security aspects of the next generation system, TS 33.501: Security architecture and procedures for 5G system
5GPPP	5GPPP Security WG	Security architecture, subscriber privacy, authentication mechanism	5G PPP Security Landscape (White Paper) June 2017
IETF	I2NSF, DICE WG, ACE WG, DetNet WG	Security solutions for massive IoT devices in 5G, user privacy, network security functions (NSFs)	RFC 8192, RFC 7744, Deterministic Networking (DetNet) Security Considerations
NGMN	NGMN 5G security group (NGMN P1 WS1 5G security group)	Subscriber privacy, network slicing, MEC security	5G security recommendations: Package 1 and 2, and 5G security: Package 3
ETSI	ETSI TC CYBER, ETSI NFV SEC WG	Security architecture, NFV security, MEC security, privacy	ETSI GS NFV SEC 010, ETSI GS NFV SEC 013 ETSI GS NFV SEC 006 and ETSI GS MEC 009

- **Roaming security**: User-security parameters are not updated with roaming from one operator to another, resulting in security compromises with roaming.
- **Denial of Service (DoS) attacks on the infrastructure**: There are visible network control elements and unencrypted control channels.
- **Signaling storms**: Distributed control systems need coordination (non-access stratum (NAS) layer of Third Generation Partnership Project (3GPP) protocols, for example).
- **DoS attacks on end-user devices**: On user devices, there are no security measures for applications, operating systems, and configuration data. The 5G design principles described by NGMN beyond radio efficiency comprise creating a common core and simplified management and operations by adopting new networking and computing technologies. Table 1 summarizes the diverse classes of security threats and attacks, the targeted elements or services in a network, and the technologies that are most prone to the attacks or threats are ticked. These security challenges are briefly described in the following sections (Sriram et al. 2019).

4 5G Security Architecture

4.1 Security Domains

The 5G security architecture comprises many network architectural elements and concepts. It is organized into an application stratum, a serving stratum, and a transport stratum. Figure 1 illustrates the following security domains,

simplified diagram of the serving stratum and the transport stratum (Jayakody et al. 2019; Mnserrat 2018).

- **Network access security (I)**: The set of security features that enables User Equipment (UE) or any devices used directly by an end-user to authenticate and access services via the network securely, including 3GPP access and non-3GPP access, and particularly to protect against attacks on radio interfaces. In addition, it includes the security context delivery from the SN to UE for access security. Therefore, UEs exchange protocol messages via the access network with the Serving Network (SN) and leverage PKI, where keys are stored in the USIM and the Home Environment (HE) (Jayakody et al. 2019; Mnserrat 2018).
- **Network domain security (II)**: The set of security features and mechanisms that enable network nodes to securely exchange signaling data and user plane data within 3GPP networks and across networks (Jayakody et al. 2019; Mnserrat 2018).
- **User domain security (III)**: The set of security features at the UE that secures the access to mobile equipment and mobile services, for this, it establishes hardware security mechanisms to prevent mobile terminals and USIMs to be altered (Jayakody et al. 2019; Mnserrat 2018).
- **Application domain security (IV)**: The set of security features that allows applications in the user domain and in the provider domain in order to exchange messages securely (Jayakody et al. 2019; Mnserrat 2018).
- **Service-Based Architecture (SBA) domain security (V)**: The set of security features for network element registration, discovery, and authorization, also for protecting the service-based interfaces, it's used to implement

Fig. 1 5G security architecture

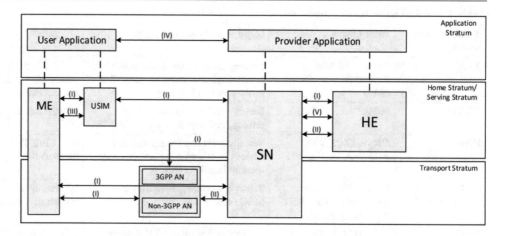

and perform business and nonbusiness functionality. SBA domain security features comprise also secure roaming, which contains the SN as well as the Home Network (HN)/HE (Jayakody et al. 2019; Mnserrat 2018).

- **Visibility and configurability of security (VI)**: The set of features that enables the user to be informed if a security feature is in operation or not (Mnserrat 2018).

5   Significance of Security and Privacy

In spite of the profits carried from 5G security architecture, it is impossible to supervise the significance of security and privacy in this novel technology. To deal with security threats to smart devices and sensitive data, four characteristics of 5G networks and their usage was defined with suggestions for security and protection is presented below (Maleh et al. 2020):

- **Modern confide models**
  Let's assume that models change over time, and since 5G will support new action plans, the trusted models will change, and the new types of gadgets will go through a wide variety of security needs and meanwhile have completely different security requirements. in order to guarantee that 5G reinforces the requirements of the new action plans and guarantee requested security, thus, the confidence display outline is redrawn (Yurish 2020).
- **New relevance transmission models**
  The use of clouds and virtualization highlights dependence on protected programming and causes diverse security impacts. Also, the decoupling of software and hardware means that telecommunications software can never depend on the specific security properties of committed telecom hardware.

For the same reason, standard interfaces to the computing/network stages are important to ensure a sensible way to manage security (SA3—Security 2020). When operators have third-party applications in their communications structure and running on hardware that is no different to local telecommunications services, there are requests for virtualization with strong detachment properties (Zuopeng 2019; Sehgal et al. 2019).

- **Emerging risk prospects**
  5G networks will play an important role as a basic foundation considerably. The quality facilitated and created by the 5G network framework has been assessed to be much higher, allowing hardware, software, and information to be considerably more attractive for different types of attacks. In addition, thinking about the imaginable results of an aggression that could seriously affect society as a whole, encourages strengthening of certain security measures, and emphasizes that Resistance to attacks should be a key ingredient in the conception new 5G protocols. However, the new risks generally underline the requirement for a quantifiable assertion and consistency of security; in other respects, it is also important to verify the presence, accuracy and adequacy of security capabilities. The set of 5G gadgets and the network will not just influence innovative attack patterns but will increase social engineering attacks (Sehgal et al. 2019).
- **Raised privacy concerns**
  The security of individual information has been reviewed within the (Equipment User), therefore, a particularly delicate resource is the user identifier. It is being checked in standardization bodies such as the 3GPP and the IETF. Conversely, the assurance of the International Mobile Subscriber Identity (IMSI) has to date only provided limited protection (Rommer et al. 2019).

6 Security Characteristics of 5G and IoT Issue

6.1 Security Characteristics

The main Basic Security Characteristics of 5G are

- **Threat prevention**: Relating to the reduction of ground problems for the majority of security incidents. Firewalls are deployed for network protection and access control to reduce user-based risk. In order to block basic 5G security threats, intrusion detection, and prevention tools are considered (Yurish 2020).
- **Stopping and fixation Advanced malware**: Going beyond signature-based tools will help to identify the attacks designed to escape basic filters. Checks based on the behavior of the endpoints—possibly using sandboxing—are important and will trigger the removal of all instances on the network once a related threat is detected (Department of Homeland Security, U.S. Government 2019).
- **Detecting anomalies**: To identify threats that are not detected by the basic filters, packet capture, big data, and machine learning are used. Once this process is inserted into the switches and routers of the 5G network, it transforms those devices into security sensors and consequently increases efficiency (Abdulhamid and Latiff 2020).
- **Incorporate DNS (Domain Name System) intelligence**: DNS activity is supervised and protected against any malicious attacks (Zuopeng 2019).
- **Making threat intelligence paramount**: Providers must deploy hackers profiles as vendors, in order to understand the malicious efforts of hackers (Tsiatsis et al. 2014).

6.2 Internet of Things Security Issues

The Internet of Things (IoT) introduces as the next big step in the evolution of the future Internet that a device with possessing computing and sensorial capabilities is able to communicate with other devices through IP-based communication protocols. A combination of Internet and emerging technologies such as embedded wireless sensor networks, wireless communications, and context awareness transforms everyday objects into connected and intelligent objects in many areas like smart network, home automation, smart cities, and many others. Also, the IoT and Mobile are one of the applications that will span a broad prospect for 5G

technology described as the first network designed to be scalable, versatile, and energy smart for the hyper-connected IoT in the world (Brij 2019).

The security technologies for IoT, supported by the industry, form the communication protocol stack that meets the criteria of reliability, Internet connectivity, and power-efficiency. The security technique is designed to protect IoT communications and to deliver a guaranteed level of protection assessed at the level of integrity, authentication, confidentiality, and non-repudiation, therefore, security schemes are required for communications IoT/sensing devices (Tsiatsis et al. 2014).

7 5G Security Topics

5G networks are the next stage in the advancement of mobile communication and will have a central empowering influence on the Networked Society. This advance generates new security situations and necessitates new security provisions. As illustrated in Fig. 2, 5G frameworks will bring a considerable number of connected devices, interoperability of new and legacy access technologies, important increase of bandwidth, and new business cases that will open up in new challenges from the security perspective (Abbas et al. 2020). 5G security will not only be specified by quantitative aspects such as latency and bit-rates but also by subjective perception (Fitzek et al. 2020).

7.1 Security Assurance

If 5G is to play an even more central role as critical infrastructure than previous generations and that security assurance will enter the picture to a higher degree, it seems clear that Common Criteria compliance could enter as an additional assurance requirement above the Security Assurance Methodology (SECAM) (Abbas et al. 2020). Conversely, in combination with cloud-based implementation (virtualization and on-demand service), it is probably necessary to separate software assurance more concretely from platform assurance and to enable on-demand measurements of assurance as part of Service Level Agreements (SLAs). In the fact that 5G is designed to be a platform for a wide range of new user groups and applications does not automatically mean that it is necessary (or even desirable) for the 5G network to assume all security responsibility. In addition, it can provide some highly valuable security services (Abbas et al. 2020).

Fig. 2 5G security architecture

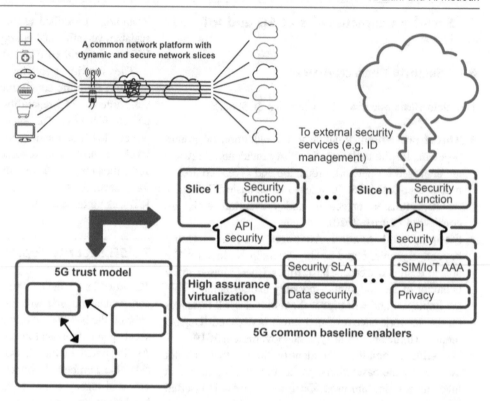

7.2   Identity Management

The use of USIM (Universal Subscriber Identity Module) on physical Universal Integrated Circuit Cards to handling identity will continue to be an essential part of 5G for reasons such as the high level of security and user friendliness. Embedded USIM card has also significantly lowered the bar for deployment issues related to machine-to-machine communication. However, there is a general tendency to bring-your-own-identity, and the 5G ecosystem would generally profit from a more open identity management architecture that permits for alternatives. Examining new ways to manage device/subscriber identities is so key considerations that enter the investigation of the new models of trust for 5G. Concepts such as network slicing can provide a tool for securely allowing different ID management solutions side-by-side by confining use to virtual and isolated slices of the network (Sahana 2019).

7.3   5G Radio Network Security

Because of the evolution of new technologies offering users low-cost alternatives to design their own devices (even at radio access level), and also the threat landscape, the attack resistance of radio networks should be a more clearly honest design consideration in 5G, analyzing threats such as denial of service from conceivably getting out of hand gadgets, and adding mitigation measures to the design of the radio protocol. With

5G radio access as a building block in, for example, industrial automation, the potential benefits of adding integrity protection seem worth exploring (Jayakody et al. 2019).

7.4   Flexible and Scalable Security Architecture

With virtualization and more dynamic configurations entering the zone for 5G, it seems logical to consider a more dynamic and flexible security for it. Security for synchronous aspects like RAN signaling (SA3—Security 2020) could have a higher level of independence from asynchronous security aspects, such as those related to the user plane, than today. New security designs with higher flexibility could also better address unnecessary conflicts between convenience and security (5G System Design: Architectural and Functional Considerations and Long Term Research).

7.5   Network Slicing Security

Network slicing not only needs the necessary security from UE accessing the slice, but also creates new security challenges. A 5G UE can simultaneously access different network slices for several services, the advantage of network slicing is that operators can deliver personalized security for each slice. Different access authentication and authorization can be provided for tenants of different network slices. Isolation should be assured for network slices, without which attackers who have access to one slice may launch an attack to other slices. Good isolation will allow integrity and confidentiality protection (Al-Dulaimi and Wang 2018).

7.6 Energy-Efficient Security

While encryption as part of the security services comes with a cost, the expense is no longer an issue for mobile phones and similar devices. The energy cost of encoding one bit is less than transmitting one bit (Marsch et al. 2018). Conversely, for the most constrained, battery dependent gadgets with a long target life time, there may be a prerequisite to consider even more lightweight arrangements, since each microjoule consumed could be of importance (Penttinen 2019).

7.7 Cloud Security

Cloud security is a critical concept that presents currently an extremely hot topic, and it has been added to the list of 5G concerns. Here is a brief list of priorities for cloud security in a 5G context (Abbas et al. 2020; Sahana 2019):

- Develop network virtualization and hypervisors and with high assurance level on isolation. So, investments in this area could pay off, as this would considerably simplify the management of various security requirements in the same infrastructure (Brij 2019).
- Set up useful ecosystems and architectures based on existing trusted computing tools and concepts for remote attestation, for example (Brij 2019).
- Offer more efficient solutions for cloud-friendly data encryption, in the occurrence, the homomorphic encryption that allowing operations on encrypted data (Brij 2019).
- Develop simple and reliable management of cloud systems and the applications that run on them. Some of these continue to be research subjects (Brij 2019).

8 Conclusion

5G will allow instantaneous connectivity to the large number of devices, the Internet of Things (IoT) and a truly connected world. With all these advantages, innovative technologies also have inherent security challenges. In terms of technologies, 5G security risks must be continuously contextualized and protocol-based security must be enhanced.

In terms of security assurance, cybersecurity requirements should be standardized and that all vendors and operators ensure that they are applicable and verifiable. In this article,

we presented the 5G security architecture that will be able to provide a trustworthy platform for this vision. And we studied security challenges in 5G technologies. Therefore, we have highlighted the main security challenges that can become more threatening in 5G. We have also presented the potential security topics, the emerging risk prospects, and key security challenges. However, the integration of IoT seems to raise more security concerns, specifically in terms of privacy. As future work, we plan to analyze the authentication protocols for 5G and IoT in the 5G security architecture.

References

5G White Paper. https://www.ngmn.org/work-programme/5g-white-paper.html. Last accessed 26 April 2020

Abbas, A., Khan, S.U., Zomaya, A.Y.: Fog Computing: Theory and Practice. John Wiley & Sons (2020)

Abdulhamid, S.M., Latiff, M.S.A.: Advanced Security Strategies in Next Generation Computing Models. IGI Global (2020)

Al-Dulaimi, A., Wang, X., Chih-Lin, I: 5G Networks: Fundamental Requirements, Enabling Technologies, and Operations Management. Wiley (2018)

Brij, B.G.: Modern Principles, Practices, and Algorithms for Cloud Security. IGI Global (2019)

Chandramouli, D., Liebhart, R., Pirskanen, J.: 5G for the Connected World. Wiley (2019)

Fitzek, F., Granelli, F., Seeling, P.: Computing in Communication Networks: From Theory to Practice. Elsevier Science & Technology (2020)

Holma, H., Toskala, A., Nakamura, T.: 5G Technology: 3GPP New Radio. Wiley (2020)

Jayakody, D.N.K., Srinivasan, K., Sharma, V. (eds.): 5G Enabled Secure Wireless Networks. Springer International Publishing, Cham (2019)

Liyanage, M., Ahmad, I., Abro, A.B., Gurtov, A., Ylianttila, M.: A Comprehensive Guide to 5G Security. Wiley (2018)

Maleh, Y., Shojafar, M., Romdhani, I., Alazab, M.: Blockchain for Cybersecurity and Privacy: Architectures, Challenges, and Applications. CRC Press LLC (2020)

Marsch, P., Bulakci, Ö., Queseth, O., Boldi, M.: 5G System Design: Architectural and Functional Considerations and Long Term Research. Wiley (2018)

Mnserrat, R.: Fundamentals of 5G Technologies. Independently Published (2018)

Penttinen, J.T.J.: 5G Explained: Security and Deployment of Advanced Mobile Communications. Wiley (2019)

Rodriguez, J.: Fundamentals of 5G Mobile Networks. Wiley (2015)

Rommer, S., Hedman, P., Olsson, M., Frid, L., Sultana, S., Mulligan, C.: 5G Core Networks: Powering Digitalization. Elsevier Science (2019)

SA3—Security. https://www.3gpp.org/specifications-groups/sa-plenary/sa3-security. Last accessed 26 April 2020

Sahana, S.: Economics and Security Implications of Cloud Computing. Educreation Publishing (2019)

Department of Homeland Security, U.S. Government: 2019 Cybersecurity and Infrastructure Security: Risks Introduced by 5G Adoption in U. S.; Insider Threat Programs for Critical Manufacturing Sector; Strategic Intent, Election Security, Urgent Threats. Independently Published (2019)

Sehgal, N.K., Bhatt, P.C.P., Acken, J.M.: Cloud Computing with Security: Concepts and Practices. Springer Nature (2019)

Sriram, P.P., Wang, H.C., Jami, H.G., Srinivasan, K.: 5G security: concepts and challenges. In: Jayakody, D.N.K., Srinivasan, K., Sharma, V. (eds.) 5G Enabled Secure Wireless Networks, pp. 1–43. Springer International Publishing, Cham (2019)

Tsiatsis, V., Karnouskos, S., Holler, J., Boyle, D., Mulligan, C.: From Machine-to-Machine to the Internet of Things: Introduction to a New Age of Intelligence. Elsevier Science (2014)

Work Groups 5G-PPP. https://5g-ppp.eu/5g-ppp-work-groups/. Last accessed 26 April 2020

Marsch, P., Bulakci, Ö., Queseth, O., Boldi, M.: 5G System Design: Architectural and Functiona Considerations and Long Term Research. Wiley (2018)

Yurish, S.: Advances in Networks, Security and Communications: Reviews, vol. 2. Lulu.com (2020)

Zuopeng (Justin), Z.: Novel Theories and Applications of Global Information Resource Management. IGI Global (2019)

An Adaptive Video Streaming Framework for Peer-To-Peer 5G Networks: Paving the Road to 5G-IMS

Adnane Ghani, El Hassan Ibn El Haj, Ahmed Hammouch, and Abdelaali Chaoub

Abstract

The IMS architecture presents many disadvantages which are centralized control, low efficiency, and low scalability in terms of core network equipment compared to network infrastructures using Cloud Computing. Cloud Computing is a new information technology paradigm, offering dynamically scalable resources, often through virtual machines and accessible as services on the Internet. IMS migration to the cloud can improve the performance of the IMS infrastructure. Video streaming in the IMS architecture of 5G technology is proposed for use in high-quality multimedia applications, the use of adaptive streaming stream is used to adapt the quality for a heterogeneous network of equipment. Knowing that 5G technology is based on cloud computing techniques and virtualization, coexistence between IMS and Peer to Peer is possible with the BitTorrent protocol which offers a solution compatible with 5G technology, by offering BitTorrent Sync a solution based on Cloud. Use of a video streaming adaptation with an architecture based on 5G technology for both to distribute the load between the elements of the P2P IMS architecture, decrease the response time for the procedures (User authentication, time required for "a Peers joined the network", time to download songs (Reduce latency in the network).

Keywords

Next generation networks • Peer to peer • Quality adaptation • Scalable video coding • 5G • Cloud computing • Virtualization

1 Introduction

The Internet Multimedia Subsystem has been standardized by the Third generation Partnership Project (3GPP) (2012) for support Peer-to-Peer architecture, is used today as architecture for triple-play services: telephony, Internet and video streaming. But streaming video requires a high cost in terms of equipment performance and bandwidth needed for better quality. Streaming in the IMS P2P network often uses data-oriented protocol in which, for example, each node periodically announces to its neighbors the blocks it has. Streaming in this network reduces the cost of infrastructure by taking advantage of customers to make content available and avoid having to set up important structures. In (Tang 2015) the authors proposed a multi-domain and multi-overlay framework referring to the RELOAD the base protocol of cloud infrastructure, the framework responds fairly well to the 3GPP standard and is intended to build the core network overlay with the SIP protocol using the P2P model, and to make elements be discovered dynamically and to re-design the original communication procedure. An end-to-end model based on the IPTV service delivery platform on the next generation of IMS/EPC network infrastructure is proposed by the authors of (Gilani et al. 2019), the proposed model is an IMS standard that is compliant, scalable, adaptive, and scalable mobile IPTV framework that has distinct key features such as support for contextual and custom broadcast (BC) and video on demand (VoD) and IMS and SIP Compliant Andriod OS-based IPTV set-top box. The authors of (Tomas and Vuksic 2012) propose convenient business model for implementation of P2P cloud architecture that is

A. Ghani (✉) · A. Hammouch
Ecole Normale supérieure de L'enseignement Technique (ENSET), Mohammed V University, Rabat, Morocco
e-mail: adnaneghani@gmail.com

A. Hammouch
e-mail: Hammouch_a@yahoo.com

E. H. Ibn El Haj · A. Chaoub
National Institute of Posts and Telecommunications (INPT), Rabat, Morocco
e-mail: ibnelhaj@inpt.ac.ma

A. Chaoub
e-mail: chaoub@inpt.ac.ma

M. Ben Ahmed et al. (eds.), *Emerging Trends in ICT for Sustainable Development*,
Advances in Science, Technology & Innovation, https://doi.org/10.1007/978-3-030-53440-0_23

both safe and reliable and it is made of multiple distributed single nodes, and to assure system reliability, and also to attract users to join large P2P Cloud network. In (Chen et al. 2010) the authors proposed NIDA (Network ID Aware) in BitTorrent P2P applications, an approach to reduce inter-network traffic effectively and to accelerate P2P applications, they showed that the P2P network built with BitTorrent standard is a random network, however, P2P network built with NIDA is sensitive to network topology, simulation results have shown that NIDA can reduce inter-network traffic, and in comparison with conventional BitTorrent can reduce download time. In (Andrade et al. 2007) the authors used BitTorrent traces and analytical modeling to control the cost of using BitTorrent compared to the benefits it can bring to the system. In (Dairaine et al. 2005), the authors described an improvement in the quality of services in Peer to Peer (P2P) networks, they used an erasure code to distribute peer information, they presented a model using the results of a P2P network on the Internet, they have shown that the quality of access service to content is guaranteed based on content replication and dissemination strategies. In (Rizk et al. 2014), the authors presented a new SVC coding evaluation framework AVIS (an Adaptive VIdeo Simulation) on Network Simulator 2 (NS2), they created two new objects on NS2, AVIS transmitter and receiver, preprocessing and post-processing tools that can be used with any SVC encoding. In this paper, we will present a new model in 5G technology which consists of exploiting the chunks selection technique offered by the BitTorrent protocol and the technique of dividing the video stream in several blocks offered by the SVC coding, to create a P2P network based on the BitTorrent protocol allowing the downloading of adaptive flow according to the performances of the peers.

2 Background

2.1 IMS in 5G

The IMS network in 5G technology is a virtualization-based cloud platform for the main IMS network. Virtualization in the telecommunications industry has become popular in recent years, moving system instances from one physical host to another is one of the most important features of virtualization. The cloud infrastructure offers the services of dynamic resource allocation, load balancing.

2.2 Bittorent Protocol

BitTorrent is arguably the most popular scalable content distribution mechanism presently. The efficiency of the protocol and some popular lightweight implementations of it

account for its success as a significant step in democratizing content publishing on the Internet. BitTorrent offloads the content provider by leveraging the resources of content consumers during download.

2.3 Zeta Protocol

Zeta is a P2P download protocol, to simulate this protocol, there is Zeta Simulator (Zetasim) a simulation framework for NS2, zeta uses non-connected mode User Datagram Protocol (UDP) instead of connected mode Transmission Control Protocol (TCP) for data transport. The PUSH system is introduced to increase transmission efficiency and reduce unnecessary data exchange. It uses a progressive decoding method to reduce network code calculation time (Wang and Han 2010).

2.3.1 Proposed Architecture

Our proposed architecture is schematized in Fig. 1, and the screen capture of the NS2 simulation nam output is schematized in Fig. 4. The main components of this architecture are Virtual Proxy-Call/Session Control Functions (vP-CSCF), Virtual Interrogating-Call/Session Control Functions (vI-CSCF), and Virtual Serving-Call/Session Control Functions (vS-CSCF). Our proposed architecture as shown in Fig. 1 is based on (xxxx), we have an virtual IMS network (which contains the virtual P-CSCF, virtual I-CSCF, virtual S-CSCF), and different BitTorrent peers, it is a model defined by 3GPP named Peer to Peer Content Distribution Services (P2P CDS) architecture, the purpose of our approach more the technical coordination between IMS, P2P and Scalable Video Coding (H.264/SVC) is to benefit from the advantages of SVC coding to provide adaptive P2P streaming for the IMS in 5G technology network and in a heterogeneous architecture containing different devices (PC, tablet, phone). After the authentication step of the different clients at the level of the IMS network, the multimedia services become possible in the network. Our simulation will include the part of virtualization of different server by using GreenCloud (Kliazovich et al. 2010) framework of the Session Initiation Protocol (SIP) signaling in the IMS network of the different peers, the creation of a P2P network in our proposed architecture is done by using the BitTorrent (https://drive.google.com/file/d/0B7S255p3kFXNaGhxNEIteC11YTg/view?usp=sharing) framework, and the part of the evaluation of the SVC coding is done by using the myevalsvc (Ke 2012) framework, in order to execute our SVC quality adaptation algorithm, we have developed our algorithm using C++ language code, to choose the streamed stream that suits the performance of the peer, then we have to connect our C++ code with the TCL (Tool Command Language) simulation on NS2. Then we use the myevalsvc framework to evaluate the transmission of SVC encoding between two sending and receiving nodes. We use the BitTorrent framework for the creation of a P2P network and the download management of the video sequence and to simulate adaptive video streaming SVC between peers, the SIP module (Prior 2007) is used for signaling in the network, where a peer plays the role of a server and other peers play the role of customers. Finally, we have created a complete architecture that can make the identification of IMS P2P users and which also allows the SVC adaptation of quality for heterogeneous devices.

$\text{Video}_{SVC} = \sum_{i=0}^{N} Li$, Where Li: layer identifier. And N: Number of layers.

As shown in Fig. 2 the encoded video is sent to the BitTorrent application, and it is processed by this application to divide the video into several chunks. The following expression defines the video at the output of the BitTorrent application:

$\text{Video}_{SVC} = \sum_{i=0}^{M} Ci$, Where Ci: chunk id. M: Number of chunks.

An adaptive level SVC will be defined as follows:

$Li\,(i, N) = \sum_{i=0}^{Nn} Ci$ Where Li (i, N) = Identifier of the layer from i = 0 until i = N andNn: Number of chunk per layer.

Chunk Mechanism in Bittorrent

A block is a piece of a file. When a file is distributed via BitTorrent, it is broken into smaller pieces or blocks. Typically the block is 256 kb in size, Breaking the file into pieces allows it to be distributed as efficiently as possible. Users get their files faster using less bandwidth (Table 1).

In Fig. 3, we present our model of quality adaptation, this model is extended from an existing model (Lahbabi et al. 2014), the main idea of the existing model is to make a comparison between the requirements of each layer of the video stream with the local static resources of a pair, it

allows to obtain the current resources and requirements of Peers, in order to match them with the achievable quality, it mainly manages the static parameters, such as screen resolution, bandwidth and currently available device power (CPU, RAM, battery life) and user preferences (screen resolution, frame rate, PSNR level), this model also aims to define the quality adequate (d, t, q) (dimensional, temporal, qualitative) with user performance. An initial quality set with the quality level parameters of the base layer d0, t0, and q0 is first filled. The extension of the adaptation model is based on two improvements which are firstly the addition of another parameter to the adaptation complexity model, which is the user's IP address and secondly the use of the chunks selection technique offered by the BitTorrent protocol to download different SVC levels, the first improvement aims to guarantee continuity of service in the IMS P2P network, by updating the list of Peers after each change of IP address, and the second improvement serves to benefit from the technique of selection of the chunks offered by BitTorrent to improve the technique of dividing the video stream into several blocks offered by SVC coding to accelerate the downloading of the different SVC levels by Peers in the P2P IMS network. We have created a C++ code which simulates this model which aims to define the adequate quality (d, t, q) (dimensional, temporal, qualitative), and we proceed as follows:

(1) Executing the spatial adaptation, extracting the SVC levels that have the resolution of the appropriate video for the user according to the resolution parameter of the user's screen.

(2) Taking into account the SVC levels already selected in Step 1, the model executes the temporal adaptation according to the user requested frame rate and the network bandwidth to extract the levels of SVC respecting these conditions.

(3) Taking into account the SVC levels already selected in Step 2, the model performs the qualitative adaptation,

Fig. 1 Proposed architecture

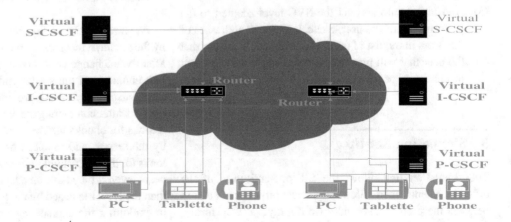

Fig. 2 The process of processing a video by our architecture

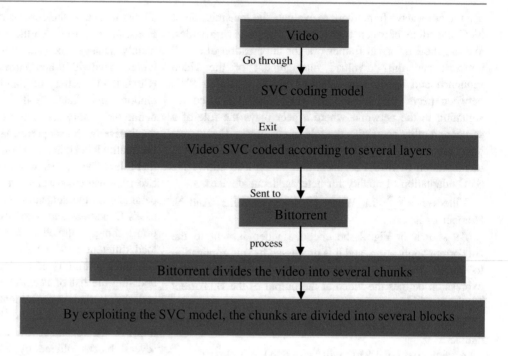

Table 1 Number of chunks versus sequence size

Sequence (MB)	Number of chunks
3	12
5	20
10	40

according to the user's preferences, does it request that the video be either with quality or not, and respect also other more complex parameters that are: Central Processing Unit (CPU) performance, Random Access Memory (RAM), battery life to finally extract a single level SVC adapted to the performance of the user.

(4) And taking into account the SVC level adapted to the performance of the user, a check of the IP addresses of the peers in the list of peers is performed, to ensure that the peer that will provide the video sequence is present in the list of peers.

(5) And taking into account the SVC level adapted to the performance of the user, a check of the IP addresses of the peers in the list of peers is performed, to ensure that the peer that will provide the video sequence is present in the list of peers.

3 Simulation Set-Up

The simulation first allows the SIP authentication of the different clients then allows the creation of the session between these clients, secondly we run myevalsvc. Thirdly,

the quality adaptation algorithm is executed, P2P clients receive decoding parameters of the video that suits with their performance so that the peer receives only the suitable layers and play the video with adequate quality (Fig. 4).

- Request 1: The peer asks the video to seeder.
- Answer 1: Answer 1: The seeder asks for peer performance.
- Request 2: the peer sends his performances to the seeder.
- Answer 2: The seeder sends to the peer the Chunks Numbers associated with the requested adaptive layer.

As shown in Fig. 5, first, the encoded SVC video is owned by the seeder, it receives the request of the video from the peer, after the exchange described above. The peer downloads the first chunk from the seeder and according to the BitTorent mechanism the peer can provide its ability to upload to the network after completing the download of the first chunk, then it looks for chunks numbers that form the adaptive level sent by the seeder, and to speed up the download time the peer looks for the other chunks either at the seeder or at the different peers, even if they have an adaptive svc level lower or higher than the level requested by the peer. Our architecture consists in exploiting the chunks selection technique offered by the

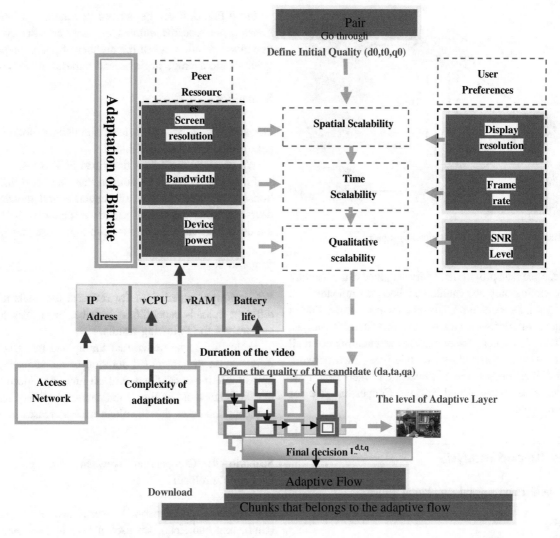

Fig. 3 The process of requesting a video by a peer

Fig. 4 Diagram describing the problem addressed

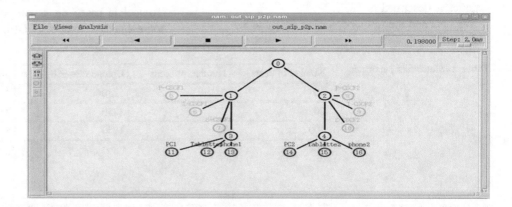

BitTorrent protocol and the technique of dividing the video stream into several blocks offered by the SVC coding, to create a p2p network based on the BitTorrent protocol allowing the downloading of an adaptive flow according to the needs of peers. In the first step the seeder owns the video SVC (containing all the layers and all the chunks). The peers join the BitTorrent P2P network and after the execution of the adaptation model the peer knows beforehand the chunks that it must download by chunk number to group the chosen SVC level and at the same time and according to the BitTorrent

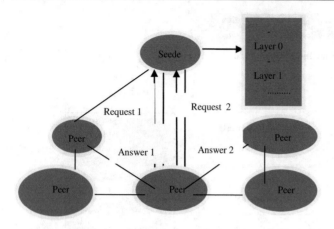

Fig. 5 The process of requesting a video by a peer

principle, a peer can provide its ability to upload to the network after completing only one chunk, so it becomes uploader and downloader at the same time. The architecture studied aims at evaluating average download Duration, first Chunk Time, last Chunk Time. This architecture contains essentially two small remote networks separated by multiple hops, each network contains different peers. We will evaluate Packets Loss Rate (PLR) for Zeta protocol and Packets Congestion rate for BitTorrent protocol.

4 Results and Analysis

We first performed several simulation scenarios:

Scenario 1:

For scenario 1, we performed simulations following the parameters described in Table 2.

We obtained the results described in Table 3.

From Fig. 6, it can be noticed that there is a linear relationship between the number of peers and the average of download duration, when the number of peers increases the value of the average of download duration increases.

Scenario 2:

For scenario 2, we performed simulations following the parameters described in Table 4.

We obtained the results described in Table 4.

From Fig. 7, it can be noticed that there is a linear relationship between the size of sequence and the download duration, when the size of sequence increases the value of the download duration increases in a proportional way.

Scenario 3:

We performed simulations for different terminals that have different uploads capabilities described in Tables 5 and 6, and we got the following results.

From Fig. 8, we notice that the upload rate has a weak influence on the download duration of the sequence, if the upload rate increases the download time is reduced weakly. On the other hand as Fig. 8 indicates, when the size of the sequence increases the download time increases in a strong way.

Scenario 4: Comparison between Zeta protocol and BitTorrent protocol

We made a comparison between two Peer protocols, (BitTorrent and Zeta), we used a Peer to Peer architecture described in Table 7.

We note that the Peer-to-Peer architecture used by Zeta protocol is based on a boot server (tracker) and peers in non-connected mode (UDP), on the other hand the

Table 2 Simulation parameter

Number of peers	Number of seeds	Upload rate (KBps)	Size of the file to download (MB)
10	2	1250	3
20	2	1250	3
100	2	1250	3

Table 3 Results of Scenario 1

Number of peers	Avearge download duration (ms)
10	14,020,641
20	18,1,192,465
100	24,204,036

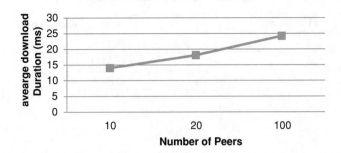

Fig. 6 Number of peers versus average download duration

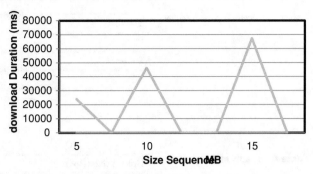

Fig. 7 Size sequence versus download duration

architecture used by the BitTorrent protocol is based on peer in connected mode (tcp), we obtained the results schematized in the graph below:

We extracted the PLR rate for Zeta packets, and we extracted the packets congestion rate for BitTorent packets.

From Fig. 9 and Table 8 we notice that the difference of the transport mode used by the two BitTorent and Zeta protocols influences on the duration of download of the sequence, and we notice that the BitTorrent protocol uses the connected mode for the transport of packets with (congestion management in the network), which justifies a low download duration compared to the mode not connected for the transport of the packets used by Zeta protocol. Most studies of the TCP protocol show that window of congestion is of the order of.

rate_congestion = $1/\sqrt{tp}$ where tp: rate of loss packets.
Expression 1: Rate of congestion.

From this expression and the results shown in Table 8, we can verify that (Table 9):

In tcp mode, we can remark that when congestion rate increases, loss rate decreases and when loss rate increases congestion rate decreases. The results show that the BitTorrent protocol is more reliable than the Zeta protocol, regarding the download.

Scenario 5: Comparison Between our adaptation model and Initial Quality Adaptation, IQA

Our model is evaluated and compared to the Initial Quality Adaptation (IQA) module (Abboud et al. 2009) (Fig. 10 and Table 10).

An improvement in download speed is remarkable between the IQA model and our adaptation scenario.

Table 4 Results of Scenario 2

Size Sequence						5 MB
Number of seeds						2
Number of peers						2
Upload rate	Peer ID	Start time (ms)	First chunk time (ms)	Last chunk time (ms)	Stop time (ms)	Download duration (ms)
10 Mbps	18	0	0	−1	2,46,573	−1
	19	8,19,076	9,12,391	2,46,573	2,46,573	1,64,666
5 Mbps	18	0	0	−1	2,47,679	−1
	19	8,19,076	9,12,907	2,47,679	2,47,679	1,65,772
2 Mbps	18	0	0	−1	25,0997	−1
	19	8,19076	9,14453	25,0997	25,0997	16,9089

Table 5 Terminal performance

Terminal	Upload rate	Download duration (ms)
Phone	2	16,4666
Tablet	5	16,5772
PC	1010	16,9089

Table 6 Results of Scenario 3

Upload rate (Bps)						250
Number of seeds						2
Number of peers						2
Size sequence	Peer ID	Start time (ms)	First chunk time (ms)	Last chunk time (ms)	Stop time (ms)	Download duration (ms)
5 MB	18	0	0	−1	23,992,8	−1
	19	8,19,076	3477,63	23,992,8	23,992,8	23,984,6
10 MB	18	0	0	−1	46,202,7	−1
	19	8,19,076	4013,16	46,202,7	46,202,7	46,194,5
15 MB	18	0	0	−1	67,334,8	−1
	19	8,19,076	3482,72	67,334,8	67,334,8	67,326,6

Fig. 8 Upload debit versus download duration

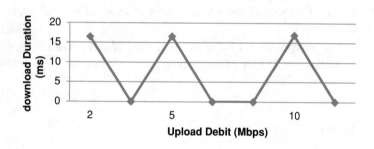

Table 7 Parameters of the simulation

Number of peers	Number of Seeder	Upload rate (KBPS)	Size sequence (MB)	Download duration (zeta) (ms)	Download duration (bitTorent) (ms)
5	1	9,8396	3	30,858	4.37

Fig. 9 Comparison between zeta and BitTorrent protocol

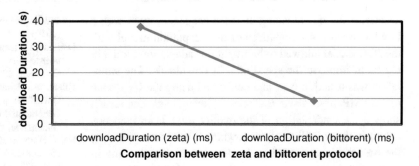

Table 8 Evaluation of transmission of the packets

Packets loss rate (zeta) (%)	Packets congestion rate (Bittorrent) (%)
28,9	0,37

Table 9 Evaluation of transmission of the packets evaluation of packet loss

Packets loss rate (zeta) (%) (UDP)	Packets loss rate (BitTorrent) (%) (TCP)
28,9	7,3

Fig. 10 Comparison between IQA model and our adaptation model

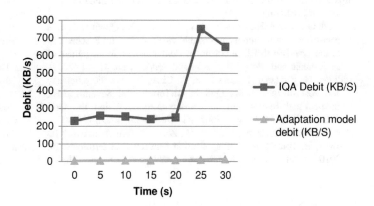

Table 10 Comparison between IQA model and our adaptation model

Time(s)	IQA Debit (KB/S)	Adaptation model debit (KB/S)
0	230	3,64
5	260	5
10	255	5,5
15	240	6
20	250	7,28
25	750	9,8396
30	650	13,4796

5 Conclusion

This article presents a quality adaptation by scalable video streaming on the virtual IMS P2P network, the video quality adaptation model proposed allowed to select the appropriate adaptive level to the preferences and performances of the Peer, in regard to the preferences of the pairs, this is the video resolution, frame rates, PSNR level, in regard to the performance of the pairs, it is about the computing power (vCPU), memory management (vRAM), and the autonomy of the battery of the Peer, thus the integration of the decentralization offers by P2P networks and the technique of selection of the chunks used by BitTorrent, allowed to accelerate the downloading of video sequences even with a low bandwidth. A comparison of our adaptation model with the IQA model allowed to highlight our adaptation model in regard to improve the speed of peer downloads. The simulation was tested in NS2 and performed using the C++ code of NS2 SIP, myevalsvc, BitTorrent, GreenCloud. The results show a clear adaptation of the quality for a heterogeneous network, the reliability of the downloading of the sequences, the best results in regard to the duration of downloading in connected mode which ensure the transmission of packets without loss, and also a low congestion of the packets, and this model allows an improvement in the duration of downloading of sequences for the case of the BitTorent protocol which uses the connected mode in comparison with a protocol which uses the unconnected mode.

References

https://drive.google.com/file/d/0B7S255p3kFXNaGhxNEIteC11YTg/view?usp=sharing

Abboud, O., Pussep, K., Kovacevic, A., Steinmetz, R.: Quality adaptive peer-to-peer streaming using scalable video coding. In: Proceedings of the 12th IFIP/IEEE International Conference on Management of Multimedia and Mobile Networks and Services, pp. 41–54 (2009)

Andrade, N., Santana, J., Brasileiro, F., Cirne, W.: On the efficiency and cost of introducing QoS in BitTorrent. In: Seventh IEEE International Symposium on Cluster Computing and the Grid (CCGrid '07), 14–17 May 2007, Rio De Janeiro, Brazil (2007)

Chen, Y., Wen, X., Zheng, W., Sun, Y., Zhao, Z.: NIDA: network ID aware BitTorrent-like P2P application between IMS terminals. In: 2010 3rd International Conference on Advanced Computer Theory and Engineering (ICACTE), 20–22 August 2010, Chengdu, China (2010)

Dairaine, L., Lacan, J., Lancérica, L., Fimes, J.: Content-access QoS in peer-to-peer networks using a fast MDS erasure code. Comput. Commun. 28(15), 15, 1778–1790 (2005)

Gilani, A., Ullah Khosa, I., Hamza, M., Qayyum, A., Bano, M.: QoENGN: a QoE Framework for Video Streaming over Next Generation Mobile Networks, ICMSSP 2019, May 10–12, 2019, Guangzhou, China (2019)

Ke, C.H.:myEvalSVC: an integrated simulation framework for evaluation of H,264/SVC transmission. KSII Trans. Internet Inform. Syst. 6(1), 378–393 (2012)

Kliazovich, D., Bouvry, P., Ullah Khan, S.: GreenCloud: a packet-level simulator of energy-aware cloud computing data centers. J. Supercomput. (2010)

Lahbabi, Y., El, I.E., Hammouch, A.: Quality adaptation using Scalable Video Coding (SVC) in Peer-to-Peer (P2P) Video-on-Demand (VoD Streaming). In: International Conference on Multimedia Computing and Systems (ICMCS) (2014)

Prior, R.: Universidade do Porto, ns2 network simulator extensions (2007)

Rizk, G.G., Zahran, A.H., Ismail, M.H.: AVIS: an adaptive video simulation framework for scalable video. In: 2014 Eight International Conference on Next Generation Mobile Applications, Services and Technologies, pp. 84–89 (2014)

Tang, Z.: A multi-domain and multi-overlay framework of P2P IMS core network based on cloud infrastructure. In: 2015 IEEE International Conference on Networking, Architecture and Storage (NAS) (2015)

The Third generation Partnership Project, 3GPP TR 23,844 V12,0,0 (2012–06)

Tomas, B., Vuksic, B.: Peer to peer distributed storage and computing cloud system. In: Proceedings of the ITI 2012 34th International Conference on Information Technology Interfaces, pp. 79–84 (2012)

Wang, J., Han, J.: Zeta: A novel network coding based p2p downloading protocol (2010). https://code,google,com/p/zetasim/

Recognition and Reconstruction of Road Marking with Generative Adversarial Networks (GANs)

Samir Allach, Mohamed Ben Ahmed,
and Anouar Abdelhakim Boudhir

Abstract

Road markings play an essential role in road safety and are one of the most important elements to guide autonomous vehicles and help the driver on the road. The recognition and detection of road markings has been very successful in recent years with the rapid development of deep learning technology. Although considerable work and progress has been made in this area, they often depend excessively on unrepresentative datasets and inappropriate conditions. In this article, to overcome these drawbacks, we propose a deep learning system for the extraction, classification, and completion of road markings which generates high-quality samples for data augmentation. For this, an in-depth learning network proposed to successfully recover a clean road marking of a fuzzy route using generative contradictory networks (GAN). The proposed data augmentation method, based on mutual information, can preserve and learn the semantic context from the actual dataset. We build and train a GAN model to increase the size of the training dataset, which makes it suitable for recognizing the target. Our system can generate clean samples from fuzzy samples and surpasses other methods, even with unconstrained road marking datasets.

Keywords

Road markings • GAN • Deep learning system • Faster R-CNN • Data augmentation

S. Allach (✉) · M. Ben Ahmed · A. A. Boudhir
Laboratory (LIST), FST of Tangier, UAE University, Tangier, Morocco
e-mail: allach.samir@gmail.com

M. Ben Ahmed
e-mail: mbenahmed@uae.ac.ma

A. A. Boudhir
e-mail: boudhir.anouar@gmail.com

1 Introduction

Departures from the road and collisions are those, which precede most of the accidents. Road departure is the cause of one of the six most serious types of accidents, and more than a third of road fatalities are due to this type of accident (Villalón-Sepúlveda et al. 2017). Thus, the detection and recognition of road markings is one of the fundamental and crucial tasks for autonomous cars. Starting from the perception system of an autonomous car, the ability to automatically recognize road markings on the road surface is one of the key steps towards understanding road conditions. Therefore, the recognition of road markings is essential to ensure safe and reliable navigation of autonomous cars on the road.

In recent years, a lot of research has been done in the area of road marking recognition. Presently, several works in the detection and recognition of road markings have been carried out with the emergence of deep convolutional neural networks.

However, the problem of recognizing road markings is always difficult in inappropriate environments such as distortion, poor lighting, low resolution, and extreme weather conditions. In addition, the existing road marking datasets are limited in diversity and quantity.

We note that most of the existing methods (Wu and Ranganathan 2012; Ahmad et al. 2017) evaluate their approaches with unrepresentative datasets, collected on extremely limited routes, they could work well only under certain controlled conditions. Adverse road environments are very common in the real world, however. For these reasons, existing methods are quite unsuitable for real-world scenarios due to inconsistent performance.

To cope with these difficulties, we propose a model of generative adversarial networks (GAN) (Goodfellow et al. 2014) capable of correcting, increasing, and classifying road markings with low error rates. There are two networks in our model: (i) a generator network and (ii) a discriminator

M. Ben Ahmed et al. (eds.), *Emerging Trends in ICT for Sustainable Development*,
Advances in Science, Technology & Innovation, https://doi.org/10.1007/978-3-030-53440-0_24

network. In the generator network, we focus on both the blurring of positive samples and the distortion of negative samples simultaneously from the unconstrained inputs. Therefore, we propose a fuzzy network to restore certain fuzzy regions for positive samples only and to decimate the entire region for negative samples. The discriminator network is made up of two branches: one distinguishes the real road marking from the fake, and the other categorizes various road markings for the object detection task. Because of the scarcity of the dataset, we propose a generative network for augmenting data, based on mutual information, to capture a much larger set of invariance and preserve the semantic context. Training the classifier with augmented data would greatly improve classification tasks. Benefiting from the augmentation module, in the test phase, the discriminator network functions like the real road marking classifier, while the generator network improves the performance of the classifier by adequately erasing positive samples and decimating any given negative sample (not road markings).

In this work, we propose an architecture of a system capable of extracting markings on the ground and generating clean images from fuzzy samples using a deep network of contradictory training for the recognition of road markings. Next, we apply a data augmentation module to create images that are different from the original samples in the dataset, improving the quality of the sample for generalization performance.

2 Related Works

We used the convolutional neural network and GANs for the extraction, classification, and completion of road markings.

Neural networks have been successfully introduced for the recognition of road markings. Kheyrollahi and Breckon

(2012) proposed an approach to this problem based on the extraction of robust road marking characteristics via a new pipeline of reverse perspective mapping and binarization at several levels. With the rise of deep neural networks and their success (Krizhevsky et al. 2012; Simonyan and Zisserman 2014) in computer vision applications.

Hu et al. (2019) formed a road marking detection model using images from NIR cameras. They modified the VGG-16 backbone of the advanced faster R-CNN algorithm using a multi-layer feature fusion technique.

Greenhalgh and Mirmehdi (2015) used a support vector machine (SVM) by extracting the HOG functionalities.

Lee et al. (2019) proposed the network to recover a clean road marking of a fuzzy route by adopting generative contradictory networks (GAN).

In addition, our work is linked to the increase in data. Ahmad et al. (2017) formed five different CNN architectures with a variable number of convolutions and fully connected layers. They introduced data augmentation to improve the size of the dataset by using the in-plane rotation of the original images. However, the standard increase in linear data only produces limited plausible alternative data.

3 Method and Architecture Proposed

In this section, we describe our approach, which includes three main steps: extraction, classification, and completion of road markings. First, we present the entire proposed architecture of our method, as shown in Fig. 1. Next, we present each part of our network in detail and represent the loss functions to form the generator network and the network discriminator, respectively, and describe how our approach improves road marking recognition performance.

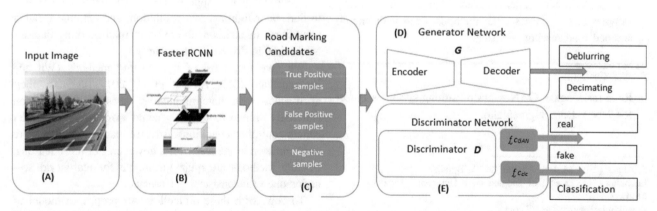

Fig. 1 The architecture of the proposed road marking recognition system. **a** The images are introduced into the network; **b** The Faster R-CNN detector; **c** The detector results are composed of positive and negative data; (**d**, **e**) The GAN is formed jointly by the loss of two branches

3.1 Road Marking Extraction

The proposed road marking detection system consists of three networks: (i) VGG-16 network for the feature extractor (ii) fast R-CNN network, which includes a RoI grouping layer for classification and regression, and (iii) the network of regional proposals (RPN).

Feature extractor (VGG-16). In our proposed method, we chose VGG-16 for the simplicity of its architecture as a functionality extractor. A typical VGG-16 network consists of thirteen convolutional layers followed by thirteen activation function (ReLU) layers, three fully connected layers, and five grouping layers. The VGG-16 has several stages: the first stage (conv1) has two convolutional layers and two ReLU layers and a grouping layer. The second step (conv2) has two convolutional layers, two ReLU layers, and a grouping layer. Step three (conv3) has three convolutional layers, three ReLU layers, and a grouping layer. The fourth step (conv4) includes three convolution layers, three ReLU layers, and a grouping layer. The last step (conv5) includes three convolutional layers, three ReLU layers, and a grouping layer (Zhu et al. 2016). At each stage, the characteristics map is preserved (Han et al. 2019; Xu et al. 2018). Once an image has passed through each stage, its characteristic card is reduced by half. After several convolutions and pooling operations, coarse characteristic maps are obtained in the last convolution step (Tian et al. 2018; Xie et al. 2019).

Merging of features. The characteristics of the deep layers have undergone several operations of convolution and subsampling. Consequently, the information in these layers becomes deeply semantic but more abstract (Yang et al. 2018; Lu et al. 2018; Peng et al. 2017). Conversely, these characteristics of the shallow layers are rich in precise positioning information for the object in the image but have a low score for the ability to represent semantic characteristics (Zhu et al. 2016). Thus, to take advantage of the advantages of the shallow and deep layers and to overcome the compromise between the spatial resolution of the lower layers and the distinctive semantic characteristics of the deep layers, the fusion of the multi-layer characteristics is imperative (Zhu et al. 2016; Xu et al. 2018). The characteristic maps obtained from the output of each convolutional stage must have sufficient semantic and positional information for an excellent performance of the model. The size of the feature map should be considered accordingly before the merger; too small a feature map provides insufficient feature information and too large a size increases the computational complexity (Xu et al. 2018). In previous research, various techniques have been used to merge the shallow and deep layers of convolutional

neural networks (Ren et al. 2018). In most of the existing publications (Zhu et al. 2016; Ren et al. 2018), the characteristics are extracted from the last convolutional layer of each step. In our approach, we concatenate feature maps from the convolutional layers of each step. Since at each convolutional stage, the characteristics map is preserved. Therefore, we have concatenated the feature maps without resizing. In addition, the dimensions of the channels have been dimensioned using a convolutional layer. This convolutional layer can be used to increase or reduce the number of filters. By reducing the number of filters, the quantity of training parameters in the network decreases. We then normalized the feature maps of conv1, conv3, and conv5 using batch normalization. This concept of batch standardization has been followed by Yan et al. (2016). After that, the maps of the characteristics of the convolutional layer of conv1, conv3, and conv5 of the VGG-16 were concatenated. The conv1 and conv3 feature cards have been subsampled to the size of the conv5 feature maps. This technique of subsampling to the size of the map of the characteristics of the last convolution step worked well for (Sun et al. 2018) in their search application. By setting the subsampling parameters accordingly, the feature maps were converted to the same size.

Detector network. Initially, R-CNN relies on a selective search algorithm to generate around 2000 regions of interest (RoI) (Yang et al. 2018). However, R-CNN is too slow for the real-time detection task. There is an improved version of R-CNN that uses RPN instead of selective search to generate regions of interest (RoI). A network of regional proposals (RPN) takes input of characteristic maps and produces thousands of proposals. A RPN is a fully convolutional kernel size layer that branches into two convolutional layers for independent class classification and regression of RoI. To generate propositions, we drag a sliding window onto the last map of shared convolutional entities obtained from VGG-16. At each sliding window position, there are nine anchors, which translate into a map of characteristics from which different sizes of proposals are obtained (Girshick et al. 2014). RoIs are sent in two convolutional layers for the box regression and box classification task (Xie et al. 2019). The maximum possible proposition for each location is indicated by k. The classification layer (cls) is assigned scores of 2 k (there is an object or no object) for each proposition. The intersection on the union between the truth delimitation area on the ground area (X) and the anchor box (Y) is used to determine the presence of a positive and negative proposition (Han et al. 2019).

The expression of intersection-on-union is given by the equation:

$$IoU = \frac{X \cap Y}{X \cup Y} \qquad (1)$$

When forming a network of region propositions, we consider two class labels (object or non-object) for each anchor. We consider two cases for positive anchors: (i) the anchor which has the highest intersection–union overlap (IoU) with the truth box on the ground or (ii) an anchor which has an intersection–union overlap more than 0.7 with any truth box on the ground. Similarly, we have assigned an intersection-on-union of less than 0.3 for negative anchors. Any anchor with an IoU greater than 0.3 but less than 0.7 is not taken into account in the training objective (Ren et al. 2017).

Anchors are available at different scales and different aspect ratios; they can be sized accordingly to suit a particular application. The proposals generated by the RPN are introduced into the pooling regions of the region of interest (RoI). The pooling layer takes a fixed size of RoI. After that, the output is introduced into the classification and regression layers for classification and regression, respectively.

3.2 The CNN Classifier

After the consolidation process on several scales, the track lines and the pedestrian crossings are deleted. The other markings are small road markings which are grouped into the following ten classes: two classes of dotted lines, four classes of text, two classes of arrows, one class of diamonds, and one class of triangular marking. A CNN classifier is designed as follows: A four-layer convolutional network, with a core size of 3×3 and a stride of one, is used to extract the characteristics. During the training process, batch normalization is used after each convolution operator. Then the activation function, ReLU, is added. After extracting features, three-dimensional feature maps are flattened into one-dimensional feature vectors. Then the vectors are sent to fully connected layers. The first two fully connected layers contain 1024 nodes, and the activation function is also ReLU. To avoid over-adjustment, an abandonment operation (ratio = 0.5) is used during training. The last fully connected layer contains ten nodes and the activation function is Softmax, the output of which is the class score. The image is classified as the category with the highest score.

A set of small marking training samples is prepared manually as templates. Models are rotated at different angles and some flaws are added to generate additional training samples.

3.3 The Generative Adversarial Networks (GAN)

The GAN framework (Radford et al. 2015) is composed of two networks: a generator network G and a discriminator network D. The generator network G and the discriminator network D are trained simultaneously by playing a minimax game with two players. Thus, the generator G attempts to minimize the differences between the real samples and the false samples generated by G to deceive the discriminator D. On the other hand, the discriminator D aims to maximize these differences to distinguish the real from the false. The GAN function can be formulated as the following minimax function:

$$min_{\theta_G} max_{\theta_D} V(D, G) = \mathbb{E}_{x \sim Pt(x)}[logD(x)] + \mathbb{E}_{z \sim Pz(z)}[log(1 - D(G(z)))] \qquad (2)$$

where Pt is the true data distribution observation of x and Pz is the false data distribution observation of a random distribution z. GAN alternately optimizes two competing networks. Therefore, the conclusion for playing the minimax game may be that the probability distribution (Pz) generated by the generator G corresponds exactly to the data distribution (Pt). After all, the discriminator D will not be able to distinguish between the sampling distribution of the generator G and the distribution of real data. At this time, for the fixed generator, the optimal discriminant function is as follows:

$$D_G^*(x) = \frac{P_t(x)}{P_t(x) + P_z(x)} \qquad (3)$$

More recently, associated with the mutual maximization of information, we have designed a GAN framework to deal with the problem of trivial markings. By retaining information about the semantic characteristics of the data, the network is able to learn more meaningful hidden representations for erasure and classification. Mutual information can be formulated as follows:

$$I(c; G(z, c)) = H(c) - H(c|G(z, c)) \qquad (4)$$

where c denotes the salient structured semantic characteristics of the data distribution by c1; c2; …; cN, the functions H () and H (|) represent, respectively, the marginal and conditional entropy, and I () denotes mutual information. We maximize mutual information because information from latent code c should not be lost in the generation process. Consequently, the minimax clearance of a common G + D network is formulated as follows:

$$argmin_{\theta_G} max_{\theta_D} V(D, G) = \mathbb{E}_{\tilde{X}, y}[logD(X, y)] + \mathbb{E}_{\left(\tilde{X}, y\right) \sim P_t + \mathcal{N}_\sigma}\left[log(1 - D\left(G\left(\frac{\tilde{y}}{X,}\right)\right))\right] - \lambda I(y; G\left(\frac{\tilde{y}}{X}\right)) \qquad (5)$$

where $\mathcal{N}$ is the noise model of the normal distribution with the standard deviation σ, X designates the candidates for road marking with a clean image, $\widetilde{X}$ represents the candidates for road marking with a fuzzy noise and y is the label of road marking. In the discriminator network D, we jointly distinguish the generated images and the classification of road markings.

3.4 Proposed GAN Network Architecture

Our framework of generative contradictory networks includes two subnets: (i) the generator network and (ii) the discriminator network. Here are the details of our network architectures.

Generator network. Inspired by the recent success of image reconstruction (Sabokrou et al. 2016), we implement the generator model G using a fully convolutional auto-encoder, as shown in Fig. 2. Since the input image sampled from the unconstrained scenes is very corrupt, the conversion to a clean image is very useful for the classification of road markings. In addition to this idea, our G network forms blur suppression mappings from corrupted positive images (is road markings) to proper images, while it learns to decimate outlier mappings (is not road marking), which facilitates classification to distinguish outliers from unrestricted samples from start to finish. Therefore, G can eliminate noise only in positive samples while preserving abundant image details. Finally, G learns the target representation to provide a property without artifacts and preserving details to positive samples and corruption to negative samples.

Discriminator network. We use a sequence of convolutional layers in our discriminator network D, as shown in Fig. 3. In the existing GAN frameworks, the discriminator network has only one branch which distinguishes the real and generated samples. Instead, in our discriminator model,

D has two parallel branches, namely, a real/fake sample classifier f_{cGAN} and a road marking classifier f_{cclc}. Figure 3 shows the details of the architecture of this network. The output of the branch f_{cGAN} is the probability that the input is a real, and the output of f_{cclc} is the softmax probability of the input being categories of road markings.

3.5 Overall Loss Function

To form the model, we compute the loss function of the common G + D network using only the positive samples. When we optimize our generator network, we adopt pixel loss instead of functionality mismatch, defined as

$$L_{MSE} = \|X - X'\|^2 \tag{6}$$

where X' is the output of G. However, the MSE optimization problem often leads to a lack of sharpening effect, which leads to an over-smoothing effect which always causes blurring. To solve this problem, we also introduce the loss of classification, which favors G to make the sharpening effect as large as possible. The formulation of the loss of classification is

$$L_{clc} = \left(\log\left(y_i - D\left(G\left(\widetilde{X}\right)\right)\right)\right) + \log(y_i - D(X))) \tag{7}$$

where y indicates the corresponding image labels and i represents the number of classes. Therefore, the loss of classification provides an improvement in the recovered images only when they are not used. Based on the above functions, the model is optimized to minimize the loss function:

$$L = L_{G+D} + \lambda L_{MSE} + L_{clc} \tag{8}$$

Fig. 2 Network architecture G, composed of encoding (first part) and decoding (second part) layers

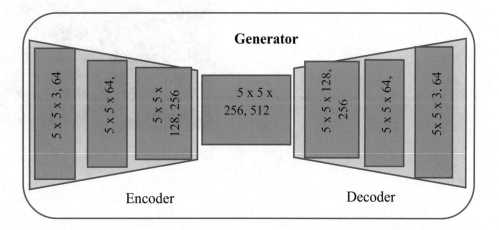

Fig. 3 The network D is composed of three convolutional layers with two parallel branches

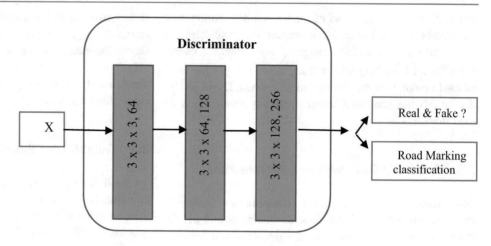

where $\lambda > 0$ is a compromise weight which controls the relative importance of the MSE loss. To find an appropriate hyperparameter λ of the training L, we conducted a number of experiments and finally introduced an appropriate value of 0.05. On the basis of this discovery, we stopped the formation of networks, when G can reconstruct its entry with a minimal error ($\|X - X'\|^2 <$ r, where r is a small positive number).

After the formation of the networks, we analyze the behavior of each entry. For any given positive sample that follows Pt, G is formed and performs the denoising task, and its output will be a clean version of the input data. On the other hand, for any given outlier that does not follow Pt, G is disturbed and cannot reconstruct it correctly. Consequently, G plays a role of denoising as well as decimation for the non-critical negative samples and, in doing so, removes the negative samples from the possibility of false positive error.

3.6 GAN-Based Data Augmentation

The data augmentation model can be used in the latent code of the G network on the basis of mutual information. Let us consider a source domain composed of data D = {X1, X2; …; XN} and corresponding target classes {y1; y2; …; yN}. Given a data point (X; y), we could get a meaningful representation of the data point, which encapsulates the information necessary to generate other related data. As a result, the latent code of G learns and preserves the class-specific semantic context, as shown in Fig. 4.

To illustrate the advantage of the increase based on InfoGAN (Chen et al. 2016), we also use the classical GAN. For the classic GAN, we form a fully convolutional generator without using labeled data. While the classic GAN optimizes its loss function, its generator visually generates almost false data, as shown in Fig. 5.

Fig. 4 Results of data augmentation using GAN based on mutual information. There is no real picture on this figure

Fig. 5 Some examples of fuzzy road markings generated by the classic GAN

4 Conclusions

In this work, we presented an architecture based on GAN, for the detection and classification of road markings. Specifically, our framework has proven to be robust to various environments without constraint. We design a network of generators to directly restore a blurred image from a blurred positive image, while distorting a decimated image from a negative image. In addition, we have proposed a two-way classification branch to the discriminating network, which can distinguish the fake/real and road marking categories, including the negative class, resulting in improved accuracy and practical evolution. In addition, applying data augmentation based on mutual information can lead to better performance than other advanced augmentation methods. We have demonstrated its general applicability by incorporating the Faster R-CNN as a reference base. Finally, our approach can be easily integrated into ADAS systems for exact location of road markings in the future.

References

Ahmad, T., et al.: Symbolic road marking recognition using convolutional neural networks. In: Intelligent Vehicles Symposium (IV). IEEE (2017)

Chen, X., Duan, Y., Houthooft, R., Schulman, J., Sutskever, I., Abbeel, P.: Infogan: interpretable representation learning by information maximizing generative adversarial nets. In: Advances in Neural Information Processing Systems. pp. 2172–2180 (2016)

Girshick, R., Donahue, J., Darrell, T., Malik, J.: Rich feature hierarchies for accurate object detection and semantic segmentation. In: 2014 IEEE Conference on Computer Vision and Pattern Recognition, pp. 580–587, Columbus, OH, USA, June (2014)

Goodfellow, I., et al.: Generative adversarial nets. In: Advances in Neural Information Processing Systems (2014)

Greenhalgh, J., Mirmehdi, M.: Automatic detection and recognition of symbols and text on the road surface. In: International Conference on Pattern Recognition Applications and Methods. Springer, Cham (2015)

Han, C., Gao, G., Zhang, Y.: Real-time small traffic sign detection with revised faster-RCNN. Multimed. Tools Appl. **78**(10), 13263–13278 (2019)

Hu, J., et al.: Near-infrared road-marking detection based on a modified faster regional convolutional neural network. J. Sens. 2019 (2019)

Kheyrollahi, A., Breckon, T.P.: Automatic real-time road marking recognition using a feature driven approach. Mach. Vis. Appl. **23** (1), 123–133 (2012)

Krizhevsky, A., Sutskever, I., Hinton, G.E.: Imagenet classification with deep convolutional neural networks. In: Advances in Neural Information Processing Systems (2012)

Lee, Y., et al.: Unconstrained road marking recognition with generative adversarial networks. In: 2019 IEEE Intelligent Vehicles Symposium (IV). IEEE (2019)

Lu, Y., Lu, J., Zhang, S., Hall, P.: Traffic signal detection and classification in street views using an attention model. Comput. Vis. Media **4**(3), 253–266 (2018)

Peng, E., Chen, F., Song, X.: Traffic sign detection with convolutional neural networks. In: Communications in Computer and Information Science, pp. 214–224. Springer Nature (2017)

Radford, A., Metz, L., Chintala, S.: Unsupervised representation learning with deep convolutional generative adversarial networks (2015). arXiv preprint arXiv:1511.06434

Ren, S., He, K., Girshick, R., Sun, J.: Faster R-CNN: towards real-time object detection with region proposal networks. IEEE Trans. Pattern Anal. Mach. Intell. **39**(6), 1137–1149 (2017)

Ren, Y., Zhu, C., Xiao, S.: Small object detection in optical remote sensing images via modified faster R-CNN. Appl. Sci. **8**(5), 813 (2018)

Sabokrou, M., Fathy, M., Hoseini, M.: Video anomaly detection and localisation based on the sparsity and reconstruction error of autoencoder. Electron. Lett. **52**(13), 1122–1124 (2016)

Simonyan, K., Zisserman, A.: Very deep convolutional networks for large-scale image recognition (2014). arXiv preprint arXiv:1409.1556

Sun, L., Chen, J., Xie, K., Gu, T.: Deep and shallow features fusion based on deep convolutional neural network for speech emotion recognition. Int. J. Speech Technol. **21**(4), 931–940 (2018)

Tian, Y., Gelernter, J., Wang, X., et al.: Lane marking detection via deep convolutional neural network. Neurocomputing **280**, 46–55 (2018)

Villalón-Sepúlveda, G., Torres-Torriti, M., Flores-Calero, M.: Traffic sign detection system for locating road intersections and roundabouts: the Chilean case. Sensors **17**(6), 1207 (2017)

Wu, T., Ranganathan, A.: A practical system for road marking detection and recognition. In: Intelligent Vehicles Symposium (IV). IEEE (2012)

Xie, Y., Dai, W., Hu, Z., Liu, Y., Li, C., Pu, X.: A novel convolutional neural network architecture for SAR target recognition. J. Sens. Article ID 1246548, 9 (2019)

Xu, Y., Zhu, M., Xin, P., Li, S., Qi, M., Ma, S.: Rapid airplane detection in remote sensing images based on multilayer feature fusion in fully convolutional neural networks. Sensors **18**(7), 2335 (2018)

Yan, Z., Zhang, H., Jia, Y., Breuel, T., Yu, Y.: Combining the best of convolutional layers and recurrent layers: a hybrid network for semantic segmentation (2016)

Yang, T., Long, X., Sangaiah, A.K., Zheng, Z., Tong, C.: Deep detection network for real-life traffic sign in vehicular networks. Comput. Netw. **136**, 95–104 (2018)

Zhu, Y., Zhang, C., Zhou, D., Wang, X., Bai, X., Liu, W.: Traffic sign detection and recognition using fully convolutional network guided proposals. Neurocomputing **214**, 758–766 (2016)

A Smart Agricultural System to Classify Agricultural Plants and Fungus Diseases Using Deep Learning

Oussama Bakkali Yedri, Mohamed Ben Ahmed, Mohammed Bouhorma, and Lotfi El Achaak

Abstract

Sustainable agriculture is the world focus of modern agriculture, to meet with actual and future generation needs, without destroying the environment, by lowering pollution level, minimizing water use, and using less pesticides with early interventions. Fungus is one of the greatest agricultural problems, it infects a large amount of plants species, benefiting from its ease in spreading on air, it can attack and destruct whole field crop production. Even though, it can be easily treated with early interventions before it's too late. The main challenge is early stages of detection in order to prevent spreading. With the help of modern technologies, Deep Learning (DL) algorithms offer us the possibility of classifying healthy and infected leaves, so we can apply early interventions to protect our plants. The purpose of this study is to build a Convolutional Neural Network (CNN) to help us to classify affected plants based on pre-learned and pre-processed features in order to take necessary interventions at early stages of infections. To do so, we use a set of high-quality images and inject them as an entry of our CNN algorithm which was chosen based on its performance for such study.

Keywords

Fungus disease • Smart agriculture • Deep learning • Convolutional neural network

O. Bakkali Yedri (✉) · M. Ben Ahmed · M. Bouhorma ·
L. El Achaak
University of Sciences and Technologies, Abdelmalek Essaadi
University, Tangier, Morocco
e-mail: bakkali.yedri.oussama@gmail.com

M. Ben Ahmed
e-mail: mbenahmed@uae.ac.ma

M. Bouhorma
e-mail: mbouhorma@uae.ac.ma

L. El Achaak
e-mail: lelachaak@uae.ac.ma

1 Introduction

Smart Agriculture is an important dimension of smart cities concept, looking attentively at its impact on the economical side of any country, considering the value of the agricultural economical source of income of the country and improving the existing sources rather than spending money on new investments.

Fungal diseases present a major problem in the agricultural field. When trees suffer from infections, it has a fatal economic impact by diminishing gravelly crop production (Moshou 2011).

A combination of practices can adapt to climate change and control the necessary supplements while having the potential to increase food production. In other hands, artificial intelligence is the theory and development of computer systems able to perform tasks normally requiring human intelligence, such as visual perception, speech recognition, decision-making, and translation between languages.

Deep learning is a type of artificial intelligence derived from machine learning where the machine is able to learn by itself. It is based on a network of artificial neurons inspired by the human brain. This network is made up of tens or even hundreds of "layers" of neurons, each receiving and interpreting information from the previous layer.

A convolutional neuronal network (CNNs) is a concrete case of deep learning neural networks, which were already used at the end of the 90 s but which in recent years have become enormously popular (McCoss 2016) when achieving very impressive results in the recognition of image, deeply impacting the area of computer vision.

CNN is a type of deep learning model for processing data that has a grid pattern, such as images, which is inspired by the organization of animal visual cortex (Yamashita et al. 2018).

In recent years, deep learning has made promising achievements in the machine learning field (Fragkiadaki et al. 2015; Yuan et al. 2015; Girshick et al. 2016). Among many deep learning algorithms, we find that Convolutional neural network (CNN) makes great success in image classification, Hu (Girshick et al. 2016) proposed a CNN that contains different layers, which are input, convolutional, max pooling, fully connected, and an output layer, for image classification. Where Makantasis (Makantasis et al. 2015) presented a deep learning-based classification method that exploits a CNN to encode the image information and a multilayer perceptron to conduct the classification task. Unlike Chen (Chen et al. 2016) who proposed a regularized 3D CNN-based feature extraction model to extract efficient spectral–spatial features for hyperspectral image classification.

In our research, we find several works focused on smart agriculture in order to classify leaf crops and also to detect infected from healthy plants, using several machine learning algorithms, each of these algorithms uses its own way to predict this classification.

In this paper, we present a comprehensive review of the application of deep learning in agriculture. A number of relevant papers are presented that emphasize key and unique features of popular ML models.

2 Theoretical Background

Deep learning is being used in multiple fields for image classification and recognition, we find multiple authors talk about classification using multiple algorithms such as SVM, CNN, RNN, etc., by proving it efficiency for image recognition, most of its technics have been replaced by deep learning for many domains (Komura and Ishikawa 2018), such as pathology, medical image analysis (Litjens 2017; Shen et al. 2017; Xing and Yang 2016), and agriculture (Hasan et al. 2019; Amara et al. 2017).

Typically, machine learning methodologies involve a learning process with the objective to learn from "experience" (training data) to perform a task. Data in ML consists of a set of examples. Usually, an individual example is described by a set of attributes, also known as features or variables. A feature can be nominal (enumeration), binary (i.e., 0 or 1), ordinal (e.g., A+ or B−), or numeric (integer, real number, etc.).

The performance of the ML model in a specific task is measured by a performance meter that is improved with experience over time. To calculate the performance of ML models and algorithms, various statistical and mathematical models are used. After the end of the learning process, the trained model can be used to classify, predict, or cluster new examples (testing data) using the experience obtained during the training process. Recently, many authors used deep learning for crop leaf classification and disease detection.

First, we find the paper entitled Leaf Classification Based on GLCM Texture and SVM (Based and on GLCM Texture and SVM 2018) the authors used classification of leaves using Gray-level Co-Occurrence Matrix (GLCM), by extracting leave's features and Support vector machines (SVM) algorithm, the author describes how artificial neural network is used to identify plant by inputting leaf image, using image processing techniques to extract leaf shape and features.

The paper has a focus on the following:

- Introducing an approach of plant classification based on the characterization of texture properties using combined "classifier learning vector quantization" and "radial basis function," the main idea is to recognize a leaf from only parts of it and its texture with no need for full leaf picture.
- Proposing a method that incorporates shape, vein, color, and texture features based on Probabilistic neural network PNN.

However, the main idea of the project is to classify different leaves species based on their texture using GLCM for texture feature extraction and SVM for classification.

Moreover, the second study is entitled Architecture-Based Classification of Plant Leaf Images. The author used a method of three main steps to classify any leaf based on its shape and boundaries (Sadeghi et al. 2018) using The Otsu, polygon decimation simplification, and centroid contour distances algorithms:

- Leaf image pre-processing: By taking only leaves shape and ignoring the color by turning it to black and white image
- Architecture-based features extraction
- Classification and species identification.

The third study, entitled Multi-class K-support Vector Nearest Neighbor (K-SVNN) for mango leaf classification, the author (Prasetyo et al. 1826) research is to classify mango varieties using K-SVNN that uses data reduction while preserving the accuracy. While K-SVNN works only for binary class, the author also proposed the entropy to calculate the Significant Degree (SD) to solve the multi-class problem (Prasetyo et al. 1826).

Entropy can measure the impurity of data class distribution so that the selection of the SD can be conducted based on the high entropy. The data with same class distribution has zero impurity, whereas data with uniform class distribution has the highest impurity.

Other studies used many other algorithms, such in this study CNN-RNN: a large-scale hierarchical.

Using image classification framework (Guo et al. 2018), we find that the author talked about his own dataset before proposing the deep learning algorithm. Traditionally, input dataset for classifiers uses small datasets, by encoding hand-crafted features (Zhang et al. 2014).

Mosin Hasan and his co-authors (Hasan et al. 2019) go for tomato leaf disease detection using transfer learning in CNN, Transfer learning, as the author says, is an approach which reduces the number of parameters, using a neural network which has already been trained. Unlike Jihen Amara (Amara et al. 2017) used CNN for banana diseases classification, the author compared the results of two types of images, RBG and grayscale image in order to identify which type of images gives us the best accuracy, while other authors (Delalieux et al. 2012; Ham et al. 2005) and (Ho 1998) went for hyperspectral analysis with supervised classification using decision tree random forest (DTRF) which develops multiple trees from samples and combines the results via voting or using maximum and support vector machines for multi-class classification who has achieved great success in various applications and is considered a stable and efficient algorithm for image classification.

Table 1 summarizes algorithms used by each article discussed above.

Deep learning methods have proven to be successful for image classification. In recent years, researchers have built various deep structures (Makantasis et al. 2015; Chen et al. 2016), and have achieved quite accurate predictions on small datasets (Guo et al. 2018).

The first contribution of this paper is a framework capable of generating hierarchical labels, by integrating the powerful CNNs and Recurrent Neural Networks (RNN). CNN is used to generate discriminative features, and RNN is used to generate sequential labels.

In recent years, deep learning methods have attracted significant attention (Fisher and Bastian 2019) and have achieved revolutionary successes in various applications (He et al. 2015). Two important structures for deep learning are CNN and RNN. CNN has proven to be successful in processing imagelike data, while RNN is more appropriate in modelling sequential data.

From the studies, we show that ML and DL models have been applied in multiple applications for crop management, mostly yield prediction and disease detection, distribution of ML applications in livestock management (19%), water management (10%), and soil management (10%). This trend in the distribution of the application reflects the data-intense applications within the crop and high use of images. Data analysis, as a scientific field, provides the basics for the development of numerous applications related to this field because ML/DL predictions can be extracted without the need for third-party resources. It is also evident from the analysis that most of the studies used multiple models. More specifically, ANNs were used mostly for implementations in the crop, water, and soil management, while SVMs were used mostly for livestock management.

Table 1 Algorithms used in each discussed article

Article title	Algorithms						
	GLCM	SVM	PNN	K-SVNN	CNN	RNN	DT RF
Leaf classification based on GLCM texture and SVM (Leaf Classification Based on GLCM Texture and SVM)	X	X	X				
Multi-class K-support Vector Nearest Neighbor for Mango Leaf Classification (Prasetyo et al. 1826)				X			
CNN-RNN: a large-scale hierarchical image classification framework (Guo et al. 2018)					X	X	
Deep Learning Precision Farming: Tomato Leaf Disease Detection by Transfer Learning (Hasan et al. 2019)					X		
A Deep Learning-based Approach for Banana Leaf Diseases Classification (Amara et al. 2017)					X		
Heathland conservation status mapping through integration of hyperspectral mixture analysis and decision tree classifiers (Delalieux et al. 2012)		X					X
Investigation of the random forest framework for classification of hyperspectral data (Ham et al. 2005)							
The random subspace method for constructing decision forests (Ho 1998)		X					X
Deep Convolutional Neural Networks for Hyperspectral Image Classification (Girshick et al. 2016)					X		

By applying deep learning to crop image analysis, crop growth rate and crop disease detection at early stages provide an ultimate crop production improvement, a real increase of production levels and bio-products quality.

3 Methodology

Our approach on classifying plant leaves images, Rust and Mildew fungus disease on mint plant focuses on the recognition of marks that appears on mint leaves as shown on Fig. 1.

The proposed model in Fig. 2 consists of capturing field images to detect any plant infection, and captured images are uploaded manually by our main user in order to run the CNN algorithm. Our algorithm is based on pre-learned features saved in flat files. So after uploading our dataset, we use these well-predicted images as a learning set to increase pre-learned features, also to increase our accuracy. In parallel to that, we notify our user alerting him with possible infection in a specific area with recommended actions to stop the infection.

Fig. 1 Rust as it appears on leaves

The main purpose of using deep learning is the ability of training classifiers in order to detect whether a leaf is infected or healthy. CNN uses many layers of convolutions followed by a pooling operation to extract images or plant features. This process creates multiple minimized images and keeps only those with higher weights for our final model.

3.1 Process Description

The proposed process in our study consists of three main steps: the first step as data pre-processing, where all input images are collected, prepared, and then split in order to fit in the following steps. Deep Learning as the second step, where we use a part of images to train our classifier and learn images features. And finally, forecasting step, which is the last step where we predict new input images. Figure 3 shows us a general overview of our model.

3.2 Data Pre-srocessing

- Input images: We have multiple input images categories, for two different purposes, starting from the first purpose, which is leaf plant recognition, we have two plant categories or types, Mint and Cannabis plants. And on the other hand, we have two fungus types, Rust and Powdery Mildew in the mint plant. Each input image is labeled with its appropriate nomination affiliation.
- Shuffle Data: in order to have a good combination of data, selected in an aleatory way, to ensure a good data variety for each part of learning and test process.
- Random splitting: the dataset is being split using shuffled data into different portions or subsets, 70% Training, 15% validation, and the last 15% for testing which will remain unknown for the algorithm by keeping it away from the learning or training process, more details are presented in Table 2.

3.3 Deep Learning

- CNN: applying our algorithm architecture, defining the number of convolutional layers to have the best possible results based on experience.
- Activation layer: defines the output of each layer and input for the next layer of our CNN model, is used at the end of a hidden layer unit to introduce non-linear complexities to the model.

Fig. 2 Proposed system architecture

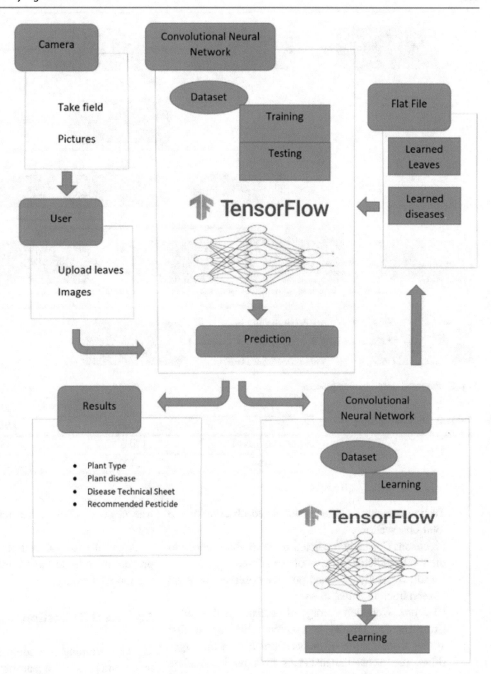

- Loss function: the main function is responsible for the network weight optimization.
- Train CNN: training our CNN model in order to describe and define relations in our model.
- Performance validation: using validation dataset, we tend to evaluate the performance of our model, to help the algorithm to update its weights. we define the validation and error accuracy in order to minimize this error and guaranty better results.

3.4 Prediction

- CNN results test: after the learning process, we calculate validation accuracy to evaluate our model performance to either validate it or improve it.
- Predict test data: we tend to use new input data which is unknown for our model
- Metrics Performance: many metrics are calculated in order to evaluate the model's performance, as follows:

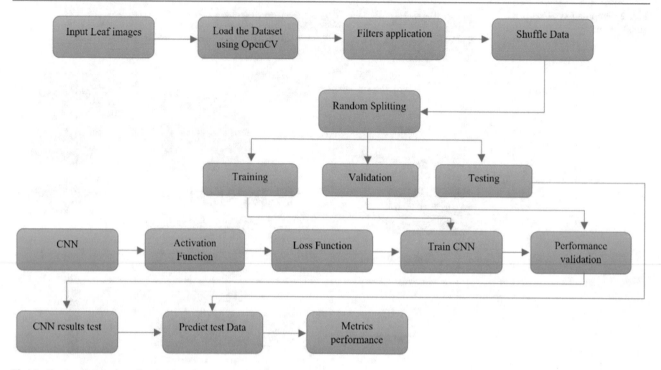

Fig. 3 Proposed deep learning process

Table 2 Random splitting details

Dataset	Training	Validation	Testing
% of total dataset	70%	15%	15
# of images	1120	240	240

- Testing accuracy: Correctly predicted observations to total observations.
- Precision: Correctly predicted positive observations to the total predicted positive observations.
- Recall: Correctly predicted positive observations to all observations in actual class.
- F1 score: Weighted average of precision and recall.
- Error matrix: Display a specific table layout that allows visualization of the performance of our algorithm, represents the instances in a predicted class while each column represents the instances in an actual class.

For further details, we have detailed mathematical formulas in Table 3, using values of true positive (TP),

true negative (TN), false positive (FP), and false negative (FN).

After metrics calculation and evaluation, we rerun the process in order to have more significant values and ensure the model stability.

3.5 Data Collection

Machine learning is a complex resilient field that needs the maximum possible of learning data in order to give significant results. To do so, we collected open source images from the internet in order to ensure the desired amount of data with different images sizes, but the main challenge is to challenge the algorithm and try to learn necessary features from a small dataset. Examples of our dataset are shown in Fig. 4.

Table 3 Performance metrics formulas

	Formula
Accuracy	TP + TN/TP + FP + FN + TN
Precision	TP/TP + FP
Recall	TP/TP + FN
F1 score	[2 * Recall * Precision]/[Recall + Precision]

Fig. 4 Captions of used input images

4 Convolutional Neural Network

The important idea of convolutional networks is that each neuron in each intermediate layer is exposed is only exposed to a receptor field, and the analysis of this receptor field is the same as analysis performed by another neuron of the same layer with its own receiver field.

We multiply the value of each pixel by a synaptic weight and calculate the sum of the weighted luminosities and hold us only the positive part of the result, which corresponds to the activation function.

The synaptic weights then form a convolution matrix (Filter/kernel), which will be applied to all regions of the image and other neurons of the layer in question, in particular, these synapses have always the synaptic weight, we talk about sharing synaptic weights (weight sharing). This allows us to drastically reduce the parameters of our neural network.

CNN's have a pre-programmed architecture for translational invariance of images. To reduce the size of the images, we use a dimensionality reduction function which amounts to summarizing the information of several neighboring neurons into a single piece of information, in order to do so, the average value of the excitations of these neurons is then calculated, we then talk about the extraction of the maximum (Max pooling) using the same convolution matrix for all the

receptive fields that automatically guarantees our network of convolutional neurons, they have a pre-programmed architecture for translation invariance of the images.

After pooling, the different sub-regions of the image were then described using the resume composed of a few canes. Then we start again any debt of similar operations convolution pooling, etc. Generally, after a few lines, we obtain a vector of much smaller dimension and which has an abstract representation of the contents of the different regions of the image, more details are shown on the next Fig. 5.

In our research, we used a CNN algorithm in order to identify three plant species from there leaves, Verbena, Mint, and Cannabis plants. To get the best results, we used five convolutional layers, with the shape of (none, 64, 64, 32), (none, 32, 32, 32), (none, 16, 16, 64), (none, 8, 8, 64), (none, 4, 4, 128) consecutively as described in Table 4.

Input images are reshaped to 128 * 128 pixels, as described above, the first convolutional layer contains 32 kernels of a size of 3 * 3, we can have an overview in Fig. 6, followed by a max-pooling layer of size 2 * 2, which chooses the maximum value of consecutive four pixels.

In order to well understand the filter, we applied it on a real image, so we can see the added value of the filter in Fig. 7.

This step shows us how our algorithm reacts and how it digs deeper to extract images features.

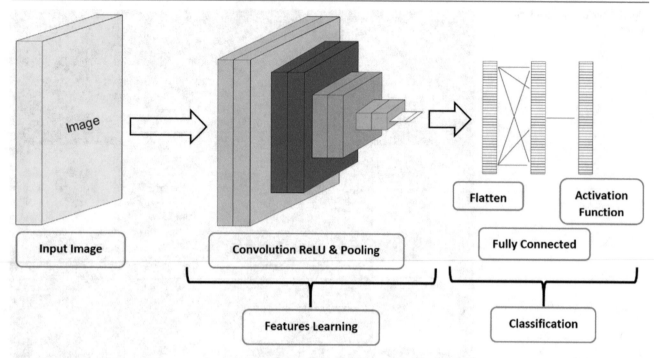

Fig. 5 Proposed CNN architecture

Table 4 Proposed CNN architecture

Layer	Output shape	Parameters
Conv2d_1 (Conv2D)	(none, 64, 64, 32)	864
Max_pooling2d_1 (MaxPooling2)	(none, 34, 34, 32)	0
Conv2d_2 (Conv2D)	(none, 32, 32, 32)	9216
Max_pooling2d_2 (MaxPooling2)	(none, 18, 18, 32)	0
Conv2d_3 (Conv2D)	(none, 16, 16, 64)	18,432
Max_pooling2d_3 (MaxPooling2)	(none, 10, 10, 64)	0
Conv2d_4 (Conv2D)	(none, 8, 8, 64)	36,864
Max_pooling2d_4 (MaxPooling2)	(none, 6, 6, 64)	0
Conv2d_5 (Conv2D)	(none, 4, 4, 128)	73,728
Max_pooling2d_5 (MaxPooling2)	(none, 2, 2, 128)	0
Flatten_1 (Flatten)		0
Dropout_1 (Dropout)		0
Dense_1 (Dense)		401,954
Dense_2 (Dense)		256

Total parameters: 541,314
Trainable parameters: 541,314
Non-Trainable parameters: 0

5 Results

For this study, we encoded the whole project with python programming language using Jupyter Notebook interactive computing. To start programming, we need many libraries for our algorithm, so we need TensorFlow for our CNN algorithm, Sci-Kit Learn for Data manipulation (normalization, splitting…), OpenCV for image loading and manipulation, Matplotlib for visualizations, and Numpy for mathematical operations. We also used an Intel Core i7-6700HQ CPU @ 2.60 GHz, and a Nvidia GeForce GTX 960 M to execute our script.

Fig. 6 Filters of the first convolutional layer

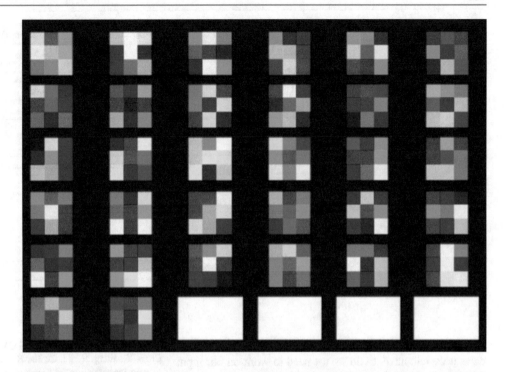

Fig. 7 Filters of the first convolutional layer applied to a real image

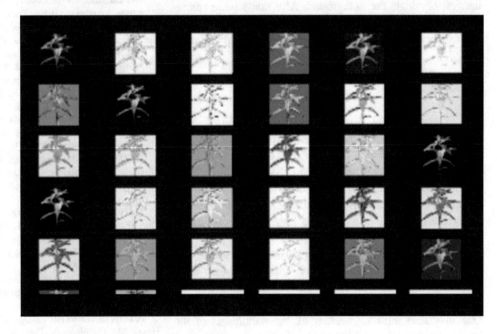

In many studies discussed in previous chapters, also in other researches (Yuan 2019), we find different results from different other algorithms, we also notice that CNN outperforms many other algorithms when it comes to image classification, as shown on next Table 5.

The main challenge for CNN algorithm is that the highest number of images for the learning stage, the better the results, in our case, we used a small Dataset containing 400 images of each plant and each disease, but focusing on selecting the best images which contain the best features.

With this approach, we used performance optimization by using many iterations, we end up having 93.8% on testing accuracy, 90.6% on validation accuracy, and 0.262 on validation loss as shown in the next Table 6.

Also as described in the next confusion matrices in Table 7, it is a table that describes the results of our prediction based on the number of properly classified examples, false positives, and false negatives.

The proposed model was built from scratch, in order to respond to our main problematic, but we need to challenge our

Table 5 Classical machine learning algorithms results

	Random Forest	Ada-Boost	SVM	Decision Trees	Artificial Neural Network
Accuracy	79.55	77.45	81.37	64.59	83.38
Precision	0.8	0.78	0.84	0.65	0.84
Recall	0.8	0.77	0.84	0.65	0.84
F1 score	0.8	0.77	0.84	0.65	0.84

Table 6 Results caption

	Validation accuracy	Testing accuracy	Validation loss	Testing loss
Proposed CNN algorithm	90.6%	93.8%	0.262	0.23

Table 7 Confusion matrix

s		Predicted infected	Predicted HEALTHY
True	Infected	362	38
	Healthy	29	371

algorithm to get better accuracy results with the smallest dataset we can have, to do so, we need to work on our input images, eliminate the background. Also many other researches (Anagnostis et al. 2020) use the Fast Fourier transform method for features selection and achieved a 96% of accuracy.

6 Conclusion

The automatic plant disease classification requires more improvement in order to have an industrial solution with the least human intervention. In this context, this paper presented a CNN algorithm that helps us to classify plants by their leaves, also to detect either a plant is infected with fungus disease or not. We have also compared our results with similar researches and specified a list of challenges and improvement that may be applied to have much better results.

Further perspectives for this study could be testing the robustness of this classification on small and noisy data. Stochastic methods both in the training algorithm and the bagging of the data have shown good results in the literature. One may also think of whether all the parameters of the images are worth being used as entries. The optimization of the training could be achieved, for example, by 'Dropping-Out' nodes we think are not very useful.

References

Amara, J., Bouaziz, B., Algergawy, A.: A Deep Learning-based Approach for Banana Leaf Diseases Classification, p. 10 (2017)

Anagnostis, A., Asiminari, G., Papageorgiou, E., Bochtis, D.: A Convolutional Neural Networks Based Method for Anthracnose Infected Walnut Tree Leaves Identification, p. 24 (2020)

Leaf Classification Based on GLCM Texture and SVM. **4**(3), 4

Chen, Y., Jiang, H., Li, C., Jia, X., Ghamisi, P.: Deep feature extraction and classification of hyperspectral images based on convolutional neural networks . IEEE Trans. Geosci. Remote Sensing **54**(10), 6232–6251 (2016). https://doi.org/10.1109/TGRS.2016.2584107

Delalieux, S., Somers, B., Haest, B., Spanhove, T., Vanden Borre, J., Mücher, C.A.: Heathland conservation status mapping through integration of hyperspectral mixture analysis and decision tree classifiers. Remote. Sens. Environ. **126**, 222–231 (2012). https://doi.org/10.1016/j.rse.2012.08.029

Fisher, D E., Bastian, B.C.: Melanoma (2019)

Fragkiadaki, K., Levine, S., Felsen, P., Malik, J.: Recurrent network models for human dynamics . In: 2015 IEEE International Conference on Computer Vision (ICCV), Santiago, Chile, p. 4346 4354 (2015). https://doi.org/10.1109/ICCV.2015.494

Girshick, R., Donahue, J., Darrell, T., Malik, J.: Region-based convolutional networks for accurate object detection and segmentation. IEEE Trans. Pattern Anal. Mach. Intell. **38**(1), 142–158 (2016). https://doi.org/10.1109/TPAMI.2015.2437384

Guo, Y., Liu, Y., Bakker, E.M., Guo, Y., Lew, M.S.: CNN-RNN: a large-scale hierarchical image classification framework. Multimed. Tools Appl. **77**(8), 10251–10271 (2018). https://doi.org/10.1007/s11042-017-5443-x

Ham, J., Chen, Y., Crawford, M.M., Ghosh, J.: Investigation of the random forest framework for classification of hyperspectral data. IEEE Trans. Geosci. Remote Sens. **43**(3), 492–501 (2005). https://doi.org/10.1109/TGRS.2004.842481

Hasan, M., Tanawala, B., Patel, K.J.: Deep learning precision farming: tomato leaf disease detection by transfer learning. SSRN J. (2019) https://doi.org/10.2139/ssrn.3349597

He, K., Zhang, X., Ren, S., Sun, J.: Deep residual learning for image recognition. arXiv:1512.03385 [cs], déc. 2015, Consulté le: déc. 19, 2019. [En ligne]. Disponible sur: https://arxiv.org/abs/1512.03385

Ho, T.K.: The random subspace method for constructing decision forests. IEEE Trans. Pattern Anal. Machine Intell. **20**(8), 832–844 (1998). https://doi.org/10.1109/34.709601

Komura, D. Ishikawa, S.: Machine learning methods for histopathological image analysis. Comput. Struct. Biotechnol. J. **16**, 34–42 (2018). https://doi.org/10.1016/j.csbj.2018.01.001

Litjens, G., et al.: A survey on deep learning in medical image analysis. Med. Image Anal. **42**, 60–88 (2017). https://doi.org/10.1016/j.media.2017.07.005

Makantasis, K., Karantzalos, K., Doulamis, A., Doulamis, N.: Deep supervised learning for hyperspectral data classification through convolutional neural networks. In: IEEE International Geoscience and Remote Sensing Symposium (IGARSS). Milan, Juill. 2015, 4959–4962 (2015). https://doi.org/10.1109/IGARSS.2015.7326945

McCoss, A.: Quantum deep learning triuniverse. JQIS **06**(04), 223–248 (2016). https://doi.org/10.4236/jqis.2016.64015

Moshou, D., et al.: Intelligent multi-sensor system for the detection and treatment of fungal diseases in arable crops. Biosyst. Eng. **108**(4), 311–321 (2011). https://doi.org/10.1016/j.biosystemseng.2011.01.003

Prasetyo, E., Adityo, R.D., Suciati, N., Fatichah, C.: Multi-class K-support vector nearest neighbor for mango leaf classification. TELKOMNIKA **16**(4), 1826 (2018). https://doi.org/10.12928/telkomnika.v16i4.8482

Sadeghi, M., Zakerolhosseini, A., Sonboli, A.: Architecture-Based Classification of Plant Leaf Images, p. 28 (2018)

Shen, D., Wu, G., Suk, H.-I.: Deep learning in medical image analysis. Annu. Rev. Biomed. Eng. **19**(1), 221–248 (2017). https://doi.org/10.1146/annurev-bioeng-071516-044442

Xing, F., Yang, L.: Robust nucleus/cell detection and segmentation in digital pathology and microscopy images: a comprehensive review. IEEE Rev. Biomed. Eng. **9**, 234–263 (2016). https://doi.org/10.1109/RBME.2016.2515127

Yamashita, R., Nishio, M., Do, R.K.G., Togashi, K.: Convolutional neural networks: an overview and application in radiology. Insights Imaging **9**(4), 611–629 (2018). https://doi.org/10.1007/s13244-018-0639-9

Yuan, L., et al.: Detection of anthracnose in tea plants based on hyperspectral imaging. Comput. Electron. Agric. **167**, 105039 (2019). https://doi.org/10.1016/j.compag.2019.105039

Yuan, Y., Mou, L., Lu, X.: Scene recognition by manifold regularized deep learning architecture . IEEE Trans. Neural Netw. Learn. Syst. **26**(10), 2222–2233 (2015). https://doi.org/10.1109/TNNLS.2014.2359471

Zhang, S., Sun, F., Liu, H.: Locality-constrained linear coding with spatial pyramid matching for SAR image classification. In: Sun, F., Hu, D., Liu, H. (eds.) Foundations and Practical Applications of Cognitive Systems and Information Processing, vol. 215, p. 867–876. Berlin, Heidelberg: Springer (2014)

Secure Data Collection for Wireless Sensor Network

Samir Ifzarne, Imad Hafidi, and Nadia Idrissi

Abstract

WSNs are growing rapidly around the world as the adoption of low-cost sensors allowed to make WSN in use for various environmental monitoring applications. Critical applications like health monitoring require high security as any loss of confidentiality, integrity, authenticity, or availability may have real and direct consequences on patient health or life. In general, any security issue may have an impact on WSN efficiency and safety. Many researchers focused on wireless systems security via disclosing vulnerabilities, detecting attacks or securing the network layers. Secure Data Collection is proposed to protect data privacy during data aggregation. Data aggregation is one critical steps as the non-leaf nodes consolidate data from children nodes and, hence, they have a major role in routing data to the base station. In this work, homomorphic encryption is used in combination with compressed sensing to protect data confidentiality without overloading the network. Compressive sensing compresses the data during the sampling which has the advantage of reducing the volume of transmitted data. The compression load is low and the data recovery which requires high computation effort is done on the sink side where there are no resource constraints. Encryption is executed at each node with a linear computation effort. During data transfer across the network, no decryption is required as the homomorphic encryption allow arithmetic operations on ciphertext without losing the useful data. Simulations show a better performance of the proposed scheme compared to a secure data aggregation method.

Keywords

Compressive sensing • Homomorphic encryption • WSN • Data aggregation security

1 Introduction

Wireless Sensor Networks allows monitoring the environment by sensing physical phenomenon. The sensor nodes collect the data such as temperature, pressure, humidity, or and allow to monitor the environment. Smart applications are processing the sensed data to enable monitoring risks like building monitoring or coordinating activities like in robotics. Generally, applications enable taking decisions and actions which can be executed manually, semi-automated, or fully automated.

Let's consider a WSN where Si is sensor node and xi is expressing the data generated by node Si ($i = 1, 2,\ldots, n$). Each node Si forwards its own reading along with the data received from the child node to the parent node. The sink completes data collection in WSN $x = (x1, 2,\ldots, xn)$ (Fig. 1).

As each node has a lifetime battery, the network lifetime will be dependent on the energy of sensors close to the sink. These sensors transmit more data and consume more energy than other children nodes.

Compressive sensing (CS) is a novel sensing approach which is very much different from the traditional compression theory where huge quantities of raw data are sensed than compression take place. CS theory highlights the fact that a sparse signal can be recovered from far fewer measurements or samples which means reducing the sampled data quantity and communication cost for data collection across the network which leads to prolong the WSN lifetime.

S. Ifzarne (✉) · I. Hafidi · N. Idrissi
National School of Applied Science: ENSA, 25000 Khouribga, Morocco
e-mail: sifzarne@gmail.com

I. Hafidi
e-mail: imad.hafidi@gmail.com

N. Idrissi
e-mail: nadia.idrissi@gmail.com

M. Ben Ahmed et al. (eds.), *Emerging Trends in ICT for Sustainable Development*,
Advances in Science, Technology & Innovation, https://doi.org/10.1007/978-3-030-53440-0_26

Fig. 1 Data collection scheme in
WSN with one route

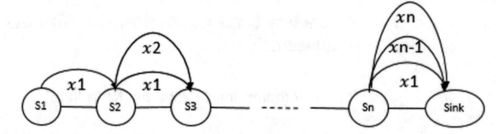

In a hostile environment, energy saving is the primary concern because the devices cannot be replaced immediately in case of failure. Data aggregation is another way to reduce the energy consumption by forwarding only the aggregated data to the base station, rather than sending all raw data directly to the base station (Xie et al. 2017). Fusing multiple sensor data related to the same physical event reduces the total amount of messages to be forwarded by the sensor devices. Thus, the primary goal of data aggregation is to reduce unnecessary energy consumption among sensor devices. The energy-efficient operations in WSNs have a significant impact on overall routing efficiency and lifetime. The central issue in integrating the WSN data aggregation with routing protocols is to extend it to the secure applications for providing an energy-efficient data aggregation and secure data transmission.

The following four techniques are employed in implementing the data aggregation: Compressive Sensing, Signal Processing, Information Theory, and Networking.

Among them, the compressive sensing can be applied in cluster-based WSNs, where each cluster member in a cluster sends its reading to the cluster head. In most of the sensing scenarios, the sensed data from several sensors are highly correlated. There is a possibility to combine the data and jointly process the collected data during data transmission to the base station. The compressive sensing scale downs the number of transmissions in the WSNs. The compressive sensing reduces both the bandwidth and energy consumption in the resource-constrained network, and therefore, it plays the leading role in WSNs. Another reason behind the energy depletion is that the sensor nodes closer to the base station have to forward a significant amount of data than others, resulting in a decline network lifetime. When dealing with data aggregation and compressive sensing for WSNs, the main challenges and constraints of WSNs are given below:

Resource Provisioning: The sensor nodes in WSNs are battery-powered, and the available energy of those sensors is limited. In most of the WSNs applications, human intervention is not desirable. Thus, it is usually impossible to recharge a sensor node. Thus, the design of energy-efficient compressive sensing is essential.

Successful Packet Delivery: In clustered WSNs, more communication traffic is induced at the cluster head,

compared with the routing without data aggregation. The limited node capacity and the interference caused by wireless communication make challenges in the design of useful compressive sensor algorithms for WSNs applications.

Scalability and Robustness: In WSN, new nodes may join a network, and existing nodes may leave the network at any time. It implies the dynamic topological structure of a WSN over time. To deal with the dynamics, an efficient distributed compressive sensing algorithm is essential.

Security Issues: Due to the open wireless environment, WSNs are susceptible to active and passive attacks. Notably, passive attacks have a severe impact on the privacy of compressive data gathering. Despite much recent interest in applying compressive sensing concept in WSN, only a few research works attempt to study secure compressive sensing by protecting the measurement matrix from the malicious attacks. However, most of the existing schemes are vulnerable to information leakage and involve high communication overhead to protect the measurement matrix.

2 Related Works

Several techniques are suggested for securing the compressive sensing over WSNs. In Rachlin and Baron (2008) the authors used the measurement matrix as a symmetric encryption key. The same key is used on one side by the node for encryption and on the other side by the base station to decrypt the received data. This technique is called symmetric encryption for compressed sensing. This protects the measurement matrix and signals against ciphertext-only attacks.

The complex deployment of WSN adds new challenges into the existing secure compressive sensing methods. Based on the one-time sensing assumption, the secret key or the measurement matrix is used only once. A one-time password was suggested in Bianchi et al. (2014, 2016) which is based on a random Gaussian matrix used for encryption. The random Gaussian matric is changing at each sensing round and each time a new matrix is used for encryption. This is a secure scheme but suffering from high cost to update the matrix. Independently, when the cluster head collects measurement results from other cluster members. Each sensing node has to submit its measurement matrix to the base station. To realize the one-time sense concept, these secret keys stored in the measurement matrix have to be synchronously updated in each measurement round. However, it is difficult under resource-constrained WSNs since it drastically increases the communication cost.

Secure communications with asymptotically Gaussian compressed encryption have been suggested in Cho and Yu (2018). This is also a cryptosystem with a compressive sensing Gaussian. A measurement matrix that does not satisfy restricted isometric property tends to a biased measurement, and it results in high compressive sensing recovery error. Moreover, the multi-hop routing concept in WSNs increases the communication cost of one-time sense that means a single-key distribution process requires several rounds of data transmission. To solve these issues, recent research works focus on the asymmetric homomorphic encryption schemes for providing security over compressive sensing in WSNs.

In Liu et al. (2019), the authors presented an adaptable secure compressive sensing-based data collection scheme for distributed wireless sensor network. To prevent passive attacks, both encryption and decryption are used, but they are computationally intensive operations. Even though the homomorphic encryption reduces the computational cost compared to others, but it still high overhead for resource-constrained WSNs. Recently, the homomorphic encryption algorithm enables each node to encrypt the sensed data but allows only the base station to decrypt the data. In Ngabo and El Beqqali (2019), an Implementation scheme of Homomorphic Encryption for Wireless Sensor Networks Integrated with Cloud Infrastructure is suggested. Notably, the base station is a powerful device in WSNs. Thus, it is essential to focus much on the energy-efficient homomorphic encryption techniques for WSNs.

The main aim of applying compressive sensing to WSN data aggregation is to reduce the communication cost. The compressive sensing working together network coding is applied in significantly reducing the number of transmissions in WSNs (Ebrahimi and Assi 2016). However, it fails in dealing with the temporal and spatial correlations simultaneously. A decentralized method in Kong et al. (2015) attempts to solve the joint problem of forwarding tree construction and link scheduling for compressive data gathering in WSNs. These schemes are suitable only for a particular geometric structure of the network with specific node locations. However, the real-time WSNs applications are mostly deployed in an area with a highly irregular shape and vulnerable environment. Thus, secure compressive sensing becomes the most crucial research topic in WSNs.

A Privacy-Preserving Compressive Sensing scheme in Xie et al. (2017) is suggested for crowdsensing-based trajectory recovery. It utilizes the homomorphic obfuscation

method in the compressive sensing framework for improving the recovery accuracy and privacy preservation simultaneously. In Hsieh et al. (2018), the homomorphic encryption is employed along with the compressive sensing, and it enables the base station can recover the original sensory reading by implementing the reconstruction algorithm on compressive sensing and by decrypting the messages received. Even though they protect the WSNs from the stalkers and eavesdroppers, they are used in crowdsensing recovery to reconstruct all trajectories of users on the basis of trajectory correlations, but they are not suitable for data gathering in clustered WSNs. In Zhang et al. (2018a), a data acquisition system by compressive sensing is designed. The measured data obtained by compressive sensing is noised as random noise is affecting the original sparse signal. Even though, the original data can be reconstructed from the noise and processed at the base station. It improves the security of data communication without increasing the communication overhead. The disadvantage is the key storage capacity and management difficulties due to the usage of the symmetric encryption key.

The combination of compressive sensing and homomorphic encryption in Zhang et al. (2018b) is suggested as a secure data collection scheme. This scheme exploits the homomorphic encryption while encrypting the measurement reading received by the compressive sensing method to achieve ciphertext calculation and to reach the purpose of privacy protection. It effectively deals with known plaintext attacks and select plaintext attacks, but it ignores the features of compressive sensing. Also, the used homomorphic encryption increases the system overhead. Most of the recently developed security algorithms use semi-homomorphic encryption algorithm, which maps only one operation of plain text with the ciphertext since the fully homomorphic encryption has practical implications in the wireless networks. The semi-homomorphic encryption is utilized in Bianchi et al. (2016). Each sensor node is allowed to encrypt the data. Each intermediate router is responsible for aggregating the encrypted data directly in cipher domain. It can avoid the creation of network bottleneck and balance the network traffic significantly. To prevent the WSN from ciphertext-only attack, the security solution in Cambareri et al. (2015) implements the compressive sensing to provide information-theoretic secrecy and but it does not provide the same secrecy under generic setting. They also hold a one-time sensing assumption, that means the measurement matrix or the secret key is employed only once. It increases the network overhead drastically. In the compressive sensing encryption schemes (Hung et al. 2015), the vector size of the transformation results is reduced compared to the original data, which means the security mechanism compresses the original data during the transformation. In Forster and Murphy (2011), the randomize transformation-based scheme is utilized, and it applies the obfuscation transformation only to the compressive measurement results. Unlike the compressive sensing techniques (Kong et al. 2015), the vector size of the measurement results in Alsheikh et al. (2014) is the same as that of the transformation result. In Di and Joo (2007), privacy-preserving compressive sensing is suggested for crowdsensing-based trajectory recovery. They do not work well under chosen-plaintext attack.

3 Compressive Sensing Overview

Traditional compression techniques suggest proceeding by sampling full information's and then get rid of redundant or non-useful data. There is a huge waste of resources to get high volume of raw data and then do the compression.

Compressed sensing senses only the useful information. When the signal is sparse, we can sample a small number of non-adaptive linear measurements of the signal which can be used to recover the original signal with very limited information's loss. The decompression phase uses different algorithms taking advantageof signal sparsity.

In Nyquist (1928), Shannon (1948), Whittaker (1915), Candès et al. (2006) authors demonstrate that signals can be recovered from a set of uniformly spaced samples taken at the so-called Nyquist rate of twice the highest frequency present in the signal.

Unfortunately, in many emerging applications such as medical imaging, video, remote surveillance, resulting Nyquist rate is so high that the data samples are too high.

The fundamental idea behind CS is rather than sampling at a high rate and then compressing the sampled data, we would like to find ways to directly sense the data in a compressed form —i.e., at a lower sampling rate. In their works, Emmanuel Candès, Justin Romberg, and Terence Tao and of David Donoho demonstrate that a finite-dimensional signal having a sparse or compressible representation can be recovered from a small set of linear non-adaptive measurements (Donoho 2006).

3.1 Compressed Signal Recovering

Recovering a signal $x \in \mathbb{R}^N$ from $y \in \mathbb{R}^M$ measurements such as $M \ll N$. Φ is a measurement matrix

$$y = \Phi x \qquad (1)$$

where M is equations and N is unknowns. There are much more unknowns than equations. So there are many solutions $x \in \mathbb{R}^N$ with $y = \Phi x$ without an additional hypothesis about x, we cannot recover x from y and Φ with $M < N$.

By a compressible representation, we mean that the signal is well approximated by a signal with only a few nonzero coefficients. To consider this mathematically, let x be a signal that is compressible in the basis Ψ:

$$x = \Psi \alpha$$

where α are the coefficients of x in the basis Ψ. The error between the true signal and its K term approximation is denoted the K-term

$$\sigma_k(x) = \arg \min_{\alpha \in \sum_k} \|x - \Psi \alpha\|_2 \qquad (2)$$

4 Homomorphic Encryption and Data Aggregation

In hop-by-hop encrypted data aggregation, each non-Leaf node is decrypting the received data and encrypting it again before sending to the next hop. This scenario presents a high risk for data confidentiality if an attacker compromises a node. The attacker can then get access to the encryption key and become able to perform the decryption process.

End-to-end encrypted data aggregation uses homomorphic encryption to apply certain aggregation functions such as addition or multiplication on the encrypted data. The encrypted reading by a node does not need decryption by the nodes during the routing path of data. Therefore, this scenario reduces significantly the workload of the sensors in the network. In addition, the asymmetric encryption ensures data confidentiality during transmission, as the decryption key is not available on the network. Only the sink has the decryption key that is the private key.

In the Pailleir & Benaloh cryptosystem, given a message m and the public key (n, g, h), the encryption function is $E(m) = g^m h^r (mod n)$ satisfies the following homomorphic property:

$$E(m_1) \cdot E(m_2) = g^{m_1} h^{r_1} (mod n) \cdot g^{m_2} h^{r_2} (mod n)$$

$$E(m_1) \cdot E(m_2) = g^{(m_1 + m_2)} h^{(r_1 + r_2)} (mod n) = E(m_1 + m_2)$$

Given a scalar t, we can dedicate that

$$E(t \cdot m) = E\left(\sum_{i=1}^{t} m\right) = \prod_{i=1}^{t} E(m) = E^t(m).$$

And $E\left(\sum_i t_i \cdot m_i\right) = \prod_i E(t_i \cdot m_i) = \prod_i E^{t_i}(m_i)$

5   Contribution

There are several attacks such as active and passive attacks which target the performance of compressive sensing over WSN. The active attacks disturb the compressive sensing-based routing functions by dropping or injecting the false messages into the network. Several secure data collection schemes have been developed in compressive sensing. However, data acquisition and processing method of compressive sensing theory raise challenges in the performance improvement of the intrusion detection method. Most of the conventional compressive sensing security schemes work under the assumption of one-time sensing where the random matrix is used only once. Without this assumption, the security can be broken easily and thus it is inadequate for the high-security scenarios, such as military application. In order to improve safety, the asymmetric encryption mechanisms are used, i.e., encrypting the data using the public key and decrypting the data using the private key. The sensor nodes only keep the public key, and the sink node maintains the private key of all the nodes. As a result, security is maintained with the ease of key distribution and management. However, the received data of the intermediate nodes is in the cipher; there is no way to perform the data aggregation without a private key. Moreover, the false data injected by the active malicious nodes are not determined accurately with the considerable overhead. To solve these issues, the proposed scheme develops a solution for detecting active and passive attacks in the network.

Aim and Objectives

The main aim and objectives of the proposed methodology are as follows:

- To develop a security scheme for compressive sensing over clustered WSNs using Homomorphic encryption.
- To resolve the challenges in the design of energy-efficient security scheme.
- To secure the compressive sensing measurement matrix by reducing the data dimension using compressive data sampling.

Proposed Methodology

The proposed methodology aims to protect data collection across the WSN and detect the active attacks using homomorphic encryption and machine learning algorithm. There are two typical secure data forwarding techniques in WSNs: encrypted data transmission through Cluster Head (CH) and hop-by-hop encrypted data aggregation at a secure routing path. The first technique attains security and data aggregation together in a clustered WSNs with considerable overhead. The data aggregators or CH nodes must aggregate the encrypted messages according to the corresponding aggregation function, and should append the deviation before forwarding it. To ensure the data reliability and avoid the active attacks in WSNs, the Benaloh and Paillier decryption algorithms are applied at the base station for data confidentiality. After decrypting the message, the Base station compares the standard deviation with the average of decrypted message. It facilitates the proposed methodology to ensure secure routing in clustered WSNs against the false injection attacks. Even if an inside attacker compromises an intermediate node to get the decryption key, the data confidentiality is not breached, since only the Base station knows the decryption key.

To mitigate both the active and passive attacks, homomorphic encryption is used in the provision of end-to-end data confidentiality. Currently, massive data processing degrades the performance of network protocols. In the phase of data acquisition, the reduction in the dimension of the data and directly getting characteristic information of network data from a small amount of data, the efficiency of the detection is greatly improved. Thus, the proposed methodology provides secure communication in WSNs with reduced overhead and less delay (Fig. 2).

Performance Evaluation

For the performance evaluation of the proposed methodology, we compare our scheme for Secure Homomorphic Data Collection (SDC) with a Cluster-based Secure Data Aggregation (CSDA) (Fang et al. 2017). The proposed methodology applies homomorphic encryption for preventing the confidentiality attacks in WSNs.

The proposed scheme is evaluated using the following metrics:

Overhead: Total number of control messages used for providing the security in WSNs.

The SDC has less overhead as requires less control messages and hence doesn't add high load on the network (Fig. 3).

Throughput: Total number of delivered bits to the base station.

SDC is slightly reducing the total size of data delivered to the base station (Fig. 4).

Delay: Total time taken by a packet to reach the base station in the network.

The data collection is happening more quickly within the SDC as the delay taken by a packet to reach the base station is less than CSDA (Fig. 5).

Fig. 2 Block diagram of the
proposed methodology

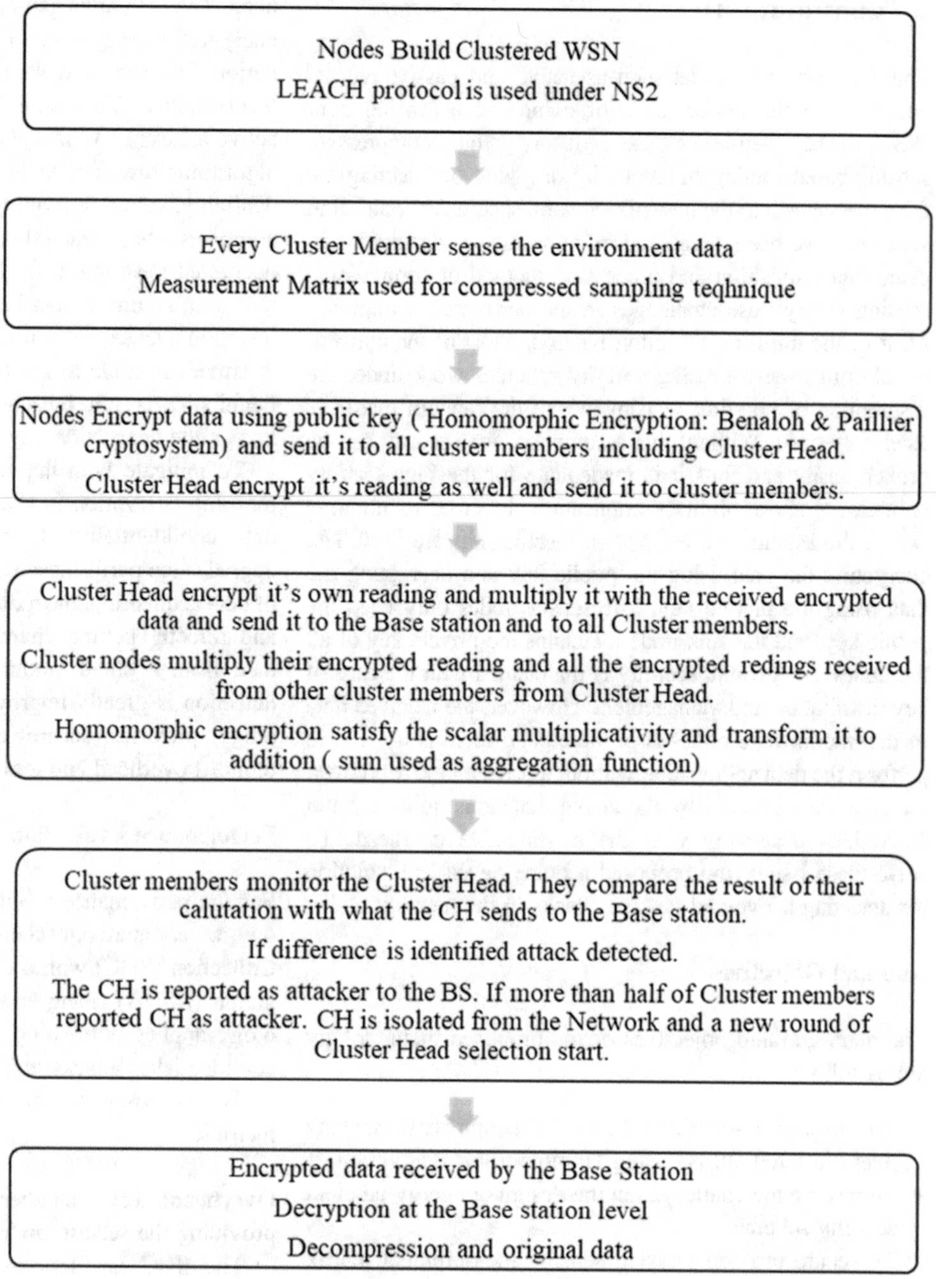

Nodes Build Clustered WSN

LEACH protocol is used under NS2

Every Cluster Member sense the environment data

Measurement Matrix used for compressed sampling teqhnique

Nodes Encrypt data using public key (Homomorphic Encryption: Benaloh & Paillier cryptosystem) and send it to all cluster members including Cluster Head.

Cluster Head encrypt it's reading as well and send it to cluster members.

Cluster Head encrypt it's own reading and multiply it with the received encrypted data and send it to the Base station and to all Cluster members.

Cluster nodes multiply their encrypted reading and all the encrypted redings received from other cluster members from Cluster Head.

Homomorphic encryption satisfy the scalar multiplicativity and transform it to addition (sum used as aggregation function)

Cluster members monitor the Cluster Head. They compare the result of their calutation with what the CH sends to the Base station.

If difference is identified attack detected.

The CH is reported as attacker to the BS. If more than half of Cluster members reported CH as attacker. CH is isolated from the Network and a new round of Cluster Head selection start.

Encrypted data received by the Base Station

Decryption at the Base station level

Decompression and original data

Fig. 3 Number of node versus overhead

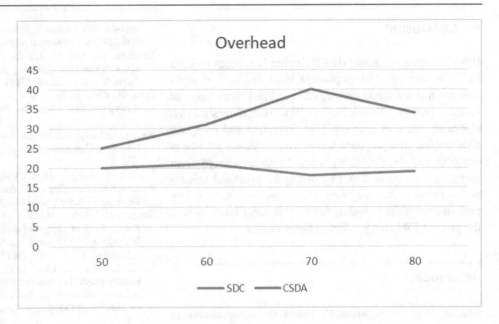

Fig. 4 Number of node versus throughput

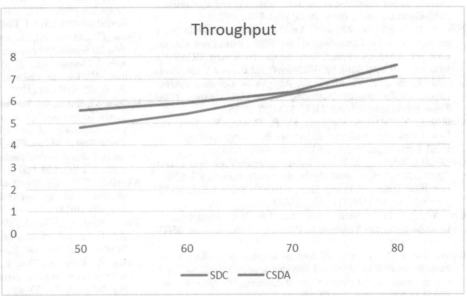

Fig. 5 Number of nodes versus delay

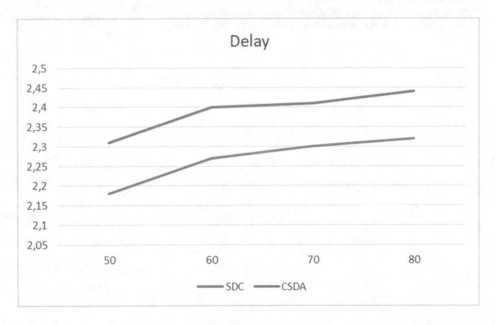

6   Conclusion

Privacy protection during data collection is a major concern for WSN security. The adoption of WSN in a critical application is a challenge as security controls bring more load and hence impact the network lifetime. This work suggests a new algorithm that uses homomorphic encryption and compressive sensing. The proposed secure data collection scheme is enabling data confidentiality while being transferred from nodes to the base station. Comparing the proposed scheme with a secure data aggregation scheme shows better network performance while providing the same security level. Hence, the proposed scheme is offering better results.

References

Alsheikh, M.A., Lin, S., Niyato, D., Tan, H.-P.: Machine learning in wireless sensor networks: algorithms, strategies, and applications. IEEE Commun. Surv. Tutor. **16**(4), 1996–2018 (2014)

Bianchi, T., Bioglio, V., Magli, E.: On the security of random linear measurements. In: Proceedings of the IEEE International Conference on Acoustics, Speech and Signal Processing (ICASSP), Florence, 4–9 May 2014, pp. 3992–3996. IEEE, New York (2014)

Bianchi, T., Bioglio, V., Magli, E.: Analysis of one-time random projections for privacy preserving compressive sensing. IEEE Trans. Inf. Forensics Secur. **11**(2), 313–327 (2016)

Cambareri, V., Mangia, M., Pareschi, F., Rovatti, R., Setti, G.: Low-complexity multiclass encryption by compressive sensing. IEEE Trans. Signal Process. **63**(9), 2183–2195 (2015)

Candès, E.J., Romberg, J., Tao, T.: Robust uncertainty principles: exact signal reconstruction from highly incomplete frequency information. IEEE Trans. Inf. Theory **52**(2), 489–509 (2006). MR 2236170. https://doi.org/10.1109/TIT.2005.862083

Cho, W., Yu, N.Y.: Secure communications with asymptotically Gaussian compressed encryption. IEEE Signal Process. Lett. **25**(1), 80–84 (2018)

Di, M., Joo, E.M.: A survey of machine learning in wireless sensor networks from networking and application perspectives. In: IEEE 6th International Conference on Information, Communications Signal Processing, pp. 1–5 (2007)

Donoho, D.L.: For most large underdetermined systems of equations, the minimal l_1-norm near-solution approximates the sparsest near-solution. Commun. Pure Appl. Math. **59**(7), 907–934 (2006). MR 2222440. https://doi.org/10.1002/cpa.20131

Ebrahimi, D., Assi, C.: On the interaction between scheduling and compressive data gathering in wireless sensor networks. IEEE Trans. Wirel. Commun. **15**(4), 2845–2858 (2016)

Fang, W., Wen, X., Xu, J., Zhu, J.: CSDA: a novel cluster-based secure data aggregation scheme for WSNs. In: Cluster Computing, pp. 1–12 (2017)

Forster, A., Murphy, A.L.: Machine learning across the WSN layers. In: Emerging Communications for Wireless Sensor Networks. InTech (2011)

Hsieh, S.-H., Hung, T.-H., Lu, C.-S., Chen, Y.-C., Pei, S.-C.: A secure compressive sensing-based data gathering system via cloud assistance. IEEE Access **6**, 31840–31853 (2018)

Hung, T.-H., Hsieh, S.-H., Lu, C.-S.: Privacy-preserving data collection and recovery of compressive sensing. In: IEEE China Summit and International Conference on Signal and Information Processing (ChinaSIP), pp. 473–477 (2015)

Kong, L., He, L., Liu, X.-Y., Gu, Y., Wu, M.-Y., Liu, X.: Privacy-preserving compressive sensing for crowdsensing based trajectory recovery. In: 2015 IEEE 35th International Conference on Distributed Computing Systems (ICDCS), pp. 31–40 (2015)

Liu, Z., Han, Y.-L., Yang, X.-Y.: A compressive sensing–based adaptable secure data collection scheme for distributed wireless sensor networks. Int. J. Distrib. Sens. Netw. (2019)

Ngabo, C.I., El Beqqali, O.: Implementation of homomorphic encryption for wireless sensor networks integrated with cloud infrastructure. J. Comput. Sci. (2019)

Nyquist, H.: Certain topics in telegraph transmission theory. Trans. AIEE. **47**, 617–644 (1928). Bibcode: 1928TAIEE..47...617N

Rachlin, Y., Baron, D.: The secrecy of compressed sensing measurements. In: Proceedings of the 46th Annual Allerton Conference on Communication, Control, and Computing, Urbana-Champaign, IL, 23–26 Sept 2008, pp. 813–817. IEEE, New York (2008)

Shannon, C.E.: A mathematical theory of communication. Bell Syst. Tech. J. **27**(4), 623–666 (1948)

Whittaker, E.T.: On the functions which are represented by the expansions of the interpolation theory. Proc. R. Soc. Edinb. **35**, 181–194 (1915)

Xie, K., Ning, X., Wang, X., He, S., Ning, Z., Liu, X., Wen, J., Qin, Z.: An efficient privacy-preserving compressive data gathering scheme in WSNs. Inf. Sci. **390**, 82–94 (2017)

Zhang, P., Wang, S., Guo, K., Wang, J.: A secure data collection scheme based on compressive sensing in wireless sensor networks. Ad Hoc Netw. **70**, 73–84 (2018a)

Zhang, P., Wang, J., Guo, K., Wu, F., Min, G.: Multi-functional secure data aggregation schemes for WSNs. Ad Hoc Netw. **69**, 86–99 (2018b)

Computing Technologies to Construct an Islamic Geometric Patterns Respecting the "Hasba" Method

Yassine Ait Lahcen, Abdelaziz Jali, Ahmed El Oirrak, and Youssef Aboufadil

Abstract

In this presentation, we put forth in the plane ornamental art, particularly in the geometric drawing. These geometric patterns are generally found in tiles that cover floors or walls of many buildings throughout the whole the Islamic world such as mosques. This form of art has evolved over centuries from simple drawing to complex geometry involving a high degree of symmetry. We describe here one method of construction of the geometric patterns encountered in the Islamic art. Artisans adopt this method called "Hasba"; it is rather suitable for sculpture and painting on wood. This method is based on geometric rules and the concept of symmetry. Here, we describe systematically the basic approach to create a new model respecting rules of the "Hasba" method by tracing grids with precise criteria of measurement.

Keywords

Islamic patterns • Symmetry • "Hasba" (measure) • Geometric art

1 Introduction

Since centuries, the Islamic world has had great decorative traditions which extends from Africa to Asia, we find artistic treasures (Aboufadil 2014).

Moroccan geometric art, in spite of the richness of its Islamic patterns and the variety of materials on which it is made (wood, Zellige (tiles), plaster, metal, etc.), remains largely unknown in the academic world. Only Thalal and Aboufadil have done interesting work on "Hasba" method (Thalal et al. 2011; Aboufadil et al. 2014). Our work aims at promoting this art in the academic world and contribute to the development of Moroccan crafts, which has experienced a long period of stagnation, using the possibilities offered today by science and new technologies and to add a fingerprint to what has been archived so far.

Several authors have published large collections of Islamic patterns, they have focused their work on classification and analysis of these patterns (Bourgoin), computer calculation (Abbas and Salman), and Castera was the only who described the craftsmen's method of construction encountered in the Moroccan "Zellij" patterns and explained how the patterns were constructed and how did the Zellij patterns evolve from the simple to the complex. In this paper, our objective is to put the light on the ornamental Arabesque specially in the method called "Hasba" as shown in Fig. 1.

The objective of this article is to solve the difficulty of creating a new Islamic geometric pattern respecting the Hasba method. In order to help the craftsmen working on wood, to realize this kind of geometric pattern in a very precise way. Moreover, the number of this geometric Islamic pattern is very limited, for this very purpose, this article is written in order to focus on this method of building these kinds of geometric patterns.

The paper is organized as follows: In Sect. 3, description of the method Hasba is given. In Sect. 4, we apply the proposed approach. Section 5 contains the conclusion and the future works.

2 Related Work

Several authors have published large collections of Islamic patterns, they have focused their work on classification and analysis of these patterns (Bourgoin 1879; El-Said and Parman 1976; Wade 1976; Paccard 1980), computer

Y. Ait Lahcen (✉) · A. Jali · Y. Aboufadil
Laboratory of Material Sciences, Faculty of Sciences of Semlalia, Marrakech, Morocco
e-mail: aitlahcen.yassine@gmail.com

A. El Oirrak
Laboratory of Engineering Information System, Faculty of Sciences of Semlalia, Marrakech, Morocco

© The Editor(s) (if applicable) and The Author(s), under exclusive license to Springer Nature Switzerland AG 2021
M. Ben Ahmed et al. (eds.), *Emerging Trends in ICT for Sustainable Development*,
Advances in Science, Technology & Innovation, https://doi.org/10.1007/978-3-030-53440-0_27

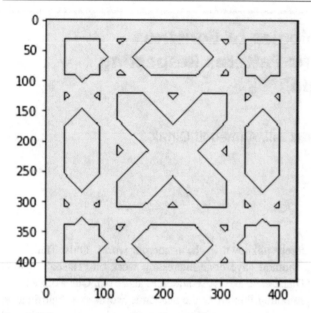

Fig. 1 Pattern constructed by method "Hasba"

calculation (Ostromouv 1998; Critchlow 1976), and group symmetry analysis (Graünbaum and Graünbaum 1986; Makovicky and Hach-Ali 1995, 1996; Makovicky et al. 1998). Moreover, we have find articles that talk about the Hasba method, which use some software to create the models respecting "Hasba" method. However, none articles talk about how to solve the difficulty of creates a new Islamic geometric pattern which takes on consideration the rules of "Hasba" method by using the computer program to generate it dynamically. Our approach is based on given the steps to follow in order to create a new pattern respecting the rules of the Hasba method. There are only some authors like Thalal (2011) who are interested in "Hasba" method in order to adopt by the Moroccan artisans ("Maâlam"), especially whose working on wood material and handed over to their disciples. In addition, there is a study of Castera (1996), which talk about another method used in the construction of Moroccan geometric patterns called "Zellij" method (fine mosaics).

3 Hasba Method

The geometric patterns (Thalal et al. 2008) are made on panels of various geometric shapes: squares, rectangles, pentagons, hexagons, octagons… These patterns can be simple or complex depending on the value of h (Hasba) which represents the multiplying coefficient below the names of the authors.

In this study, the work focused only on square patterns. Patterns with octagonal, triangular, and other shapes generally deduced in their design from the square.

The Hasba method consists of evaluating the dimensions of the frame, on which the geometric pattern will be produced, in a multiple quantity of a unit of measurement or Qasma q.

A geometric pattern from the "Hasba" method will be produced on a square frame of dimension L = h * q. h is a specific measure of the pattern called the multiplying coefficient, see Fig. 3. The type of pattern produced strongly depends on h which can be an integer or a rational greater than or equal to 8. In this study, we will adapt the value of h = 16.

4 Proposed Approach

In this section, we will present a new approach, which describes the steps to follow to create a new pattern respecting the rules of the Hasba method.

4.1 Construction of the Grid

Construction of the pattern grid respecting the "Hasba" method is done according to the following steps:

4.1.1 Grid with Concentric Circles

Draw the frame of dimension L = h * q, on which we will make a Hasba pattern. Depends on the value of coefficient h (Hasba), which can take many value for exemple 8, 16 or 24, for this article we have take like value h=16, and q (Qasma). In this case, we will take as value of (Hasba) h = 16 and

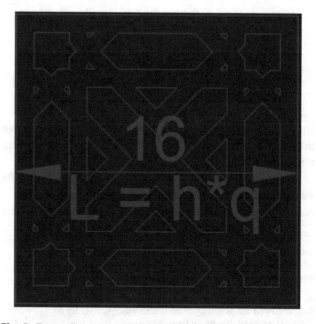

Fig. 2 Frame dimension with h = 16 and q = 1

(Qasma) q = 60 and drawing eight concentric circles of respective diameter 4 * q and 2q located at the four corners, and at the middle of each side of the graduated frame, as shown in the Fig. 3.

4.1.2 Tangent for Each Circle

In this step, we will trace the different tangents for the different circles. We need to draw four sets of parallel lines: horizontal, vertical, and diagonal lines with the directed coefficient $\alpha = 1$ and $\alpha = -1$.

Determination of Horizontal and Vertical Lancets

In order to draw the vertical tangents, we will use the equation of the line x = n * q with, $n \in [0\text{–}16]$. For the horizontal tangents, we have used the equation y = n * q with, $n \in [0\text{–}16]$ as shown in Fig. 4.

Determination of Diagonal Tangents

In this step, we will trace the different tangents for the different circles. We need to draw four sets of parallel lines: horizontal, vertical, and diagonal lines with the directed coefficient $\alpha = 1$ and $\alpha = -1$.

- **Step 1: Determination of the coordinates of the intersections points**

To get the different tangent lines for our different circles, first, we need to define points of intersection of tangents to circles and the equation of tangents. They are four points noted: M, N, P, and Q as shown in Fig. 5.

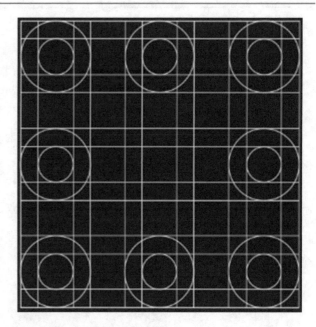

Fig. 4 Horizontal and vertical tangents

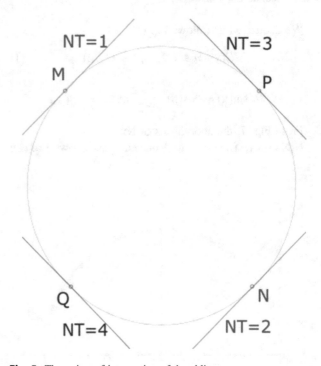

Fig. 5 The points of intersection of the oblique tangents

We must first find the coordinates of these points to find the equation of each tangent line (https://en.wikipedia.org/wiki/Tangent). From Fig. 6, we deduce the coordinates of M as follows: M $(X_0 - d_1, Y_0 + d_2)$. With $d_1 = r * \sin(\beta)$, $d_2 = r * \cos(\alpha)$ and $d_1 = d_2$ because ($\alpha = \beta = 45°$) which implies $\cos(\alpha) = \cos(\beta)$ and $\sin(\alpha) = \sin(\beta)$. So

$$M(X_0 - r * \sin(\beta), Y_0 - r * \sin(\beta))$$

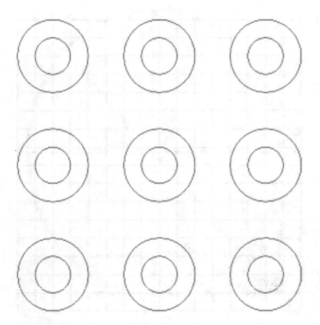

Fig. 3 Grid of a pattern of size L = h * q

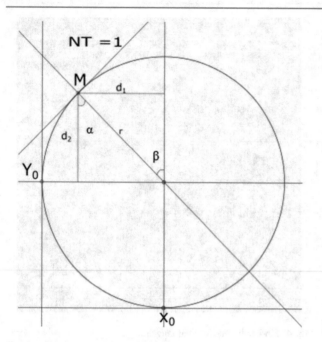

Fig. 6 Coordinates of the points M

We deduce M as follows Eq. (1):

$$M(X_0 - q * \sqrt{2},\ Y_0 + q * \sqrt{2}) \tag{1}$$

With $\sin(\alpha) = \cos(\beta) = \dfrac{\sqrt{2}}{2}$ and $r = 2 * q$.

From Fig. 7, the coordinates of N:
$N(X_0 + d_1,\ Y_0 - d_2)$. So deducted N as follows Eq. (2):

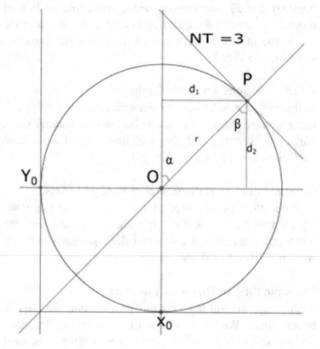

Fig. 8 Coordinates of the points P

$$N(X_0 + q * \sqrt{2},\ Y_0 - q * \sqrt{2}) \tag{2}$$

From Fig. 8, the coordinates of P:
$P(X_0 + d_1,\ Y_0 + d_2)$. So deducted P as follows Eq. (3):

$$P(X_0 + q * \sqrt{2},\ Y_0 + q * \sqrt{2}) \tag{3}$$

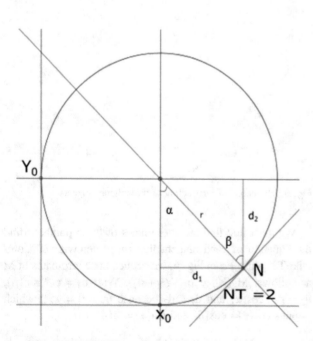

Fig. 7 Coordinates of the points N

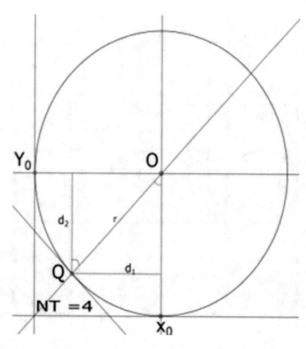

Fig. 9 Coordinates of the points Q

From Fig. 9, the coordinates of Q:
$Q(X_0 - d_1, Y_0 - d_2)$. So deducted Q as follows Eq. (4):

$$Q(X_0 - q * \sqrt{2}, Y_0 - q * \sqrt{2}) \tag{4}$$

- **Step 2: Determination of the Tangents equations**.

The equation of a line in the plane is given in Eq. (5). The coefficient of a line (Weisstein 2016) of the tangents NT = 1, NT = 2 equal $\alpha = 1$. Moreover, for the two tangents NT = 3 and NT = 4 equal $\alpha = -1$.

$$Y = ax + b \tag{5}$$

So for the two tangents NT = 1 and NT = 2: y = x + b. Then b = y − x.

– **For the tangent NT = 1**

The point M (Xm, Ym) is the point of intersection of the circle with the tangent, then its coordinates verify the equation y = x + b:

1. $b = y_m - x_m$
2. $b = y_0 + q\sqrt{2} - x_0 + q\sqrt{2}$
3. $b = y_0 - x_0 + 2q\sqrt{2}$

Then, the equation of the tangent NT = 1 Eq. (6):

$$y = x + y_0 - x_0 + 2q\sqrt{2} \tag{6}$$

– **For the tangent NT = 2**

1. $b = Y_n - X_n$
2. $b = y_0 - q\sqrt{2} - x_0 - q\sqrt{2}$ (Eq. (2))
3. $b = y_0 - x_0 - 2q\sqrt{2}$

Then, the equation of the tangent NT = 2 Eq. (7):

$$y = x + y_0 - x_0 - 2q\sqrt{2} \tag{7}$$

– **For the tangent NT = 3**

In the tangent NT = 3, y = −x + b, $\alpha = -1$ so

1. $b = y_p + x_p$
2. $b = y_0 + q\sqrt{2} + x_0 + q\sqrt{2}$ (see Eq. (3))
3. $b = y_0 + x_0 + 2q\sqrt{2}$

Then, the equation of the tangent NT = 3 Eq. (8):

$$y = -x + y_0 + x_0 + 2q\sqrt{2} \tag{8}$$

– **For the tangent NT = 4**

1. $b = y_q + x_q$
2. $b = y_0 - q\sqrt{2} + x_0 - q\sqrt{2}$
3. $b = y_0 + x_0 - 2q\sqrt{2}$

Then, the equation of the tangent NT = 3 Eq. (9):

$$y = -x + y_0 + x_0 - 2q\sqrt{2} \tag{9}$$

- **Step 3: Determination of segments Tangent to Circles**

We have determined the equations of the lines tangent to the circles. In practice for the layout, we need the segments not the lines so we must determine the ends of these segments.

The ends are calculated from the intersection between the lines and the edges of the grid. For the edges of the grid, we have the equations Eq. (10), Eq. (11), Eq. (12) ET Eq. (13).

$$y = 0 \tag{10}$$

$$x = 0 \tag{11}$$

$$y = 16 * q \tag{12}$$

$$x = 16 * q \tag{13}$$

We see that the circles C_{ij}, such that the sum of i and j is the same, have the same segment corresponding to the tangents NT = 1 and NT = 2. For example, C_{12} have the same tangent line to the circle C_{21}, because i + j = 3 in the two circles as shown in Fig. 10.

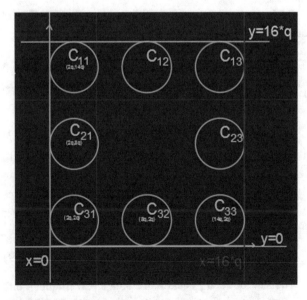

Fig. 10 Grid with the edges equations

To determine the interactions of tangents with the edges of the grid, we will treat the two cases:

– **In the case of the coefficient α = 1:**

Starting from the upper left we have

The sum of i and j equals 4 Eq. (14), so the line NT = 1 is intersected with the two tangent lines, x = 0 Eq. (11) and y = 16 * q Eq. (12).

$$\sum(i+j) = 4\,(NT = 1) \tag{14}$$

The sum of i and j less strictly than 4 Eq. (15), so the two tangents NT = 1 and NT = 2 are intersected with the two tangent lines x = 0 Eq. (11) and y = 16 * q Eq. (12).

$$\sum(i+j) < 4\,(NT = 1\,and\,NT = 2) \tag{15}$$

The sum of i and j equal 4 Eq. (16), so the line NT = 2 is intersected with the two tangent lines, x = 16 * q Eq. (13) and y = 0 Eq. (10).

$$\sum(i+j) = 4\,(NT = 2) \tag{16}$$

The sum of i and j superior strictly than 4 Eq. (17), so the two tangents NT = 1 and NT = 2 are intersected with the two tangent lines x = 16 * q Eq. (13) and y = 0 Eq. (10).

$$\sum(i+j) > 4\,(NT = 1\,and\,NT = 2) \tag{17}$$

– **For the case of the coefficient α = −1:**

Starting from the upper right we have, we will consider the circles Cij.

Such as $\sum(i+j) = n,$ $\sum(i'+j') = n'$ and $\sum(i''+j'') = n''.$

With n' = (n + 2) and n'' = (n' + 2).

The sum of i and j equal 4 Eq. (18), so the lines NT = 3 and NT = 4 are intersected with the two tangents lines, x = 16 * q Eq. (13) and y = 16 * q Eq. (12).

$$\sum(i+j) = 4\,(NT = 3\,and\,NT = 4) \tag{18}$$

The sum of i and j equal 3, the sum of i' and j' equal 5 (n ' = n + 2).

With n = $\sum(i+j)$, Eq. (19), so the two tangents lines NT = 3 and NT = 4 are intersected with the two lines tangents x = 16 * q Eq. (13) and y = 16 * q Eq. (12).

$$\sum(i+j) = 3\,and\,\sum(i'+j') = 5 \tag{19}$$

The sum of i and j equal 2, the sum i' and j' equal 4 and the sum i'' and j'' equal 6 Eq. (20), so the tangent line NT = 3 is intersected with the two tangents lines x = 16 * q Eq. (13) and y = 16 * q Eq. (12).

$$\sum(i+j) = 2,\ \sum(i'+j') = 4\,and\,\sum(i''+j'') = 6 \tag{20}$$

The sum of i and j equal 2, the sum i' and j' equal 4 and the sum i'' and j'' equal 6 with (n'' = n' + 2) Eq. (21), so the tangent line NT = 4 is intersected with the two tangents lines x = 0 and y = 0.

$$\sum(i+j) = 2,\ \sum(i'+j') = 4\,and\,\sum(i''+j'') = 6 \tag{21}$$

The sum i and j equal 3, the sum i' and j' equal 5(n' = n + 2).

With n = $\sum(i+j)$ Eq. (22), so the tangents lines NT = 3 et NT = 4 are intersected with the two tangents lines x = 0 et y = 0.

$$\sum(i+j) = 3,\ \sum(i'+j') = 5\,(NT = 4\,and\,NT = 3) \tag{22}$$

The sum i and j equal 4 Eq. (23), so the two lines NT = 3 and NT = 4 are intersected with the two tangents lines, x = 0 and y = 0.

$$\sum(i+j) = 4\,(NT = 4\,and\,NT = 3) \tag{23}$$

- **Step 4: Optimization of the parameters of the segments Tangents**

To extract the equations of the lines, we deduced that the best solution is to choose an optimal number of circles to draw the tangents by treating only a family of circles L (Figs. 11 and 12).

When we are going through the 2L instead of treating 5 cases our objective is to reduce the number of cases from 5 to 3 cases only.

For the coefficient α = 1, we have the circle C_{11} represent the 1st case $(\sum(i+j) < 4)$, C_{12}, C_{21} the 2nd case $(\sum(i+j) < 4)$, C_{13}, C_{22} and C_{31} the 3rd case $(\sum(i+j) = 4)$, for the circles C_{23}, C_{32} represent the 4th case $(\sum(i+j) > 4)$ and C_{33} the 5th case $(\sum(i+j) > 4)$. For the (L) right the 3 cases are highlighted via the line which has the equation: y = x. We will deal with the cases of circles whose centers are above and below or on the line y = x, see Fig. 13.

So for the case of "L" right, we have the same treatment for circles C_{11} and C_{21}, because they have the intersection with the same edges equations y = 16 * q and x = 0.

For the circles C_{32} and C_{33} are also the same treatment because they have intersection with the same edges with the equations y = 0 and x = 16q. And for the circle C_{31} has special treatment because it intersects with the 4 edges for NT = 1, NT = 2, NT = 3, and NT = 4.

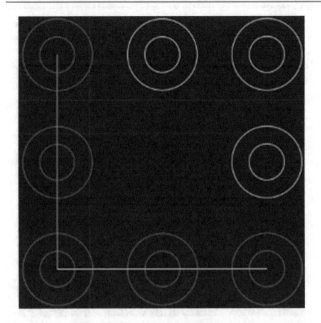

Fig. 11 Circles forming "L" chosen to plot NT = 1 and NT = 2

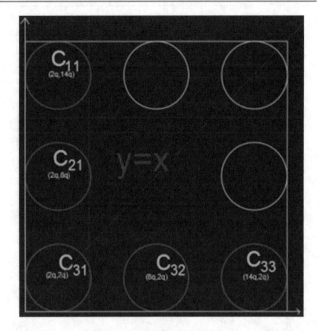

Fig. 13 Circles treated with the line y = x

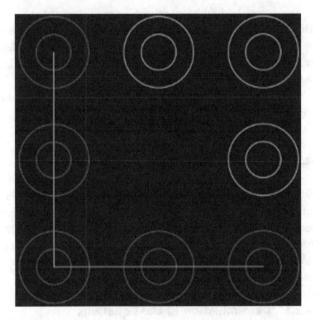

Fig. 12 Circles forming "L" chosen to plot NT = 3 and NT = 4

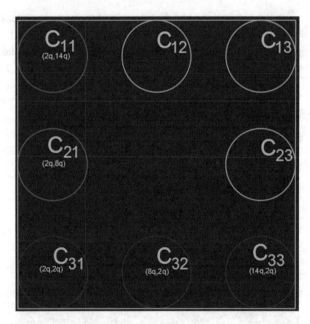

Fig. 14 Family of circles with "L" right

The circles belong to right "L" which have the same colors have the same treatment Fig. 14.

For the circles C21, C32 represent the 4th case $(\sum(i+j) = 3, \sum(i'+j') = 5)$ and C31 the 5th case $(\sum(i+j) = 4)$. For the coefficient $\alpha = 1$, we have the circle C_{13} represent the 1st case $(\sum(i+j) = 4)$, C_{12}, C_{23} the 2nd case $(\sum(i+j) = 3$ and $\sum(i'+j') = 5)$, C11, C22 and C33 the 3rd case $((\sum(i+j)=2, \sum(i'+j') = 4, \sum(i''+j'') = 6)$, for the circles C_{21}, C_{32} represent the 4th case

$(\sum(i+j) = 3, \sum(i'+j') = 5)$ and C_{31} the 5th case $(\sum(i+j) = 4)$.

For the (L) left the 3 cases are highlighted via the line which has the equation y = −x + 16q. We will deal with the cases of circles whose centers are above and below or on the line y = −x + 16q, see Fig. 15.

So for the case of "L" left, we have the same treatment for circles C_{13} and C_{23} because they have the intersection with the same edges equations y = 16 * q and x = 16 * q.

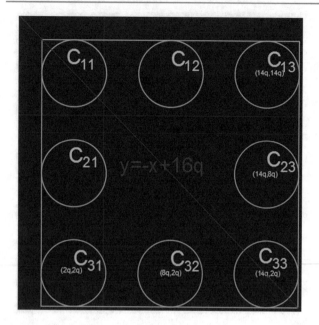

Fig. 15 Circles treated with the line y = −x + 16q

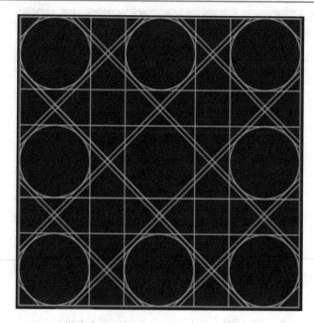

Fig. 17 Grid with all tangents

For the circles C_{32} and C_{31} are also the same treatment because they have intersection with the same edges with the equations y = 0 and x = 0. And for the circle C_{33} has special treatment because it intersects with the 4 edges for NT = 1, NT = 2, NT = 3 and NT = 4.

The circles belong to left "L" which have the same colors have the same treatment Fig. 16.

So, we deduced that instead of working for each L with the 5 cases, we will work with 3 cases by considering the

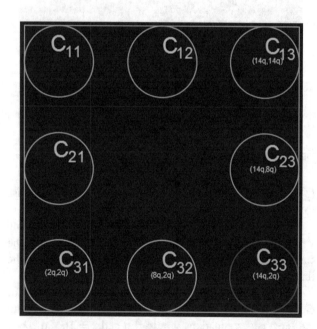

Fig. 16 Family of circles with "L" left

right y = x for "L" right and y = −x + 16q for "L" left. We will treat the 3 cases according to the position of the circles relative to one of the two lines either below or above. The result of this study is displayed in Fig. 17.

We applied the same formalism and the same steps to build the small circles and its tangents Fig. 18.

4.2 Determination of the Symmetric Region

We are going to determinate a zone, which we call fundamental zone see Fig. 19, and we will find inside all the tangents. Thereafter, we will extract all the points of intersection between the different segments in our fundamental zone (https://en.wikipedia.org/wiki/Line%E2%80%93line_intersection).

4.2.1 Determination of the Equations of the Fundamental Zone

In this step, we will determine the equations of the segments delimiting our fundamental zone as shown in Fig. 20.

The first segment is delimited by the two points: a1 (0, 16q) and b1 $\left(\frac{16q}{2}, \frac{16q}{2}\right)$. So the coefficient α is equal: $\left((Yb_1 - Ya_1)\big/(Xb_1 - Xa_1)\right) = -1$. Therefore,

$$y = -x + 16 * q \qquad (24)$$

For the 2nd segment, we have the coordinates of the two points: a_2 (0, 16q) and b_2 $\left(\frac{16q}{2}, 16q\right)$, so the coefficient α is equal: $\left((Yb_2 - Ya_2)\big/(Xb_2 - Xa_2)\right) = 0$.

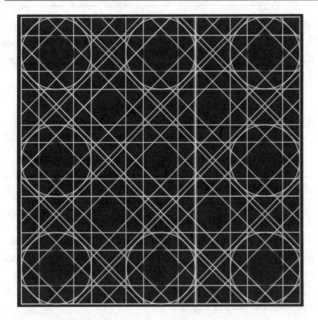

Fig. 18 Grid with small circles and its tangents

Fig. 20 Equations of the fundamental zone

4.2.2 Determination of Intersection Points

To find the points of intersection between the different tangents in our fundamental zone, we opt to solve the system of each equation of the tangent line of our grid, with the equations of the segments of the edge of the fundamental zone Fig. 21.

For example to find the point of intersection between the tangent of equation: $y = -x + 16q$ and the segment equation $x = \frac{16q}{2}$ Eq. (26), we must solve the following system:

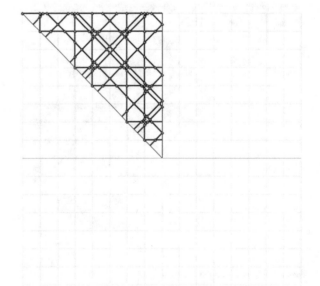

Fig. 19 Symmetric region (fundamental zone)

Therefore,

$$y = 16 * q \tag{25}$$

For the 3rd segment, the coordinates of the two points are $a_3\left(\frac{16q}{2}, \frac{16q}{2}\right)$ and $b_3\left(\frac{16q}{2}, 16q\right)$. Therefore,

$$x = \frac{16q}{2} \tag{26}$$

Fig. 21 Intersection points

$$\begin{cases} y + x = 16q \\ x = \frac{16q}{2} \end{cases}$$

Substitution method:

1. We express an unknown as a function of the other from one of the two equations, here the first equation gives us: $x = \frac{16q}{2}$.
2. We replace this unknown by its expression in the other equation. We get with the second equation: y + 16q/2 = 16q.
3. We solve the equation with an unknown factor obtained

$$y + 16q/2 = 16q \text{ Is } y = 16q/2.$$

4. The system, therefore, has a single solution: $x = \frac{16q}{2}$ and y $= \frac{16q}{2}$

4.2.3 Determination of the Segments of the Fundamental Zone

In this step, we will extract a few segments from our fundamental zone. We started by giving numbers to our intersection points already extracted as shown in Fig. 22.

Then, we build an array of a few segments since the number of segments is very large. With for each segment,

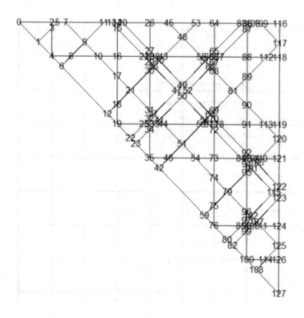

Fig. 22 Number of intersection points

we have indicated the type of segment (vertical: 0 Horizontal: 1 oblique: 2), the numbers of the two vertices delimiting each segment, and the coordinates of each vertex Table 1.

4.3 Draw Geometric Shapes in the Symmetric Region

After determining all the segments in the fundamental zone, we try to draw some forms of the "Hasba" method.

4.3.1 Draw Part of the Geometric Shapes
According to the segments found in the fundamental zone, we managed to draw a part of the geometric shape, respecting the distance, which must be fixed and equal to q (Qasma), between the already drawn shape and the other neighboring shapes Fig. 23.

4.3.2 Symmetry with Respect to a Straight Line
In our proposed approach, the symmetry (Thalal et al. 2010) is relative to the axis $x = \frac{16q}{2}$. Then, the 1st part of the geometric shape is the image of the 2nd part by symmetry with respect to the equation of the line Fig. 24.

4.3.3 Rotation Around a Point
In our case, we will make successive rotations of 90°, 180°, 270° around a Point C (8q, 8q). For each point of our segment plot in the asymmetric zone, we will apply one of the equations Eqs. 27, 28, and 29 depending on the angle of rotation.

For rotation, $\theta = 90°$ the general formula is as follows:

$$\begin{aligned} x_r &= \left(-y - c_y\right) + c_x \\ y_r &= x - c_x + c_y \end{aligned} \tag{27}$$

With

p(x, y) indicates the starting point.
$p_r\ (x_r, y_r)$ image of P by rotation of 90°
c (c_x, c_y) Center of rotation.

For rotation, $\theta = 180°$ the general formula is as follows:

$$\begin{aligned} x_{r1} &= 2 * c_x - x \\ y r_1 &= 2 * c_x - y \end{aligned} \tag{28}$$

For rotation, $\theta = 270°$ the general formula is as follows:

$$\begin{aligned} x_{r2} &= y - c_y + c_x \\ y_{r2} &= c_x - x + c_y \end{aligned} \tag{29}$$

Table 1 Segments with coordinates of the vertices limiting

N° S_1	(x_1, y_1)	N° S_2	(x_2, y_2)	Type
4	(60, 900)	8	(95.14, 900)	1
8	(95.14, 900)	9	(120, 924.86)	2
9	(120, 924.86)	10	(144.85, 900)	2
10	(144.85, 900)	16	(180, 900)	1
16	(180, 900)	17	(180, 864.86)	0
17	(180, 864.86)	21	(204.85, 120)	2
21	(204.85, 120)	18	(180, 815.15)	2
18	(180, 815.15)	19	(180, 180)	0

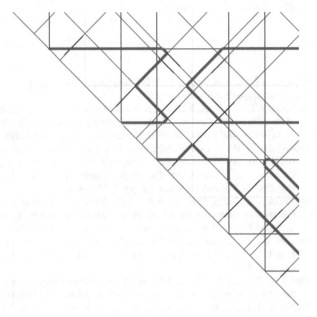

Fig. 23 Part of the geometric shapes

Fig. 24 Symmetry about the axis x = 16q/2

4.4 Display of the Different Pattern Hasba

After applying these formulas, we obtained the results displayed in Fig. 25 which displays a pattern respecting the rules of the "Hasba" method (Aboufadil et al. 2010; Thalal et al. 2017).

5 Conclusion

Many works have used this method to show us the different patterns respecting the Hasba method only in some paper or by using some drawing software like Inkscape, for example. Our proposed approach gives us the details gradually on how to create this sort of pattern respecting the Hasba method.

This paper presented an approach to solve the problem of how to create a valid pattern respecting the Hasba method, in order to help the artisans of the ministry of tourism, crafts, air transport, and social economy to create a valid pattern.

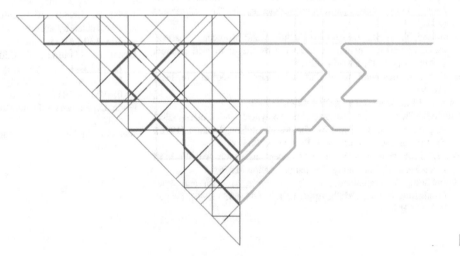

Fig. 25 Patterns made with the Hasba method

The modern dialog between artisans and researchers, in combination with modern science and technology, has allowed the formalization of the method of constructing Moroccan patterns, found new patterns. We are going to add new moments and apply our approach to create an application to automatically generate this type of Islamic pattern, because the number of patterns generated so far is very limited and realized statically, no computer program to generate it dynamically, we need to generate other patterns which must be different from what exists today.

References

Aboufadil, Y.: Crystallography in Moroccan geometric ornamental art. Ph.D. thesis, Cadi Ayyad University, Marrakesh. Morocco (2014)

Aboufadil, Y., Thalal, A., Benatia, J., Jali, A.: Classification of the Moroccan ornamental patterns constructed by the "Hasba" method (2010)

Aboufadil, Y., Thalal, A., Elidrissi Raghni, M.A.: Moroccan ornamental quasi periodic patterns constructed by the multi grid method. J. Appl. Cryst. **47**, 630–641 (2014)

Bourgoin, J.: Les élements de l'art arabe. Firmin-Didot et Cie, Paris (1879)

Castera, J.M.: Arabesques. ACR Edition (1996)

Critchlow, K.: Islamic Patterns: An Analytical and Cosmological Approach. First published by Thames & Hudson, London (1976). First paper back edition 1983, Reprinted 2004

El-Said, I., Parman, A.: Geometrical Concepts in Islamic Art: World of Islam Festival. Publishing Company, London (1976)

Graünbaum, B., Graünbaum, Z.: Symmetry in Moorish and other ornaments. Comput. Math. Appl. **12B**(¾),641–653 (1986). Printed in Great Britain

https://en.wikipedia.org/wiki/Line%E2%80%93line_intersection
https://en.wikipedia.org/wiki/Tangent

Makovicky, E., Hach-Ali, P.F.: The twin star mosaic from niches of the "Sala de la Barca", "Salon de Comares", Alhambra, Granada, Boltin de la Sociedad Española de Mineralogica, vol. 18, pp. 1–6 (1995)

Makovicky, E., Hach-Ali, P.F.: Mirador de Lindaraja: Islamic ornament patterns based on quasiperiodic octagonal lattices in Alhambra, Granada, and Alcazar, Sevilla, Spain, Boltin de la Sociedad Española de Mineralogica, vol. 19, pp. 1–26 (1996)

Makovicky, E., Perez, R.F., Hach-Ali, P.F.: Dodecagonal patterns in the Islamic ornament art of Spain and Morocco, Boltin de la Sociedad Española de Mineralogica, vol. 21, pp. 107–127 (1998)

Paccard, A.: Traditional Islamic Craft in Moroccan Architecture, vols. 1 and 2. Published by Editions ateliers 74, 74410 Saint-Jorioz, France (1980)

Thalal, A., Benatia, J., Jali, A.: Analysis of the craftsman's approach to the Moroccan geometric patterns. Acta Cryst. **A64**(C635) (2008)

Thalal, A., Benatia, M.J., Jali, A., Aboufadil, Y., Elidrissi Raghni, M.A.: Islamic geometric patterns constructed by craftsmen working on wood. Accepted for publication in the Special Issue on Tessellations Symmetry J. Hung. (2010)

Thalal, A., Benatia, M.J., Jali, A., et al.: Islamic geometric patterns constructed by artisans working on wood. Symmetry Cult. Sci. **22**, 103–130 (2011)

Thalal, A., Aboufadil, Y., Elidrissi Raghni, M.A., Jali, A., Oueriagli, A., Ait Rai, K.: Symmetry in art and architecture of the Western Islamic world. Crystallogr. Rev. (2017). https://doi.org/10.1080/0889311X.2017.1343306

Wade, D.: Patterns in Islamic art. Cassell & Collier Macmillan, London (1976)

Weisstein, E.W.: Slope. MathWorld—A Wolfram Web Resource (2016)

Technology for Sustainable Development: Solar Adsorption Cooling System Cold Room Modelization Using Python

Hanane Abakouy, Hanae El Kalkha, and Adel Bouajaj

Abstract

Solar adsorption cooling system (SACS) is an innovative process that produces cold using solar thermal heat energy, and it is considered a technology for sustainable development as it uses environmentally friendly materials. The target of this study is to maximize the performance of SACS by defining the optimal insulation thickness for the cold room, which is the element where the products are stored to be cooled down. A mathematical model has been developed and tested with different kinds of materials. A 29% decrease in heat losses through walls is achieved by using 5 cm of polyurethane and 21% decrease is achieved by using 6 cm of polystyrene. The simulation was carried out using python. This research shows the validity of this model to reduce heat losses and the possibility to explore different materials in order to choose the best insulation thickness that helps to reduce heat transferred to the cold room.

Keywords

Adsorption • Modelization • Cold room • Insulation thickness • Cooling • Solar energy

1 Introduction

The solar adsorption cooling system (SACS) is the process that produces cold using solar thermal heat energy, and this process is based on the phenomenon of adsorption that occurs when a balance is established between a working pair called adsorbate/adsorbent. The advantages of this system are that it works with a low-temperature heat source, and it is considered as zero ozone depletion potential, and zero warming global potential because it uses environmentally friendly materials.

Researches in solar adsorption cooling system were developed and carried out since 1848, before the conventional refrigeration systems; it was first demonstrated by Michael Faraday who developed a system using ammonia and silver chloride as the working principle (Meunier and Douss 1990). Afterward, Plank and Kuprianoff described a practical adsorption system that used the working pair methanol and activated carbon (Critoph 1994). But researches in this field decreased since the introduction of chlorofluorocarbons and the development of the conventional electrical compressors (Hassan and Mohamad 2012a, b). The traditional vapor compression machines are dominating electricity consumers, and their operation causes high electricity peak loads (Hassan and Mohamad 2013). Besides, there are problems related to environmental pollution.

Several studies have been made in this field of solar adsorption cooling, and some researchers have paid attention to the improvement of the COP values while others are focusing on the system cooling load and the improvement of mass and heat transfer. The evaporative cooling system technique is proposed in developing countries, mostly as a replacement for conventional refrigerators based on vapor compression (Wang et al. 2014). A lot of researchers have published their work: Sumanthy et al., Clauss et al., and Alam et al. studied air-conditioning system (Sumathy et al. 2009; Clausse et al. 2008; Alam et al. 2013). Others examined the desalination system (Mitra et al. 2014; Wakil

H. Abakouy (✉)
Laboratory of Innovative Technologies, National School of Applied Sciences, Abdelmalek Essaâdi University, Tangier, Morocco
e-mail: hananeabakouy@gmail.com

H. El Kalkha · A. Bouajaj
Department of Industrial and Electrical Engineering, National School of Applied Sciences, Abdelmalek Essaâdi University, Tangier, Morocco
e-mail: elkalkha_hanae@yahoo.fr

A. Bouajaj
e-mail: dbouajaj@yahoo.fr

© The Editor(s) (if applicable) and The Author(s), under exclusive license to Springer Nature Switzerland AG 2021
M. Ben Ahmed et al. (eds.), *Emerging Trends in ICT for Sustainable Development*,
Advances in Science, Technology & Innovation, https://doi.org/10.1007/978-3-030-53440-0_28

et al. 2014) and hot water plants (Young-Deuk et al. 2015); Meunier and Mugnier 2013). Wu and Uyun studied heat recovery of solar-driven adsorption chiller with cascading adsorption refrigeration cycle (Wu et al. 2002) and a continuous adsorption heat pump (Uyun et al. 2009); a recent study was made for the improvement of COP with the heat recovery scheme for solar adsorption cooling system by Ariful Kabir et al. (2018). And for desalination purposes, some studies were carried out recently by Rezk et al. (2019), Du et al. (2017), and Elsayed et al. (2017).

The literature is abundant with experimental work as well as theoretical investigation, although the solar adsorption cooling system is not yet competitive enough to replace electricity-driven refrigerators. Nonetheless, researches in this field are still on-going to increase the overall performance of the system, by the development of new adsorbents, advanced components, and optimized cycles.

This work is a contribution to the optimization of the solar adsorption cooling system. In this paper, the authors paid attention to the cold room as it is a part of the system, and the cold room is the element where the cold is produced. The main objective of this work is the research of the optimal insulation thickness to reduce heat transfer and improve system efficiency.

In the second section, the description of the whole system is introduced. A short discussion about the modelization process is provided in Sect. 3. Section 4 is devoted to the simulation results followed by a conclusion in the last section.

2 System Description

The solar adsorption refrigerator is built by four main elements: the collector-adsorber, the condenser, and the evaporator located inside the cold room:

- The collector-adsorber is a reactor (adsorber) enclosed inside the solar collector, and the collector captures the solar rays; converts it to thermal energy, which is transferred into the reactor to heat the adsorbate, this element is where the phenomenon of adsorption and desorption occurs.
- The condenser: where the refrigerant is liquefied when the maximum temperature is reached, in desorption process, the heated adsorber reactor releases the refrigerant from its adsorbent and pushes it into the condenser, and when the hot gas meets the cooler air of the outside, it becomes liquid, which is at its high pressure.
- The evaporator: this is the element where the refrigerant evaporates extracting heat from its surrounding, which is the process that produces cold.

- The cold room: this part contains the evaporator which allows its cooling, when the adsorbent cools down as it flows into the evaporator, then it absorbs the heat inside the cold room, cooling down the air.

3 Modelization

3.1 Cold Room Heat Sources

In this study, three mean heat sources were taken into account, the transmission loads which occurs due to the difference of the temperature between the cold room and its surrounding as the heat always passes from the hot side to the cold side, and its importance varies with the type of material used; the product loads account when products are introduced in the cold room, and the infiltration loads which occurs when the door opens and the heat is transferred inside through the air.

3.2 Thermodynamic Modelization

The cooling load may be defined as the maximum hourly cooling output required from the refrigeration to establish or maintain the design's internal conditions when the outside environment is at the design outside condition.

The accurate determination of each of the components of the heat load would be a most tedious calculation and the number of simple correction factors that are used in totaling the cooling load will be described later in this study.

The design outside condition for cooling is, therefore, the outside condition at the time of maximum ambient temperature, so the system can work in extreme conditions.

The assumptions that have been taken into account to model the refrigerator are

- All the walls are at the same temperature;
- The ground temperature is 10 °C;
- The temperature between the wall and the atmosphere is 5 °C.

In the following, the equations of heat losses (Sakadura 2000) for each heat source:

- Model of heat loss by air

$$Q_{air} = V \times \Delta h \times \varphi \times n \tag{1}$$

With

V   cold room volume (m^3).

n   cold room opening number.

Δh the enthalpy difference between the ambience in the cold room and the atmosphere (Wh/kg).

φ air density (kg/m^3).

- Model of the quantity of heat to be extracted from products

$$Q_{product} = P \times C_s \times \Delta T + P \times 1.4 \qquad (2)$$

With

P   weight of the products (kg).

Cs   specific heat of the products (Wh/kgK).

ΔT Difference between the temperature at the arrival of the products and the temperature of storage (K).

- Model of heat loss through the walls

$$Q_w = kS\Delta T \qquad (3)$$

With

S   cold room area (m^2).

ΔT Difference between the air temperature inside the room and the ambient external air temperature (K).

k   Overall heat transfer coefficient (W/m^2 K),

For the greatest effectiveness, the thermal resistance of the structure must be high, and this may be achieved by the enhancement of insulation.

Of the terms of Eq. (3), the only item which may be varied is the thermal transmittance (k). This is defined by

$$k = \frac{1}{\frac{1}{h_0} + \frac{1}{h_i} + \frac{E_p}{\lambda}} \qquad (4)$$

where

h_0 Heat transfer coefficient on the outside surface (W/m^2 K).

h_i Heat transfer coefficient on the inner surface (W/m^2 K).

I_{th} Insulation thickness (m).

λ Thermal conductivity of the insulator (W/mK)

$$Q_w = \frac{24 \times ((L + 2I_{th})(l + 2I_{th})(T_{amb} + 10) + 2 \times T_{amb}(h + 2I_{th})(l + L + 4 \times I_{th}))}{\frac{1}{h_0} + \frac{1}{h_i} + \frac{I_{th}}{\lambda}} \qquad (5)$$

The main parameters that were highlighted in the general model are

- The ambient temperature
- The insulation thickness
- The thermal conductivity
- The cold room dimensions

It is recommended that the cold room must be placed away from the direct influence of sunshine; however, if it is exposed to direct solar radiation, then the radiation loads must be taken into account. So, an allowance of 5 K above the outside temperature is assumed in calculating the heat gain through the exposed walls and roof (Ibe and Anyanwu 2013).

3.3 Coefficient of Performance

The performance of the solar adsorption cooling system can be evaluated concerning the thermal coefficient of performance *TCOP* and the solar coefficient of performance *SCOP*.

The thermal coefficient of performance is given by the following equation:

$$TCOP = \frac{Q_e}{E_u}$$

where

Q_e The cold produced at the evaporator (kJ).

E_u The necessary irradiation to perform the desorption process (kJ/m^2).

The solar coefficient of performance is given by the following equation:

$$SCOP = \frac{Q_e}{E}$$

where

E   The total irradiation received by the front of the collector (kJ/m^2).

4 Results and Discussion

The prescribed mathematical model was carried out using Python 3.7. And the parameters used in this study are summarized in Table 1.

Table 1 Design parameter used in the simulation

Parameter	Value	Unit
Glass wool conductivity	0.036	W/m K
Polystyrene conductivity	0.028	W/m K
Polyurethane conductivity	0.025	W/m K
Convection coefficient	8	W/m^2
Convection coefficient in touch with the air	30	W/m^2
Cold room volume	56.35	l
Cold room total area	0.889	m^2

4.1 Climate Data

The results of numerical simulation are obtained using the hourly climate data (ambient temperature) corresponding to different days from March 13 to April 15 in Rabat (Morocco, 34°00′47 N, 6°49′57 W) (Lemmini and Errougani 2005). From the monthly metrological data, the design months are selected to test the experiment with different sky state. Figure 1 shows the daily average ambient temperature for each day.

4.2 Model

To test the model, we have used a real data obtained from the solar adsorption cooling system built in Rabat Morocco and this system is constituted by the collector, condenser, and the evaporator inside the cold chamber; and the cold room was insulated with glass wool. This experiment has been already described in previous work (Lemmini and Errougani 2005). The following model describes the behavior of the system:

$$Q_t = \frac{\left((0.76 + 7.88 \times I_{th} + 20 \times I_{th}^2) \times T_{amb} + 12 + 14 \times I_{th} + 40 \times I_{th}^2\right) \times 24 + 0.186 \times T_{amb}}{0.158 + \frac{l_a}{\lambda}}$$

(6)

4.3 Results

Figure 2 shows the variation of numerical results of the heat losses with the insulation thickness:

The solar radiation intensity affects directly the heat lost by the cold room. Indeed, for a well sunny day, the quantity of cold produced should increase to cool down the air inside the cold room, and this creates a big temperature difference between the atmosphere and the cold room, and so much more heat losses.

For clear days we added 5 K to the temperature used in the simulation. As it can be seen in Fig. 3, the heat loss increases when the sky is clear and the solar radiation is intense.

The heat transferred to the cold room from solar heat occurs throughout the year even on totally overcast days, so

Fig. 1 Daily average ambient temperature used in the simulation (Lemmini and Errougani 2005)

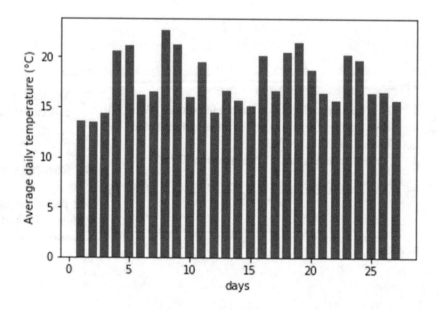

Fig. 2 Variation of heat losses with the insulation thickness using glass wool

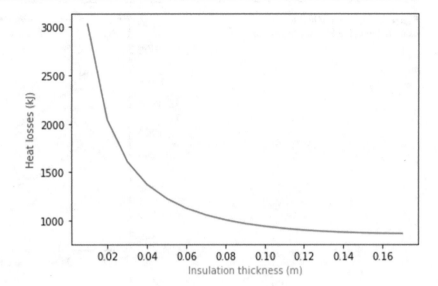

Fig. 3 The heat losses taking into account direct sunlight

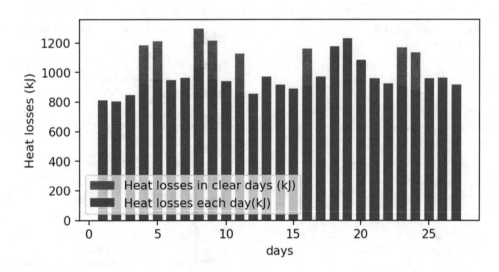

to minimize the heat loss, it is recommended to place the cold room away from direct sunlight and the collector on the roof; or as a suggestion, the direct solar radiation may be excluded by devices such as canopy or roller blinds, which can be positioned to cut off the direct radiation.

4.4 Test of Different Materials

In the experiment, the glass wool was used to minimize the heat transfer, in this section and with the same dimensions of the cold room used in the experiment we tested different kinds of insulators: the wood wool, the polyurethane, and polystyrene, and Fig. 6 shows the results obtained with the polyurethane and polystyrene; the heat losses decrease with the enhancement of the insulation thickness.

It can be seen from Fig. 4 that the heat losses transferred to the cold room can be decreased with the polyurethane and polystyrene with less than 10 cm of insulation thickness.

With 10 cm of the glass wool, we have 938.33 kJ of heat losses. Indeed with 10 cm of polyurethane, heat losses decrease by 29%, from 939.33 to 665.66 kJ. And we can use just 5 cm, with this insulation the heat losses are 881.57 kJ, so we can decrease heat losses by 6%.

With 10 cm of polystyrene, heat losses decrease by 21%, from 938.33 to 740.92 kJ. And with 6 cm, we obtain 895.46 kJ, which means that heat losses will decrease by 4.5%.

4.5 Coefficient of Performance

The performance of the SADC system is evaluated concerning the solar coefficient of performance (COP). Figure 5

Fig. 4 Comparison of heat losses
using different kinds of insulators

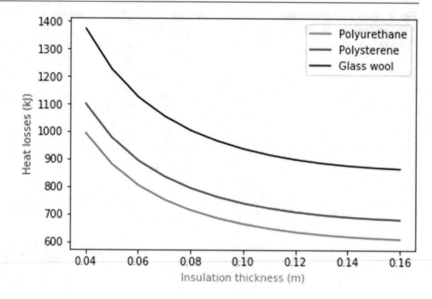

Fig. 5 Solar coefficient of
performance

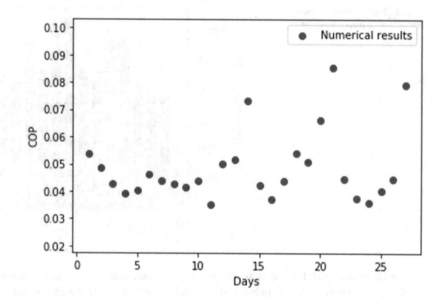

Fig. 6 Thermal coefficient of
performance

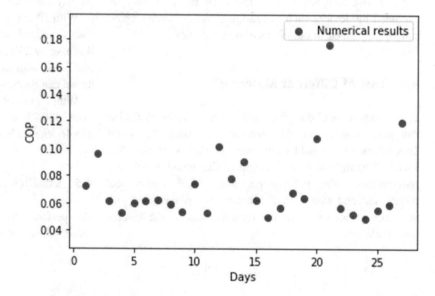

shows that with polyurethane (polystyrene), the COP vary from 3 to 9% and increase by 11% compared to the results obtained by the glass wool.

From Fig. 6, we can see that with the polyurethane, the COP vary from 5 to 17% and increase by 29% compared to the results obtained using the glass wool.

5 Conclusion

Solar adsorption cooling system is an innovative process that produces cold using low solar thermal heat energy, and this system can be built by four main elements, the collector-adsorber, the condenser, the evaporator, and the cold room—which is the main element where the cold is produced.

Improving the performance of the solar adsorption cooling system is the main goal of this research; by the enhancement of the cold room insulation thickness, a mathematical model has been developed, and the authors have compared different kinds of insulator with different conductivity values. A 29% decrease in heat losses through walls is achieved by using 5 cm of polyurethane and 21% decrease is achieved by using 6 cm of polystyrene.

This research shows the validity of this model to reduce heat losses and the possibility to explore different materials in order to choose the best insulator that helps to reduce heat transferred to the cold room. And this study shows also the influence of direct sunlight on the heat transferred to the cold room, so we recommend to place the cold room sheltered from the direct sunlight, in order to minimize the temperature difference between the inside of the cold room and its surrounding. The goal is to improve system efficiency.

As a future work, we can enhance the simulation by introducing more climate data with a time step of one hour at least that takes into account the initial temperature of the morning, and this temperature is an important factor that influences the system performance.

References

Alam, K.C.A., Saha, B.B., Akisawa, A.: Adsorption cooling driven by solar collector: a case study for Tokyo solar data. Appl. Therm. Eng. **50**, 1603–1609 (2013). https://doi.org/10.1016/j.applthermaleng.2011.09.028

Ariful Kabir, K.M., Rouf, R.A., Sarker, M.M.A., Amanul Alam, K.C., Saha, B.B.: Improvement of COP with heat recovery scheme for solar adsorption cooling system. Int. J. Air Cond. Refrig. **26**(02), 1850016 (2018). https://doi.org/10.1142/S2010132518500165

Clausse, M., Alam, K.C.A., Meunier, F.: Residential air conditioning and heating by means of enhanced solar collectors coupled to an adsorption system. Sol. Energy **82**, 885–892 (2008). https://doi.org/10.1016/j.solener.2008.04.001

Critoph, R.E.: An ammonia carbon solar refrigerator for vaccine cooling. Renew. Energy **5**, 502–508 (1994). https://doi.org/10.1016/0960-1481(94)90424-3

Du, B., Gao, J., Zeng, L., Su, X., Zhang, X., Yu, S., Ma, H.: Area optimization of solar collectors for adsorption desalination. Sol. Energy **157**, 298–308 (2017). https://doi.org/10.1016/j.solener.2017.08.032

Elsayed, E., Raya, A.D., Mahmoud, S., Anderson, P.A., Elsayed, A., Youssef, P.G.: CPO-27 (Ni), aluminium fumarate and MIL-101 (Cr) MOF materials for adsorption water desalination. Desalination **406**, 25–36 (2017). https://doi.org/10.1016/j.desal.2016.07.030

Hassan, H.Z., Mohamad, A.A.: A review on solar cold production through absorption technology. Renew. Sustain. Energy Rev. **16**, 5331–5348 (2012a). https://doi.org/10.1016/j.rser.2012.04.049

Hassan, H.Z., Mohamad, A.A.: A review on solar-powered closed physisorption cooling systems. Renew. Sustain. Energy Rev. **16**, 2516–2538 (2012b). https://doi.org/10.1016/j.rser.2012.02.068

Hassan, H.Z., Mohamad, A.A.: Thermodynamic analysis and theoretical study of a continuous operation solar-powered adsorption refrigeration system. Energy **61**, 167–178 (2013). https://doi.org/10.1016/j.energy.2013.09.004

Ibe, C.A., Anyanwu, E.E.: Principles of Tropical Air Conditioning. Author House (2013)

Lemmini, F., Errougani, A.: Building and experimentation of a solar powered adsorption refrigerator. Renew. Energy **30**(13), 1989–2003 (2005). https://doi.org/10.1016/j.renene.2005.03.003

Meunier, F., Douss, N.: Performance of adsorption heat pumps. Active carbon-methanol and eolith-water. ASHRAE Trans. **2**, 267–274 (1990)

Meunier, F., Mugnier, D.: La climatisation solaire. In: Dunod (ed.) Competition Between Solar Hot Water and Solar A/C, vol. 1, pp. 235–239 (2013)

Mitra, S., Srinivasan, K., Kumar, P., Murthy, S.S., Dutta, P.: Solar driven adsorption desalination system. Energy Procedia **49**, 2261–2269 (2014). https://doi.org/10.1016/j.egypro.2014.03.239

Rezk, H., Alsaman, A.S., Al-Dhaifallah, M., Askalany, A.A., Abdelkareem, M.A., Nassef, A.M.: Identifying optimal operating conditions of solar-driven silica gel based adsorption desalination cooling system via modern optimization. Sol. Energy **181**, 475–489 (2019). https://doi.org/10.1016/j.solener.2019.02.024

Sakadura, J.F.: Initiation aux Transferts Thermiques (Introduction to Thermal Transfers), Edition Techniques et Documentation (2000)

Sumathy, K., Li, Y., Muller, S.H., Kerskes, H.: Performance analysis of a modified two-bed solar-adsorption air-conditioning system. Int. J. Energy Res. **33**, 675–686 (2009). https://doi.org/10.1002/er.1503

Uyun, A.S., Miyazaki, T., Ueda, Y., Akisawa, A.: High performance cascading adsorption refrigeration cycle with internal heat recovery driven by a low-grade heat source temperature. Energies **2**, 1170–1191 (2009). https://doi.org/10.3390/en20401170

Wakil, S.M., Thu, K., Saha, B.B.: An emerging hybrid multi-effect adsorption desalination system. Evergr.: Jt. J. Nov. Carbon Resour. Sci. Green Asia Strategy **1**, 30–36 (2014). https://doi.org/10.5109/1495161

Wang, D., Zhang, J., Yang, Q., Li, N., Sumathy, K.: Study of adsorption characteristics in silica gel-water adsorption refrigeration. Appl. Energy **113**, 734–741 (2014). https://doi.org/10.1016/j.apenergy.2013.08.011

Wu, J.Y., Wang, R.Z., Xu, Y.X.: Dynamic analysis of heat recovery process for a continuous heat recovery adsorption heat pump. Energy Convers. Manag. **43**, 2201–2211 (2002). https://doi.org/10.1016/S0196-8904(01)00158-3

Young-Deuk, K., Thu, K., Ng, K.C.: Evaluation and parametric optimization of the thermal performance and cost effectiveness of active-indirect solar hot water plants. Evergr.: Jt. J. Nov. Carbon Resour. Sci. Green Asia Strategy **2**, 50–60 (2015). https://doi.org/10.5109/1544080

Satellite Big Data Ingestion for Environmentally Sustainable Development

Badr-Eddine Boudriki Semlali and Chaker El Amrani

Abstract

Currently, many environmental applications take advantage of remote sensing techniques, particularly air quality monitoring, climate changes overseeing, and natural disasters prediction. However, a massive volume of remote sensing data is generated in near-real-time; such data are complex and are provided with high velocity and variety. This study aims to confirm that satellite data are big data and proposes a new big data architecture for satellite data processing. In this paper, we mainly focused on the ingestion layer enabling an efficient remote sensing big data preprocessing. As a result, the developed ingestion layer removed eighty-six percent of the unnecessary daily files. Moreover, it eliminated about twenty percent of the erroneous and inaccurate plots, therefore, reducing storage consumption and improving satellite data accuracy. Finally, the processed data was efficiently integrated into a Hadoop storage system.

Keywords

Remote sensing big data • Ingestion layer • Data preprocessing • Data integration

1 Introduction

The considerable rise of industrial, transport, and agricultural activities has led to many environmental issues. Notably, the outdoor Air Pollution (AP) due to the emission of many anthropogenic pollutants such as the Monoxide of Carbon (CO), Dioxide of Carbone (CO_2), Nitrogenous of Oxide (NO_x), Methane (CH_4), and so on Nowak et al. (2018). Consequently, AP can seriously affect human health and catalyze climate change. For this reason, Air Quality (AQ) currently deserves special attention from several scientific communities. Indeed, continuous AQ monitoring is one of the proposed solutions helping to decision makers (Boudriki Semlali et al. 2019). It enables a Near-Real-Time (NRT) monitoring of the Aerosol Optical Depths (AOD), it provides a potential input data for AQ models, tracks the pollutant plumes emitted from industrial and agricultural areas (Boudriki Semlali et al. 2019), and estimates the Ozone precursor.

Generally, the Remote Sensing (RS) technique refers to the use of the Satellite Sensors (SS) to measure the ocean, Earth, and atmospheric components through Electromagnetic Energy (EME) without making physical contact with it (Ma et al. 2015). Nowadays, there are more than three thousand satellites in orbits (Lavender 2018), used for many purposes, such as military, earth observation weather, and forecasting support. All these satellites are equipped with many sensors within different temporal, spatial, and spectral resolutions ranging from low to high (Zhu et al. 2018). Satellites sensors measure data and then they transmit data into ground stations through downlink channels. In our survey, we collect data from the European Organization for the Exploitation of Meteorological Satellites (EUMETSAT) via the Mediterranean Dialogue Earth Observatory (MDEO) ground station installed at the Abdelmalek Essaâdi University of Tangier in Morocco (Boudriki Semlali and Amrani 2019b). We also obtained RS data from the Earth Observation System Data and Information System (EOSDIS) of the National Aeronautics and Space Administration (NASA), the Infusing Satellite Data into Environmental Applications (NESDIS) of the National Oceanic and Atmospheric Administration (NOAA), and the Copernicus platform operated by the European Space Agency (ESA).

B.-E. Boudriki Semlali (✉) · C. El Amrani
LIST Laboratory, Faculty of Sciences and Techniques of Tangier, Abdelmalek Essaâdi University, Tetouan, Morocco
e-mail: Badreddine.boudrikisemlali@uae.ac.ma

C. El Amrani
e-mail: ch.elamrani@fstt.ac.ma

The RS data gathered come from many polar and geostationary satellites and various sensors.

Moreover, these data are stored in a complicated scientific file format precisely: The Binary Universal Form for the Representation of meteorological data (BUFR) (Karhila 2010), the Network Common Data Form (NetCDF) (Rew and Davis 1990), the Hierarchical Data Format (HDF5) (Gosink et al. 2006), and so on. The daily volume of the received RS data reaches sixty Gigabits (GB) and sumps up twenty Terabits (TB) per year. Furthermore, the speed with which data is collected is fast, with a rate of fifty thousand files per day. According to the attribute definition (Venue, Volume, Variety, Veracity, Velocity, and so on) of Big Data (BD) (Hu et al. 2014), we will confirm that satellite data are BD (Boudriki Semlali et al. 2020a). Consequently, the processing is challenging and also take a vital execution time. For this goal, we have designed a novel BD architecture to split and facilitate problem-solving the problems of RSBD (Boudriki Semlali et al. 2020). In this paper, we will show only the mechanisms and results of the developed ingestion layer. Thus, this phase is very critical because it is responsible for collecting unprocessed RS data, for handling an enormous volume of input data and for extracting and filtering it. As a result, the ingestion layer has proceeded efficiently and obtained potential values with high accuracy, low volume, and reasonable execution time. Besides, this layer has performed all steps automatically and processed global RS global data in NRT.

2 Background

RS techniques have become widely employed in several environmental applications, including AP monitoring and climate change monitoring (Navalgund et al. 2007). Satellites are the principal instrument of the measurement. They use sensors within different temporal, spatial, and spectral resolutions. Satellites always pass by unique orbits. This orbit can be polar or geostationary (Boain' 2004). In our study, we apply RS techniques to supervise the AQ of Morocco and monitor climate changes in NRT (Boudriki Semlali and Amrani 2019a). We acquired data from four organizations, which are the MDEO, NASA, the NOAA, and the ESA. Thus, the most used satellites in our investigation are polar excepting the Meteosat Second Generation (MSG) is geostationary (Schmetz et al. 2002). Also, we exploited many active and passive SS, such as (IASI, AMSU, VIIRS, GOME-2, ASCAT, TROPOMI, and so on).

Table 1 details all the measured variables of each satellite sensor. We notice that all these instruments are for the earth observation mission, and 50% of them measure atmospheric variables, 42% estimate tropospheric variable. The rest calculate only land and sea targets such as LST, SST vegetation

cover soil moisture, and so on. Besides, the majority of the used sensors measure whether variable particularly, temperature, humidity, pressure, and wind speed. Twenty percent of them notably the IASI, GOME-2, MLS, OMI, and TROPOMI estimate the TC, VCD, and VMR of TC of CO, NO_2, CO_2, SO_2, CH_4, and so on. AOD properties are also measured by five satellite sensors, especially the GOME-2, MODIS, MISR, VIIRS, and the OLCI. Forest fire is monitored by three instruments which are: the SEVIRI, MODIS, and the VIIRS. Ten percent of the used satellite measure the cloud priorities, precipitation, and forecasting such as SEVIRI, MHS, and MWR. Besides, some sensors measure other variables like the emissivity, altitude, the concentration of chlorophylls, and so on.

In our case study, RS data come with a high velocity reaching forty thousand files per day with an average latency of thirty minutes. These data continuously increase the storage space with fifty GB per day. The collected data are stored in scientific file format, particularly the NetCDF, HDF5, and BUFR. Consequently, RSBD turns out to be extremely challenging problems to be treated. The processing chain of the RSBD includes many challenges. Firstly, satellite data are transmitted into ground stations, so the big protest is how to collect these data in real-time to protect data freshness. These data should be preprocessed to remove erroneous, inaccurate, and unneeded datasets to keep only data of interest and also integrate them into a distributed and scalable storage platform. Also, satellites generate pervasive data with high velocity, which is not possible to be continuously hosted inside a standard storage system (Ma et al. 2015). So, it is compulsory to build a model deciding which RSBD to keep and which one to discard. Finally, RSBD processing demands some mathematical knowledge in probability and statistics to employ deep learning, machine learning, and neural network algorithms to unlock new insights. Consequently, many kinds of research have been conducted on different architectures to solve RS data processing issues. These investigations aim mainly to use parallel computing by the integration and the exploitation of the hardware's capacity (Wei et al. 2014) to store and process RSBD inside distributed clusters such as the Hadoop (Wang et al. 2015). To optimize algorithms and the processing patterns (Jin et al. 2010), and to process RS data in streaming (Ranjan 2014).

3 Method

We have proposed a BD architecture enabling an efficient RS processing. Figure 1 illustrates the proposed architecture. It is composed of six layers which are: the data sources, the ingestion layer, the Hadoop storage, the processing and management layer, and finally, the visualization layer

Table 1 The measured variables with the used satellite sensors

Sensor	Measured variables	Layer
MetOp		
SEVIRI	LST, clouds, winds, water vapor, O_3 and fires	Atmosphere and surface
IASI	Humidity, temperature, CO, NO_2, CO_2, and CH_4	Atmosphere
MHS	Surface temperature, clouds, precipitation, and water vapor	Troposphere
GOME-2	VCD and TC of CO, NO_2, CO_2, SO_2 and CH_4, AOD	Atmosphere
ASCAT	Wind speed, soil moisture, ice snow	Atmosphere and surface
AVHRR	Snow ice, cloud, SST and LST	Troposphere and surface
HIRS	Vertical profile temperature, clouds, water vapor, O_3, N_2O, LST	Atmosphere
JASON-2 & 3		
AMR	SST, water vapor, clouds, altitude	Atmosphere and surface
POSEIDON	Winds, altitude, oceanography	Atmosphere and surface
TERRA		
ASTER	LST	Surface
MISR	Clouds and AOD	Atmosphere
MOPITT	TC of CO	Troposphere
AIRS	Vertical temperature and humidity profiles	Troposphere
AMSU	Vertical temperature and humidity profiles	Troposphere
AQUA		
HSB	Vertical humidity profile	Atmosphere
MODIS	Vegetation cover, LST, ocean chlorophyll, AOD, and fires	Atmosphere and surface
MLS	Temperature, VMR of CO, NO_2, SO_2, and CH_4 and so on	Atmosphere
AURA		
OMI	VCD of O_3, NO_2, SO_2	Atmosphere
TES	VCD of O_3	Troposphere
SUOMI NPP		
VIIRS	Clouds, AOD, SST, LST, and fires	Atmosphere and surface
OMPS	TC of O_3	Atmosphere
SENTINEL-5P		
TROPOMI	VCD of CO, NO_2, SO_2 and CH_4	Troposphere
SENTINEL-3		
SLSTR	LST, SST	Surface
SRAL	Sea surface topography, winds, sea ice, height and thickness	Troposphere and surface
MWR	Water vapor, clouds, and earth emissivity	Troposphere and surface

(Boudriki Semlali et al. 2020c). In this paper, we focus our clarifications on the ingestion layer. The ingestion layer is regarded as a potential part of the proposed BD architecture. It is responsible for preprocessing RS data (Erraissi et al. 2018). Thus, the acquired data are firstly stored and then processed. As a result, data staging is batch processing. Before beginning the design and the development of this layer, we have done many pieces of research to understand the nature and the characteristics of satellite data. The developed ingestion layer acquires, decompresses, filters, converts, and extracts refined information and datasets from enormous RSBD input (Boudriki Semlali et al. 2020b). JAVA, Python, and Shell were the principal programming languages used in development.

Figure 1 illustrates the different stages of the ingestion layer. The acquisition is the first step in the satellite data processing. In this study, we collect datasets from four sources which are: the MDEO ground station through the EUMETCast protocol (Amrani et al. 2012), the EOSDIS of NASA (EOSDIS 2017), the NESDIS of the NOAA (NOAA National Environmental Satellite, Data, and Information Service (NESDIS) 2018), and the Copernicus platform (esa 2019). Table 2 contains all the specifications of the used RS data, such as the name of the product received in the downlink channel, the file format, the library for excitation, the daily volume, velocity, and the latency of the collected data. The total size sumps up 60 Gb and 50 000 files with a latency ranging between 35 to 180 min.

These data come compressed from the ground station or datacenters. Consequently, the next step decompresses satellite data automatically to be ready to use. A BASH script decompresses all the extension of the received RSBD (Tar, Zip, and bz2). After the decompression, the number of files increases 240 times, and the size up to forty percent.

According to these results, we confirm that RS data consume more storage space and become more complex to be processed after the decompression step. Generally, the HDF5, NetCDF, the BUFR, and the BIN file's format are used to store RS data. However, exploring this data demands a conversion from the scientific file's format to a text or XML. We have used two Python libraries, which are the BUFRextract (BUFREXC) (Breame 2018) and the pybu-fr_ecmwf (ECMWF) (Siemen et al. 2007). We have developed a BASH script which filters files based on the hours of a scan of every satellite sensor with the studied zone countries (Austria, China, Morocco, Qatar, Spain, and the USA). Table 3 includes the coverage hours of each satellite sensor in Morocco. We notice that big countries have extended

Fig. 1 The general architecture of RS data preprocessing

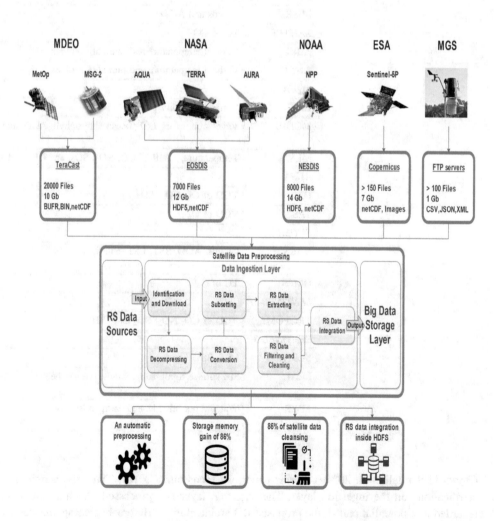

Table 2 The specification of the used satellite sensors data

Channel	Sensors	Product	Library	Format	Size/day	File/day	Latency
AIRS2SUP_NRT.006	AIRS	L2.RetSup_IR	pyhdf	HDF5	800	240	75–140
ASCATSM	ASCAT	ASCAT_C_EUMP	h5py	NetCDF	330	12	180
Data_Channel_3	SEVIRI	MPEF-AMV	BUFREXC	Native	10	60	35
Data_Channel_3	SEVIRI	MPEF-ASR	ECMWF	Native	10	60	35
Data_Channel_3	SEVIRI	MPEF-CLA	BUFREXC	Native	10	60	35
Data_Channel_3	SEVIRI	MPEF-CSR	BUFREXC	Native	14	60	35
Data_Channel_3	SEVIRI	MPEF-GII	ECMWF	Native	15	60	35
Data_Channel_3	SEVIRI	MPEF-TH	ECMWF	Native	18	60	35
Data_Channel_3	SEVIRI	MPEF-TOZ	BUFREXC	Native	19	60	35
EPS-Africa	ASCAT	ear_o	ECMWF	BUFR	200	40	35
EPS-Africa	ASCAT	eps_o_125_ssm	BUFREXC	Bin	17	36	35
EPS-Africa	ASCAT	eps_o_250_ssm	BUFREXC	Bin	17	36	35
EPS-Africa	IASI	eps_o_clp	BUFREXC	Bin	34	480	35
EPS-Africa	IASI	eps_o_cox	BUFREXC	Bin	54	480	35
EPS-Africa	IASI	eps_o_ems	BUFREXC	Bin	35	480	35
EPS-Africa	IASI	eps_o_ozo	BUFREXC	Bin	164	480	35
EPS-Africa	IASI	eps_o_so2	BUFREXC	Bin	122	480	35
EPS-Africa	IASI	eps_o_trg	BUFREXC	Bin	29	480	35
EPS-Africa	IASI	eps_o_twt	BUFREXC	Bin	322	480	35
EPS-Global	HIRS	ATOVS_C_EUMP	BUFREXC	Bin	96	480	35
IASICSP	IASI	IASI_C_EUMP	h5py	NetCDF	950	12	180
MCDAODHD	MODIS	MCDAODHD	pyhdf	HDF5	4	4	60–125
ML2CO_NRT.004	MLS	L2GP-CO	h5py	HDF5	24	90	45–140
ML2H2O_NRT.004	MLS	L2GP-H2O	h5py	HDF5	24	90	40–140
ML2HNO3_NRT.004	MLS	L2GP-HNO3	h5py	HDF5	24	90	40–140
ML2N2O_NRT.004	MLS	L2GP-N2O	h5py	HDF5	24	90	40–140
ML2O3_NRT.004	MLS	L2GP-O3	h5py	HDF5	24	90	40–140
ML2SO2_NRT.004	MLS	L2GP-SO2	h5py	HDF5	24	90	50–140
ML2T_NRT.004	MLS	L2GP-Temperature	h5py	HDF5	24	90	50–140
MOD11_L2	MODIS	MOD11_L2	pyhdf	HDF5	3030	105	60–125
MOD14	MODIS	MOD14	pyhdf	HDF5	4320	240	60–125
MOP02R_NRT	MOPITT	MOP02R	h5py	HDF5	520	40	180
MSGAMV	SEVIRI	SEVI-MSGAMVE	BUFREXC	BUFR	3	6	180
MSGTOZ	SEVIRI	SEVI-MSGTOZN	BUFREXC	BUFR	1300	24	180
NMTO3NRT	OMPS	NMTO3-L2	h5py	HDF5	250	28	180
NPBUVO3-L2-NRT	OMPS	NPBUVO3-L2	h5py	HDF5	15	25	180
NPP-3	SEVIRI	L3U_GHRSST-SST	h5py	NetCDF	2730	334	35
NPP-3	SEVIRI	VRSYCW	h5py	NetCDF	2880	144	35
NPP-3	OMPS	OMPS_C_EUMP_	ECMWF	Bin	12	2000	35
NPP-3	VIIRS	VIIRS_C_EUMP_	ECMWF	Bin	39	2000	35
NUCAPS-Product	AMSU	NUCAPS-EDR	h5py	NetCDF	7500	2600	60–180
OMPS-Ozone	OMPS	V8PRO-EDR	h5py	NetCDF	140	1000	60–180
OMSO2NRTb	OMI	L2-OMSO2NRT	h5py	HDF5	520	20	180
OMTO3	OMI	L2-OMTO3	h5py	HDF5	900	40	100–165

(continued)

Table 2 (continued)

Channel	Sensors	Product	Library	Format	Size/day	File/day	Latency
PMSAOP	GOME-2	GOME-GOMPM	h5py	NetCDF	150	6	180
Precipitation	AMRS2	AMSR2-PRECIP	h5py	NetCDF	240	6	180
Precipitation-and-Surface	MISR	NPR-MIRS-IMG	h5py	NetCDF	1700	5200	60–180
SAF-Africa	SEVIRI	MSG_LST_NAfr	h5py	HDF5	870	60	35
SAF-Africa	GOME-2	O3-NO2-NO2Tropo-	h5py	HDF5	7000	60	35
SAF-Africa	GOME-2	O3M_GOME_NA	h5py	HDF5	8000	60	35
SAF-Africa	GOME-2	GOME_O3-NO2-SO2	ECMWF	BUFR	20	250	35
SAF-Europe	SEVIRI	MSG_ET_NAfr	h5py	HDF5	407	200	35
SAF-Europe	GOME-2	O3-NO2-NO2Tropo	h5py	HDF5	367	300	35
SAF-Europe	GOME-2	GOME2_C_EHDB	ECMWF	BUFR	800	200	35
Sentinel-5P	TROPOMI	L2__CO	h5py	NetCDF	400	21	120–180
Sentinel-5P	TROPOMI	L2__NO2	h5py	NetCDF	400	21	120–180
Sentinel-5P	TROPOMI	L2__O3	h5py	NetCDF	400	21	120–180
Sentinel-5P	TROPOMI	L2__SO2	h5py	NetCDF	400	21	120–180
Soil Moisture	AMRS2	AMSR2-SOIL	h5py	NetCDF	440	15	100–165
VIIRS-Active-Fire	VIIRS	AF_	h5py	NetCDF	129	1000	60–180
VIIRS-AOD	VIIRS	JRR-AOD	h5py	NetCDF	27000	560	60–180

Table 3 The time coverage of the used satellite sensors over Morocco

Country	Satellite	Sensors	Hours of cross
Morocco	AQUA	AIRS	2-3-4-5-13-14-15-16-17
Morocco	AQUA	AIRS	3-14
Morocco	NPP	AMSU	3-14
Morocco	MetOp	ASCAT	9-10-11-20-21-22
Morocco	MetOp	GOME-2	09-10-2011
Morocco	MetOp	HIRS	4-5-6-9-10-11-16-17-21-22
Morocco	MetOp	IASI	9-10-11-20-21-22
Morocco	AURA	MLS	3-4-5-14-15-16
Morocco	AURA	OMI	6-13-14-15-16-17
Morocco	MSG	SEVIRI	Permanent
Morocco	NPP	VIIRS	12-13-14-15

coverage time, reaching 18 h daily. For instance, the MetOp satellite regularly scans Morocco twice a day between 9 to 11 h and 20 to 22 h.

After the conversion, datasets are ready to be extracted. Thus, the total size of data remains approximately the same after this operation. The collected data come from polar satellites flying in a Low Earth Orbit (LEO) with an altitude of 800 km and making sixteen orbits daily. Besides, processing all data of the globe will be challenging in terms of preprocessing. As a result, we developed a Python script that filters satellite data by countries using the longitude and latitude as the main parameters. Table 4 shows the minimum and the maximum longitudes and latitudes of the six studied countries. Accordingly, we find that big countries such as the USA, China, and Australia have a high number of files reaching more than seven hundred files per day. However, the smallest state, which is Qatar, contains only about fifty files, similarly in data size (MB). We have filter data also by the minimum and maximum accepted values delimiting the measured value in a logical range, for instance, the minimum and the maximum value of the pressure must be between 0 atm to 1120 hPa. Then, we have also filtered the data by their quality. RS files usually include some indicators useful to classify the quality of the received data. Tables 5 and 6 contain, respectively, the quality flags used by NASA, NOAA, ESA, and the MDEO files.

The final step is extraction. It allows the selection of the needed variables; for instance, we were interested in twelve variables, including the temperature, humidity, pressure, wind speed, AOD, the VCD of trace gasses, and so on. We developed a Python script performing an automatic data extraction, subset, and filters of inaccurate data.

A Comma-Separated Values (CSV) output file will be generated containing all the refined satellite data. It stores a 3D model of altitude ranging between two hundred meters to seven Kilometers, Fig. 6, shows the final output files containing the Vertical Mixing Ration (VMR) of the CO of Morocco in 2017/25/06, and Fig. 7 shows a map visualization of CO in morocco in 2018/06/11.

4 Results and Discussion

The ingestion layer performed an automatic download, decompression, filter, conversion, subset, and extraction efficiency. As a result, and as shown in Fig. 2, we notice that the total daily size acquired was thirty-five GB. An increase of forty percent happened after the decompression step because the ground station compresses RS data to facilitate data transmission. After the conversion step, the total size remained the same; however, after the subset operation, data decrease significantly due to the elimination of unneeded data. Globally, this ingestion layer helps to gains storage space by eighty-six percent. Accordingly, this result could be regarded as a solution to the satellite data perversity.

Table 7 details the benchmarking results particularly the number of processed files (P-Files), the execution time (E-Time) the Central Processing Unit (CPU), Random Access Memory (RAM), CPU per second, total files (T-Files), and the temperature of the CPU obtained during all the preprocessing steps.

In our analysis, we made in use a standard laptop equipped with an i5 in the processor (Generation 2), 16 GB in Random Access Memory (RAM), 1 GB in the Graphical Processing Unit (GPU) with an Internet bandwidth of 2 Megabits/s. Generally, the preprocessing of the RSBD takes an important execution time. From Fig. 3, we remark that the download phase took approximately five minutes, this number could be reduced by increasing the Internet speed or/and changing the Internet Protocol (IP) protocol from the Transmission Control Protocol (TCP) to the User Datagram Protocol (UDP). The decompressing's execution time took an average of ninety minutes. Thus, the conversion is the most prolonged operation reaching about eight hours, and then the subset needs more than an hour. Finally, the extraction processes data took an average of thirty minutes. As a result, the total execution time is about five hours.

The extraction is the final step of the ingestion layer, and it is also a significant stage, as explained in the previous section. Figure 4 shows the daily total number of plots of the six countries. We notice that after the sub-setting, the number of plots decreases potentially to keep only values covering the countries' zone of interest. Then, the quality, the minimum, and maximum filter eliminate about twenty percent of inaccurate and erroneous datasets, as shown in Fig. 4. Finally, refined and final plots were stored into associated CSV output files that can be integrated into an HDFS. Accordingly, the extraction step reduces the number of uninteresting datasets up to twenty percent (Figs. 5, 6 and 7).

5 Related Works and Comparison

Despite the existing robust architectures, platforms, and systems from prominent organizations such as NASA, NOAA, EUMETSAT, and the ESA. We can find some limits and challenges of the processing; besides, sometimes their tech-

Table 4 The longitudes and latitudes limits of the studied countries

Country	Long_min	Long_max	Lat_min	Lat_max
Morocco	−17	−1	21	36
Spain	−10	4	36	44
USA	−125	−63	25	48
Qatar	50	52	24	26
China	74	134	21	49
Australia	113	153	−44	−10

Table 5 The list of the data quality variables of NASA, NOAA, and ESA satellite sensor

Name of product	Name of variable	Quality indicator
AMSR2-PRECIP	Convection	Rain_Rate_QC_Flag
AMSR2-SOIL	Soil Moisture	Soil_Moisture_QA
JRR-AOD	AOD-FineModWgt	QCAll
L2.RetSup_IR	VMR-CO2	CO2ppmv_QC
L2.RetSup_IR	VMR-CH4	CH4VMRLevSup_QC
L2.RetSup_IR	TC-CH4	CH4_total_column_QC
L2.RetSup_IR	VCD-CH4	CH4CDSup_QC
L2.RetSup_IR	TC-CO	CO_total_column_QC
L2.RetSup_IR	VMR-CO	COVMRLevSup_QC
L2.RetSup_IR	VMR-CO	COVMRSurf_QC
L2.RetSup_IR	VCD-H2O	H2OCDSup_QC
L2.RetSup_IR	MMR-H2O	H2OMMRLevSup_QC
L2.RetSup_IR	MMR-H2O	H2OMMRSurf_QC
L2.RetSup_IR	TC-O3	O3CDSup_QC
L2.RetSup_IR	VMR-O3	O3VMRLevSup_QC
L2.RetSup_IR	VMR-CO	O3VMRSurf_QC
L2.RetSup_IR	Humidity	RelHumSurf_QC
L2.RetSup_IR	Temperature	TAirSup_QC
L2.RetSup_IR	Temperature	TSurfAir_QC
L2__CO	TC-CO	qa_value
L2__NO2	TC-NO2	qa_value
L2__O3	TC-O3	qa_value
L2__SO2	VCD-SO2-DU	qa_value
L3U_GHRSST-SST	TemperatureSea	quality_level
L3U_GHRSST-SST	WindSpeed	quality_level
MLS-Aura_L2GP-CO	VMR-CO	Quality
MLS-Aura_L2GP-N2O	VMR-N2O	Quality
MLS-Aura_L2GP-O3	VMR-O3	Quality
MLS-Aura_L2GP-SO2	VMR-SO2	Quality
MLS-Aura_L2GP-Temperature	Temperature	Quality
MOD11_L2	Temperature	QC
NUCAPS-EDR	VCD-CH4	Quality_Flag
O3M_GOME_NA	TC-O3	QualityInput
O3M_GOME_NA	Altitude	QualityInput
O3M_GOME_NA	TC-O3	QualityProcessing
O3-NO2-NO2Tropo-	VCD-SO2	VCD_Error
O3-NO2-NO2Tropo-	VCD-NO2	VCDTropo_Error
OMI-Aura_L2-OMSO2NRT	VCD-SO2-DU	QualityFlags_PBL
OMI-Aura_L2-OMTO3	TC-O3	MeasurementQualityFlags
OMI-Aura_L2-OMTO3	TC-O3	QualityFlags
OMPS-NPP_NMTO3-L2	TC-O3	MeasurementQualityFlags
OMPS-NPP_NMTO3-L2	TC-O3	QualityFlags
OMPS-NPP_NPBUVO3-L2	VMR-O3	O3MixingRatioError
OMPS-NPP_NPBUVO3-L2	TC-O3	ProfileTotalO3Error
OMPS-NPP_NPBUVO3-L2	TC-O3	TotalO3ErrorFlag

Table 6 The list of the data quality variables of the MetOp satellite sensor

Name of Product	Quality Indicator	Accepted limit
eps_o_ozo	AMSUBAD	0
eps_o_250_ssm	ASCATKP	0
eps_o_250_ssm	ASCATSD	1
eps_o_250_ssm	ASCATSO	1
eps_o_250_ssm	ASCATTP	1
eps_o_ozo	FLAG_CLDNESS	3
eps_o_trg	FLAG_CLDNESS	3
eps_o_twt	FLAG_CLDNESS	3
eps_o_trg	FLAG_IASIBAD	1
eps_o_twt	FLAG_IASIBAD	1
eps_o_ozo	FLAG_MHSBAD	0
eps_o_twt	FLAG_MHSBAD	0
eps_o_ozo	FLAG_SATMAN	1
eps_o_trg	FLAG_SATMAN	1
eps_o_twt	FLAG_SATMAN	1
eps_o_ozo	FLAG_SUNGLINT	0
eps_o_clp	FLAG_SUNGLINT	0
eps_o_trg	FLAG_SUNGLINT	0
eps_o_twt	FLAG_SUNGLINT	0
eps_o_ozo	FLAGIASIBAD	1
OMPS_C_EUMP_	POQ	0
eps_o_250_ssm	RR	8
OMPS_C_EUMP_	SGI	0
OMPS_C_EUMP_	TOQ	0

Fig. 2 Total daily size of the RSBD (Gb) of Morocco, Spain, USA, Qatar, China, and Australia during the processing in the ingestion layer

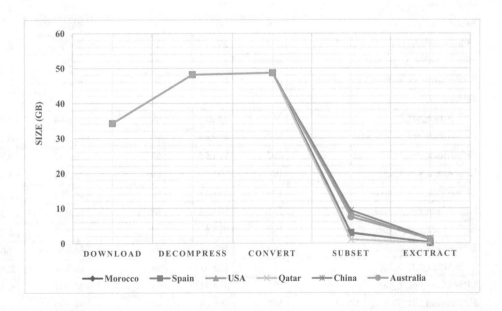

Table 7 The summary of the benchmarking statistic results of the processing

Product	P-Files	E-Time (s)	CPU (%)	RAM (Mb)	CPU/s	T-Files	Temp (°C)
Decompression							
Data_Channel_3	273	7	11	3120	1	273	44
EPS-Africa	8682	72	10	4984	7	8682	46
EPS-Africa_ear_o	36	7	21	5016	0	147	52
EPS-Global	954	6	11	3120	1	954	45
NPP-3	789	77	4	3040	3	789	47
NUCAPS	2623	662	14	3092	15	2794	46
OMPS-Nadir	1016	2	41	3092	0	1114	50
Precipitation	15	1	77	3064	0	15	48
Precipitation	24,227	462	26	6718	17	25,883	46
SAF-Africa	1765	326	38	3854	5	3530	49
SAF-Europe	4745	264	4	3660	11	4745	44
Sentinel-3	41	1051	20	3084	32	41	33
Soil Moisture	15	3	12	3064	0	15	46
VIIRS-Active	1008	5	36	3022	0	1164	45
VIIRS-Aerosol-Optical-Depth	565	2054	19	3028	67	670	43
Orbit filter							
EPS-Africa_eps_o_125_ssm	189	25	9	5068	1	336	46
EPS-Africa_eps_o_250_ssm	190	25	8	5016	1	526	47
EPS-Africa_eps_o_clp	171	17	12	5088	1	697	49
EPS-Africa_eps_o_cox	178	24	9	5016	1	875	51
EPS-Africa_eps_o_ems	179	14	15	5024	1	1054	53
EPS-Africa_eps_o_ozo	179	21	10	5024	1	1233	54
EPS-Africa_eps_o_so2	186	27	9	5012	1	1419	58
EPS-Africa_eps_o_trg	179	16	14	5084	1	1598	61
EPS-Africa_eps_o_twt	179	26	9	5024	1	1777	57
EPS-Global_ATOVS_C_EUMP	230	27	7	3608	1	242	51
NPP-3_VIIRS_C_EUMP_	60	6	5	3228	0	64	58
SAF-Europe_GOME2_C_EHDB	64	6	8	3648	0	69	49
Conversion							
EPS-Africa_ear_o	36	37.21	99	52,984	3.01	1779	58
EPS-Africa_eps_o_125_ssm	202	489.43	99	41,272	4.76	1779	62
EPS-Africa_eps_o_250_ssm	202	178.65	99	39,352	3.79	1779	63
EPS-Africa_eps_o_clp	191	104.03	99	29,344	3.36	1779	66
EPS-Africa_eps_o_cox	190	531.71	99	26,388	27.42	1779	63
EPS-Africa_eps_o_ems	191	164.38	98	33,008	3.93	1779	73
EPS-Africa_eps_o_ozo	191	917.6	84	26,692	5.49	1779	60
EPS-Africa_eps_o_so2	190	148.22	95	24,624	7.62	1779	64
EPS-Africa_eps_o_trg	191	123.51	98	28,784	2.85	1779	63
EPS-Africa_eps_o_twt	191	988.11	98	27,504	6.4	1779	67
EPS-Global_ATOVS_C_EUMP	242	388.08	97	31,260	5.11	242	64
NPP-3_VIIRS_C_EUMP_	64	27.36	93	37,400	1.99	64	61
SAF-Africa_GOME_O3-NO2-SO2	40	19.97	97	35,004	1.81	40	61
SAF-Europe_GOME2_C_EHDB	69	266.18	99	61,704	8.05	69	65

(continued)

Table 7 (continued)

Product	P-Files	E-Time (s)	CPU (%)	RAM (Mb)	CPU/s	T-Files	Temp (°C)
Subset							
Aerosol-Optical-Depth	8	1557.78	89	182,992	7.27	8	64
AIRS2SUP_NRT.006	3	132.87	10	71,032	4.99	3	57
EPS-Africa_ear_o	36	6.16	29	65,332	0.32	1743	51
EPS-Africa_eps_o_125_ssm	190	26.43	84	103,572	2.32	1553	54
EPS-Africa_eps_o_250_ssm	190	8.8	79	70,620	1.09	1363	55
EPS-Africa_eps_o_clp	171	3.93	52	66,892	0.51	1192	55
EPS-Africa_eps_o_cox	178	19.55	77	94,796	2.14	1014	52
EPS-Africa_eps_o_ems	179	10.35	76	73,088	1.34	835	52
EPS-Africa_eps_o_ozo	179	39.2	85	174,896	5.46	656	55
EPS-Africa_eps_o_so2	186	2.23	49	60,292	0.42	470	54
EPS-Africa_eps_o_trg	179	5.96	78	67,032	0.79	291	53
EPS-Africa_eps_o_twt	179	45.08	85	197,536	7.27	112	55
EPS-Global_ATOVS_C_EUMP	230	18.25	80	95,780	2.88	12	53
MIRS-Precipitation-and-Surface	299	160.14	60	71,508	8.52	299	51
ML2CO_NRT.004	2	2.11	50	66,948	0.19	2	48
ML2H2O_NRT.004	2	1.98	49	67,048	0.15	2	48
ML2HNO3_NRT.004	2	2.19	44	66,996	0.14	2	48
ML2N2O_NRT.004	13	4.32	29	67,996	0.28	13	51
ML2O3_NRT.004	2	2.16	46	67,148	0.19	2	48
ML2SO2_NRT.004	2	1.99	50	67,180	0.14	2	49
ML2T_NRT.004	2	2.1	48	67,124	0.12	2	48
MOD11_L2	5	59.08	67	72,528	0.65	5	54
MOD14	5	888.58	87	240,232	6.97	5	62
MOP02R_NRT	3	6.93	23	70,392	0.4	3	58
NMTO3NRT	1	2.44	48	67,796	0.18	1	58
NPBUVO3-L2-NRT	1	1.1	64	67,580	0.14	1	58
NPP-3_VIIRS_C_EUMP_	60	2.73	63	68,792	0.32	4	52
NUCAPS	19	49.27	22	67,744	1.76	19	47
OMPS-Nadir-Profile-Ozone	7	8.2	49	67,904	0.53	7	55
OMSO2NRTb	2	4.73	64	72,760	0.24	2	57
OMTO3	3	6.84	55	72,740	0.43	3	56
Precipitation	2	38.77	79	205,172	1.01	2	57
SAF-Africa_GOME_O3-NO2-SO2	40	1.29	68	60,876	0.19	0	52
SAF-Europe_GOME2_C_EHDB	64	5.45	67	89,816	0.85	5	51
Sentinel-5P	2	7.91	60	74,208	0.29	2	56
Soil Moisture	2	19.72	77	141,096	0.56	2	59
Extraction							
AIRS2SUP_NRT.006_L2.RetSup_IR	2360	2709	49,094.35	0.09	30.50	2,493,232	53.12
EPS-Africa_ear_o	132	226	64,501.33	0.13	2	276	52
EPS-Africa_eps_o_125_ssm	4004	437	160,600	0.67	4.20	8688	58
EPS-Africa_eps_o_250_ssm	1032	404	84,605.60	0.27	7.20	22,764	60.40
EPS-Africa_eps_o_clp	60	343	67,849	0.11	10	20,928	58
EPS-Africa_eps_o_cox	196	79	102,988	0.27	11	5428	56

(continued)

Table 7 (continued)

Product	P-Files	E-Time (s)	CPU (%)	RAM (Mb)	CPU/s	T-Files	Temp (°C)
EPS-Africa_eps_o_ems	576	70	89,988	0.48	12	6004	59
EPS-Africa_eps_o_ozo	2552	177	244,780	0.86	13.50	15,836	57.50
EPS-Africa_eps_o_trg	2396	304	85,806	0.51	15.75	40,228	60.50
EPS-Africa_eps_o_twt	3272	242	272,444	0.99	19	39,136	60.67
EPS-Global_ATOVS_C_EUMP	1556	398	110,818.40	0.34	22.20	74,428	60.40
L2GP-CO	4	115	37,246	0.05	35.50	147,844	71
L2GP-HNO3	4	119	37,234	0.04	36	147,848	65.50
L2GP-N2O	16	153	92,726	0.21	36.50	147,864	65.50
L2GP-O3	4	98	37,404	0.03	37	147,868	65.50
L2GP-SO2	4	111	37,282	0.04	37.50	147,872	64.50
L2GP-Temperature	4	110	37,660	0.05	38	147,876	64.50
L2-OMSO2NRT	1052	555	78,987.33	0.43	49.50	866,144	71.50
L2-OMTO3	1976	744	96,674	0.60	51.25	1,160,756	74.25
MCDAODHD	4	585	36,444.80	0.04	35	739,200	70.40
MIRS-Precipitation	45,608	2065	83,650.73	0.86	29	6,156,268	72.77
MOD11_L2_MOD11_L2	138,040	197	1,895,500	7.12	38.50	285,916	72.50
MOD14_MOD14	8	189	1,457,186	1.21	39	285,924	75
MOP02R_NRT_MOP02R	1504	1089	57,176.57	0.18	41	2,007,748	73.43
NMTO3NRT_OMPS-NPP_NMTO3-L2	452	729	39,130.40	0.06	44	1,438,500	72.50
NPBUVO3-L2-NRT	28	785	36,842	0.04	47	2,015,272	70.93
NPP-3_L3U_GHRSST-SST	8	206	36,260	0.03	36.25	66,036	71.50
NPP-3_VIIRS_C_EUMP_	316	156	70,346	0.21	25.50	32,032	60
NUCAPS	3604	3176	43,662.11	0.08	36.50	11,163,396	71.79
OMPS-Nadir-Profile	8	210	37,163	0.05	25.75	1,079,868	67
Precipitation	15,240	569	583,834	0.72	21.50	903,168	61.17
SAF-Africa_O3M_GOME_NA	12	347	36,384.67	0.04	41.50	99,180	71
SAF-Africa_O3-NO2-NO2Tropo	32	893	36,300.25	0.03	38.75	264,304	73.44
SAF-Europe_GOME2_C_EHDB	384	219	91,540	0.24	28	49,028	57.33
Sentinel-5P_S5P_NRTI_L2__NO2	4	107	36,314	0.03	2.50	7612	68
Sentinel-5P_S5P_NRTI_L2__O3	7588	185	74,142	0.24	3	15,200	73
Sentinel-5P_S5P_NRTI_L2__SO2	4	112	36,190	0.03	3.50	15,204	73
Sentinel-5P_S5P_OFFL_L2__CO	4	118	36,204	0.04	4	15,208	73
Soil Moisture-containing	348	180	353,798	0.43	22.50	310,164	63.50
VIIRS-Aerosol-Optical-Depth	229,764	981	2,846,265.60	9.89	24	2,237,000	70.90

Fig. 3 Time execution of the RSBD (Min) of Morocco, Spain, USA, Qatar, China, and Australia during the processing in the ingestion layer

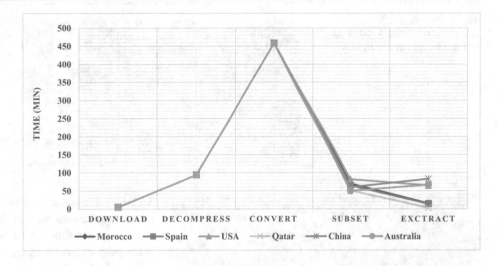

Fig. 4 Daily number of plots of Morocco during the MinMax and Quality filter

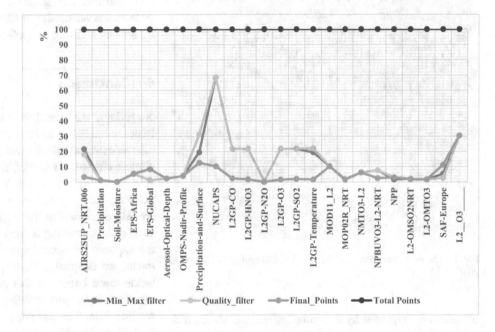

Fig. 5 Total daily number of plots (Millions) of Morocco, Spain, USA, Qatar, China and Australia during the extraction step

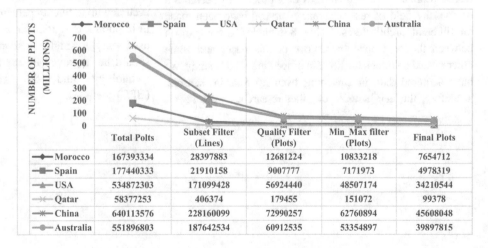

	Total Polts	Subset Filter (Lines)	Quality Filter (Plots)	Min_Max filter (Plots)	Final Plots
Morocco	167393334	28397883	12681224	10833218	7654712
Spain	177440333	21910158	9007777	7171973	4978319
USA	534872303	171099428	56924440	48507174	34210544
Qatar	58377253	406374	179455	151072	99378
China	640113576	228160099	72990257	62760894	45608048
Australia	551896803	187642534	60912535	53354897	39897815

Fig. 6 The output CSV file containing the VMR (ppb) of the CO of Morocco in 2017/25/06

EpochTime	Y	M	D	H	M.1	Latitude	Longitude	LevelGround	Level0	Level1	Level2	Level3	Level4	Level5	Level6	Level7	Level8	Level9	Level10	Level11	Level12	Total_column
1498355700	2017	6	25	3	55	35.85	-1.07	-0.0	-0.0	0.1	0.1	0.1	0.1	0.09	0.09	0.09	0.08	0.07	0.06	0.06	0.05	0.0
1498355700	2017	6	25	3	55	35.91	-1.56	-0.0	-0.0	0.1	0.1	0.1	0.1	0.09	0.09	0.09	0.08	0.07	0.06	0.06	0.05	0.0
1498355700	2017	6	25	3	55	33.26	-4.48	-0.0	-0.0	-0.0	-0.0	-0.0	-0.0	-0.0	-0.0	0.11	0.1	0.1	0.1	0.1	0.09	-0.0
1498355700	2017	6	25	3	55	33.32	-5.15	-0.0	-0.0	-0.0	-0.0	-0.0	-0.0	-0.0	-0.0	0.12	0.11	0.11	0.1	0.1	0.09	-0.0
1498355700	2017	6	25	3	55	33.39	-5.88	-0.0	-0.0	-0.0	-0.0	-0.0	0.1	0.1	0.1	0.1	0.09	0.09	0.09	0.08	0.08	-0.0
1498355700	2017	6	25	3	55	33.46	-6.72	-0.0	-0.0	-0.0	-0.0	0.1	0.1	0.1	0.1	0.1	0.1	0.09	0.09	0.09	0.08	-0.0
1498355700	2017	6	25	3	55	33.54	-7.68	-0.0	-0.0	-0.0	0.11	0.1	0.1	0.1	0.1	0.1	0.09	0.09	0.09	0.09	0.08	-0.0
1498355700	2017	6	25	3	55	33.62	-8.81	-0.0	-0.0	-0.0	0.1	0.1	0.1	0.1	0.1	0.09	0.09	0.09	0.08	0.08	0.08	-0.0
1498355700	2017	6	25	3	55	34.1	-3.08	-0.0	-0.0	-0.0	-0.0	-0.0	0.11	0.1	0.1	0.1	0.09	0.09	0.09	0.09	0.09	-0.0
1498355700	2017	6	25	3	55	34.16	-3.65	-0.0	-0.0	-0.0	-0.0	0.11	0.11	0.11	0.11	0.11	0.1	0.1	0.1	0.1	0.09	-0.0
1498355700	2017	6	25	3	55	34.22	-4.27	-0.0	-0.0	-0.0	-0.0	0.11	0.11	0.1	0.1	0.1	0.1	0.1	0.1	0.09	0.09	-0.0
1498355700	2017	6	25	3	55	34.29	-4.94	-0.0	-0.0	-0.0	-0.0	0.11	0.11	0.11	0.11	0.1	0.1	0.1	0.1	0.1	0.09	-0.0
1498355700	2017	6	25	3	55	34.36	-5.69	-0.0	-0.0	-0.0	0.11	0.1	0.1	0.1	0.1	0.1	0.09	0.09	0.09	0.08	0.08	-0.0
1498355700	2017	6	25	3	55	34.43	-6.53	-0.0	-0.0	-0.0	0.12	0.11	0.11	0.11	0.11	0.11	0.1	0.09	0.09	0.09	0.08	-0.0
1498355700	2017	6	25	3	55	34.51	-7.5	-0.0	-0.0	-0.0	0.1	0.1	0.1	0.1	0.09	0.09	0.09	0.08	0.08	0.08	0.08	-0.0
1498355700	2017	6	25	3	55	34.59	-8.64	-0.0	-0.0	-0.0	0.1	0.1	0.1	0.1	0.09	0.09	0.09	0.08	0.08	0.08	0.08	-0.0
1498355700	2017	6	25	3	55	33.39	-1.22	-0.0	-0.0	-0.0	-0.0	-0.0	-0.0	0.13	0.13	0.12	0.12	0.11	0.11	0.1	0.1	-0.0
1498355700	2017	6	25	3	55	33.5	-2.16	-0.0	-0.0	-0.0	-0.0	-0.0	-0.0	0.11	0.11	0.11	0.1	0.1	0.09	0.09	0.09	-0.0
1498355700	2017	6	25	3	55	33.56	-2.66	-0.0	-0.0	-0.0	-0.0	-0.0	-0.0	0.11	0.11	0.11	0.1	0.1	0.1	0.09	0.09	-0.0
1498355700	2017	6	25	3	55	33.62	-3.2	-0.0	-0.0	-0.0	-0.0	-0.0	-0.0	0.11	0.11	0.1	0.1	0.09	0.09	0.09	0.09	-0.0
1498355700	2017	6	25	3	55	33.68	-3.76	-0.0	-0.0	-0.0	-0.0	-0.0	-0.0	-0.0	0.11	0.11	0.1	0.1	0.09	0.09	0.09	-0.0

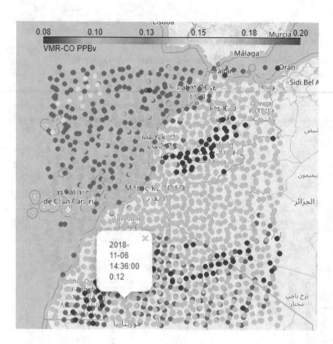

Fig. 7 The map visualization of the VMR of CO (ppb) of Morocco in 2018/06/11

SAT-ETL-Integrator preprocesses RS data for all environmental applications such as forest fire monitoring or climate changes supervision and so on. The developed software is a batch proposed solution consumes a big part of RAM and CPU and take a long execution time. The proposed solution generates small streams or CSV outputs that can be consumed by another software or application.

6 Conclusion

Nowadays, the world is witnessing many environmental issues, notably AP and climate change. Thus, RS techniques play an essential role in monitoring the AQ and supervise the climate changes. However, data provided by SS are pervasive, complex, and have a considerable size and high velocity. As a result, we have confirmed that satellite data are BD according to the eight salient of BD. Accordingly, such data processing is very challenging and goes beyond the capacity of current systems and architectures. For this intent, we adopted a Hadoop BD architecture that should tackle these issues. In this paper, we focused only on the ingestion layer part enabling an efficient preprocessing of satellite data. The developed layer allowed a daily storage gain of eighty-six percent and improved satellite dataset's accuracy up to twenty percent. However, the processing took in important execution time, reaching ten hours because we used only a single standard computer. This architecture should be improved by the integration of cloud computing technology and the High-Performance Computing (HPC) methods.

nologies are surpassed by the complication, and the considerable volume of the acquired RS data processing becomes a noteworthy field of research; many surveys have been made on different architectures. Table 8 includes a comparison between the developed framework of this paper and some other related solutions for RS data processing. In this study, we have acquired data in streaming from 26 Satellite sensors. Therefore, the data sources of other researches were fewer.

Table 8 The table of comparison

Feature/Study	This study	Akram et al. (2018)	Garcia-de-Prado et al. (2018)	Díaz et al. (2018)	Manogaran et al. (2018)	Almeer (2012)
RS data input						
Input Type	NetCDF, HDF5, BUFR, GRIB, BIN	BUFR, Grib	Steam, CSV, HTML	Stream, CSV	Stream, Batch, Database	Images
Sources	16 SS	1 SS	61 MGS	100 MGS	32 MGS	1 SS
Size (Mb)	60.000	120	1	1	–	982
Velocity (Rows/day)	17 million	28.000	50.000	50.000	50 million	200
RS data application						
AP	Yes	Yes	Yes	Yes	No	No
Climate change	Yes	No	No	No	Yes	No
Other	Yes	No	No	No	No	Yes
RS data processing						
Streaming	No	No	Yes	Yes	Yes	No
Batch	Yes	Yes	No	No	Yes	Yes
Single	Yes	Yes	Yes	Yes	No	Yes
Distributed	Yes	No	No	No	Yes	Yes
Technologies/Tools	Hadoop	–	–	Esper CEP, CPN Tools	Hadoop	Hadoop
Languages	Java, Python, BASH	Python	Android Firebase	EPL	MR, SQL, Spark	MR, Java
Benchmarking						
Execution time (Min)	200	1	0.5	–	1	4
Used RAM (Gb)	16	4	3	–	–	20
Used CPU (%)	100	100	–	–	–	100
RS data output						
Output Type	CSV	CSV	Push alerts	Stream	Stream	Image
Size (Mb)	600	–	–	–	–	–
GUI	No	No	Yes	Yes	No	No

References

Akram, M., Amrani, M., El, C.: Air2Day: an air quality monitoring adviser in Morocco. Int. J. Comput. Appl. **181**, 1–6 (2018)

Almeer, M.H.: Cloud Hadoop map reduce for remote sensing image analysis. J. Emerg. Trends Comput. Inf. Sci. **3**, 8 (2012)

Boain', R.J.: A-B-Cs of Sun-Synchronous Orbit Mission Design, p. 20 (2004).

Boudriki Semlali, B.E., El Amrani, C.: Towards remote sensing datasets collection and processing. Int. J. Embed. Real Time Commun. Syst. **10**, 49–67 (2019a)

Boudriki Semlali, B.-E., El Amrani, C.: Towards remote sensing datasets collection and processing. In: Hameurlain, A., Wagner, R., Dang, T.K. (eds.) Transactions on Large-Scale Data- and Knowledge-Centered Systems XLI, pp. 286–294. Springer, Berlin, Heidelberg. https://link.springer.com/10.1007/978-3-030-11196-0_26 (2019b). Accessed 22 Feb 2019

Boudriki Semlali, B., El Amrani, C., Denys, S.: Development of a Java-based application for environmental remote sensing data processing. Int. J. Electr. Comput. Eng. (IJECE) **9**, 1978–1986 (2019c)

Boudriki Semlali, B.E., El Amrani, C., Ortiz, G.: Hadoop paradigm for satellite environmental big data processing. Int. J. Agric. Environ. Inf. Syst. **11**, 24–47 (2020a)

Boudriki Semlali, B.E., El Amrani, C., Ortiz, G.: SAT-ETL-Integrator: an extract-transform-load software for satellite big data ingestion. J. Appl. Remote Sens. (JARS) **14**, 28 (2020b)

Boudriki Semlali, B.-E., El Amrani, C., Ortiz, G.: Adopting the Hadoop Architecture to Process Satellite Pollution Big Data, p. 8 (2020c)

Breame, F.: BUFRextract_User_Guide_v30.pdf (2018)

Díaz, G., Macià, H, Valero, V., Boubeta-Puig, J., Cuartero, F.: An Intelligent Transportation System to control air pollution and road traffic in cities integrating CEP and Colored Petri Nets. Neural Comput. Appl. https://link.springer.com/10.1007/s00521-018-3850-1 (2018). Accessed 14 Mar 2019

El Amrani, C., Rochon, G.L., El-Ghazawi, T., Altay, G., Rachidi, T.: Development of a Real-Time Urban Remote Sensing Initiative in the Mediterranean Region for Early Warning and Mitigation of Disasters, pp. 2782–2785. IEEE. https://ieeexplore.ieee.org/document/6350855/ (2012). Accessed 17 May 2018

EOSDIS: Wikipedia. https://en.wikipedia.org/w/index.php?title=EOSDIS&oldid=808578344 (2017). Accessed 6 Jan 2019

Erraissi, A., Belangour, A., Tragha, A.: Meta-modeling of data sources and ingestion big data layers. SSRN Electron. J. https://www.ssrn.com/abstract=3185342 (2018). Accessed 30 Dec 2018

esa: Overview. European Space Agency. https://www.esa.int/Our_Activities/Observing_the_Earth/Copernicus/Overview3 (2019). Accessed 6 Jan 2019

Garcia-de-Prado, A., Ortiz, G., Boubeta-Puig, J., Corral-Plaza, D.: Air4People: a smart air quality monitoring and context-aware notification system. J. Univers. Comput. Sci. 24, 18 (2018)

Gosink, L., Shalf, J., Stockinger, K., Wu, K., Bethel, W.: HDF5-FastQuery: Accelerating Complex Queries on HDF Datasets using Fast Bitmap Indices, pp. 149–158. IEEE. https://ieeexplore.ieee.org/document/1644309/ (2006). Accessed 7 June 2018

Hu, H., Wen, Y., Chua, T.-S., Li, X.: Toward scalable systems for big data analytics: a technology tutorial. IEEE Access 2, 652–687 (2014)

Jin, H., Lu, X., Liu, H.: Large-scale terrain realistic rendering based on programmable GPU hardware. Geomat. Inf. Sci. Wuhan Univ. (2010)

Karhila, V.: BUFR: A Meteorological Code for the 21st Century, p. 5 (2010)

Lavender, A.: How Many Satellites are Orbiting the Earth in 2018?, p. 3 (2018)

Ma, Y., Wu, H., Wang, L., Huang, B., Ranjan, R., Zomaya, A., et al.: Remote sensing big data computing: challenges and opportunities. Future Gener. Comput. Syst. 51, 47–60 (2015)

Manogaran, G., Lopez, D., Chilamkurti, N.: In-Mapper combiner based MapReduce algorithm for processing of big climate data. Future Gener. Comput. Syst. https://linkinghub.elsevier.com/retrieve/pii/S0167739X17324639 (2018). Accessed 30 May 2018

Navalgund, R.R., Jayaraman, V., Roy, P.S.: Remote sensing applications: an overview. Curr. Sci. 93, 20 (2007)

NOAA National Environmental Satellite, Data, and Information Service (NESDIS): https://www.nesdis.noaa.gov/ (2018). Accessed 6 Jan 2019

Nowak, D.J., Hirabayashi, S., Doyle, M., McGovern, M., Pasher, J.: Air pollution removal by urban forests in Canada and its effect on air quality and human health. Urban For. Urban Green. 29, 40–48 (2018)

Ranjan, R.: Streaming big data processing in datacenter clouds. IEEE Cloud Comput. 1, 78–83 (2014)

Rew, R., Davis, G.: NetCDF: an interface for scientific data access. IEEE Comput. Graphics Appl. 10, 76–82 (1990)

Schmetz, J., Pili, P., Tjemkes, S., Just, D., Kerkmann, J., Rota, S., et al.: An introduction to Meteosat Second Generation (MSG). Bull. Am. Meteorol. Soc. 83, 977–992 (2002)

Siemen, S., Lamy-Thepaut, S., Li, F., Russell, I.: The next generation of ECMWF's meteorological graphics library—Magics++. ECMWF. https://www.ecmwf.int/node/17714 (2007). Accessed 22 Feb 2019

Wang, C., Hu, F., Hu, X., Zhao, S., Wen, W., Yang, C.: A Hadoop-based distributed framework for efficient managing and processing big remote sensing images. ISPRS Ann. Photogramm. Remote Sens. Spat. Inf. Sci. II-4(W2), 63–66 (2015)

Wei, J., Liu, D., Wang, L.: A general metric and parallel framework for adaptive image fusion in clusters: a general metric and parallel framework for adaptive image fusion. Concurr. Comput.: Pract. Exp. 26, 1375–1387 (2014)

Zhu, L., Suomalainen, J., Liu, J., Hyyppä, J., Kaartinen, H., Haggren, H.: A review: remote sensing sensors. In: Rustamov, R.B., Hasanova, S., Zeynalova, M.H. (eds.) Multi-purposeful Application of Geospatial Data. InTech. https://www.intechopen.com/books/multi-purposeful-application-of-geospatial-data/a-review-remote-sensing-sensors (2018). Accessed 15 Dec 2018

A New Approach for Estimating Monthly Global Solar Irradiation Based on Empirical and Artificial Neural Networks Models: A Case Study of Al-Hoceima Province, in Morocco North Region

Badr Benamrou, Mustapha Ouardouz, Imane Allaouzi, and Mohamed Ben ahmed

Abstract

Knowledge of the monthly global solar irradiation is crucial for the planning, design, and dimensioning of solar photovoltaic systems. However, global horizontal solar irradiation is not readily available at all sites. In this regard, the authors proposed a hybrid model that combines an empirical model with an artificial neural network technique to estimate the monthly global solar irradiation for the case study of the city of Al-Hoceima, Morocco. To this end, 17 empirical models are calibrated and evaluated in order to obtain the best empirical model, where the output of this empirical model is combined with one or more parameters such as sunshine duration (S), extraterrestrial radiation (H_o), mean, maximum and minimum temperatures (T_{mean}), (T_{max}), (T_{min}), the ratio (T_{min}/T_{max}), and the difference (T_{max}-T_{min}). These combinations of parameters are used as inputs to different MLP models that were developed and evaluated in order to find the model with the highest overall performance. The simulation results demonstrated that the proposed hybrid approach provides the best results in terms of RMSE, R^2, and MAE and outperforms MLP or empirical models used individually. Also, the results indicated that the proposed model gives higher performance than five machine learning algorithms including SVR, Random Forest, Xgboost, decision tree, and K-nearest neighbors.

Keywords

Solar energy • Monthly global solar irradiation • Empirical models • Machine learning

1 Introduction

Knowledge of global solar radiation is essential for the study, planning, and design of solar energy systems. Likewise, global solar radiation is a key parameter for estimating the electrical energy produced by a solar photovoltaic (PV) system.

Estimating the monthly solar radiation in a given area allows the design and size of PV systems to estimate their production and conduct economic feasibility studies for the mentioned systems. The stations dedicated for continuous recording of solar radiation are extremely rare, which makes measurements of this parameter not available at all sites and difficult to acquire due to the high cost of measuring instruments and equipment (pyranometer). Indeed, it is indispensable to develop models that predict global solar radiation based on the most readily available meteorological data.

There are different methods for estimating global monthly solar radiation, such as physical models based on physical laws that describe the interactions between solar radiation and atmospheric components. Empirical models based only on mathematical formulas and do not use any type of physical laws to relate solar radiation with available weather variables. And machine learning methods that model complex non-linear regression between solar radiation and meteorological parameters.

In this work, we calibrated and evaluated 17 empirical models based on accessible meteorological data such as duration of sunshine (SD), average air temperature (T_{avg}), maximum air temperature (T_{max}), minimum air temperature (T_{min}), and extraterrestrial solar radiation (H_o). Thereafter, the most precise empirical model among the 17 calibrated models

B. Benamrou (✉) · M. Ouardouz
Department of Mechanical Engineering, Research Group: "Mathematical Modelling and Control (MMC), Faculty of Sciences and Techniques, University Abdelmalek Essaadi, Route Boukhalef, BP 416, Tangier, Morocco
e-mail: badr.benamrou@gmail.com

I. Allaouzi · M. Ben ahmed
Department of Computer Science, LIST Laboratory, Faculty of Sciences and Techniques, University Abdelmalek Essaadi, Route Boukhalef, BP 416, Tangier, Morocco

is combined with the MLP Feed-Forward neural network in order to improve the performance of the monthly global solar irradiation estimation for Al-Hoceima city, Morocco.

2 Background

2.1 Review of Empirical Models

The use of empirical models is more widespread due to their greater operability. These models are constructed using mathematical expressions that relate solar radiation (dependent variable) to different meteorological variables. The coefficients of these models are obtained by regression adjustments using statistical techniques.

According to scientific literature, several studies have used experimental models to estimate global solar radiation, the most important of which use available meteorological data such as.

2.2 Empirical Models Based on Air Temperature

Models that use temperature to estimate solar radiation are widely used due to their easy availability of data and their simple and robust measurement. These models use as inputs the average air temperature (T_{avg}), maximum air temperature (T_{max}), and minimum air temperature (T_{min}).

Hargreaves and Samani (1982) proposed a mathematical model that employs as inputs the difference between T_{min} and T_{max} in order to estimate daily GHI. From this model, many methodologies have been developed such as Bristow and Campbell (1984), Chen et al. (2004), and Jahani et al. (2017).

In other studies, the authors proposed empirical models that employed as inputs the ratio of T_{min} to T_{max} to estimate monthly clearness index, such as (Pandey and Katiyar 2010, Okonkwo and Nwokoye 2014). While in (Falayi et al. 2008), the authors have proposed mathematical models which correlate the clarity index with the mean temperature and the ratio of T_{min} to T_{max}.

2.3 Empirical Models Based on Sunshine Duration

Sunshine duration is considered one of the most important parameters for estimating global horizontal solar radiation. Among the most significant sunshine, duration-based models is that of Ångström–Prescott (1924), which estimates global solar radiation from a simple linear regression equation.

Over the years, several studies have modified the Angstrom models by developing non-linear regression models

based only on sunshine duration including quadratic, cubic, exponential, logarithmic, and power (Ogelman et al. 1984; Bahel et al. 1987; Elagib and Mansell 2000; Ampratwum and Dorvlo 1999; Tougrul and Tougrul 2002).

In Almorox and Hontoria (2004), the authors compared quadratic, cubic logarithmic, and exponential functions in order to estimate global solar radiation from sunshine duration for 16 meteorological stations in Spain, the results demonstrated that the third-degree model outperformed other models. In Betti et al. (2020), the authors compared 17 models based on sunshine duration using data collected from 7 meteorological stations in Croatia. The results illustrated that the linear model is the most efficient. In Fan et al. (2019), the authors employed 12 empirical models and 12 machine learning models for estimating GHI in different climatic zones of China. The results revealed that the machine learning methods outperform generally the empirical models. However, almost nine empirical models have given good results in terms of prediction accuracy.

2.4 Machine Learning Models

Machine learning is a subdivision of artificial intelligence, the main characteristic of which is the ability to "learn" through algorithms that analyze data and generate predictions. In recent years, several studies in the literature have shown that machine learning techniques estimate global solar radiation with great accuracy in different places of the world.

In Alsina et al. (2016), the authors used Artificial Neural Network (ANN) to estimate monthly average daily global solar radiation in different locations in Italy. This model was trained in a dataset of 17 locations and tested on a subset of 28 locations. In Çelik et al. (2016), the authors developed an optimized ANN model to estimate monthly global solar radiation with high accuracy for the Eastern Mediterranean Region of Turkey. In Hejase et al. (2014), an optimal MLP technique is proposed to estimate monthly GHI for three cities of the United Arab Emirates (UAE). This model used as inputs four meteorological parameters collected from the UAE Solar Atlas. In Belaid and Mellit (2016), the authors applied a support vector machine (SVM) for the prediction of daily and monthly GHI in Ghardaia, Algeria. The results showed that the SVM method outperformed ANN techniques. In (Shamshirband et al. 2015), the authors carried out a comparative study between Extreme Learning Machine (ELM), Genetic Programming (GP), SVM, and ANN in order to find the best model that estimates monthly GHI for the city of Shiraz, Iran. The results showed that the ELM model based on several parameters is more precise than SVM, GP, and ANN.

3 Materials and Methods

3.1 Data Description

In this paper, monthly global horizontal solar radiation (Mj/m^2), air maximum and minimum temperatures (°C), mean temperature (°C), and sunshine duration (hours) were collected from the meteorological station of Al-Hoceima between 2011 and 2017. Al-Hoceima is a city located in the north of Morocco with a Mediterranean climate where the winters are mild and moderately rainy, while summers are hot and sunny.

3.2 Empirical Models

There are many empirical models in the literature that have been developed to estimate the monthly global horizontal radiation. In this work, we have developed 17 empirical models based on the sunshine duration, the minimum and maximum air temperatures, and the average temperature. These models were selected based on the availability of their input parameters and their applicability in different climates and zones.

- **Model 1: Angstrom–Prescott Model (linear model)**

Angstrom–Prescot made the first contribution to estimate monthly or daily global solar radiation by proposing a simple linear model that relates the global solar radiation to the sunshine duration using the following formula:

$$\frac{H}{Ho} = a + b \times \frac{S}{So} \quad (1)$$

where:

    a and b are the regression coefficients.
    H: monthly average daily global solar radiation.
    Ho: extraterrestrial radiation.
    S: Measured sunshine.
    So: monthly average day length.

- **Model 2: Ögelman, Ecevit and Tasdemiroglu**

On the basis of Angstrom model, (Ogelman et al. 1984) proposed a second-order function that relates the monthly average daily global solar radiation and the sunshine duration as shown by the following equation:

$$\frac{H}{Ho} = a + b \times \frac{S}{So} + c \times \left(\frac{S}{So}\right)^2 \quad (2)$$

- **Model 3: Bahel model**

In Bahel et al. (1987), the authors established a cubic correlation of 48 meteorological stations around the world that relates the monthly average daily global solar radiation to the sunshine duration by using the following formula:

$$\frac{H}{Ho} = a + b \times \frac{S}{So} + c \times \left(\frac{S}{So}\right)^2 + d \times \left(\frac{S}{So}\right)^3 \quad (3)$$

- **Model 4 and 5: Almorox and Hantoria model**

In Almorox and Hontoria (2004), the authors proposed exponential and logarithmic models to estimate monthly global horizontal irradiation in different zones of Spain. These models are given by the following formulas:

$$\frac{H}{Ho} = a \times exp(b \times \frac{S}{So}) \quad (4)$$

$$\frac{H}{Ho} = a + b \times \log(\frac{S}{So}) \quad (5)$$

- **Model 6: Torgul model**

An exponential correlation was proposed by Tougrul and Tougrul (2002), and was tested in 16 sites of Sudan:

$$\frac{H}{Ho} = a \times b^{S/So} \quad (6)$$

- **Model 7: Hargreaves model**

In Hargreaves and Samani (1982), the authors suggested a simple model where the relationship between the monthly average daily solar radiation and the difference between the maximum and minimum air temperatures is given as follows:

$$\frac{H}{Ho} = a \times (T_{max} - T_{min})^{0,5} \quad (7)$$

- **Model 8 and 9: Chen model**

In Chen et al. (2004), the authors proposed the following equations to estimate the monthly global solar radiation:

$$\frac{H}{Ho} = a \times [ln(T_{max} - T_{min})] + b \quad (8)$$

$$\frac{H}{Ho} = a + b \times (S/So)^c + d \times ln(T_{max} - T_{min}) \quad (9)$$

- **Model 10 and 11: Pandey and Katiyar model**

In Pandey and Katiyar (2010), the authors suggested a second and third-order function to estimate global solar radiation based on the ratio of maximum temperature T_{max} to minimum temperature T_{min}. The model is presented in the following formulas:

$$\frac{H}{Ho} = a + b \times \frac{T_{max}}{T_{min}} + c \times \left(\frac{T_{max}}{T_{min}}\right)^2 \quad (10)$$

$$\frac{H}{Ho} = a + b \times \frac{T_{max}}{T_{min}} + c \times \left(\frac{T_{max}}{T_{min}}\right)^2 + d \times \left(\frac{T_{max}}{T_{min}}\right)^3 \quad (11)$$

- **Model 12 and 13: Jahani model**

In Jahani and Dinpashoh (2017), the authors proposed a polynomial correlation between global solar radiation and the difference between the maximum and minimum temperatures using the following equations:

$$\frac{H}{Ho} = a + b \times (T_{max} - T_{min}) + c \times (T_{max} - T_{min})^2 + d \times (T_{max} - T_{min})^3 \quad (12)$$

$$\frac{H}{Ho} = a + b \times (T_{max} - T_{min})^{0,5} + c \times (T_{max} - T_{min})^{1,5} + d \times (T_{max} - T_{min})^{2,5} \quad (13)$$

- **Model 14: Olomiyesan model**

A multi-linear two parameters model was developed by Olomiyesan et al. (2014), it relates global solar radiation to S/So and the difference between the maximum and minimum temperatures:

$$\frac{H}{Ho} = a + b \times \left(\frac{S}{So}\right)^{0,5} + c \times \left(\frac{T_{max} - T_{min}}{S}\right) \quad (14)$$

- **Model 15: Hassan model**

In Hassan et al. (2016), the authors proposed a model to estimate monthly global solar radiation based on means temperature, maximum and minimum temperatures using the following equation:

$$\frac{H}{Ho} = (a + b \times T_{mean}) \times (T_{max} - T_{min})^c \quad (15)$$

- **Model 16: Fourier model**

In the Fourier model, the global solar radiation is related to sunshine duration using Fourier series:

$$\frac{H}{Ho} = a \times cos\left(\frac{S}{So} + b\right) + c \times cos\left(2 \times + \left(\frac{S}{So}\right)d\right) + e \quad (16)$$

- **Model 17: Weibull model**

A model based on the Weibull equation was proposed by Achour et al. (2017), it correlates H/Ho and S/So using the following formula:

$$\frac{H}{Ho} = a + (b - a) \times \left(exp\left(-c \times \left(\frac{S}{So}\right)^d\right)\right) \quad (17)$$

- **Extra-terrestrial radiation**

The monthly average daily extraterrestrial radiation Ho is computed by the formula given below (Krenker et al. 2011):

$$Ho = \frac{86400}{\pi} Isc\left(1 + 0,033cos\left(2\pi.\frac{n}{365}\right)\right) \\ (cos\varphi.\cos\delta.\cos\omega s + \xi s.\sin\varphi.\sin\delta) \quad (18)$$

$$\delta = 23,45.\sin\left(2\pi.\frac{284 + n}{365}\right) \quad (19)$$

$$\omega s = \arccos\left[-\tan(fi)\tan(\delta)\right] \quad (20)$$

$$So = 2.\omega s/15 \quad (21)$$

where
   Isc = 1367 w/m^2 is the solar constant.
   n is the day of the year starting from 1 January.
   φ is the latitude of location.
   δ is the declination angle.
   ωs is the sunset hour angle.
   So is the maximum possible sunshine duration.

3.3 Artificial Neural Network

An MLP model (Duffie and Beckman 1991) is made up of a layer of N input neurons, a layer of M output neurons, and one or more hidden layers. In this type of architecture, the connections between neurons are always forward, that is, the

Table 1 Input combinations to ANN models

Models	Inputs	Models	Inputs
M15-MLP1	$H_{est-emp}$	MLP10	S/So, Ho
M15-MLP2	$H_{est-emp}$, S	MLP11	S/So, $T_{max}-T_{min}$, Ho
M15-MLP3	$H_{est-emp}$, S, T_{mean}	MLP12	$T_{max}-T_{min}$, Ho
M15-MLP4	$H_{est-emp}$, S, T_{min}, T_{max}	MLP13	S/So, S, $T_{max}-T_{min}$, Ho
M15-MLP5	$H_{est-emp}$, Ho, S, T_{min}, T_{max}	MLP14	T_{max}/T_{min}, Ho
M15-MLP6	$H_{est-emp}$, S/So	MLP15	S/So, T_{max}/T_{min}, Ho
M15-MLP7	$H_{est-emp}$, T_{max}/T_{min}	MLP16	T_{max}, T_{min}, S, Ho
M15-MLP8	$H_{est-emp}$, S/So, $T_{max}-T_{min}$	MLP17	T_{mean}, T_{max}, T_{min}, Ho
M15-MLP9	$H_{est-emp}$, S/So, $T_{max}-T_{min}$, T_{max}/T_{min}	MLP18	S/So, $T_{max}-T_{min}$, T_{max}/T_{min}

outputs of a certain layer are used as inputs for the neurons of the next layer.

The output, $z_i^{(k+1)}$, of a neuron i in layer k is defined by

$$z_i^{(k+1)} = f_{k+1}\left(\sum_{j-1}^{D(k)}\left(\sum_{j=1}^{D(K)}\left(w_{j,i}^{(k)}z_j^{(k)}\right)+w_{0.i}^{(k)}\right)\right) \quad (22)$$

where

f_{k+1}: is the activation function of layer $k + 1$.

$D(k)$: is the number of neurons in layer k.

$w_{0.i}^{(k)}$ is a bias parameter.

$w_{j,i}^{(k)}$ is a weight parameter.

The authors got inspired by empirical models to create different combinations of parameters that will be used as inputs for ANN models in order to estimate the monthly GHI with the highest accuracy.

The combinations of inputs as shown in Table 1 consist of the output ($H_{est-M15}$) of the best empirical model and other meteorological parameters that were used as inputs for the 17 empirical models such as T_{min}, T_{max}, T_{mean}, S, Ho, T_{max}/T_{min}, $T_{max}-T_{min}$, and the S/So ratio. In total, 18 models were created, as shown in Table 1, the hybrid models that combine the best empirical model with MLP technique were named M15–MLP1 to M15–MLP9, while the stand-alone MLP models were named MLP10–MLP18.

4 Evaluation Metrics

To evaluate the performance of the estimation models, the authors employed the root mean squared error (RMSE), the Mean Absolute Error (MAE), and the coefficient of determination (R^2). They are defined as follows:

- **Mean Absolute Error:**

$$MAE = \frac{1}{N} \times \sum_{i=1}^{N} |(\hat{y}(i) - y(i))|$$

- **Root mean square error:**

$$RMSE = \sqrt{\frac{1}{N} \times \sum_{i=1}^{N} (\hat{y}(i) - y(i))^2}$$

- **Coefficient of determination:**

$$R^2 = 1 - \frac{\sum_{i=1}^{N} (y(i) - \hat{y}(i))^2}{\sum_{i=1}^{N} \left(y(i) - \overline{\hat{y}(i)}\right)^2}$$

where ŷ is the predicted value, y is the actual value.

5 The Proposed Methodology

Figure 1 illustrates the different steps of the proposed methodology. As can be seen, the dataset was collected from the meteorological station during the period 2011–2017. This dataset was divided into two groups: the first group of data measured during the period 2011–2016 was used to calibrate the 17 selected empirical models, while the data measured during 2016–2017 was used for the testing process in order to obtain the empirical model with the best results.

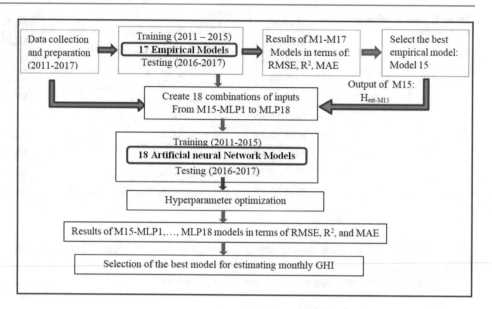

Fig. 1 The proposed methodology

Then the best empirical model is used to estimate the global solar radiation ($H_{est-M15}$), which will be combined with parameters including Ho, S/So, S, T_{min}, T_{max}, T_{max}-T_{min}, and T_{max}/T_{min} to create different sets of inputs for ANN. In total, 18 models were established and divided into two groups: the first group consists of hybrid models that combine the empirical model M15 with the MLP technique, using 9 different input combinations. While the second group consists of stand-alone MLP models using eight different input combinations.

Then, each model was trained using Adam (Diederik and Jimmy Ba 2014) algorithm, thereafter, the hyper-parameter tuning is used to test different architectures in order to build the optimal architecture for each model. Finally, among the optimized MLP models, we have selected the most accurate model which gives the best results in terms of RMSE, R^2, and MAE.

6 Simulation Results and Discussion:

6.1 Regression Coefficients and Performance of Empirical Models

The dataset measured during 2010–2015 were used to calibrate the empirical models in order to obtain the regression coefficients (a,b,c,d) of each model. These coefficients were fitted using Minitab 19 software. The period between 2016 and 2017 was used to validate the empirical models on the basis of statistical errors such as RMSE, R^2, and MAE.

The calibrated coefficients and the performances of each model are illustrated in Table 2. As can be seen, the RMSE is ranged from 1.39 Mj/m^2 to 2.40 Mj/m^2, R^2 from 0.68 to 0.93,

and MAE from 1.10 Mj/m^2 to 1.91 Mj/m^2. It is clear from the table that the calibrated model which presents the best correlation with 0.93 and the lowest RMSE and MAE with 1.39 Mj/m^2 and 1.10 Mj/m^2, respectively is the model M15 followed by the model M14 with R^2 of 0.89. While the model M11 presents the lowest correlation with 0.68, the highest errors with RMSE of 2.40 Mj/m^2, and MAE of 1.91 Mj/m^2.

The best empirical model 'M15' obtained in this study is based only on the following readily available data including maximum temperature, minimum temperature, and average temperature.

6.2 Simulation Results of Proposed Models

In this case, 18 models were created. Each model consists of a hybrid model that combines M15 with an MLP model and uses as input a combination of the (H_{est}-M15) global solar radiation estimated by the empirical model M15 and one or more of the following parameters Ho, S/So, S, T_{min}, T_{max}, T_{max}−T_{min}, T_{max}/T_{min}. A total of 18 artificial neural networks models were investigated to build the best hybrid model.

All the 18 MLP models have been optimized by adjusting hyper-parameters such as the number of neurons, the optimizer, the number of hidden layers, epochs, and the activation function. The performances of optimized models are presented in Table 3, as can be seen, all the models yield satisfactory results with R^2 varying between 0.9 and 0.95. However, it should be noted that the results of proposed hybrid models (from M15–MLP1 to M15–MLP9) outperform and are more accurate than MLP or empirical models used individually. This indicates that combining the best empirical model M15 with MLP technique has a positive

Table 2 Regression coefficients and performances of empirical models

Model	a	b	c	d	RMSE	R2	MAE
M1	0 0.184	0.488	–	–	1.90	0.83	1.56
M2	−0.045	1.224	−0.578	–	1.88	0.83	1.54
M3	0.877	−3.25	6.524	−3.686	1.80	0.85	1.50
M4	0.267	0.96	–	–	1.93	0.82	1.59
M5	0.65	0.62	–	–	1.87	0.83	1.54
M6	0.63	0.305	–	–	1.84	0.84	1.52
M7	0.169	–	–	–	2.18	0.73	1.78
M8	0.34	0.077	–	–	2.15	0.76	1.74
M9	0.98	−0.48	0.11	–	2.02	0.83	1.66
M10	2.38	−2.6	1.157	−0.164	2.06	0.82	1.68
M11	4.76	−1.34	0.14	0.0047	2.40	0.68	1.91
M12	14.4	−8.44	0.58	−0.017	2.29	0.72	1.83
M13	0.33	0.52	−0.077	–	1.93	0.83	1.57
M14	0 0.064	0.63	−0.058	–	1.65	0.89	1.33
M15	**0.11**	**0.003**	**0.49**	**–**	**1.39**	**0.93**	**1.14**
M16	3.34	0.021	−1.33	−1.78	1.79	0.85	1.48
M17	0.57	0 0.35	−8.78	−4.55	1.78	0.85	1.48

Table 3 Performance of hybrid models and MLP models

Model	RSME	R^2	MAE
M15-MLP1	1.346	0.948	1.077
M15-MLP2	1.339	0.949	1.069
M15-MLP3	1.343	0.949	1.069
M15-MLP4	1.351	0.948	1.074
M15-MLP5	1.418	0.94	1.150
M15-MLP6	1.346	0.948	1.072
M15-MLP7	1.373	0.945	1.124
M15-MLP8	1.344	0.948	1.074
M15-MLP9	**1.339**	**0.95**	**1.072**
MLP10	1.793	0.91	1.527
MLP11	1.827	0.906	1.546
MLP12	1.793	0.908	1.529
MLP13	1.791	0.91	1.523
MLP14	1.673	0.921	1.459
MLP15	1.784	0.91	1.520
MLP16	1.785	0.91	1.521
MLP17	1.839	0.90	1.556
MLP18	1.786	0.91	1.522

effect in increasing the prediction accuracy of monthly global solar irradiation (R^2 ranging from 0.94 to 0.95 and RMSE from 371.9 to 393.78).

The model M15-MLP9 (see, Fig. 2 which employed input combinations of $H_{est-M15}$, S/So, T_{max}-T_{min}, and T_{max}/ T_{min} categorized as the most accurate model with the highest correlation of 0.95, the lowest RMSE of 1.346 MJ/m2, and the lowest MAE of 1.072 MJ/m2 (see Table3.

Figure 3 illustrates the comparison between the monthly global solar irradiance measured and estimated during the

Fig. 2 Architecture of the best-proposed model: M15-MLP9

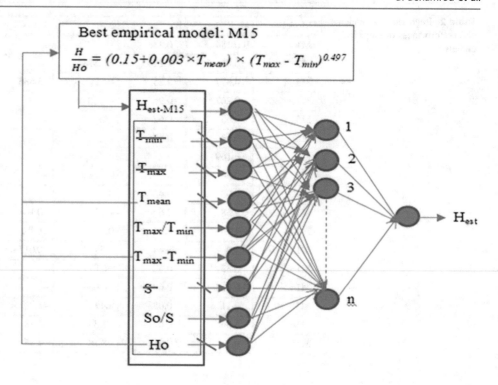

Fig. 3 Measured and estimated monthly solar radiation by the best-proposed model (M15-MLP9) on the training and testing data

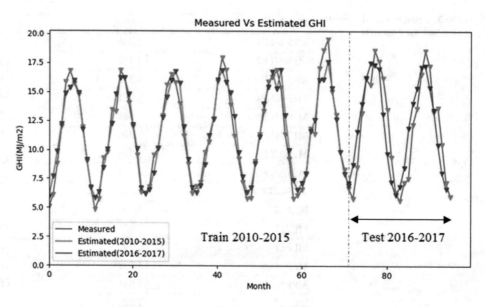

training and testing process. It is clear from this graph that there is a good correlation between the estimated and measured values in most months and during the two phases of training and testing.

6.3 Comparison with Other Machine Learning Technique

In this part, we compared our proposed models with the best-performing Machine Learning (ML) methods used to tackle the aforementioned tasks. For this end, we selected five machine learning approaches including Support Vector Regression (SVR) Smola and Scholkopf (2004), Random Forest (RF) Breiman (2001), Decision Tree (Ahmad et al. 2017), Xgboost (Chen and Guestrin 2016), and K-nearest Neighbors (Yang et al. 2015).

As shown in Table 4, the five machine learning techniques provide satisfactory results. However, the comparative study proved that the proposed model M15–MLP9 outperformed SVR, RF, Xgboost, Decision-tree, and K-nearest Neighbors in terms of RMSE, R^2, and MAE.

Table 4 Comparison of performance of the proposed model with the best empirical model and other machine learning methods

Models	RSME	R^2	MAE
SVR	2,15	0,84	1,64
RF	1,38	0,94	1,18
Xgboost	1,89	0,9	1,56
Decision-tree	1,79	0,91	1,45
k-nearest Neighbors	1,53	0,93	1,30
The proposed Model M15-MLP9	**1,33**	**0,95**	**1,072**
Best empirical M15	1,41	0,93	1,19

7 Conclusion

As previously mentioned the monthly global solar radiation is an essential parameter for the planning and dimensioning of photovoltaic systems. In this article, the authors have proposed an accurate model to estimate the monthly global solar irradiation for the city of Al-Hoceima, Morocco. To this end, 17 empirical models were established to obtain regression coefficients by curve fitting. Then the calibrated models were tested to find the best empirical model. The output of this empirical model has been combined with other parameters such as T_{min}, T_{max}, T_{mean}, S, Ho, T_{max}/T_{min}, T_{max}-T_{min}, and S/So. A total of 18 combinations of inputs were used to create nine hybrid models and nine MLP neural network models. These models were trained during the 2010–2015 period and were tested during the 2016–2017 period using the following evaluation metrics R^2, RMSE, and MAE.

The obtained results showed that the hybrid models, which combine MLP with the empirical model, outperform the individual models. This proves the validity of the basic hypothesis which assumes that the combination of the empirical model with the MLP model will improve the prediction accuracy. In addition, the results showed that the proposed model M15–MLP9 outperformed other machine learning methods include SVR, RF, decision tree, Xgboost, and K-nearest neighbors.

By analyzing the simulation results obtained in this paper, the authors concluded that calibrated 17 empirical models, machine learning methods, and all proposed models give acceptable results. However, the most accurate model for estimating monthly global solar irradiation for Al-Hoceima city, Morocco, is the hybrid model M15–MLP9 that uses as inputs $H_{est-M15}$ estimated by the best empirical model as well as S/So, Tmax-Tmin, and Tmax/Tmin.

Acknowledgments This work has been supported by MESRSFC and CNRST under the project PPR2-OGI-Env, reference PPR2/2016/79.

References

Achour, L., Bouharkat, M., Assas, O. et al.: Smart model for accurate estimation of solar radiation. Front. Energy (2017)

Ahmad, M.W., Mourshed, M. Rezgui, Y.: Trees vs. Neurons: comparison between random forest and ANN for high-resolution prediction of building energy consumption. Energy Build **147** (Supplement C), 77–89 (2017)

Almorox, J., Hontoria, C.: Global solar radiation estimation using sunshine duration in Spain. Energy Convers. Manage. **45**(9–10), 1529–1535 (2004)

Alsina, E., Bortolini, M., Gamberi, M., Regattieri, A.: Artifcial neural network optimisation for monthly average daily global solar radiation prediction. Energy Convers. Manage. **120**, 320–329 (2016)

Ampratwum, D.B., Dorvlo, A.S.S.: Estimation of solar radiation from the number of sunshine hours. Appl Energy **63**, 161–167 (1999)

Ångström-Prescott, A.: Solar and terrestrial radiation. Q. J. R. Meteorol. Soc. **50**(210), 121–125 (1924)

Bahel, V., Srinivasan, R., Bakhsh, H.: Solar radiation for Dhahran, Saudi Arabia. Energy **11**(10), 985–989 (1987)

Belaid, S., Mellit, A.: Prediction of daily and mean monthly global solar radiation using support vector machine in an arid climate. Energy Convers. Manag. **118**, 105–118 (2016)

Betti, T., Zulim, I., Brkić, S., Tuka, B.: A comparison of models for estimating solar radiation from sunshine duration in croatia. Int. J. Photoenergy 14 (2020)

Breiman, L.: Random forests. Mach. Learn. **45**, 5–32 (2001)

Bristow, K.L., Campbell, G.S.: On the relationship between incoming solar radiation and daily maximum and minimum temperature. Agric. For. Meteorol. **31**, 159–166 (1984)

Çelik, Ö., Teke, A., Yildirim, H.: The optimized artifcial neural network model with Levenberg-Marquardt algorithm for global solar radiation estimation in Eastern Mediterranean Region of Turkey. J. Clean. Prod. **116**, 1–12 (2016)

Chen, T., Guestrin, C.: XGBoost: a scalable tree boosting system. In: Proceedings of the 22nd ACM SIGKDD International Conference on Knowledge Discovery and Data Mining. KDD '16. ACM, New York, NY, USA. pp 785–794 (2016)

Chen, R.S., Ersi, K., Yang, J.P., Lu, S.H., Zhao, W.Z.: Validation of five global radiation models with measured daily data in China. Energy Convers. Manage. **45**, 1759–1769 (2004)

Diederik, K., Jimmy Ba, A.: A method for stochastic optimization. (2014). arXiv:1412.6980

Duffie, J.A., Beckman, W.A.: Solar engineering of thermal processes. Wiley, New York (1991)

Elagib, N.A., Mansell, M.G.: New approaches for estimating global solar radiation across Sudan. Energy Convers. Manag. **41**, 419–434 (2000)

Falayi, E.O., Adepitan, J.O., Rabiu, A.B.: Empirical models for the correlation of global solar radiation with meteorological data for Iseyin Nigeria. Int. J. Phys. Sci. **3**, 210–216 (2008)

Fan, J., Wu, L., Zhang, F., Cai, H., Zeng, W., Wang, X., Zou, H.: Empirical and machine learning models for predicting daily global solar radiation from sunshine duration: a review and case study in China. Renew. Sustain. Energy Rev. **100**, 186–212 (2019)

Hargreaves, G.H., Samani, Z.A.: Estimating potential evapotranspiration. J. Irrig. Drainage Eng. **108**(IR3), 223–230 (1982)

Hassan, G.E., Youssef, M.E., Mohamed, Z.E., Ali, M.A., Hanafy, A. A.: New temperature based models for predicting global solar radiation. Appl Energy **179**, 437–450 (2016)

Hejase, H., Al-Shamisi, M.H., Assi, A.: Modeling of global horizontal irradiance in the United Arab Emirates with artifcial neural networks. Energy **77**, 542–552 (2014)

Jahani, B., Dinpashoh, Y.: Nafch, AR evaluation and development of empirical models for estimating daily solar radiation. Renew. Sustain. Energy Rev. **73**, 878–891 (2017)

Krenker, A., Bešter, J., Kos, A.: Introduction to the artificial neural networks, artificial neural networks. Methodological Advances and Biomedical Applications. Kenji Suzuki, IntechOpen. (2011).https://doi.org/10.5772/15751

Ogelman, H., Ecevit, A., Tasdemiroglu, E.: A new method for estimating solar radiation from bright sunshine data. Solar Energy **33**, 619–25 (1984)

Okonkwo, G.N., Nwokoye, A.O.C.: Estimating global solar radiation from temperature data in minna location. Eur. Sci. J. **10**, 1857–7431 (2014)

Olomiyesan, B.M., Oyedum, O.D., Ugwuoke, P.E., Ezenwora, J.A.: Abdullahi, estimation of mean monthly global solar radiation for minna using sunshine hours. J. Sci. Technol. Math. Educ. **10**(3), 15–22 (2014)

Pandey, C.K., Katiyar, A.K.: Temperature base correlation for the estimation of global solar radiation on horizontal surface. Int. J. Energy Environ. **1**(4), 737–744 (2010)

Shamshirband, S., Mohammadi, K., Yee, L., Petković, D., Mostafaeipour, A.: A comparative evaluation for identifying the suitability of extreme learning machine to predict horizontal global solar radiation. Renew. Sustain. Energy Rev. **52**, 1031–1042 (2015)

Smola, A.J., Scholkopf, B.: A tutorial on support vector regression. Stat. Comput. **14**, 199–222 (2004)

Tougrul, I.T., Tougrul, H.: Global solar radiation over Turkey: comparison of predicted and measured data. Renew. Energy **25**, 55–56 (2002)

Yang, D., Sharmaa, V., Ye, Z., Lim, L.I., Zhao, L., Aryaputera, A.W.: Forecasting of global horizontal irradiance by exponential smoothing using decompositions. Energy **81**, 111–119 (2015)

Computing Technologies for Sustainable Development

Efficient Congestion Management for Sustainable Wireless Mesh Networks

Kaoutar Bazi and Bouchaib Nassereddine

Abstract

Sustainable development is a lever of innovation allowing rethinking models. It must be equitable, effective, and tolerable. In the world of IT networks, sustainability is to make smarter decisions in order to provide the resources necessary to meet user needs, to manage huge quantity of data, and to improve the methods and models available in order to guarantee better services in the future. As we all know, when we have transport entities on many machines, and that send too many packets too quickly in the network, it will be congested and suffer performance losses as the packets are lost and/or delayed. As part of sustainable development, this congestion must be managed. It is the responsibility of the transport layer, which must slow down its rate of sending packets to the network layer, by regulating its rate of sending while ensuring good management of bandwidth. The transport layer must also deploy reliable and robust congestion control mechanisms, ensuring good management of data exchanges between the different entities sharing the network, without high delay, loss, or alteration. These mechanisms must also ensure fair, just and optimal management of the network bandwidth without being congested. This paper deals with the congestion problem of Wireless Mesh Networks, then it focuses on the TCP protocol as a reliable transport protocol implementing a panoply of congestion control mechanisms (Tahoe, Reno, New Reno, Vegas, Cubic, Sack, Fack, Scalable …), to then compare their behavior in congested networks, thanks to simulations carried out with the NS2 network simulator. Toward the end, conclusions are drawn, making it possible to distinguish the mechanisms which are delay-based (requiring a data transfer without loss or alteration) like TCP Vegas, from those which are loss-based (do not support a long transmission delay) such as Tahoe, Reno, New Reno, Fack, and Scalable, in order to conclude which one is suitable for which type of network. And this is based on the three metrics throughput, drop delay and latency.

Keywords

Mesh • IEEE 802.11s • TCP • Congestion control • ns2 • Tahoe • Reno • New reno • Vegas • Fack • Sack • Cubic • Scalable

1 Introduction

Wireless mesh networks (WMNs) (Akyildiz et al. 2005; Camp and Knightly 2008) are an optimal solution for providing a communication backbone to a large number of users with very high speed, WMNs networks are ultra-fast, provide an optimal signal and require no cables. Ordinary networks are unable to meet user requirements. There are some who want to watch streaming videos, another who plays online, a third who starts a video meeting from his computer, others who want to surf online and follow the news… So the network, saturated by all these requests, is unable to serve all users, the link becomes slow with interruptions. Hence, the need for an optimal solution providing high speed with large bandwidth to a very large number of users, a mesh network is made up of a number of network devices, connected to each other to collaborate and distribute the load over large wireless areas network. This means that a mesh network is made up of several wireless transmitters ensuring optimal coverage. Such a wireless transmitter or node is also called mesh station; it can establish links between them using a routing protocol. The interconnection and relaying of the frames are ensured by nodes called mesh points, which can also be Mesh Access

K. Bazi (✉) · B. Nassereddine
Faculty of Sciences and Technologies,
Hassan First University of Settat, Settat, Morocco
e-mail: k.bazi@uhp.ac.ma

B. Nassereddine
e-mail: nassereddine_bouchaib@yahoo.com

Point when they act as an access point. A node can also combine another functionality and link the mesh network to another mesh of a different nature (Ethernet, IEEE 802.3…), it is called in this case Mesh Portal Point. The nodes of a mesh network work together to mesh and distribute the load over large areas in order to ensure optimal coverage.

Mesh networks are perfect for places without an Ethernet cable and where the Wifi signal is not very strong. Failure problems are rare since when one transmitter is deactivated, another takes over, the system automatically recognizes the most appropriate node, and establishes the link without the user noticing. As a result, the network maintains optimal continuity without loss of speed and without signal interruption. A mesh wireless network has other advantages such as integrated protection against cybercrime, adjusting bandwidth in the presence of a flow requiring a large part than others, the installation of filters in order to secure navigation…

WMNs 802.11s (Chan and Ramjee 2005; Wang and Lim 2008) inherit 802.11 WLANs (Wireless Local Area Networks) with the coverage of a large geographical area, and this is thanks to a multi-hop service where the relaying of a data flow is done through several intermediate nodes before reaching the destination. However, there are no problems with neither energy consumption nor mobility as is the case in MANETs (Mobile Ad hoc Networks). It should be noted that (roughly) the main objective of WMNs is to face the problem of mobility, as well as to improve the network capacity in terms of bandwidth and flow, instead of minimizing the use of energy.

The adoption of WMNs makes it possible to improve several network properties, among others we quote:

- Throughput: To serve a large number of users and pass a very large amount of data, the Wireless Mesh Networks must have the capacity to support a large load.
- The delay: WMNs must also reduce the waiting time by avoiding overloading the buffers of the routers, since the delay is critical for most of the applications currently used.
- The fairness: Another requirement is the fairness; WMNs must ensure equitable management of competing flows crossing a bandwidth, especially in the presence of greedy flows which devour the bandwidth shared with other flows.

In addition to the endless user demands, WMNs face another problem which is network congestion. Indeed, the continuity of the services required by the sources, which flood the network with large amounts of data, causes the saturation of the network buffers and subsequently a delay and even a loss of packets. Therefore, in this way, the network loses its performance. To deal with the problem of network congestion, various works have been developed in this direction; among others, we find the management of network congestion for the TCP protocol.

2 The Protocol TCP

TCP (Transmission Control Protocol) is a transport protocol for the transport layer of the TCP/IP stack. It is designed to reliably transport data from end to end without loss, high transmission delay or alteration, by adapting dynamically to the different variations that a network can experience. It has a set of congestion control mechanisms. As we all know, when the load injected into the network exceeds its carrying capacity, there is a phenomenon of congestion. The network layer detects congestion when queues grow at routers. So, it informs the transport layer which encourages it to slow down the traffic it injects into the network. TCP plays a major role in congestion control as well as in reliable transport. During its implementation, several congestion control mechanisms (Verma et al. 2019) were brought into play at the transport layer; here, we will describe the way in which hosts can regulate their packet transfer rate in the network since the internet is essentially based on the transport layer for this operation. Specific algorithms have been included for this purpose in TCP as well as other protocols.

2.1 Bandwidth Allocation

The objective is not only to avoid congestion but also to achieve an optimal allocation of bandwidth for transport entities using the network. A good allocation will ensure good performance by using all the available bandwidth while avoiding congestion, and this by giving fair treatment to the different transport entities and adapting quickly to changes in traffic demand. In fact, efficiency and power in allocating bandwidth remain the main criteria for congestion control. As the load increases, the payload increases at the same rate, but the closer it approaches the maximum link capacity, the more this increase diminishes. This drop is due to the fact that bursts of traffic can sometimes cause losses at the level of certain network buffers since the delay cannot stretch indefinitely in the buffers of the routers, and thereafter the packets will be lost after being subject to the maximum buffering delay. If the transport protocol is poorly designed and retransmits packets that have been delayed but not lost, the network may become saturated. The sharing of bandwidth between the different entities must take into account the fact that if the network allocates to a transmitter a certain amount of bandwidth, the latter should only be able to use this amount for its activities. However, it is frequent that the networks do not follow this limitation for each flow or each connection. They sometimes apply it for certain flows if the quality of service is guaranteed. But many connections will seek to use all of the available bandwidth or benefit from a common bandwidth allocation with other connections. In this case, it is the congestion management mechanism that allocates bandwidth to them. Another criterion is the equitable distribution of charges within a network. It is fairly straightforward to get with N streams along with a single link, in which case each stream can get 1/N of bandwidth, although the performance rules specify that they use a little less if the traffic is burst. But if one of the streams takes multiple paths, then it will consume more network resources, and it would therefore be fairer to allocate less bandwidth to it than to those on a link. We should certainly be able to support more flow to a link by reducing the bandwidth allocated to the flow with several links. This point indicates an intrinsic conflict between equity and efficiency. On the other hand, allocating an equal fraction of bandwidth is relatively complicated since different connections will take different paths across the network, and these paths will themselves be characterized by different lengths and capacities. In this case, it is possible that flow is blocked on a link and borrows a smaller portion of an upstream link than the other flows. Reducing the bandwidth of other streams would slow them down but would not help the blocked stream in any way. Note that with TCP it is possible to open multiple connections and encourage more aggressive competition between them. This tactic is used by bandwidth-intensive applications for point to point file sharing.

In order to have the desirable bandwidth allocation (Chan and Ramjee 2005), TCP regulates the rate of data transmission. Indeed, limiting the sending rate depends on two main factors. The first is flow control; it concerns the traffic between a transmitter and a receiver, it is a question of forcing a powerful transmitter to lower its frequency of sending data, in order to not saturate the less powerful recipient who cannot adopt the same high rhythm as the transmitter. The second factor is congestion; it concerns the general condition of the network, ensuring that it is able to carry the traffic that arises, in this case, all the hosts are concerned.

2.2 Flow Control Using the TCP Window

As already mentioned, the role of flow control (Miler et al. 2005) is to prevent a fast sender from driving a slower recipient to saturation. In TCP, window management dissociates the acknowledgment of the good reception of the segments and the allocation of buffers at the recipient level. For example, assuming that the recipient has a buffer memory of X bytes. If the sender transmits a segment of X/2 bytes, the recipient, if it receives it correctly, acknowledges

it. However, since now it has only X/2 bytes free in its memory buffer (until the application removes data from it), it announces a window of X/2 bytes which starts at the next expected byte. The sender then sends again X/2 bytes, which are acknowledged, but the window becomes null. The sender removes data from the memory buffer, which then allows TCP to advertise a larger window and allow the transfer of new data. When the window is zero, the sender must not send segments, except in two cases. First, if it is urgent data to allow, for example, a user to stop an application running on the remote machine. Second, if it's a one-byte segment to force the recipient to announce the next expected byte and the window size again. This packet is called a window sensor. The TCP standard explicitly provides this option to avoid blocking in the event that a window update does not arrive at its destination. In order to reduce the load placed on the network by the recipient, TCP implements the principle of duplicated acknowledgments and window updates for 500 ms in the hope of acquiring data on which the indications of service (which will be transported free of charge). However, a sender sending multiple small packets continues to operate inefficiently. The Nagle algorithm makes it possible to remedy this situation.

2.3 TCP Timer Management

To do its job, TCP relies on RTO (Retransmission Timeout). When a segment is sent, a retransmission timer is started. If the segment is acknowledged before expiring, the timer is stopped. If, however, the timer expires before the acknowledgment arrives, the segment is retransmitted and the timer starts again at 0. To determine the length of time before expiration, TCP uses a dynamic algorithm that continuously adapts to the expiration of the timer, based on continuous monitoring of network performance. The algorithm generally used by TCP works as follows: for each connection, TCP maintains a variable SRTT (smoothed Round Trip Time) (Kim et al. 2014), which is the best current estimate of the time of round trip to the destination in question. When a segment is sent, the timer is started, both to see how long the acknowledgment takes and to trigger retransmission if it takes too long, if the accused returns before the end of the expiration period. TCP records the time it took to arrive, for example, R and then updates the SRTT variable according to the following formula:

$$SRTT = \alpha SRTT + (1 - \alpha)R \tag{1}$$

where α is a smoothing factor which determines the speed at which the old values are forgotten. In general, $\alpha = 7/8$.

Even with good SRTT value, choosing an appropriate delay before retransmission is not easy. Early versions of

TCP used 2 * RTT, but experience has shown that a constant value is too rigid because it does not respond when the variance increases.

In particular, the random traffic queuing models (for example, along the fish curve) predict that when approaching carrying capacity, the delay becomes large and very variable. This can cause the retransmission of a packet and the timer to start, while the original is still in the network. This phenomenon is all the more likely to occur as the load conditions are high, and thereafter it is no longer the time to send additional packets over the network. To solve this problem, it has been proposed to make the expiration time-sensitive to the variance of the round trip times as well as to the smooth curve of this variable. This change requires the monitoring of another smoothed variable, RTTVAR (Round Trip Time Variation), which is updated using the following formula:

$$RTTVAR = \beta RTTVAR + (1 - \beta)|SRTT - R| \quad (2)$$

where generally $\beta = 3/4$.

The RTO retransmission time is fixed at

$$RTO = SRTT + 4 * RTTVAR(RFC2988) \quad (3)$$

The retransmission timer is kept to a minimum of one second, regardless of estimates, it is a value chosen with precision, in order to avoid erroneous retransmissions based on the measurements.

One problem that occurs with collecting round trip time samples R, is what to do when a segment expires and is returned. When the acknowledgment arrives, it is unclear whether it refers to the original transmission or a later version, and misinterpretation can seriously affect the timing of retransmission. This problem was first brought to light by an amateur radio enthusiast, Karn, who was interested in transmitting TCP/IP packets by radio. His idea was simple: do not update the estimates for the segments that have been retransmitted. In addition, double the expiration time on each successive retransmission until the segments pass the first time. This fix is called the Karn algorithm. And most TCP implementations incorporate it.

3 TCP Congestion Control

In addition to flow control, the TCP transport protocol must also apply another remedy to prevent congestion. It is congestion control. Congestion control is one of the key functions of TCP. When the load injected into the network, whatever it is, is greater than its carrying capacity, a phenomenon of congestion is formed. When the queues get bigger at the routers, the network layer detects congestion and informs the transport layer to encourage it to slow down the traffic it injects into the network layer. In the Internet,

TCP plays a key role in congestion control, as well as in reliable transport. That's why this protocol is so important.

3.1 Congestion Control Mechanisms

Studies have concluded that the Additive Increase Multiplicative Decrease (AIMD) algorithm is the appropriate control mechanism to arrive at the point of fair and efficient operation. When the sum of the allocations reaches 100%, i.e. the capacity of the link, the global allocation is effective. AIMD also provides network stability, making it easier to generate than solving congestion within a network. So the speed of increment should always be slower than the speed of reduction. But that is not entirely fair, strictly speaking, because TCP connections adjust the size of their window by a value given to each RTT. TCP congestion control implements this method using a window and using packet loss as a congestion signal. To do this, TCP maintains a congestion window cwnd whose size corresponds to the number of bytes that the sender can have in circulation in the network at any time. In the AIMD rule, it is a question of alternating phases of additive increment where we increase linearly the cwnd as long as there is no congestion, and phases of multiplicative reduction where we multiplie the cwnd by a number < 1. The TCP cwnd congestion window is maintained in addition to the flow management window, which specifies the number of bytes that the receiver can buffer. The two windows are managed in parallel, and the number of bytes that can be sent is determined by the smaller of the two windows. All TCP algorithms assume that packet loss is due to congestion, so they monitor timeouts and other signals. A good retransmission timer is needed to detect packet loss signals accurately and in real time. All we have to do is monitor the congestion window using sequence numbers and acknowledgments and adjust the congestion window using a congestion management mechanism. The acknowledgments return to the sender at about the same speed as that at which the packets cross the slowest link along the way. If new packets are injected into the network at this rate, they will be sent as fast as the slow link allows. But they won't clutter any routers placed on the way. This rhythm is given by an entity called the ack clock. This is an essential part of TCP, thanks to which it smooths traffic and avoids unnecessary queues at the router level. A second consideration to take into account is that the AIMD rule will take a long time to reach an adequate operating point within fast networks if the congestion window starts with a low value. An improvement to this situation has been proposed, it is a mixture of linear and multiplicative increment when a connection is established, the sender initializes the congestion window at a low initial value of four segments maximum (Hiertz et al. 2007). In addition, the use of four segments is an increase in the previous initial value of an experience-based segment. The sender then sends the previous initial window of a segment taking into account the experience. The sender then sends the initial window. Packets will only be acknowledged after an RTT. For each segment acknowledged, before the retransmission timer expires, the sender adds a number of bytes equivalent to one segment to the congestion window. Because for each segment acknowledged, there is one segment less in the network. The result is that each acknowledged segment allows two additional segments to be sent, and therefore the congestion window is doubled on each RTT. This algorithm is known as Slow Start (Kim et al. 2014). It is a question of starting to send the segments slowly then more quickly in an exponential way. When this happens, queues form in the network, which when full cause the loss of one or more packets. To curb this undesirable phenomenon, the sender maintains a threshold for connection called the SSThresh (slow start threshold). Initially, this value is arbitrarily high equivalent to the size of the flow control window, so it does not limit the connection. TCP continues to increment the congestion window until a timeout expires or the congestion window exceeds the defined threshold (or the recipient window becomes full). Whenever a packet loss is detected, for example, due to the expiration of a timeout, the SSThresh is set at half the congestion window, and the whole process begins again. The size of the congestion window is therefore readjusted to its initial value and the slow start resumes. Each time the Slow Start Threshold is crossed, TCP switches to Congestion Avoidance mode (Mathis and Mahdavi 1996). The congestion window is then incremented by a single segment at each RTT. Like slow start, this increment is generally based on acknowledgments rather than on RTTs. If we call the congestion window cwnd and the maximum segment size MSS, a common approximation is to increase cwnd by (MSS*MSS)/cwnd for each of the cwnd/MSS packets acknowledged. This increment does not have to be fast. The idea is to maximize the time that a TCP connection spends with an optimal congestion window size is either not too small (not to curb excessively), or too large (not to cause congestion). In congestion avoidance mode, at the end of each RTT, the sender's congestion window is large enough for it to inject an additional packet into the network. Compared to the slow start, the linear growth rate is much slower. After losing a packet, the recipient cannot continue to send acknowledgments. So the acknowledgment number does not change any more. The sender is therefore no longer able to send new packets over the network because his congestion window remains full. This situation can continue for a relatively long period of time until the timer starts and the lost packet is retransmitted. At this point, TCP begins a new slow start. There is a quick way for the sender to recognize that one of their packets has been lost. As the packets following

the lost packet reach the recipient, they trigger acknowledgments all bearing the same number, we talk about duplicated acknowledgments. Whenever the sender receives such an acknowledgment, it knows that probably another packet has arrived at the recipient before the lost packet has reappeared. Since packets can take different paths across the network, they can arrive out of sequence which triggers duplicated acknowledgments even if no packets have been lost. However, this is usually uncommon on the internet. When reorganization takes place along a multi-link path, packets usually don't get out of order too much. Thus, TCP somewhat arbitrarily presumes that three duplicate acknowledgments mean that a packet has been lost. The identity of this packet can also be inferred from the acknowledgment number, as it is the next packet in the sequence. This packet can then be retransmitted immediately before the retransmission time expires. This heuristic method is called Fast Retransmission.

3.2 Congestion Control Algorithms

After triggering, the slow start threshold is always set at half of the current congestion window, just like after the expiration of the delay. You can restart slow start by adjusting the size of the congestion window to a packet. With this window size, a new packet will be sent after the round trip time required acknowledging the retransmitted packet as well as all the data that had been transmitted before the loss was detected. This version of the TCP protocol is called TCP Tahoe published in 1988. The maximum segment size here is 1 KB. Initially, the congestion window was 64 Kb, but a timeout has occurred, so the threshold is fixed at 32 Kb and the congestion window at 1 Kb for zero transmission. This is incremented exponentially until reaching the threshold (32 Kb), each time a new acknowledgment of receipt arrives rather than continuously, which leads to a progression curve in the form of stairs. Once the threshold is crossed, the window continues to progress linearly, incremented by a segment at each RTT. When one of the packets gets lost in the network, detected on the arrival of three duplicated acknowledgments; at this time, the lost packet is retransmitted, the threshold is fixed at half of the current window, and the slow start procedure is restarted. TCP Tahoe (Sikdar et al. 2004), which included good retransmission timers, provided a functional congestion control algorithm to solve the problem of congestion collapse. But realized that we could do even better, at the time of the fast retransmission, the connection has a too large congestion window, but it still has an ack clock. Each other duplicated acknowledgment that arrives probably signals a new packet loss.

A new algorithm is generated; it's TCP Reno (Miler et al. 2005). This is the equivalent of TCP Tahoe to which the Fast

Recovery has been added. After an initial slow start, the congestion window is incremented in a linear way until the detection of a packet loss via duplicated acknowledgments. The lost packet is retransmitted and quick retrieval is used to keep the ack clock running until the retransmission is acknowledged. At this time, the cwnd resumes from the new threshold of slow start rather than 1. This behavior continues indefinitely, and the connection spends most of the time with its cwnd close to the optimal value of the product bandwidth-delay waiting. With these cwnd tuning mechanisms, TCP Reno has formed the basis of TCP congestion control for more than two decades. Most of the modifications made in the following years are minor adjustments to these mechanisms, for example, the modification of the choice of the initial window is the removal of various ambiguities. Some improvements have been made to ensure the recovery of more than one packet from the same window. As described in RFC 3782, the TCP New Reno version (Floyd et al. 2004) thus uses, after each retransmission, a partial increment of the acknowledgment number to find and repair another loss. Since the mid-1990s, several variants have emerged, based on the principle that we have described using slightly different control mechanisms. We also talk about TCP Vegas (Brakmo and Peterson 1995), which is an improvement of TCP Reno in order to minimize the lost packet rate, and this by making three changes, the first is to modify the slow start by lowering the rate of increment the cwnd, the second concerns retransmission where we check the RTO of each segment without waiting for duplicate acknowledgments, and in the third, we improve the congestion avoidance phase.

There is also TCP CUBIC (Poojary and Sharma 2019) implemented in the last versions of Linux and Windows, it is designed for bandwidth and high latency networks. In Cubic, we do not rely on acknowledgments to manage the cwnd, it is set up based on information from the latest congestion phenomenon. As a result, cubic ensures equity between the streams since its window is adjusted independently of the RTT values.

Two other important changes have also affected TCP implementations. One is concerned with the complexity of TCP, much of which is due to the fact that one should seek to deduce from a duplicate acknowledgment stream the packets that have arrived and those that have been lost. The cumulative acknowledgment number does not provide this information. A simple solution is to use the TCP Selective Acknowledgment or SACK option (Mathis et al. 1996), which lists a maximum of three ranges of bytes that have been received. Using this information, the sender can more directly determine the packets to be retransmitted and follow the packets in circulation in order to respect congestion. With SACK, TCP can more easily recover from two situations in which multiple packets are lost roughly simultaneously because the TCP sender knows which packets were not received. This option, described in RFC 2883, is widely deployed today. SACK-based TCP congestion control is described in RFC 3517. We must also mention TCP Fack (Mathis and Mahdavi 1996) which is an alternative of TCP Sack guaranteeing good data management by controlling the quantities of data circulating in the network based on SACK's loss recovery algorithm.

Another TCP congestion control variant that we have to mention is TCP Scalable (Wen et al. 2009), the main objective is to achieve very large throughput values even when the network is congested, it is designed to be a robust algorithm improving the performance of high-speed networks. In case of management, instead of dividing the cwnd by 2 each time a packet is lost, it is decremented by a factor of 1/8 until the losses stop. In this way, the throughput does not experience a sudden fall and its value does not decrease significantly.

4 Proposed Work

In the above, we talked about a set of TCP variants, we discussed the advantages and limits of each, and we also mentioned the different congestion control techniques that have been implemented in it. But, to better understand reality in its entirety, we must study their behavior closely by examining their reactions under the same circumstances. To do this, we have chosen to make network simulations in which the different TCP flavors are subjected to the same conditions to compare their behavior in normal situations (in the absence of congestion), and critical situations where the network becomes congested (bottleneck strangulation). These simulations will serve as a guideline to draw the strengths and limits of each of our TCP variants.

The simulations were carried out using the network simulator NS2. We chose to test the behavior of the eight variants (TCP Tahoe, TCP Reno, TCP New Reno, TCP Vegas, TCP Fack, TCP Sack, TCP Scalable, and TCP Cubic), each variant is tested separately, to configure the state of the network we applied a CBR flux whose values were varied from 1 MB up to 12 MB with a step increment of 0.5 MB in order to have congestion. In this way, 24 scenarios were applied to each TCP variant. In total, we had 192 scenarios. For the other parameters of the simulation, we used a mesh topology, we chose a value of 10 MB between every two adjacent nodes. The total duration of the simulations is 720 s. The results collected from the simulations were collected so that they could be sorted and then filtered in order to bring out the information necessary for our study. The comparison is based on the three metrics throughput, latency, and packet drop rate. For this, we used the data already collected to calculate these parameters according to the following equations:

Fig. 1 Variation of the throughput according to CBR values

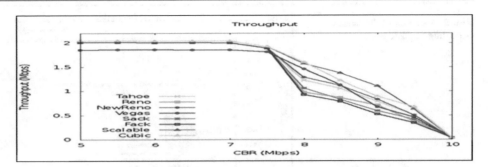

$$\text{Throughput} = \frac{\text{Total Received By Receiver Node}}{\text{First Packet Sent Time} - \text{Last Packet Received Time}} \quad (4)$$

$$\text{Latency} = \frac{\sum (\text{Ack Time} - \text{Sent Time})}{\text{Total Packet Number}} \quad (5)$$

Packet Delivery Ratio
$$= \frac{(\text{Total Sent Packets Number} - \text{Total Received Packets Number}) * 100}{\text{Total Sent Packets Number}} \quad (6)$$

To better clarify the results obtained from the three equations, we used the GNUplot tool from NS2 to transform the numerical values into graphs that are easy to interpret (Figs. 1 and 2).

These two graphs translate, for each TCP variant, the variation of the throughput as a function of the value of the CBR. At the beginning of the connection, all the variants adopt the same ehaviour since we are still in normal mode (there is no congestion), they start with almost the same value of throupghput except for TCP Cubic which started with a value a bit high compared to the rest, and TCP Vegas which adopted from the start a minimum value. Once the network was congested, the values of the throughput decreased significantly for all the variants except that TCP Scalable, TCP Cubic, and TCP Vegas experienced a slight

decrease compared to the other variants, while the most significant decrease was that from TCP Fack and TCP Reno.

The rhythm of the variations is well confirmed in the graph representing the average of the throughput where the lowest average is that of TCP Vegas (adopts minimum values from the beginning), while TCP Scalable keeps the highest average since it has kept the highest values from the start (Fig. 3).

This graph represents the variation of the latency as a function of the CBR, we notice that in the absence of congestion, all the variants adopt almost the same values of latency, but in the presence of the congestion these values differed in a remarkable way, especially for TCP Vegas which has experienced a very high increase, so we can deduce that it is a delay-based algorithm (Figs. 4 and 5).

This graph represents the variation of the packet loss rate as a function of the CBR values. At the beginning, everything is normal, the rate of packet loss is zero since there is no congestion. But as soon as the network becomes congested, all variants begin to experience losses. This graph also confirms what has already been said, TCP Vegas is the variant which has known minimum loss values (it is delay-based, recommended for applications which require data transfer without loss or alteration), and TCP Scalable next to TCP Fack have experienced the highest loss values with a very significant loss increase for TCP Scalable (it is

Fig. 2 Average throughput of TCP Variants

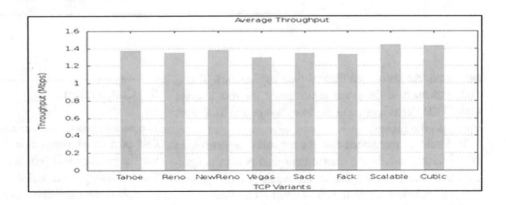

Fig. 3 Variation of the latency according to CBR value

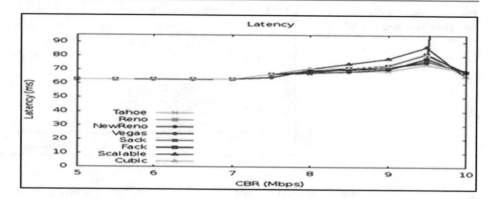

Fig. 4 Variation of the drop rate according to CBR values

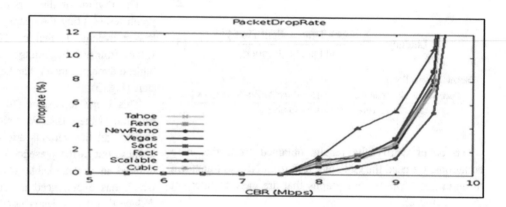

Fig. 5 Average packets losses of the TCP Variants

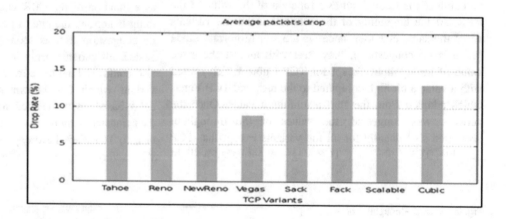

loss-based, recommended for real-time application). This is well confirmed in the graph representing the loss average where TCP Vegas represents the low average and TCP Fack represents the highest.

It should be mentioned that the networks implementing TCP Scalable remain the ones that reacted best to congestion, and this is very clear in the results of the graphs. Although it is loss-based (favoring delay rather than packet loss), it does not allow very large data loss and this is visible in the last graphs.

5 Conclusion

The various works and analyzes carried out over time in the field of network congestion control, give us the impression that the problem is definitively resolved, but on the contrary, the problem persists and the acceleration of networks still prompts many researchers to take another look on this topic. This is why we tried in this paper to discuss the limitations of Wireless Mesh Network due to the problem of network congestion. We focused on the TCP protocol as a reliable transport protocol implementing a variety of variants, each ensuring congestion control in its own way depending on what each situation requires. The simulations carried out allowed us to conclude that some variants are delay-based (requiring data transfer without loss or alteration) like TCP Vegas. Others are loss-based (do not support long transmission delay) such as Tahoe, Reno, New Reno, Fack, Sack, Cubic, and Scalable. It has also been deduced that an improvement is required concerning the allocation of bandwidth. In fact, the objective is not only to avoid congestion but also to achieve an optimal allocation of bandwidth for the transport entities using the network, by giving fair treatment to the different transport entities, and by adapting quickly to changing traffic demand. In fact, this is the case for some implementations like TCP Reno and Scalable which devour the bandwidth shared with the other flows (the other variants), which is translated by the latter as false congestion. The challenge then is to implement a new variant capable of adapting quickly to changes that a network may experience, and which is not based on packet loss as a congestion signal, especially since it is very rare to serve as a useful signal. However, we need to detect the problem as soon as possible before our network loses its performance. One possibility is to think of a mechanism that absolutely does not use packet loss as a signal. The signal can be provided by the round trip time which increases as the network becomes congested.

References

Akyildiz, F., Wang, X., Wang, W.: Wireless mesh networks: a survey. Comput. Netw. ISDN Syst. **47**, 445–487 (2005)

Brakmo, L., Peterson, L.: TCP Vegas: end to end congestion avoidance on a global Internet. IEEE J. Sel. Areas Commun. **13**(8), 1465–1480 (1995). https://doi.org/10.1109/49.464716

Camp, J.D., Knightly, E.W.: The IEEE 802.11s extended service set mesh networking standard. IEE Commun. Mag. **46**, 120–126 (2008)

Chan, M.C, Ramjee,R.: TCP/IP performance over 3G wireless links with rate and delay variation. Wireless Netw. **11**(1–2), 81–97, Springer (2005)

Floyd, S., Henderson, T., Gurtov, A.: RFC3782 the NewReno modification to TCP's fast recovery algorithm. Internet Eng. Task Force (2004). https://tools.ietf.org/html/rfc3782

Hiertz, G.R., Max, S., Zhao, R., Denteneer, D., Berlemann, L.: Principles of IEEE 802.11s. In: Proceedings of 16th International Conference on Computer Communications and Networks (ICCCN), Honolulu, Hawaii, USA (2007). https://ieeexplore.ieee.org/abstract/document/4317949

Kim, M., Klein, T., Soljanin, E., Barros, J., Médard, M.: Modeling network coded tcp: Analysis of throughput and energy cost. Mobile Netw. App. **19**(6), 790–803, Springer (2014)

Mathis, M., Mahdavi, J.: Forward acknowledgement: refining TCP congestion control. SIGCOMM Comput. Commun. Rev. **26**(4), 281–291 (1996). https://doi.org/10.1145/248157.248181

Mathis, M., Mahdavi, J., Floyd, S., Romanov, A.: RFC2018 TCP selective acknowledgment options. Internet Eng. Task Force (1996). https://tools.ietf.org/html/rfc2018

Miler, B.M., Avrachenkov, K.E., Stepanyan, K.V., Miller, G.B.: Flow control as a stochastic optimal control problem with incomplete information. Prob. Inf. Trans. **41**(2), 150–170, Springer (2005)

Poojary, S., Sharma, V.: An asymptotic approximation for TCP CUBIC. Queue. Syst. **91**, 171–203, springer (2019)

Sikdar, B., Kalyanaraman, S., Vastola, K.S.: Analytic models for the latency and steady-state throughput of TCP Tahoe, Reno, and Sack. IEEE/ACM Trans. Netw. **11**, 959–971 (2004)

Verma, L.P., Verma, I., Kumar, M.: An adaptive congestion control algorithm. Model. Meas. Control A **92**(1), 30–36 (2019)

Wang, X., Lim, A.O.: IEEE 802.11s wireless mesh networks: Framework and challenges. Ad Hoc Netw. **6**, 970–984 (2008). https://www.sciencedirect.com/science/article/abs/pii/S157087050001370

Wen, S., Fang, Y., Sun, H.: Differentiated bandwidth allocation with TCP Protection In Core Routers. In: IEEE Transactions on Parallel and Distributed Systems, vol. 20, no.1, pp. 34–47 (2009)

Classifying Security Attacks in IoT Using CTM Method

Hind Meziane, Noura Ouerdi, Mohammed Amine Kasmi, and Sanae Mazouz

Abstract

Internet of Things (IoT) is an active research area. Because in Springer database, the number of "IoT attacks" (Articles & Chapter) indicates 7625 Results, that means a trend that research on IoT is becoming increasingly popular. Security and privacy are challenging tasks. Hence, it is essential to pay attention during the basic stages in planning and building the IoT, to avoid any danger or risk due to its widespread in sensitive areas in most countries such as national defense sites, various industries, and military sites… and in order to develop new reliable applications and systems. Moreover, "safety" is a basic criterion for IoT quality and a prerequisite for trusting and ensuring its viability. Security in IoT is more important and essential because IoT devices share sensitive data with other devices, peoples. So, the leading companies in this field should consider cybersecurity as a top priority. In this chapter, we survey and describe different attack surfaces in IoT. To do this, it seems necessary to start with an analysis of existing classifications of IoT attacks and to present their strengths and weaknesses. Then, it would be judicious to study the criteria of each taxonomy in order to extract what may be relevant to our study. Indeed, each of the existing classifications has been developed for a specific purpose. We will thus eliminate the least useful criteria and we will suggest others; the goal being to arrive at a general classification of attacks in IoT. To achieve this goal, we propose a new classification of attacks on IoT, based on Classification Tree Method (CTM). In short, we use the CTE (Classification Tree Editor) tool to generate and select the test cases.

Keywords

IoT • Taxonomy • Attacks • CTM • Challenges • Security

1 Introduction

What exactly is IoT? There is no standard, unified, and shared definition of IoT since an accurate and precise definition is not a simple task.

IoT is a vast domain which contains physical objects, networks communication, technologies, hardware (devices, computers), protocols, electronics, platforms, and applications that interconnects anything from the physical environment (physical objects, animals, places, plants, machines, and people…) to the internet in order to exchange data without any human interaction.

IoT was first used by Kevin Ashton in 1999 (Bilal 2017), he has also invented the term Radio-Frequency Identification (RFID) that tags the physical objects to the internet for the purpose of counting and tracking of goods without any human interference. IoT is the network of interconnected things/devices which are embedded with sensors, software, network connectivity, and necessary electronics that enables them to collect and exchange data making them responsive.

The IoT technology is in continuous progress, and soon all devices in homes and companies will be connected to the network, which makes them vulnerable to cyberattack and theft in the absence or weak protection. The absence of work and the real intention to develop and improve ways to protect IoT will lead to the users refraining from engaging in it

H. Meziane (✉) · N. Ouerdi
Faculty of Sciences, ACSA Laboratory, Oujda, Morocco
e-mail: wwhindmeziane94@gmail.com

N. Ouerdi
e-mail: n.ouerdi@ump.ac.ma

M. A. Kasmi
Faculty of Sciences, LARI Laboratory, Oujda, Morocco
e-mail: a.kasmi@ump.ac.ma

S. Mazouz
Faculty of Sciences, SmartICT Laboratory, Oujda, Morocco
e-mail: s.mazouz@ump.ac.ma

for fear of violating its privacy, which maybe will limit the user base. It is true that this interconnection of devices will provide us with a comfortable and affluent lifestyle, but this requires us not to lose sight of the need to preserve our information stored on these devices. This places the government with all its institutions and devices in front of a major challenge to face cyber threats, not only by enacting laws and regulations but also by carrying out procedures and measures and using the techniques and tools necessary to protect the safety of networks and programs and centers of devices and data from electronic attacks or damage or unauthorized access to them.

Classification is a process of categorizing a given set of data into classes. In other words, it is a predictive technique because it makes possible to predict if an element (test case) is a member of a given class (precisely a sheet) according to the characteristics of this test case. Artificial Neural Network (ANN) classifier, Bayesian classification, Deep Learning (DL), and Support Vector Machine (SVM) are employed as classification methods, but as a start of work, we will use CTM method.

The goal of this chapter is to try to classify attacks in IoT by employing CTM method.

This chapter is structured as follows: We can start with introducing the idea, vision, definition of some ambiguous terms, components, and challenges of IoT. Then we are going to present an overview about some previous research by critically analyzing the strengths and weaknesses of each classification. In the fourth section, we propose, as result, our new classification based on CTM method and then, we compare with the previous research. Finally, we conclude this chapter.

2 IoT Components and Challenges

The principle of IoT is to make the elements of the physical environment able to be aware of the world in which they are found.

The main goal is to provide the best services for human being, to allow many smart new applications in the medical, industrial, economic, educational, and even individual daily life levels.

The elements of the physical environment are isolated and its non-communicating space (i.e., no information flow).

Integration (fusion) of the digital world into the physical world (or real world), where real-time information can make our life much easier and simpler. The five necessary characteristics for the elements of the physical environment in order to be represented virtually (to have a virtual personality) are given below:

- Identify: recognize each element of the physical environment in a unique way by the physical label for the remote access including 6LowPan, RFID, GPS, URI, Barcode, EPC (Electronic Product Code), uCode (ubiquitous Code), IPv6 (for communicating; and accessing to smart objects/devices (sensors, actuators, and communication devices (Xiao et al. 2014)), UNSPSC and ecl@ss (as device catalogs) are two standards introduced in IoT in order to categorize devices (Xiao et al. 2014);
- Connected with any other elements in a flexible and transparent way, through the use of a technology to connect to networks as a technique ZigBee, Wi-Fi, Bluetooth…. in order to exchange data;
- Autonomous, it can do treatments and sometimes even make decisions without any human intervention, i.e., they must be equipped with intelligent algorithms to perceive and act;
- Intelligent, the elements of physical environment must have a certain capacity of calculation and memorization;
- Interacting with any other things, devices, or humans related to the Internet by IoT components which is represented in (it means) sensing, actioning, and coordinating.

M2M is an example of a connected object in which two or more machines connected to each other (either wired or wireless network) can exchange information without human intervention.

RFID (Radio-Frequency IDentification) allows remote identification of elements of the physical environment including stationary or moving individuals and the exchange of data of all kinds. RFID system includes a reader which sends an electromagnetic wave carrying a signal, and a label ("tag") is fixed on the objects reacting to the reception of the signal sent by the reader by sending back to the latter the requested information. The reader receives, in return, the information returned by these objects. This system can be supplemented by a computer making it possible to process and store the information transmitted by the reader.

IoT is more likely to be attacked than the Internet since billions of more devices will produce and consume services. Furthermore, there are several obstacles to the development of IoT such as heterogeneity, interoperability, scalability (a large number of IoT devices), Big Data, no standardization (in definition, architecture, attacks…). Besides, resource constrains in terms of energy, storage space, and computing capacity. On the other hand, the lack of encryption in the cloud should also be considered. In addition to this, IoT devices are uniquely identifiable and are mostly characterized by limited processing, small memory, and low power, and also due to the great diversity of IoT devices and the rapid emergence of new technology and elements on the market. Another point is security and privacy that means to provide the necessary protection for data and maintain the privacy of users, this is the biggest challenge of IoT, how IoT manufacturers protect their products against risks, threats, and vulnerability.

However, there is currently no security standards that describes in detail the attacks and vulnerabilities of IoT. Furthermore, if security standards are not set for IoT networks, devices (PCs, smartphones Mobile), hardware (Routers, Cables (the protocol on the cable)), software, layers, technologies of communications, protocols (especially protocols in each IoT layers), servers, clouds…, companies and users will be susceptible to cyberattacks as we saw on October 21, 2016 (ScienceDaily and October 2016), a DDoS (distributed denial-of-service) attack crippled a large domain name system (DNS) company called Dyn, resulting in the outage of popular websites such as Amazon, Twitter, PlayStation Network, Reddit, and many others for almost a day.

3 Previous Research

For (Vikas Hassija, 2019) (Hassija et al. 2019), any IoT application can be divided into four layers: (1) sensing layer; (2) network layer; (3) middleware layer; (4) application layer. Even though this approach describes security attacks in each layer, it lacks a comprehensive set of protocol attacks of each layer. For example, the authors identify security attacks against IoT network at high level without analyzing the attacks against each network protocol. Furthermore, the authors limited the first layer to sensors and actuators. Similarly, the authors were interested more in IoT solution rather than attacks.

According to (Mukrimah Nawir, 2016) (Nawir et al. 2016), attacks can be classified to eight categories (Device Property, Information Damage Level, Location, Strategy, Host-based, Access Level, Communication Stack Protocol, Protocol-based), this author did not identify all attacks (the most dangerous or possible attacks) for each level. For example, only two or three types of attacks are described on Physical, Transport, Data Link, and Application layers, like Jamming and Tampering, have been identified in the Physical layer. Moreover, authors identify information damage level attacks into six categories (Interruption, Eavesdropping, Alteration, Fabrication, Massage Replay, MITM (Man in The Middle)), we can generalize from this classification a new class for attacks against the concept/objective of security.

For (Jyoti Deogirikar, 2017) (Deogirikar and Vidhate 2017), IoT technology works on three layers (perception, network, and application layers), and the authors classify the attacks in four categories (physical, network, software, and encryption attacks). For each category, Jyoti and all considered one attack that is most dangerous from all the attacks of that category, which is insufficient to have a general classification for most attacks.

According to (Amir Abdullah, 2020) (Abdullah et al. 2020), there are four levels on the IoT architecture, based on this architecture and we add another layer named "Business layer" (Burhan et al. 2018), which is fundamental for the real success of IoT technology, it can be used to consume the data obtained from the application layer to build business models, graphs, and flowcharts, which are useful in evaluating the new technology of IoT.

Another interesting work described in (Arsalan Mosenia, 2016) (Mosenia and Jha 2016), authors have surveyed vulnerabilities specific to the IoT in brief, authors detailed vulnerabilities and countermeasures against only the edge side layer of IoT. Nevertheless, this approach lacks in-depth analysis of cloud and fog computing issues in the IoT space.

4 New Classification

4.1 Different Attacks in IoT

This section discusses our most important proposed taxonomy of IoT attacks.

Opposing with (Hassija et al. 2019), problems of security are classified not only by layers of the IoT architecture but also many categories such as attacks based on data, on control, on layers, on software, on hardware, on the technology of communication, on devices, on encryption, on protocol application, on security concepts on vulnerability, on impact, on sources, on target, on detectability, on prevention, and on the goal of attackers.

Attacks based on Layers: Our proposed model contains five layers: Physical layer, Network layer, Middleware layer, Application layer, and Business layer.

According to an in-depth study, the first layer is not limited to sensors and actuators (Hassija et al. 2019), but we can generalize this first layer's name of the IoT model "sensing layer" by "physical layer" because "sensors" can only measure and gather information about the environment, it can't report or monitor and protocols of each layer are different. As I see, this layer comprises sensors, sensor nodes, actuators, coordinator, devices, objects, machines, RFID tags and reader, Barcode, Infrared, WSN, camera, tools, circuits, peoples, plants, clothes, environments, buildings, animals, and controllers. These devices include mobile devices such as smartphones or tablets and microcontroller units. The connected devices are the real endpoint for IoT. As devices get smarter, they can react better to our needs, wishes, and even moods, e.g., (personality type; sleep patterns; stress levels; happiness; levels of exercise). Billions of things (sensors and computing devices…) are continually sensing, collecting, consolidating, and analyzing the significant amount of our personal information, such as location, contact list, browsing patterns, health and fitness information. This layer's main responsibility is to collect useful information from things, then process, and digitize the data (Bilal 2017). The main purpose of objects is unique address identification and communication between short-range technologies (Chen et al. 2018). This layer covers how a device is physically connected to a network via wired or wireless mechanisms. Finally, the first layer is "physical layer" also known as perception, recognition (Gan et al. 2011). The main function is to identify, generate, and collect data from the physical world to perceive their environment by detecting changes using sensors. And the components of this first layer are Tag, sensors/actuators (photoelectric, photochemistry, solid state, infrared, GPS, Gyroscope), coordinator and network (LAN (Wi-Fi, Ethernet), PAN (UWB, ZigBee, Bluetooth, 6LowPan, wired)).

Sensors: collect information from their environment (temperature, position, force, pressure, movement, shock, gas, rate…). A sensor captures a physical quantity to be measured as a usable quantity, it is a translator that translates a physical quantity into an electrical signal (logic (TOR), analog, digital).

Actuators to perform actions and an intelligent function.

In (Nawir et al. 2016; Hossain et al. 2015), a coordinator is a device which acts as a device manager, it also sent a report aggregating to the IoT service provider about their event and activities.

The first layer attack is performed when the attacker is in a close distance of the device. The main challenge for this layer is the malicious attack on the sensor and the identification technology, which includes attacks that disable RFID tags, as well as relay attacks, (RFID) technologies, have currently attracted a lot of attention. To get access to physical devices, social engineering is one of the most prominent methods where the attackers access the devices and perform the real attack. Furthermore, the attacker can start further attacks, such as spoofing attack and tag cloning: tags are vulnerable due to the deployment of tags on different objects. An attacker could easily capture these tags and build a replica of them, which look like original ones to compromise an RFID system by deceiving even the RFID readers. Besides, this layer is targeted by several attacks such as Node Capture, DoS (Denial-of-Service) Attack, Denial-of-Sleep Attack, DDoS, Fake Node, Replay Attack, Routing Threats, Side-Channel Attack, Mass Node Authentication, Unauthorized Access to the Tags, Eavesdropping, RF Jamming, Spoofing Attack, Sleep Deprivation Attack, Worm Hole, and Timing attack.

The network layer contains (Router, Switches, Gateway, Internet) and must support massive volumes of IoT data produced by wireless sensors and smart devices, which is responsible for the transmission or to transfer the data between different layer by through network by using wired or wireless. The goal of this layer is to transfer, transmits data from the first layer to the middleware layer. The attacks are referred at routing, data, traffic analysis, spoofing, and launching MITM attack. Besides, Sybil attacks are also possible at the network layer where fake identities/Sybil identities are used to create illusions in the network. Several attacks on this layer are identified: DoS, Denial-of-Sleep Attack, Sinkhole attacks, Eavesdropping/sniffing, Spoofing, Routing attacks (Black Hole, Gray Hole, Worm Hole, Hello Flood, Sybil), Selective forwarding, Malicious code injection, RFIDs interference.

The middleware layer contains (Cloud, Datacenter, API, data analysis, and data visualization …), stores, analyses, and processes a huge amounts of data, employs databases, cloud computing, big data processing modules, classification, and polymerization (Gan et al. 2011). It receives the data from the second layer and store in the database. The main challenges are database security and cloud security, there are many attacks such as DoS, Flooding Attack in Cloud, Cloud Malware Injection, Signature Wrapping Attack, Web Browser Attack, and SQL Injection Attack.

The application layer, responsible for delivering application-specific services to the user and for all industries. The IoT application covers "intelligent" environments/spaces in areas such as transport, building, city, lifestyle, commerce, agriculture, factory, supply chain, emergency, health, environment, and energy. Attackers exploit the vulnerabilities of application and programs (Buffer Overflow, Code Injection), Phishing Attack, DoS, Authentication and Authorization, Data Access and Authentication, Social engineering, Sniffing Attack, and Cross-Site Scripting.

The business layer manages the whole IoT system including the control applications, business, making decision processes, and profit models of the IoT also user's privacy, based on the received data from the Application layer. The goal is to build flowcharts, graphs, and a business model. It is supposed to monitor design, develop implement analyze, and evaluate IoT system-related elements (Vandana and Chikkamannur 2016). The attacks that can be found are Business Logic Attack and Zero-Day Attack.

Attacks based on Communication Technologies: This section presents a new classification of communication technologies attacks, based on two categories: (1) "Short-range technologies" and (2) "Long-range technologies."

The first class includes, but not limited to RFID, NFC, WSN, 6LowPan, Z-Wave, Zigbee, Bluetooth, Wi-Fi, BLE, LR-WPANs, IEEE 802.11 ah, IEEE 802.15.4, many attacks defined in this category such as Spoofing, MITM,

Eavesdropping, Replay, Sniffing, Bluejacking, Bluebugging, Bluesnarfing, Dictionary Attack, Phishing attacks.

There are many LPWAN technologies, among them are Sigfox, LoRa/LoRaWAN, NB-IoT, Satellite, LTE/LTE-A, 3G/4G, GSM, GPRS, GPS. This class contains many attacks such as replay attack, worm hole attack, and jamming attack.

Attacks based on Security Concepts: Security solutions must help meet at least the following criteria: Availability, integrity, confidentiality, and to these criteria, we can add authentication, non-repudiation. This class is represented by five following criteria:

Availability: A resource must be accessible, with acceptable response time, against DoS attacks.

Integrity: makes it possible to certify that data, processing or services have not been modified, altered or destroyed, whether intentionally or accidentally. Attacks related to integrity are message alteration attack and message fabrication attack, in addition to this, active eavesdropping on the network which can modify the intercepted data.

Confidentiality is the maintenance of the secrecy of information. Its objective is to ensure that information can only be read by authorized persons. Unauthorized access is an example of an attack.

Authentication process can be described in two distinct phases—identification and authentication. The identification can be seen as a simple login on a system. Authentication can be a password known only to the user. Among the authentication attacks, there are impersonation attacks where an attacker pretends to be another entity and Sybil attack where the attacker uses different identities at the same time.

Non-repudiation is the fact of not denying or rejecting that an event (action, transaction) has taken place.

Attacks based on Protocol Application: The most popular protocols are MQTT (Message Queuing Telemetry Transport) is a publish-subscribe messaging protocol based on the TCP/IP protocol, and CoAP (Constrained Application Protocol) is a web transfer protocol optimized for constrained devices and networks used in wireless sensor networks to form the IoT.

After several searches, and based on (Abdul-Ghani et al. 2018), this class can be classified into two subclasses, the first one is about attacks on protocols themselves and the second is for poor SSL implementation. The first subclass contains:

MQTT attacks (MITM, Buffer Overflow, and DoS),

CoAP attacks (sniffing, pre-shared key, DoS/DDoS, and MITM),

XMPP attacks (Authentication attack, XMPPloit, DoS/DDoS, and MITM).

The second categories based on poor implementation SSL include MQTT authentication attack, XMPP bomb, XMPPDeamon crash.

Attacks based on Software: Based on (Abdul-Ghani et al. 2018), this section identifies possible attacks on IoT software, including IoT applications located either in IoT objects or in cloud, Firmware, and operating systems.

Application-based attacks: the attacks targeting web application are Exploitation of a misconfiguration, Malicious code injection, Path-based DoS attack, Reprogram attack, Malware, and DDoS.

Operating system-based attacks: the attacker can indirectly try to install malicious applications (plugins, updates, etc.) in the equipment, such as phishing attack, Backdoors, Virus, worm attack, Brute-force search attack, and unknown attack.

Firmware-based attacks: the cause of this attack is the failure to receive updates from IoT systems, this subclass include Malware, Eavesdropping, Reverse Engineering, and Control hijacking.

Attacks based on Impact: According to (Ali et al. 2019), we can generate a new classification, based on the impact of the attack, this class is divided into three subclasses named, respectively ,"Low," "Medium," and "High."

In the first subclass "Low," we can find the following attacks: RFIDs interference, Node jamming in WSN, and Eavesdropping.

The second subclass named "Medium" includes Side-Channel Attack, Physical damage, Protecting Sensor Data, Interoperability and Portability, Business continuity and Disaster Recovery, Cloud Audit, Network Congestion problems, and Phishing Attacks.

The last subclass "High" contains Malwares attack, Data Access and Authentication, Node Tempering, Fake Node, Malicious Code injection, Mass Node authentication, Heterogeneity problem, DoS, RFID Spoofing, Routing attacks, Sybil Attack, and Data Security.

Attacks based on Devices: Antivirus company Kaspersky reveals that attacks on IoT Devices have increased nine-fold, from 12 million attacks in 2018 to 105 million in 2019. Most often, they are an uncorrected flaw or a simple password. The author reaches this conclusion by using lures, called "honeypot" which imitate IoT Devices to attract hackers. Kaspersky recommends, upon purchase, to change the default password. It's obvious to avoid an overly simple password like strings, for example. The antivirus vendor also recommends that updates be made as soon as they are available. These updates very often contain files relating to the security of your devices.

Reports issued by the Gartner's company indicate, that in 2021, the number of IoT devices will be 25.1 billion (Middleton et al. 2017), 80 billion in 2025 estimated by IDC (Velocity Business Solutions Limited 2025), and as many as 125 billion connected IoT devices by 2030 according to HIS (Jenalea Howell 2030).

The classification of (Harbi et al. 2019) based on availability and trust of the IoT devices. Trust management is divided into two main categories: Deterministic and non-deterministic trust. The first includes certificate-based mechanisms (determined using public or private keys and digital signatures) and policies (use a set of policies to identify trust), while the second includes systems based on recommendations (use prior information to define trust), on reputation (use the global reputation of entities), on predictions, and on social networks (take into account the social reputation of entities). The availability of IoT networks should be achieved at the hardware level (means the existence of all devices at all times) and software (this is the ability to provide services anywhere and at any time). The IoT devices may face several attacks as DoS and DDoS.

Attacks based on Target: According to many researches, the targets can be operating system, an application, a network, a local computer, and user's personal information. This category can be classified into

Information: It means stealing sensitive data (personal or business data) or hacking of video cameras.

Devices control (Spamming, Botnet (Mirai, Reaper, Hajime …)).

The last one is causing physical harm to the user such as disabling of medical life support equipment.

Attacks based on Encryption: The purpose of this type of attack is to break the encryption process. According to many researches, we can classify this class into two subclasses:

The first one is for Cryptanalysis Attack: its purpose is to recover the encryption key to break the encryption function in the IoT system, some examples of this type of attack are Chosen Ciphertext attack, Ciphertext Only attack, Chosen Plaintext attack, Known Plaintext attack, and MITM.

The second type of attack is Side-channel Attack: Attacker can find the encryption key to access hacked data using techniques such as electromagnetic and power analysis.

Attacks based on Vulnerability: One of the most important elements in cybersecurity is the deployment of updates on systems. As soon as a vulnerability is made public, hackers very quickly try to exploit it (as we have seen with the WannaCry malware and the EternalBlue vulnerability). Generally, this class can be divided into two subclasses:

Configuration: the misconfiguration happens when no secure configuration has been applied to the frameworks, application server, web server, database server, or the platform of the application. For example, opening a number of vulnerable ports.

Implementation: A bad implementation of the application can make the system vulnerable to SQL injection attacks. Owing to insufficient authorization and authentication, insecure web interfaces and lack of encryption mechanism, IoT devices are vulnerable. Besides, a system permits the users to choose weak passwords to have a vulnerability in its implementation.

Attacks based on Location (Source): Depending on the location of the attacker in relation to the network, two distinct categories of attack can be identified: external attacks and internal attacks.

Internal attacks ("Insider") come from inside the IoT network, attacker resides in the close proximity of the IoT devices to execute his own malicious code and gets unauthorized access to the native network (Malicious node, Black hole, insider eavesdropping).

External attacks ("Outsider") originate from outside the IoT network, an attacker tries to randomly, and remotely access IoT smart devices outside the network (Outsider eavesdropping, DoS).

Attacks based on Behavior: Active and passive attack are two possible subclasses:

Active attacks: smart devices and data can be destroyed (Sybil, DoS, MITM, Spoofing, Jamming, Selective Forwarding, and Hole attack).

Passive attacks consist of listening without modifying the data or the functioning of the network (eavesdropping, traffic analysis).

Attack based on Detectability: This class concern not only detection of IoT attacks but also botnets, vulnerability, anomalies, intrusion, and malware. It mainly depends on the mechanism of predicting the behavior of the system, object…. In my view, this class is the most important and can be divided into two subclasses: attacks with easy detectability and attacks with difficult detectability.

Easy: The attack can be detected via antivirus such as worm attack or IDS.

Difficult: Such as Botnet because it spreads malware and it will not affect your computer in any visible way, I can include this subclass attacks based on algorithm detectability.

Attack based on Prevention: The principal goal is to prevent intrusion or threats. As I see, we can classify this class into two subcategories:

Human by using some simple preventive procedures such as changing weak, guessable password, and don't use the same password for multiple services, apply software updates when necessary, carefully read the permissions before installing apps, make sure a website is secure before you enter personal information, avoid logging into your important accounts on public computers, avoiding suspicious files attached or sites in worm attack, or protecting you by antivirus.

Technique/Method including IPS provide Node Authentication in Sinkhole attack, using preventive method or mechanism.

Attack based on Attackers Goal: (or Attack based on the goal of attack). Depends on attacker's motivation; in other words, what the attacker(s) want to achieve, we must first understand one of the main problems is "What is the goal of the attacker?". The possible subcategories in our view can be

Cryptocurrency: The primary goal of this subclass is the financial gain to earn money quickly and easily. Ransomware (BrickerBot, Silex, wannacry...) is an example of this subclass due to their Bitcoin goal. It is a type of malware.

Spying is one of the most important international problems in the world today, such as phishing, zero-day exploits, and spoofing...

DoS means it denies its service to a legitimate user. The primary goal of a this type of attack is not to steal data but to take down a web site, which may affect all sectors from big/mid enterprises (Government, Banking, e-gaming...). Ping of death, Flooding, and teardrop attacks are examples of such attacks.

4.2 Our New Classification

Classification Tree Method (CTM): CTM is a supervised classification tool, developed by Grotchmann and Grimm in the field of software engineering in 1993. This method graphically represents the partitions of the input domain in the form of a tree.

The goal is to be able to form test cases by combining classes belonging to different dimensions. The field of test entries is considered at first according to various aspects, for each aspect, complete and separate classifications are formed. The resulting classes are, in turn, divided into subclasses. We obtain a grid which is erected below the tree. Each column of the grid contains the leaves of the classification tree.

The CTM has several advantages. First, the identification of all possible cases as well as the selection of relevant test cases are done in a systematic way, which facilitates its management and helps to reduce or eliminate certain errors.

Classification Tree Editor (CTE): The CTE tool is based on the CTM method. Once all possible test cases have been generated, relevant test cases must be identified. For this, we used the CTE tool (version 3.2.9). This tool allows the automation of the generation and identification of test cases. Thus, from our classification scheme and using the CTM, CTE tool generates all possible combinations of the subclasses. These combinations represent the possible attack test cases.

Our Taxonomy: The objective of this chapter is to classify IoT attacks based on the CTM. Once our classification of attacks in IoT is ready in the form of a tree, have a clear and structured vision, this approach should be applied to known attacks, in order to characterize and categorize

them and to select the most relevant test cases for the system studied. Figure 1 shows a taxonomy of security attacks on IoT. There are 15 categories classified for attackers to attack the IoT system. For the test, we chose Sybil attack.

An attacker conducting a Sybil attack will attempt to break into the network. To do this, he will usurp the identity of legitimate nodes. It will then copy the identity of one or more nodes to the same node physical or multiple hardware in order to gain access to a large part of the network and participate in communications. He wants to be able to monitor traffic or even influence the data collected. The node whose address has been spoofed no longer has access to the data which was previously sent to it.

Sybil attack is defined as a malicious device illegally taking multiple identities. The first test case concerns the Sybil attack:

- Sybil attack is an "Insider attack" because the adversary has connections to legitimate identity;
- The Sybil attack is a massive destructive attack;
- It is an "Active attack" and also a high-impact attack;
- Belongs to the "Network Layer";
- Present privacy and security threats to users and devices,
- Communication to an illegal node results in data loss and becomes dangerous in the network;
- Sybil attacks degrade the integrity of data, security, and resource utilization;
- Attackers can violate user privacy, breach information security, and spread spam or mislead popularity;
- It is necessary to interact with trusted IoT devices to prevent unwanted actions conducted by malicious nodes;
- Sybil's technique is to propagate malware to a website;
- Sybil attack is one of the most attacks linked to 6LowPan;
- Sybil activity is identified with the application of the CAM-PVM algorithm and is detected in the network (Dhamodharan and Vayanaperumal 2015);
- For prevention of Sybil activity, another MAP algorithm is applied along with CAM-PVM (Dhamodharan and Vayanaperumal 2015).

Our proposed taxonomy is based on the following characteristic:

- Clear and Simple because attacks are easily categorized;
- General: Because it can classify the most possible attacks (attacks can be classified by their characteristic);
- Performance: For each class, an attack belongs only to one sheet. In other words, an attack can be classified by one subclass from every class or dimension;
- Pertinence: Repetition of the process of categorizing a particular attack must always lead to the same results.

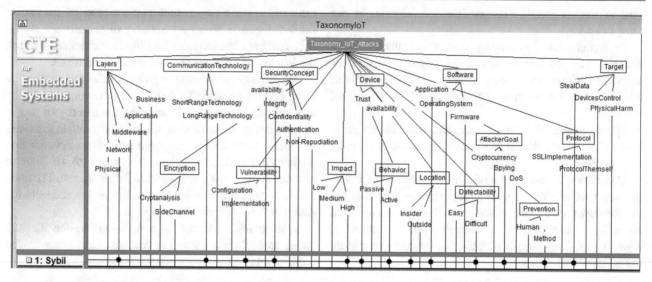

Fig. 1 Taxonomy of IoT attacks and generated test cases via CTE tool

5 Comparison and Synthesis

There are several surveys on the taxonomy of security attacks. Each research has its own advantages and drawbacks; this research depends on the vision of each author. Table 1 shows a comparison and synthesis of previous surveys. In this comparison, we show the advantages and disadvantages of related surveys.

We have changed their limits to remedy the disadvantages of the classification described by authors (Hassija et al. 2019; Nawir et al. 2016; Deogirikar and Vidhate 2017; Abdullah et al. 2020; Mosenia and Jha 2016), and in order to avoid the previously drawbacks, we proposed a new classification for IoT attacks based on CTM.

6 Conclusion

The main goal of this chapter is to provide an explicit overview of the most important challenges and aspects of IoT and to help readers to have a simple and clear idea/vision of IoT, definition, components, layers. In addition to this, we propose a classification of attacks in IoT using CTM method.

Today, we live in a new phase called the IoT stage, in which we witness an accelerated technological change, with which we have begun putting leadership in the hands of devices that carry our important and sensitive information, in which these devices are interconnected and communicate with each other at high speeds, and make decisions

Table 1 Survey comparison

Survey	Strength	Weakness
Hassiji et al. (2019)	Authors have discussed the existing and upcoming solutions to IoT security threats including Blockchain, Fog Computing, Edge Computing, Machine Learning	They have been discussed only on the attacks in IoT layers, which is insufficient for a clear vision
Nawir et al. (2016)	Attacks can be classified into eight categories	Their classification is not based on a graphical representation of the relevant aspects of the test
Deogirikar and Vidhate (2017)	Authors give an overview of IoT architecture and attacks on IoT	This approach has a poor scalability
Abdullah et al. (2020)	Authors explore IoT's various architecture layer and major security concern in each layer	It has not a comprehensive look
Mosenia and Jha (2016)	Authors surveyed vulnerabilities and countermeasures against only the edge side layer of IoT	Lacks in-depth study of attacks in cloud, fog computing in the IoT space

according to the indicators they see. Our personal information is distributed and stored in more than one database and on more than one device.

The IoT is about connecting devices to each other and the Internet. We can conclude that it aims to build a relationship between people, people and things, and things.

IoT technology proves that we will live in smart homes, smart cities (AMSTERDAM, TOKYO, BARCELONE) and perhaps in a smart world-smart planet in the coming years.

To summarize, it is necessary verification by the higher authorities (government) to test the products/services and verify their security before being introduced on the market and used by consumers (users, companies...), and more precisely, after its finishing phase. Furthermore, it is therefore necessary to set up a security strategy, "secured by design" which consists of the builder for including the concept of risk in his project from the design phase.

In this paper, we first stressed the need to have a simple and new classification of attacks as well as the CTM method, and, we proposed to use the CTE tool for the generation and semi-automatic selection of test cases. In the future work, we will explore machine and deep learning techniques used for classifying IoT attacks in order to make a comparative study between classification methods (classifiers) to improve security and privacy in the IoT domain, and also to evaluates the accuracy and the effectiveness of several learning classification techniques used for IoT security.

References

Abdul-Ghani, H.A., Konstantas, D., Mahyoub, M.: A comprehensive IoT attacks survey based on a building-blocked reference model. Int. J. Adv. Comput. Sci. Appl. (IJACSA) 9(3), 355–373 (2018)

Abdullah, A., Kaur, H., and Biswas, R.: Universal Layers of IoT Architecture and Its Security Analysis. In: New Paradigm in Decision Science and Management (pp. 293–302). Springer, Singapore. (2020) https://doi.org/10.1007/978-981-13-9330-3_30

Ali, I., Sabir, S., and Ullah, Z.: Internet of things security, device authentication and access control: a review. (2019) arXiv preprint arXiv:1901.07309

Bilal, M.: A review of internet of things architecture, technologies and analysis smartphone-based attacks against 3D printers. (2017) arXiv preprint arXiv:1708.04560

Burhan, M., Rehman, R., Khan, B., Kim, B.S.: IoT elements, layered architectures and security issues: A comprehensive survey. Sensors 18(9), 2796 (2018). https://doi.org/10.3390/s18092796

Chen, K., Zhang, S., Li, Z., Zhang, Y., Deng, Q., Ray, S., Jin, Y.: Internet-of-things security and vulnerabilities: taxonomy, challenges, and practice. Journal of Hardware and Systems Security 2(2), 97–110 (2018). https://doi.org/10.1007/s41635-017-0029-7

Deogirikar, J., and Vidhate, A.: Security attacks in IoT: A survey. In: 2017 International Conference on I-SMAC (IoT in Social, Mobile, Analytics and Cloud) (I-SMAC) (pp. 32–37). IEEE. (2017) https://doi.org/10.1109/I-SMAC.2017.8058363

Dhamodharan, U. S. R. K., and Vayanaperumal, R.: Detecting and preventing sybil attacks in wireless sensor networks using message authentication and passing method. The Scientific World Journal, 2015. (2015) https://doi.org/10.1155/2015/841267

Gan, G., Lu, Z., and Jiang, J.: Internet of things security analysis. In: 2011 international conference on internet technology and applications (pp. 1–4). IEEE. (2011) https://doi.org/10.1109/itap.2011.6006307

Harbi, Y., Aliouat, Z., Harous, S., Bentaleb, A., and Refoufi, A.: A Review of Security in Internet of Things. Wireless Personal Communications, 1–20. (2019) https://doi.org/10.1007/s11277-019-06405-y

Hassija, V., Chamola, V., Saxena, V., Jain, D., Goyal, P., Sikdar, B.: A survey on IoT security: application areas, security threats, and solution architectures. IEEE Access 7, 82721–82743 (2019). https://doi.org/10.1109/ACCESS.2019.2924045

Hossain, M. M., Fotouhi, M., and Hasan, R.: Towards an analysis of security issues, challenges, and open problems in the internet of things. In 2015 IEEE World Congress on Services (pp. 21–28). IEEE. (2015) https://doi.org/10.1109/services.2015.12

Howell, J.: "Number of Connected IoT Devices Will Surge to 125 Billion by 2030, IHS Markit Says," Press Release 24 Oct, 2017. [Online]. Available: https://technology.ihs.com/596542/number-of-connected-iot-devices-will-surge-to125-billion-by-2030-ihs-markit-says (Consulted 27-03-2020)

Mosenia, A., Jha, N.K.: A comprehensive study of security of internet-of-things. IEEE Trans. Emerg. Top. Comput. 5(4), 586–602 (2016). https://doi.org/10.1109/TETC.2016.2606384

Nawir, M., Amir, A., Yaakob, N., and Lynn, O. B.: Internet of Things (IoT): Taxonomy of security attacks. In: 2016 3rd International Conference on Electronic Design (ICED) (pp. 321–326). IEEE. (2016) https://doi.org/10.1109/ICED.2016.7804660

Peter Middleton, D. R., Tsai, T., Yamaji, M., Gupta, A.: "Forecast: Internet of Things — Endpoints and Associated Services, Worldwide, 2017," Gartner, 2017. [Online]. Available: https://www.gartner.com/doc/3840665/forecast-internetthings–endpoints (Consulted 27-03-2020)

Stanford University. "Massive cyberattack poses policy dilemma." ScienceDaily. ScienceDaily, 25 October 2016 (consulted 18-03-2020) <www.sciencedaily.com/releases/2016/10/161025215400.htm>

Vandana, C. P., and Chikkamannur, A. A.: IOT future in Edge Computing. International Journal of Advanced Engineering Research and Science, 3(12) (2016)

Velocity Business Solutions Limited, "IDC: 80 billion Connected Devices in 2025 for generating 180 trillion GB of Data and IoT Opportunities," 15 Feb 2018, 2018. [Online]. Available: http://www.vebuso.com/2018/02/idc-80-billion-connecteddevices-2025-generating-180-trillion-gb-data-iot-opportunities/. (Consulted 27-03-2020)

Xiao, G., Guo, J., Da Xu, L., Gong, Z.: User interoperability with heterogeneous IoT devices through transformation. IEEE Trans. Industr. Inf. 10(2), 1486–1496 (2014). https://doi.org/10.1109/TII.2014.2306772

Exploring the Power of Computation Technologies for Entity Matching

Youssef Aassem, Imad Hafidi, and Noureddine Aboutabit

Abstract

Sustainable development consists of a set of goals (SDG) associated with a myriad of targets. The aim is to achieve those underlying goals between 2015 and 2030. Although none of the goals refers directly to information and communication technologies (ICT) that latter can accelerate the development of human being and bridge the digital gaps, so as to build modern communities. Data cleaning is a field of computing technologies which aims to extract meaningful information that can be used in many areas in order to help the community. In particular, entity matching is a crucial step for data cleaning and data integration. The task consists of grouping similar instances of the real-world entities. The main challenge is to reduce the number of required comparisons, since a pairwise comparison of all records is time consuming (quadratic time complexity). For this reason, the use of indexing techniques such as Sorted Neighborhood Methods and blocking is indispensable considering that they divide data into partitions such manner that only records within some block are compared with each other. In this work, we propose two novel hybrid approaches which use a varying block. Indeed, reducing the similarity distance can improve the number of detected duplicates with a smaller number of comparisons. We prove theoretically the efficiency of our algorithms. The computational tests are performed by evaluating our technique on two real-world datasets and compare them with three baseline algorithms, and the results show that our proposed approach performs in efficient way.

Y. Aassem (✉) · I. Hafidi · N. Aboutabit
National School of Applied Siences, Khouribga, Morocco
e-mail: aassem.youssef@gmail.com

I. Hafidi
e-mail: imad.hafidi@gmail.com

N. Aboutabit
e-mail: n.aboutabit@usms.ma

Keywords

Entity matching • Sorted neighborhood methods • Blocking

1 Introduction

In many applications, a real-world entity may show up in many data sources as it may have quite different descriptions. For example, a person's name or mailing address may be represented in several ways. Thus, it is inevitable to identify the records that refer to the same real-world entity, which is called Entity Resolution (ER) Christen and Goiser (2007). ER is a fundamental problem in data cleaning, and it occurs in many applications such as information integration and information retrieval. The challenge in the entity resolution process is the huge amount of data which is expensive in terms of comparisons. The naive algorithm consists of comparing each record with all remaining ones. Therefore, its complexity time is quadratic in the number of records. Hence, it is indispensable to make intelligent guesses about which records have a high probability of matching the same real-world entity. These assumptions are often expressed as data partitioning in the hope that duplicate records will appear in individual partitions. Many algorithms have been suggested over the years (Steorts et al. 2014; Cohen and Richman 2002; Winkler 2006). Two most prominent ER families are Blocking and windowing (Kolb et al. 2012, Christen 2012).

Blocking algorithms consist of choosing a Blocking Key Value (BKV) and to cluster data based on its value to blocks or clusters. Then, comparisons between tuples within the same block are done and duplicates are marked.

Windowing (Sorted Neighborhood Methods) algorithms are executed following three steps:

– Key building: In which, a new blocking value is generated for each record. The BKV result from the

concatenation of certain parts of attribute values (e.g., 3 first letters of the first name | first 4 digits of phone number).
- Sorting: All records are sorted by the sorting key.
- Windowing: A fixed size window slides over data, and records within same window are compared.

Problems that occur with Sorted Neighborhood Method (Hernández and Stolfo 1995) concern key building and windowing steps. As long as key building is based on accurate and significant fields, the result of algorithm will be satisfied. However, issues with data or significant field will cause data to be far in the sorting order. Consequently, some comparisons will not be made, and the recall will be low. Concerning the windowing step, the choice of window size and nature (fixed or variable size) has a crucial importance in detecting few or many duplicates.

Draisbach et al. (2012) overcame this problem by proposing two approaches that extend the window or shrink it based on the number of detected duplicates per number of the conducted comparison.

Duplicate Count Strategy (DCS and DCS++) (Draisbach et al. 2012) algorithms represent an enhancement of the traditional SNM algorithm. The principle consists in varying the window size based on the number of identified duplicates. Due to this variation, the compared records differ from the original SNM (Hernández and Stolfo 1998). The DCS and DCS++ algorithms differ only in the way of adapting the window size when duplicates are found.

Being inspired by the two algorithms, we propose a hybrid approach called Enhanced Duplicate Count strategy. Our algorithm takes advantage of using the last found record as a representative of the current window (block). Thus, we reduce the number of comparisons and minimize the similarity distance (Christen 2006) between records which will lead to higher chances to match some additional duplicates. Our algorithms show when evaluated significant improvements in terms of number of comparisons, precision, and thus F-Score as well.

In the following section, we discuss related work. Our approaches are described and elaborated in more detail in section III. An experimental evaluation is presented in the fourth section and finally we conclude and discuss our future work in section V.

2 Related Work

Pointing at improving the quality and effectiveness of entity resolution (Christen 2012; Hernández and Stolfo 1998), some authors suggested many traditional approaches. Blocking is a term that refers to approaches that reduce the search space for duplicate detection. The idea is to divide records into groups or blocks, and then compare all records that occur only within the same block, supposing that records in different blocks are unlikely to represent the same entity. The generated blocks can be disjointed or overlapping. Examples for overlapping blocking methods are SNM (Hernández and Stolfo 1995), canopy clustering (McCallum et al. 2000), suffix array-based blocking (Aizawa and Oyama 2005), q-gram indexing (Christen 2012), spark-based Entity resolution (Mestre et al. 2017) and MFI blocking (Kenig and Gal 2013).

In the SNM family, the traditional SNM suffers from the fixed block size solution, which makes the process either costly expensive when choosing a high window size, including unnecessary comparisons, or inaccurate with low recall by using a low window size and missing true duplicates. Yan et al. (2007) have talked about adaptivity of record linkage utilizing SNM as an example. They took advantage of the windowing technique to build non-overlapping blocks. They assume that the distance between a record and its successors in the sort arrangement is monotonically increasing in a little neighborhood. For that, a hypothesis was that the distance between two records within the same window is smaller than the distance between one record within the window and the other one is outside. Then, the following may hold

$$dis(r_i, r_j)\phi \leq dist(r_i, r_k) \tag{1}$$

where φ is the minimum distance threshold, records r_i and r_j are in the same window, while r_k is outside.

Therefore, after executing blocking ideally, records outside block must be different from the records within the block. Consequently, finding blocks will necessitate only to find the boundary records between adjacent blocks. They present two algorithms and compare them with the traditional SNM. Incrementally Adaptive-SNM (IA-SNM) (Yan et al. 2007) is an algorithm that increases the window size if the distance between the first and last record of the window is smaller than a given threshold. The expansion of the window size depends on the current window size. Accumulative Adaptive-SNM (AA-SNM) Yan et al. (2007), on the other hand, creates windows with one overlapping record. Using transitivity one block can be built by grouping multiple adjacent windows. For this to be done, the distance between the last record of a window and the last record in the next adjacent window should be smaller than a specific threshold.

After IA-SNM and AA-SNM algorithms have been proposed as variants of the traditional SNM, Draisbach et al. (2012) propose an alternative approach that expands a window if a certain minimum number of records are classified as matches within the current window. DCS is based on a heuristic that assumes the window size based on the number of already detected duplicates. The more duplicates of a tuple are found within a window, the larger it is. In contrast, when not finding any duplicates within the current window, we assume that duplicates do not exist or that they are very far away in the sorting order. In the beginning, we choose a starting window size ω, which is domain dependent as it is the case for the SNM. The first record within the window is compared with its ω − 1 successors. The window elements can be expressed by W (i, i + ω − 1). At the end of comparisons executed in the window, we compare the fraction given by the Eq. (1) $\frac{numberofduplicate}{numberofrecords}$ with a given threshold defined by the user φ. This fraction represents the average number of detected duplicates per comparison. When it has a high value, there is a huge probability for the next record to be a duplicate of the record. If no duplicate can be found within this window, we do not increase the window. At the end of the algorithm, the transitive closure is calculated. It is explained by the following example: if t_i and t_j are duplicates, and t_j is duplicate of t_m then t_i and t_m are duplicates. The second strategy, also known as Multiple record increase or DCS++ (Draisbach et al. 2012), is an improvement of the duplicate count strategy DCS. The difference is when finding a duplicate; the next ω − 1 adjacent records of that duplicate are added to the current window. These records are added only once to that window. After that, we calculate the transitive closure to save some of the comparisons. Let us assume that the pair (t_i, t_j) and (t_i, t_m) are duplicates, with i < j < m. Calculating the transitive closure returns the additional duplicate pair (t_j, t_m). Consequently, the window W(j, j + ω − 1) does not need to be checked. The difference between this algorithm and DCS is the check whether a record should be skipped or not and adding adjacent records to the current window when finding a duplicate. Draisbach et al. (2012) have shown that the DCS ++ method saves between 0 and ω − 2 comparisons. In the machine learning techniques (Yin et al. 2007; Reyes-Galaviz et al. 2017) usually learn from a classifier using an example set and use the classifier to classify pairs of records as match or non-match. However, those methods are expensive because they need to enumerate n2 pairs. Moreover, they need some labeled data in order to do the training of the used classifier.

Some active works (Mishra et al. 2013; Mishra et al. 2017; Mishra et al. 2014a; Mishra et al. 2014b; Mishra et al. 2016) have restricted the study of entity matching techniques for bibliographic datasets due to their attribute's evolution over time and the extensive growth of number of records. Based on its work in which EMTBD algorithm was proposed (Mishra et al. 2013), Mishra et al. have proposed a heuristic algorithm based on genetic algorithms (Mishra et al. 2017). In order to assess the quality of the partitioning of the records, some cluster validity indices were used as

objective function (Mishra et al. 2014a; Mishra et al. 2014b, Deb and Kalyanmoy 2001). Mishra et al. (2016) have used multi-objective evolutionary algorithms (MOEA) (Mishra et al. 2016) using Non-Dominated Sorting Algorithm (NSGA-II) to resolve the problem based on multiple indices to optimize simultaneously. Some further researches were focused on proposing some cluster validity indices (Mishra et al. 2014a; Mishra et al. 2014b; Mishra et al. 2016) to assess the quality of quality the resulted partitioning. Other works have considered the problem as a CPP problem (Tauer et al. 2019) where the goal is to partition the records into cliques such that the similarity is maximized. The latter algorithm considers refreshing the classification once a new entity has arrived.

3 The Proposed Enhanced Duplicate Count Strategy

In this section, we present our two proposed algorithms, Enhanced DCS and Enhanced DCS++. Both algorithms belong to the SNM Family, we consider them as an enhancement of the traditional DCS and DCS++. Keeping the same three steps principle as these latter, we propose a new idea to the windowing step. Hence, we remind the principle of traditional DCS and DCS++, which will be used to explain how our algorithms are designed. Moreover, we demonstrate the efficiency of both proposed algorithms when compared to DCS and DCS++.

3.1 Duplicate Count Strategy vs Enhanced Duplicate Count Strategy

To elaborate the enhanced DCS algorithm's principle, we preview the logic of DCS algorithm first. As illustrated in Fig. 1, within the window w_i, all elements are compared with x_i. At the end of comparisons, we calculate the ratio mentioned in Eq. (1) and if it is significant (i.e., the ratio is bigger than a given threshold), the window will be enlarged. The process of comparison is as what follows: Initially, we compare x_i with its adjacent records within the same window w_i. x_i is compared with x_{i+1}. Whether they are duplicates or not, the next comparison will be between x_i and x_{i+2} and so on. Consequently, the string similarity distance begins to increase as we move further and the chances that two records are a match begins to decrease. Thus, true matches may be missed. To improve results quality, we propose the following scenario: when comparing x_i and x_{i+1}, the next comparison will be done between the last duplicate encountered x_m where $m \leq i + 1$ and one of its successors x_j (j > m). Therefore, we minimize the distance between compared

Fig. 1 Initial situation (example)

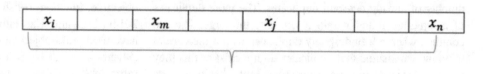

Window W_i

elements in order to increase the probability of being matched.

In our algorithm, we employ a structure called *Matrix Matches* in order to save comparisons. The pseudo-code of this variant can be found in algorithm 1.

Firstly, we sort records by key. Next, we populate the window with the first w elements of the records. Then, we slide a window over data, and we compare tuples within the same window. We use a variable *last Dup* to save the last duplicate element encountered. In order to save matches between records that will be used in the transitive closure we use a 2D array dubbed *Matrix Matches*. Comparisons made are marked in a 2D matrix called *Comparisons Matrix* (line 20). When a comparison is done between two records of indexes i and j, respectively, the value of the cell having coordinates (i, j) in the Comparisons Matrix is changed to 1 to mark the comparison as done in order to avoid repetitive ones.

When proceeding to a comparison, the function Comparison Already Done checks the matrix and returns true if the comparison is already done or false otherwise. Thus, we avoid repeating comparisons for same pair. This added functionality allows to reduce the huge number of comparisons. Let us assume that we found a duplicate x_m for x_i, the lastDup variable holds the index of x_m and next comparison will be between x_m and x_{m+1}. At the end, we calculate the value of Transitive closure using Warshall algorithm (Warshall 1962) to find the remaining duplicates.

3.2 DCS++ vs Enhanced DCS

As Enhanced DCS is an improvement of the DCS algorithm, we propose also an improvement of DCS++ algorithm. This latter differs from DCS in the enlargement strategy by using multiple record increase when finding a duplicate x_m. Consequently, the current window is enlarged by w − 1 next adjacent records from that duplicate, so we skip its window latter. The enhanced DCS++ main goal is to minimize the distance between compared elements by holding a reference to the last duplicate found within the window. That way, after the enlargement step next record will be compared with the duplicate found and so on. The pseudo-code of this algorithm can be found in algorithm 2.

Algorithm 1 EDCS(records, sorting key *key*, initial window size *w*, threshold ϕ)

```
 1: Sort records by key
 2: Populate window win with first w records of records
 3: for j = 1 to records.length − 1 do
 4:     useDup ← false
 5:     dup ← 0
 6:     comp ← 0
 7:     k ← 2
 8:     while k ≤ win.length do
 9:         current ← win[0]
10:         if (useDup) then
                current ← lastDup
11:         end if
12:         if ComparisonAlreadyDone(current, win[k]) then
13:             if Matches(current, win[k]) then
14:                 lastDup ← win[k]
15:                 useDup ← true
16:             end if
17:             k ← k + 1
18:             continue
19:         else
20:             mark comparison pair (current, win[k])
21:         end if
22:         if isDup(current, win[k]) then
23:             useDup ← true
24:             numDuplicates ← numDuplicates + 1
25:             lastDup ← win[k]
26:             mark pair duplicate (current,win[k])
27:         end if
28:         numComparisons ← numComparisons + 1
29:         if k = win.length and j + k < records.length and
            (numDuplicates/numComparisons) ≥ φ then
30:             win.add(records[j + k + 1])
31:         end if
32:         k ← k + 1
33:     end while
34:     win.remove(1)
35:     if win.length ≤ w and j + k ≤ records.length then
36:         win.add(records[j + k + 1])
37:     else
38:         while win.length > w do
39:             win.remove(win.length)
40:         end while
41:     end if
42:     j ← j + 1
43: end for
44: calculate transitive closure
```

Algorithm 2 EDCS++(records, sorting key *key*, initial window size *w*, threshold ϕ)

```
1:  Sort records by key
2:  Populate window win with first w records of records
3:  for j = 1 to records.length − 1 do
4:      useDup ← false
5:      dup ← 0
6:      comp ← 0
7:      k ← 2
8:      while k ≤ win.length do
9:          current ← win[0]
10:         if (useDup) then
11:             current ← lastDup
12:         end if
13:         if ComparisonAlreadyDone(current, win[k]) then
14:             if Matches(current, win[k]) then
15:                 lastDup ← win[k]
16:                 useDup ← true
17:             end if
18:             k ← k + 1
19:             continue
20:         else
21:             mark comparison pair (current, win[k])
22:         end if
23:         if isDup(current, win[k]) then
24:             useDup ← true
25:             numDuplicates ← numDuplicates + 1
26:             lastDup ← win[k]
27:             mark pair duplicate (current, win[k])
28:             skipRecords.add(win[k])
29:             while win.length < k + w − 1 and j + win.length < records.length
                do
30:                 win.add(records[j + win.length + 1])
31:             end while
32:         end if
33:         numComparisons ← numComparisons + 1
34:         if k = win.length and j + k < records.length and
                (numDuplicates/numComparisons) ≥ φ then
35:             win.add(records[j + k + 1])
36:         end if
37:         k ← k + 1
38:     end while
39:     win.remove(1)
40:     if win.length ≤ w and j + k ≤ records.length then
41:         win.add(records[j + k + 1])
42:     else
43:         while win.length > w do
44:             win.remove(win.length)                    1
45:         end while
46:     end if
47:     j ← j + 1
48: end for
49: calculate transitive closure
```

3.3 Demonstration of Efficiency

To compare EDCS and EDCS++ to the state of art algorithms, we have to examine the difference b between added comparisons a due to increasing the window size, and the saved comparisons s due to the skipped windows, and already done comparisons.

Proposition 1 . EDCS is at least efficient as DCS.

Let *d* be the number of duplicates found within the window w_i a is the number of added comparisons.
   s is the number of saved comparisons.
   b is the difference between a and s.

$$a = \frac{numDup}{numComp} \tag{2}$$

3.3.1 Duplicate Count Strategy DCS

Case 1. If d = 0, then no duplicates are found within the window. Consequently, the window will not be increased, and we will not save any comparisons.

$$b = a - s = 0. \tag{3}$$

Case 2. If d > 0, then we distinguish between two cases:

- If α < Threshold. So, the window will not be increased. The number of added comparisons and saved ones, respectively, are a = 0 and s = 0.

$$b = 0 \tag{4}$$

- If α > Threshold. So, the window will be increased. The number of added comparisons and saved ones, respectively, are a = k and s = 0.

$$b = -k \qquad (5)$$

where k is the number of times the same window has been enlarged with one record.

3.3.2 Enhanced Duplicate Count Strategy DCS

Case 1. If d = 0, then no duplicates are found within the window. Consequently, the window will not be increased, and we will not save any comparisons. Thereof

$$b = a - s = 0. \qquad (6)$$

Case 2. If d > 0, then we distinguish two cases:

- If α < Threshold. So, the window will not be increased. The number of added comparisons a = 0. The number of saved comparisons s = w − m where m is the index of first duplicate encountered.

$$b = m - w \qquad (7)$$

- If α > Threshold. So, the window will be increased. The number of added comparisons and saved ones respectively are a = k and s = w + k − m. Thus,

$$b = m - w \qquad (8)$$

where k is the number of times the same window has been enlarged with one record.

Proposition 2 EDCS++ is at least efficient as DCS++.

3.3.3 Duplicate Count Strategy DCS++

Case 1. We distinguish the case in which d = 0 (No duplicates are found). We will not enlarge the window. Consequently, the number of added comparisons a = 0. As well as the number of saved comparisons s = 0.

$$b = a - s = 0. \qquad (9)$$

Case 2. If d > 0, then we distinguish two cases:

- If α < Threshold. So, the window will not be increased. The number of added comparisons a = k, where k is the index of last duplicate found within the window. The number of saved comparisons s = d (w − 1).

$$b = k - d(w - 1). \qquad (10)$$

- If α > Threshold. So, the window will be enlarged by one element each time it reaches its end, and the significance of α is high. Consequently, the number of added comparisons and saved ones, respectively, are a = k + r and s = d (w − 1).

$$b = k + r - d(w - 1) \qquad (11)$$

where k is the index of last duplicate found within the window and r is the number of times we enlarged the same window by one record.

3.3.4 Enhanced Duplicate Count Strategy EDCS+ +

Case 1. We distinguish the case in which d = 0 (no duplicates are found). The values of a, s, and b are the same as in DCS++.

Case 2. In case d > 0, we have two cases to consider:

- If α < Threshold. So, the window will not be increased. The number of added comparisons a = k, where k is the index of last duplicate found within the window. The number of saved comparisons s = d (w − 1) + w − m, where m is the index of first duplicate found within the window. We then have for b.

$$b = k - d(w - 1) - w + m. \qquad (12)$$

- If α > Threshold. So, the window will be enlarged by one element each time we reached its end, and the significance of α is high. Consequently, The number of added comparisons and saved ones, respectively, are a = k + r and s = d (w − 1) + w − m, where m is the index of first duplicate found within the window, r is the number of times we enlarged the same window by one record, and D is the set of indexes of duplicates found within the window. The difference b will be expressed as follows.

$$b = k + r - d(w - 1) - w + m. \qquad (13)$$

4 Experimental Evaluation

To evaluate our algorithm, we conduct experiments using Java language on two datasets: The Cora Citation Matching (McCallum 2017) and DBLP dataset. The first one is a well-known dataset used for benchmarking Duplicate Detection algorithms. As used in previous works, we concatenate the first author's last name and the year of publication as sorting key. The second dataset is a famous computer science bibliography that indexes many publications. We use for the blocking key value the concatenation of four first letters of title, four letters of author's name, and three first characters from the venue. We consider four criteria to compare the performance of the proposed algorithms with the state-of-the-art ones: number of comparison, execution time, precision (fraction of correctly detected duplicates and all detected duplicates), and F-Score (harmonic mean of precision and recall).

4.1 Number of Comparisons

Tables 1 and 2 show the number of comparisons for different window sizes. A comparison means the execution of a complex similarity function. We see that due to window size increase (by one record) combined with skipped comparisons, EDCS outperforms the other approaches. However, EDCS++ has more comparisons due to the multiple record increase. But it is still better than both DCS and DCS++ due to the advantage of using the last duplicate as a representative of the window and save comparisons for the transitive closure.

Table 1 Number of comparisons required by algorithms for Cora dataset

Window size	DCS	EDCS	DCS++	EDCS++
3	29,238	2517	13,034	13,153
10	119,568	7171	72,778	65,571
20	181,663	12,177	74,766	69,833
60	287,259	24,603	89,131	77,070
100	336,042	32,967	92,919	80,417
200	401,592	47,614	93,263	80,761
500	474,136	70,019	93,861	81,359

Table 2 Number of comparisons required by algorithms for DBLP dataset

Window size	DCS	EDCS	DCS ++	EDCS ++
3	2920	2064	1908	1908
10	13,157	8877	8351	8354
20	35,969	18,172	20,249	20,411
60	102,145	50,782	308,941	171,006
100	153,989	80,066	345,713	335,742
200	244,001	143,203	360,324	338,342
500	406,834	273,718	362,320	350,856

4.2 Execution Time

In Figs. 2 and 3, we observe the execution time of the all studied algorithms based on the used window size. For very small window size, the algorithms take approximately the same time to detect duplicates. However, for larger window sizes, EDCS requires less time due to its reduced number of comparisons. EDCS++ basically uses the same mechanism to reduce the amount of comparisons. Even so, the extension mechanism using large window sizes causes some additional comparisons, which impact the execution time. Concerning DCS++, it outperforms DCS by the strategy of ignoring windows of already detected duplicates, which is reflected in the execution time.

4.3 Precision

Figures 4 and 5 present the variation of precision values obtained by applying the four concerned algorithms on both datasets. The detection of duplicates is mainly relative to the search strategy used by the algorithm and by distribution of duplicates across the dataset. For Cora dataset, EDCS++ outperforms other algorithms by using wide extensions in regions where true duplicates exist. And by using last encountered duplicate as a representative of the current window, EDCS detects more accurately duplicates than DCS ++ and DCS. In the second dataset, the proposed algorithms perform better than DCS and DCS++. However, EDCS and EDCS++ strategy vary slightly depending on duplicates distribution in the dataset, which result in EDCS being more accurate than EDCS++ for this dataset.

4.4 F-Score

Figures 6 and 7 show the measured F-Score of different algorithms for using different window sizes. They present approximately the same behavior across the two data sets. Using small window size permits DCS+ + to outperform EDCS and DCS by covering more true positives within the same window. Meanwhile EDCS++ still have better values than DCS++ due to the implemented comparison strategy. As the window size increases, EDCS++ keeps covering more true positives (true matches). Meanwhile EDCS algorithm outperforms DCS++ due to its enlargement strategy which doesn't enlarge window by unnecessary elements in region with low similarity. Consequently, false positives are avoided. DCS algorithm has the lowest F-Score value because of missing some true positives due the used search strategy.

Fig. 2 Execution time for Cora dataset

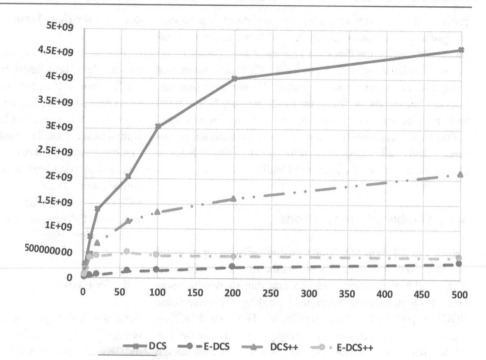

Fig. 3 Execution time for DBLP dataset

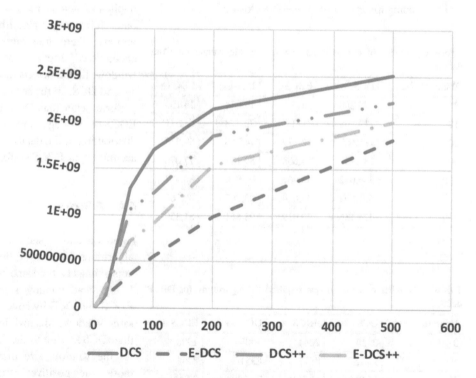

5 Conclusion

With the growth of dataset sizes, powerful entity matching techniques have become more and more important (Zaharia et al. 2012). Traditional Sorted Neighborhood Method is the first family algorithm based on which the adaptive windowing idea has emerged. This adaptivity proposed in Duplicate Count Strategy has permitted improving record linkage quality. Varying slightly the windowing step of the latter strategy permits to efficiently respond to different similarity regions for different cluster sizes within a dataset. In this paper, we have suggested two different strategies to adapt the window size, with the proposed algorithms as the

Fig. 4 Precision values for Cora dataset

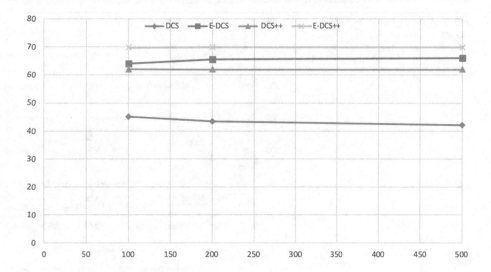

Fig. 5 Precision values for DBLP dataset

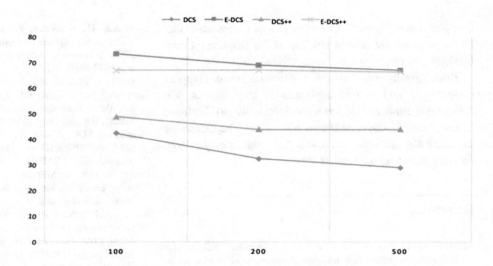

Fig. 6 F-Score values for Cora dataset

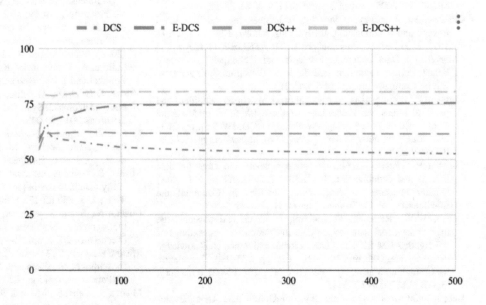

Fig. 7 F-Score values for DBLP dataset

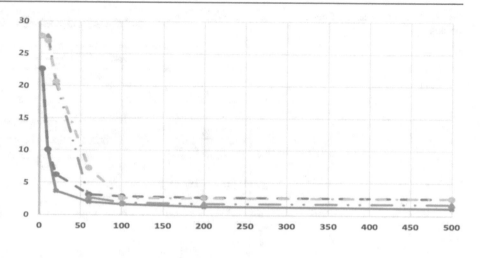

best performing. In Sect. 3 we have proven theoretically that our approaches are at least efficient as the Duplicate Count Strategy and can save more comparisons.

Both algorithms utilize transitive closure to save complex comparisons and to find duplicates in large blocks. We believe that enhanced DCS is a good alternative to Duplicate Count Strategy. The experiments have shown the increasing in efficiency allowing to search for duplicates in large datasets within an appropriate time.

References

Aizawa, A., Oyama, K.: A Fast Linkage Detection Scheme for Multi-Source Information Integration, International Workshop on Challenges in Web Information Retrieval and Integration. Tokyo, pp. 30–39 (2005). https://doi.org/10.1109/WIRI.2005.2

Christen, P.: A survey of indexing techniques for scalable record linkage and deduplication. IEEE Trans. Knowl. Data Eng. **24**(9), 1537–1555 (2012). https://doi.org/10.1109/TKDE.2011.127

Christen, P.: Data matching: concepts and techniques for record linkage. Entity resolution, and duplicate detection. Springer Publishing Company Incorporated. (2012)

Christen, P.: A comparison of personal name matching: techniques and practical issues. In: Proceedings of the Sixth IEEE International Conference on Data Mining—Workshops (ICDMW '06), pp. 290–294. IEEE Computer Society, USA (2006). https://doi.org/10.1109/ICDMW.2006.2

Christen P., Goiser, K.: Quality and complexity measures for data linkage and deduplication. In: Guillet, F.J., Hamilton, H.J. (eds.) Quality measures in data mining. Studies in Computational Intelligence, vol 43. Springer, Berlin, Heidelberg (2007)

Cohen, W.W., Richman, J.: Learning to match and cluster large high-dimensional data sets for data integration. In: Proceedings of the eighth ACM SIGKDD international conference on Knowledge discovery and data mining (KDD '02), pp. 475–480. Association for Computing Machinery, New York, NY, USA (2002). https://doi.org/10.1145/775047.775116

Deb, K., Kalyanmoy, D.: Multi-Objective Optimization Using Evolutionary Algorithms. Wiley, USA (2001)

Draisbach, U., Naumann, F., Szott, S., Wonneberg, O.: Adaptive windows for duplicate detection. In Proceedings of the 2012 IEEE 28th International Conference on Data Engineering (ICDE '12), pp. 1073–1083. IEEE Computer Society, USA 2012). https://doi.org/10.1109/ICDE.2012.20

Hernández, M.A., Stolfo, S.J.: Real-world data is dirty: data cleansing and the merge/purge problem. Data Min. Knowl. Disc. **2**, 9–37 (1998). https://doi.org/10.1023/A:1009761603038

Hernández, M.A., Stolfo, S.J.: The merge/purge problem for large databases. SIGMOD Rec. **24**(2), 127–138 (1995). https://doi.org/10.1145/568271.223807

Kenig, B., Gal, A.: MFIBlocks: An effective blocking algorithm for entity resolution. Inf. Syst. **38**, 908–926. (2013) https://doi.org/10.1016/j.is.2012.11.008

Kolb, L., Thor, A., Rahm, E.: Load balancing for MapReduce-based entity resolution. In: Proceedings of the 2012 IEEE 28th International Conference on Data Engineering, ICDE12, pp. 618–629. IEEE Computer Society (2012)

McCallum, A., Nigam, K., Ungar, L.H.: Efficient clustering of high-dimensional data sets with application to reference matching. In: Proceedings of the sixth ACM SIGKDD International Conference on Knowledge Discovery and Data Mining (KDD '00), 169–178. Association for Computing Machinery, New York, NY, USA (2000). https://doi.org/10.1145/347090.347123

McCallum, A.: Cora dataset. ICPSR—Inter University Consortium for Political and Social Research. (2017)

Mestre, D.G. et al.: An efficient spark-based adaptive windowing for entity matching. J. Syst. Softw. **128**, 1–10 (2017). https://doi.org/10.1016/j.jss.2017.03.003.

Mishra, S., Mondal, S., Saha, S.: Entity matching technique for bibliographic database. In: Database and expert systems applications, pp. 34–41. Springer (2013)

Mishra, S., Saha, S., Mondal, S.: Gaemtbd: Genetic algorithm based entity matching techniques for bibliographic databases. Appl. Intell. **47**(1), 197–230 (2017). https://doi.org/10.1007/s10489-016-0874-z

Mishra, S., Saha, S., Mondal, S.: Cluster validation techniques for bibliographic databases. In: Students Technology Symposium (TechSym), 2014, pp. 93–98. IEEE (2014a)

Mishra, S., Saha, S., Mondal, S.: On validation of clustering techniques for bibliographic databases. In: 2014 22nd International Conference on Pattern recognition (ICPR), pp. 3150–3155. IEEE (2014b)

Mishra, S., Saha, S., Mondal, S.: A multiobjective optimization based entity matching technique for bibliographic databases. Expert Syst.

Appl. **65**, 100–115 (2016). https://doi.org/10.1016/j.eswa.2016.07.043

Reyes-Galaviz, O.F., Pedrycz, W., He, Z., Pizzi, N.J.: A supervised gradient-based learning algorithm for optimized entity resolution. Data Knowledge. Eng. **112**, 106–129 (2017). https://doi.org/10.1016/j.datak.2017.10.004

Steorts, R.C., Ventura, S.L., Sadinle, M., Fienberg, S.E.: A comparison of blocking methods for record linkage. In: Domingo-Ferrer, J. (eds.) Privacy in statistical databases. PSD 2014. Lecture Notes in Computer Science, vol, 8744. Springer, Cham (2014)

Tauer, G., Date, K., Nagi, R., Sudit, M.: An incremental graph-partitioning algorithm for entity resolution. Inform. Fusion **46**, 171–183 (2019)

Warshall. S.: A theorem on boolean matrices. J. ACM **9**(1), 11–12 (1962). https://doi.org/10.1145/321105.321107

Winkler, W.E.: Overview of record linkage and current research directions technical report RR2006. US Bureau of the Census (2006)

Yan, S., Lee, D., Kan, M-Y., Giles, L.C.: Adaptive sorted neighborhood methods for efficient record linkage. In: Proceedings of the 7th ACM/IEEE-CS Joint Conference on Digital libraries (JCDL '07), pp. 185–194. Association for Computing Machinery, New York, NY, USA (2007). https://doi.org/10.1145/1255175.1255213

Yin, X., Han, J., Yu, P.: Object distinction: Distinguishing objects with identical names. In: IEEE 23rd International Conference on Data Engineering, ICDE 2007, pp. 1242–1246, IEEE (2007)

Zaharia, M., Chowdhury, M., Das, T., Dave, A., Ma, J., McCauley, M., Franklin, M.J., Shenker, S., Stoica, I.: Resilient distributed datasets: a fault-tolerant abstraction for in-memory cluster computing. In: Proceedings of the 9th USENIX conference on Networked Systems Design and Implementation (NSDI'12). USENIX Association, USA (2012)

Smart Sustainable Cities: A Chatbot Based on Question Answering System Passing by a Grammatical Correction for Serving Citizens

Bghiel Afrae, Ben Ahmed Mohamed,
and Boudhir Anouar Abdelhakim

Abstract

The smart city concept has been developed as a strategy for working with cities as they become systematically more complex through interconnected frameworks and increasingly rely on the use of Information and Communication Technology to meet the needs of their citizens. At this end, as users struggle to navigate the wealth of on-line information now available, the need for automated question answering systems becomes more urgent. Most QA systems use a wide variety of linguistic resources. We focus instead on the redundancy available in large corpora as an important resource. QA System that described in this paper, built using a seq2seq model, BiLSTM as encoder, a LSTM as decoder, and Attention mechanism for boosting the model performance whether Grammatical Correction Model or Question Answering Model, we have proposed to build a grammatical model first to pass a corrected question to the second system in order to improve QA system results.

Keywords

Sustainability • Chatbot • Question and answer • Grammatical correction • Seq2Seq • BiLSTM • Attention mechanism • Natural language • Bleu score • Smart city

1 Introduction

Sustainability and sustainable development concepts generate awareness of the production and use of resources required for residential, industrial, transportation, commercial, or recreational processes. Applying sustainable development in a strategic manner is achieved through systems thinking approach, in order to build an intelligent environment for serving citizens basing on their needs, where communication and searching are the most important things that can be serving people to reach informations; chatbots and question answering systems can help citizens to reach answers and informations they need just by click, where they can write their question in natural language and wait QA System to generate the answer.

A.I. scientists have, for decades, underestimated the complexity of human language, in both comprehension and generation. The obstacle for computers is not just understanding the meanings of words, but understanding the end-less variability of expression in how those words are collocated in language use to communicate meaning. Nonetheless, decades later, we can find an abundance of natural language interaction with intelligent agents on the Internet, where dialogue or conversational systems including chatbots, and personal assistants are becoming ubiquitous in modern society.

Conversation is defined as an exchange between two or more sides in which thoughts, feelings, and ideas are expressed, questions are asked and answered, or news and information are exchanged, and while many Chatbots today offer the possibility of answering question, and attending customers' requests, few of them are able to provide these services without the need to limit user's input by command-based UI menus.

Thanks to the big data revolution and advanced computational capabilities, companies have never had such a deep access to customer data. This is allowing organizations to interpret, understand, and forecast customer behaviors as

B. Afrae (✉) · B. A. Mohamed · B. A. Abdelhakim
List Laboratory FSTT UAE, Tangier, Morocco
e-mail: afraebghiel1995@gmail.com

B. A. Mohamed
e-mail: mbenahmed@uae.ac.ma

B. A. Abdelhakim
e-mail: boudhir.anouar@gmail.com

© The Editor(s) (if applicable) and The Author(s), under exclusive license to Springer Nature Switzerland AG 2021
M. Ben Ahmed et al. (eds.), *Emerging Trends in ICT for Sustainable Development*,
Advances in Science, Technology & Innovation, https://doi.org/10.1007/978-3-030-53440-0_34

never before; and with the AI technologies, we are able to build a question answering system that is described in this paper, which is based on deep learning. We develop a set of deep learning models for natural language retrieval and generation—including recurrent neural networks, sequence-to-sequence models, and attention mechanism—and evaluate them in the context of natural language processing. Thus, we have used those technologies in a grammatical correction system, and a question answering system by passing the first model result as input to the second one, in order to improve the result.

In this chapter, we dive into the concept of Smart Sustainable Cities, where our document aims to build a question answering systems that help customers to get information about products, and it is organized as follows: in Sect. 2, we present a brief historical overview, in Sect. 3. We discuss chatbot's platforms and related works to those two models, and, in Sect. 4., we summarize the main of proposed model. Finally, in Sect. 5., we present results and then some conclusions and point to future challenges.

2 Historical Overview

The first chatbots were ELIZA (Bradeško and Mladenić 2012), PARRY (Bradeško and Mladenić 2012), and SHRDLU (Blom and Thorsen 2013). ELIZA was created in 1963, It was created by Joseph Weizenbaum (https://fr.wikipedia.org/wiki/Joseph_Weizenbaum), of the Massachusetts Institute of Technology (MIT), and took only 3 pages in SNOBOL language. She made great use of the echolalia technology. It started by asking a neutral question "Hello. Why are you coming to see me?" To the person tested, then analyzed each time the answer to try to ask a question related to it, if a question was asked, it asked why we asked her the question, if a sentence contained the word computer, it asked: "Do you say that because I am a machine?".

Chatbot competitions are organized each year to promote emulation in this area. Currently, we are witnessing the birth of many chatbots, partly thanks to the Loebner Prize, which tries to be a kind of Turing test. During these competitions, in order to evaluate chatbots, the jury can dialogue either with a human or with a chatbot through a keyboard/screen interface. After a certain time, the jury must assess whether the candidate tested is a machine or not. The most daring chatbots are those who try to extract knowledge from their conversations like ECTOR (in English, learning bots), but they are also those whose results are currently the least impressive.

The first chatbots were developed to perform improved notification processes. However, later, the new chatbots deployed respond to more complex requests such as financial advice, savings, or meeting planning (https://fr.wikipedia.org/wiki/Chatbot).

3 Related Work

The application of chatbot and its technology has seen a drastic improvement since the very first chatbot was created in 1966; the attempt a lot of research has been done to make a developed chatbot, and to explore new approaches; and here we specify the seq2seq model development, for different similar problems to question answering system as machine translation, grammatical correction, etc.

In Cho et al. (2014) propose a comparison between different RNN such LSTM and GRU for seq2seq modeling, where they well established in the field that the LSTM unit works well on sequence-based tasks with long-term dependencies. In the same context, a research team (Zaremba et al. 2014) have used a multi-layered Long Short-Term Memory (LSTM) to map the input sequence to a vector of a fixed dimensionality, and then another deep LSTM to decode the target sequence from the vector, where they achieved a BLEU score of 34.8. Grammatical error correction is one such application of potential utility as a component of a writing support tool. Much of the recent work in grammatical error identification and correction has made use of hand-tuned rules and features that augment data-driven approaches, or individual classifiers for human-designated subsets of errors (Schmaltz et al. 2016).

Most of these approaches are based on classification or statistical machine translation (SMT), where Ng and al proposed a combination between the output from a classification-based system and an SMT-based system to improve the correction quality (Susanto et al. 2014).

4 Models

In this section, we describe the details of our proposed methods, including data preprocessing, neural networks, and ensemble strategy.

4.1 The Proposed Model

Here, we introduce our approach for question answering tasks, we propose an additional task for a question answering system which is a grammatical correction for the question in input, where the question first to the grammatical correction model, which is based on the same approach of question answering system for generating the corrected Question.

Given a correct question Q from the first model, we first preprocess the question, encode it and pass it to embedding model—in our case we have used pre-trained model glove for obtaining vector of 100 dim for each word—then runs the input through a BiLSTM or pre-attention RNN that comes before the attention mechanism to get back [<h0 > , <h1> , ..., <ht>], where we use <h1> = [h1–>, h1 <–] to represent the concatenation of the activations of both the forward and backward directions of the pre-attention BiLSTM. The usual RNN described in reads an input sequence x in order starting from the first symbol × 1 to the last one xTx.

However, in the proposed scheme, we would like the annotation of each word to summarize not only the preceding words, but also the following words. Hence, we propose to use a bidirectional RNN.

The output of bi-LSTM for each word pass to attention mechanism concatenated with the repeated previous hidden state of the second LSTM (si) till the max length of question for calculating the attention variables α-i, which are used to compute the context variable < C > for each time step in the output (i = 0, ..., question max length), then we Give context vector to the post-attention LSTM cell which comes after the attention mechanism, and pass in the previous hidden state \$s i−1 and cell-states cell i−1 of this LSTM using initial state = [previous hidden state, previous cell state]. Get back the new hidden state \$$st$ and the new cell state cell i. Finally, we apply a Softmax layer to get the output and save it till adding others outputs to generate the answer.

4.2 Seq2seq Model

In this study, our parser is built upon a seq2seq model (Zaremba et al. 2014), which encodes the input sentence forward and backward, and predicts the head position for each word in the sentence. Our model contains three main components: (1) an encoder that processes the input sentence and maps it into some hidden states that lie in a low dimensional vector space Rh, (2) a decoder that incorporates the hidden states and the previous prediction to generate head position of the current word, and (3) an attention layer that encodes the context for each focused word. Figure 1 illustrates our model. Recurrent neural networks (RNNs) (Elman 1990) are commonly used to build block for seq2seq model (Fig. 2).

Specifically, our model employs the Long Short-Term Memory (LSTM) (Hochreiter and Schmidhuber 1997)

variant of RNN, which uses numbers of gates to address the problem of vanishing or exploding gradient when trained with back-propagation through time. A standard LSTM unit consists of an input gate it, a forget gate fi, an output gate oi, a memory cell ci, and a hidden state hi, where the subscript t denotes the time step.

4.2.1 Encoder Model

The encoder of our model is a BiLSTM, which processes an input sequence in both directions by incorporating a stack of two distinct LSTMs.

Given an input sentence X = $\{x_1, \cdots, x_t\}$, the $i - th$ element xi is encoded by first feeding its embedding representation and into two distinguish LSTMs: the forward LSTM and the backward LSTM, to obtain two hidden state vectors: h_{fi} and $h_{bi,}$ respectively, and then concatenating the two vectors as $h_i = [h_{fi}; h_{bi}]$.

Question Representation (GLOVE Embedding)

Given a vocabulary V, each individual word $w_i \in$ V is mapped into a real-valued vector (word embedding) w $\in$ Rm where m = 100 is the dimension. We employ the Glove (Pennington et al. 2014) to generate the pre-trained word embedding, obtaining a distinct embedding for each work; so, the adopted word representation of our model is E = [e1..., e100].

Glove (Pennington et al. 2014), Global Vectors, is an unsupervised learning algorithm for obtaining vector representations for words. Training is performed on aggregated global word-word co-occurrence statistics from a corpus, and the resulting representations showcase interesting linear substructures of the word vector space (Rezaeinia et al. 2017) (Fig. 3).

BiLSTM

(BRNN) (Schuster and Paliwal 1997; Pollastri et al. 1999) is used to present each training sequence forwards and backwards to two separate recurrent nets, both of which are connected to the same output layer; what can overcome problem of the standard RNNs is that they have access to past but not to future context. The amount of context

Fig. 1 General architecture of the model based

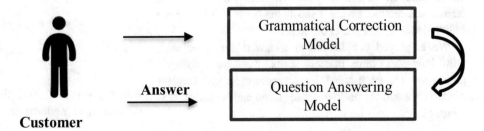

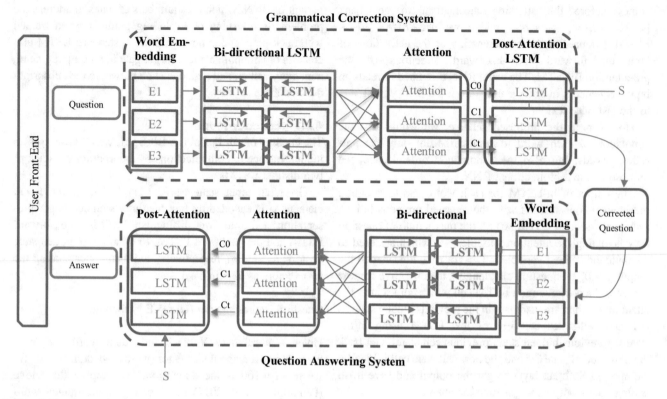

Fig. 2 The proposed model

Fig. 3 Word embedding (Glove)

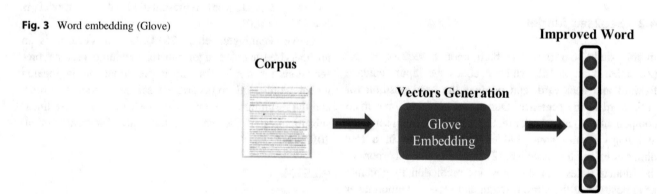

information that the network actually uses is learned during training and does not have to be specified beforehand. Thereby, forward and backward context are learned independently from each other.

However, often a small buffer is enough in order to profit from bidirectional context, so that bidirectional networks can also be applied for causal systems whenever a short output latency is tolerable. Figure 4 shows the structure of a simple bidirectional LSTM (Li et al. 2018).

We employed a bidirectional recurrent neural network with long short-term memory units, then we obtain an annotation for each word by concatenating the forward hidden state and the backward one (Jason and Nichols 2015) (Fig. 5).

4.2.2 Decoder Model

The decoder is often trained to predict the next word w_{t0}, given the context vector c and all the previously predicted words $\{y_1, ..., y_t \, 0 - 1\}$. In other words, the decoder defines a probability over the translation y by decomposing the joint probability into the ordered conditionals:

$$p(y) = \prod_{t=1}^{T} p(y_t | \{y_1, ..., y_{t-1}\}, c) \qquad (1)$$

where $y = y1, ..., yt$. With an RNN, each conditional probability is modeled as

Fig. 4 BiLSTM

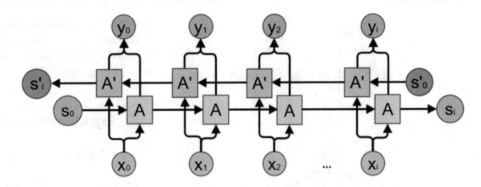

Fig. 5 BiLSTM Work

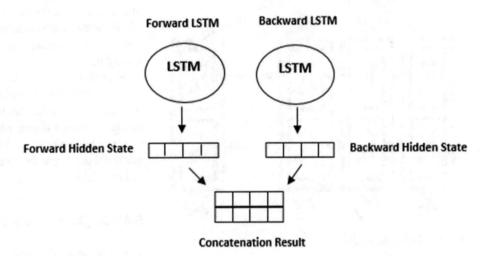

But in our model we can explain the previous equation by:

$$p(y_t|\{y_1,\ldots,y_{t-1}\},x) = g(y_{t-1},s_t,c) \tag{2}$$

And the hidden state can be writed as

$$S_i = f(S_{i-1},y_{i-1},C_i) \tag{3}$$

where g is a nonlinear, potentially multi-layered, function that outputs the probability of yt, and ht is the hidden state of the RNN. It should be noted that other architectures such as a hybrid of an RNN and a de-convolutional neural network can be used (Kalchbrenner and Blunsom 2013).

Here the probability is conditioned on a distinct context vector ci for each target wordy_i, The context vector ci depends on a sequence of annotations $(h_1,\ldots, h_{Tx})$ to which an encoder maps the input sentence. Each annotation hi contains information about the whole input sequence with a strong focus on the parts surrounding the i-th word of the input sequence. We explain in detail how the annotations are computed in the next subsection where we will detail the attention mechanism. By letting the decoder have an attention mechanism, we relieve the encoder from the burden of having to encode all information in the source sentence into a fixed length vector. With this new approach the information can be spread throughout the sequence of annotations, which can be selectively retrieved by the decoder accordingly (Fig. 6).

4.2.3 Attention Mechanism

As the graphical illustration shows, the proposed model is trying to generate the t-th target word y_t given a source sentence $(x_1,x_2,\ldots, x_T)$.

The attention mechanism is applied which allows the decoder to look at the input sequence selectively, so we employed an attention layer (Luong et al. 2015) to encode the context for each word, The context vector ci at the t-step of the decoding process is calculated as the weighted sum of the hidden states of the input sequence:

$$C_i = \sum_{j=1}^{Tx} \alpha_{ij}h_j \tag{4}$$

The weight α_{ti} of each annotation h_i is computed by

$$\alpha_{ij} = \frac{\exp(e_{ij})}{\sum_{k=1}^{Tx} \exp(e_{ik})} \tag{5}$$

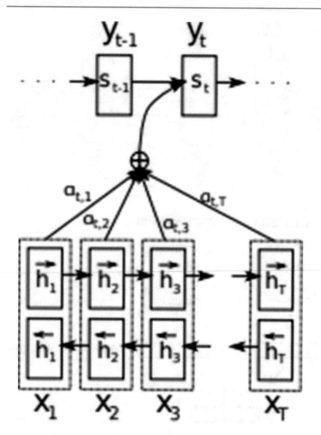

Fig. 6 Attention mechanism [18]

$$e_{ij} = a\left(s_{i-1}, h_j\right) \tag{6}$$

where (6) is an alignment model which scores how well the inputs around position j and the output at position i match. The score is based on the RNN hidden state s_{i-1} (just before emitting yi, Eq. (4)) and the j-mh annotation h_i of the input sentence. w_e parametrize the alignment model as a feedforward neural network which is jointly trained with all the other components of the proposed system.

The alignment is not considered to be a latent variable. Instead, the alignment model directly computes a soft alignment, which allows the gradient of the cost function to be backpropagated through. This gradient can be used to train the alignment model as well as the whole question answering model jointly. We can understand the approach of taking a weighted sum of all the annotations as computing an expected annotation, where the expectation is over possible alignments. Let α_{tj} be a probability that the target word y_i is aligned to, or translated from, a source word x_i. Then, the t-mh context vector ci is the expected annotation over all the annotations with probabilities α_{ij}. The probability α_{ij}, or its associated energy e_{ij}, reflects the importance of the annotation h_i with respect to the previous hidden state s_{i-1} in

deciding the next state s_t and generating y_i. Intuitively, this implements a mechanism of attention in the decoder. The decoder decides parts of the source sentence to pay attention to Li et al. (2018).

5 Experiments and Evaluation

5.1 Dataset

We have used a part of lang-8 corpus, the largest publicly available parallel corpus to train our grammatical Correction Model, where we have used 10,000 pair (sentence, corrected sentence) for training our model and about 2000 pair for testing it;

For training our question answering system we have used Amazon question answering open data (https://registry. opendata.aws/); where we have used about 20,000 questions and answers for training and 5000 pair for testing our model.

First, we preprocess data using different Natural Language Processing techniques for passing them to models for training them.

5.2 Seq2seq Model

The grammatical correction system and the question answering system are evaluated by using different key performance measures such as accuracy, and Blue Score where:

BLEU Score is a metric used to evaluate a sentence generated in a sentence of reference, proposed by Papineni et al. (2001) in 2002. A perfect match gives a score of 1.0, whereas a perfect discordance gives a score of 0.0. The Blue Score allows us to automatically calculate a score that measures the resulting phrase quality (Papineni et al. 2001).

And Accuracy (Shalev-Shwartz and Ben-David 2014) is defined as the number of times that the model predicts the correct answer for inputs, the accuracy parameter determines how far the output can be from the optimal one (this corresponds to the "approximately correct"):

$$Accuracy = \frac{(TP + TN)}{(TP + TN + FP + FN)} \tag{8}$$

5.3 Training and Results

We trained Our models conducting several experiments for different optimizers such RMSprop (Ruder 2017), Adam (Ruder 2017), RAdam (https://medium.com/@ lessw/new-state-ofthe-art-ai-optimizer-rectified-adamradam 5d854730807b), (Liu et al. 2019), and SGD (Ruder 2017);

Table 1 Training results

Model	Optimizer	Bleu score	Accuracy
Grammatical correction	ADAM	0.34	0.47
Grammatical correction	RMSprop	0.39	0.49
Grammatical correction	RAdam	0.38	0.35
Grammatical correction	**SGD**	**0.57**	**0.78**
Question answering	ADAM	0.39	0.47
Question answering	RAdam	0.4	0.45
Question answering	RMSprop	0.31	0.71
Question answering	**SGD**	**0.51**	**0.68**

and other hyper parameters, where our results are summarized in Table 1.

We noticed that Adam and RAdam give approximately same results although the last one provides an automated, dynamic adjustment to the adaptive learning rate based on their detailed study into the effects of variance and momentum during training, also RMSprop, while SGD has shown some good results that exceeded the 50% as bleu score.

6   Conclusion

Smart Sustainable Cities is an aggregate concept. In this chapter, we have shown that each of the constituent concepts —smart, sustainable, and cities—is important in its own right. Cities can't be made sustainable without the use of smart technology especially with the AI and Big data revolution, and smart technologies can also be used for sustainable development in venues other than cities.

Citizens and Users are rapidly turning to social media, assistant robots, and websites to request and receive customer service; however, a majority of these requests were not addressed timely or even not addressed at all. To overcome the problem, we create a new conversational system to automatically generate responses for user's requests. Our system is integrated with state-of-the-art deep learning techniques, where we have used recurrent neural networks, sequence-to-sequence models, and attention mechanism— and evaluate them in the context of natural language processing. Thus, we have used those technologies in a grammatical correction system, and a question answering system

by passing the first model result as input to the second one, in order to improve the result.

References

Blom, A., Thorsen, S.: A sentiment-based chat bot (2013)

Bradeško, L., Mladenić, D.: A survey of chabot systems through a loebner prize competition (2012)

Cho, K., van Merrienboer, B., Universite de Montr ´ eal ´, Bahdanau, D.: On the properties of neural machine translation: encoder–decoder approaches (2014)

Elman, J.L.: Finding structure in time. Cogn. Sci. **14**, 179–211 (1990)

Hochreiter, S., Schmidhuber, J.: Long short-term memory (1997)

https://fr.wikipedia.org/wiki/Chatbot

https://fr.wikipedia.org/wiki/Joseph_Weizenbaum

https://hackernoon.com/attention-mechanism-in-neural-network-30aaf5e39512

https://medium.com/@lessw/new-state-ofthe-art-ai-optimizer-rectified-adamradam5d854730807b

https://medium.com/@raghavaggarwal0089/bi-lstm-bc3d68da8bd0

https://registry.opendata.aws/

Jason, P.C., Nichols, E.: Named entity recognition with bidirectional LSTM-CNNs. Chiu University of British Columbia (2015)

Kalchbrenner, N., Blunsom, P.: Recurrent continuous translation models (2013)

Li, Z., Cai,J., He, S., Zhao, H.: Seq2seq dependency parsing (2018)

Liu, L., Jiang, H., He, P., Chen, W., Liu, X., Gao, J., Han, J.: On the variance of the adaptive learning rate and beyond (2019)

Luong, M-T., Pham, H., Manning, C.D.: Effective Approaches to Attention-based Neural Machine Translation (2015)

Papineni, K., Roukos, S., Ward, T., Zhu, W.-J.: Bleu: a method for automatic evaluation of machine translation (2001)

Pennington, J., Socher, R., Christopher, D.: Manning: global vectors for word representation (2014)

Pollastri, G., Baldi, P., Fariselli, P., Casadio, R.: Improved prediction of the number of residue contacts in proteins by recurrent neural networks (1999)

Rezaeinia, S.M., Ghodsi, A., Rahmani, R.: Improving the accuracy of pre-trained word embeddings for sentiment analysis (2017)

Ruder, S.: An overview of gradient descent optimization algorithms (2017). arXiv:1609.04747v2.

Schmaltz, A., Kim, Y., Rush, A.M., Shieber, R.M.: Sentence-level grammatical error identification as sequence-to-sequence correction (2016)

Schuster, M., Paliwal, K.K.: Bidirectional recurrent neural networks (1997)

Shalev-Shwartz, S., Ben-David, S.: Understanding machine learning: from theory to algorithms (2014)

Susanto, R.H., Phandi, P., Ng, H.T.: System combination for grammatical error correction (2014)

Zaremba, W., Sutskever, I., Vinyals, O.: Recurrent neural network regularization (2014)

Enhancing Wireless Transmission Efficiency for Sensors in Wireless Body Area Sensor Networks

Rahat Ali Khan, Shahzad Memon, and Qin Xin

Abstract

Wireless Body Area Sensor Network (WBASN) has gained much significance in recent time and a lot of researches are being carried in this field because it is related to the betterment of humans. Sensing devices called sensors are used on the human body to observe several physiological parameters. These sensors are easily carried on the human body because of their size being very tiny and able to communicate the observed data wirelessly. Wireless communication has advantages over wired communication being very less complex to be carried. Transmission being wireless has to experience path loss which makes it weaker and at the destination end, the signal is not what needs to be, so in this work, we present a new scheme for wearable on-body sensors so that the path loss can be reduced. The proposed scheme is based on transposing of the sensors used on the human body. Path loss is reduced based on its distance parameter.

Keywords

WBASN • Sensors • Energy • Transposition • Distance • Path loss

R. A. Khan (✉)
Department of Telecommunication Engineering, Faculty of Engineering and Technology, University of Sindh, Jamshoro, Pakistan
e-mail: rahat.khan@usindh.edu.pk

S. Memon
Department of Electronic Engineering, Faculty of Engineering and Technology, University of Sindh, Jamshoro, Pakistan
e-mail: shahzad.memon@usindh.edu.pk

Q. Xin
Faculty of Science and Technology, University of the Faroe Islands, Tórshavn, Faroe Islands
e-mail: qinx@setur.fo

Abbreviations

BAN	Body Area Networks
BS	Base Station
BSN	Body Sensor Networks
CF	Cost Function
CH	Cluster Head
Co-LAEEBA	Cooperative Link-Aware and Energy Efficient protocol for wireless Body Area networks
CSI	Channel State Information
dB	Decibels
ECG	ElectroCardioGram
EDA	Euclidean Distance Algorithm
EE	Energy Efficiency
EECBR	Even Energy Consumption and Backside Routing
EEG	ElectroEncephaloGram
E-HARP	Energy-efficient Harvested Aware clustering and cooperative Routing Protocol for WBAN
EH-RCB	Energy Harvested-aware Routing protocol with Clustering approach in Body area networks
ELR-W	Energy aware Link efficient Routing approach for WBANs
EMG	ElectroMyography
EMSMO	Enhanced Multi-objective Spider Monkey Optimization
EOCC	Energy Optimized Congestion Control
ICT	Information and Communication Technology
IoMT	Internet of Medical Things
IoT	Internet of Things
MEMS	Micro Electro Mechanical Systems
MH	Multi Hop
MHz	Mega Hertz
MICS	Medical Implant Communication Service

M. Ben Ahmed et al. (eds.), *Emerging Trends in ICT for Sustainable Development*,
Advances in Science, Technology & Innovation, https://doi.org/10.1007/978-3-030-53440-0_35

NEMS	Nano Electro Mechanical Systems
OMNeT++	Objective Modular Network Testbed in C++
pH	Potential Hydrogen
SDN	Software Defined Networking
SH	Single Hop
SN	Sensor Node
SNR	Signal-to-Noise Ratio
SO	Square Odd
TARA	Temperature Aware Routing Algorithm
TDMA	Time Domain Multiple Access
WBAN	Wireless Body Area Network
WBASN	Wireless Body Area Sensor Network
WBSN	Wireless Body Sensor Network
WSN	Wireless Sensor Networks

1   Introduction

Developing trend in Information and Communication Technology (ICT) have been so rapid that within a small time researches have emerged to develop small scaled devices (Jafri et al. 2011, 2017). These small scaled devices known as Micro Electro Mechanical Systems (MEMS) and Nano Electro Mechanical Systems (NEMS), are very small in size so that they have to be fitted on a piece of cloth. These devices have multiple abilities like they can observe parameters, process the observed data, and communicate the data wirelessly. These devices can be used to form Wireless Sensor Network (WSN). WSN is a type of network requiring small infrastructure or with no infrastructure. WSN comprises of a large number of these devices which form a network whenever there is a need (Pathan et al. 2006).

Wireless Body Area Sensor Network (WBASN) introduced by Van Dam (2001) is a subfield of WSN (Dam et al. 2001). WBASN are sometimes alternatively named as Wireless Body Area Network (WBAN), Wireless Body Sensor Network (WBSN), Body Area Networks (BAN), and Body Sensor Networks (BSN) (Khan et al. 2018,2016). The WBASN is related to those sensors which can be used to monitor physiological changes. These sensors can be categorized as either being invasive or noninvasive (Jung et al. 2014; Yang et al. 2015; Al Rasyid et al. 2015; Muramatsu et al. 2015; Hess et al. 2015; Figueiredo et al. 1504; Sajatovic et al. 2015; Kölbl et al. 2014; Rathbun et al. 2015; Sugiura et al. 2017; Rao and Chiao 2015; Peter et al. 2016; Zheng et al. 2016; Xu and Hua 2016) as shown in Table 1.

The noninvasive sensors are also known as wearable sensors comprising of ECG, EEG, EMG, and other sensors while pacemaker, deep brain stimulator, and others come into the category of invasive sensors. Invasive sensors are also known as implantable sensors, which means a sensor can be inserted into the human body by surgery.

The invasive sensors are those devices which can be placed inside the human body while noninvasive sensors are those devices which can be placed on the human body. These sensors are required to be placed on a specific point so that they may observe the physiological changes in the human body as shown in Fig. 1. It can be observed in Fig. 1, that there are places where sensors can be placed for monitoring. These sensors have capabilities of sensing, processing, and then communicating the data through a wireless medium, making them easier for use, because initially we had wired sensing devices which were complicated to use and carry around.

WBASN architecture (as shown in Fig. 2), works as: sensors communicate the sensed data to a device Base Station (BS) or sink node. The base station collects data from entire sensing nodes and communicates to the entities which can further handle the data like if data is critical then it can send a signal to ambulance service. Simultaneously, the data is sent to the doctor so that any timely action may be performed on arrival and also to the database to store the data.

These sensors operate on batteries which they carry within their hardware. As the size of the sensors is very small so they contain small batteries with them. As these sensors are related to human health and charging of the batteries is somehow a complicated process. So there is a need to design an algorithm in such a way that the sensors may operate for a longer span of time. In this chapter transposing is performed using Euclidean Distance Algorithm (EDA) to design and

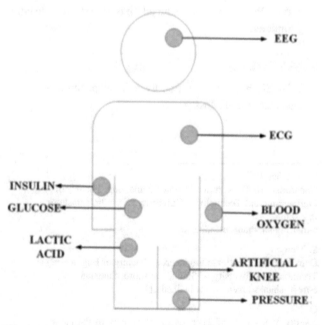

Fig. 1 Placement of sensor on the human body

Table 1 Non-invasive/wearable sensors invasive/implantable sensors

Invasive/Implantable sensors	Non-invasive/Wearable sensors
Pacemaker	Electrocardiogram (ECG)
Wireless capsule endoscope (electronic pill)	Glucose sensor
Deep brain stimulator	Electromyography (EMG)
Retina implants	Electroencephalogram (EEG)
Cochlear implants	Temperature
Electronic pill for drug delivery	Pulse oximeter
Implantable defibrillators	Blood pressure Oxygen, pH value

develop an algorithm to make the sensors work efficiently by saving the maximum amount of energy.

WBASN can also be considered as the Internet of Medical Things (IoMT). Internet of Medical Things is a technology in which the internet is the backbone for medical devices. Whatever the devices record or observe data is sent using the internet to the authorities like in WBASN.

Further, this chapter is organized as: Related work is briefly described in Sect. 2, Mathematical modeling is presented in Sect. 3

2 Related Work

Researchers in (Liu et al. 2017), proposed Time Domain Multiple Access (TDMA) based protocol. Time is allocated to data in order to use energy in a better managed way. TDMA adjusts transmission. This minimization optimizes energy consumption.

Researchers in (Sahndhu et al. 2015), propose a routing protocol in which they have utilized intermediate nodes.

These nodes have been used to gather data from sensor nodes and after gathering data aggregation is performed. This aggregated data is then sent to the sink node. Threshold value has been set for the sensor nodes. If any of the sensors depletes its energy and its energy is approaching threshold then it will only transmit critical data instead of sending normal data.

Authors in (Soontornpipit 2014), have measured data for radio propagation frequencies of 402 megahertz (MHz) to 405 megahertz. They have used the Medical Implant Communication Service (MICS) band for evaluating path loss. Their simulation results are based on measurement for two scenarios. One is with furnished and second is with the unfurnished room.

In (Kumari et al. 2019), researchers have focused on the cognitive fatigue level monitoring of vehicle drivers. They have named some parameters which they have described as contributors towards increasing fatigue levels. These parameters are driving comfort, type of vehicle being driven, health of drivers, drowsiness or insufficient sleep of drivers, continuous driving, road conditions, and traffic on road.

Fig. 2 WBASN architecture

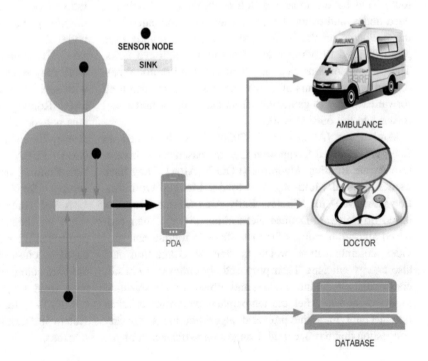

These parameters affect driving and driver's behavior towards driving. The authors have tried to propose a method to detect a cognitive fatigue level so that the accidents caused by drivers during driving vehicles. They have used the K-mean algorithm in the proposed system. They divided testing into two classes as pre driving mode and post driving mode. In the proposed algorithm Square Odd (SO) scanning technique is used for detecting an object. SO saves a significant amount of power as it is efficient and effective in detecting objects. It also switches sensors used between two modes, i.e., sleep mode and awake mode which is very useful is saving sensor power.

A new energy-efficient routing protocol for wireless body area sensor network has been proposed by researchers in (Yang et al. 2020). Their proposed system has involved scheduling of the data transmission also. The proposed system has taken into account several parameters like path loss, type of traffic, node energy and others. In order to make a reduction in traffic flooding, they have proposed to use channel competition. The proposed system first initializes, then there is a setup phase in which routes for communication are selected and they are selected on the basis of having abundant energy. In this way, the sensor nodes consume lesser energy, hence increasing network lifetime.

Rahat et al. in (Khan et al. 2018), have proposed a novel energy-efficient routing protocol for wireless body area sensor networks in which they have proposed to use eight wearable sensor nodes. These sensor nodes are given coordinates in their x and y coordinates and sink node is placed at the center of the body. They have compared their proposed protocol with the one that was already existing protocol. They have compared path loss, throughput, number of dead nodes, and residual energy. They have compared both the protocols on the basis of the number of rounds each protocol need to perform all of the aforementioned tasks. The comparison was clear that their proposed protocol performed better in terms of all the parameters. They also used forwarder nodes to gather data from other sensor nodes and forward to the outside world.

Authors in (Ahmed et al. 2020), have developed an Energy Optimized Congestion Control based on Temperature Aware Routing Algorithm (EOCC-TARA). They have used Enhanced Multi-objective Spider Monkey Optimization (EMSMO) for Wireless Body Area Sensor Networks based on Software-Defined Networking (SDN). Their proposed algorithm reduces thermal effects in routing and provides communication which is free of congestion and also energy efficient. Their proposed algorithm first considers thermal effects into routing and provides route selection to sensor nodes. After the temperature parameter is taken into account then the proposed algorithm looks for any congestion that has occurred. Congestion avoidance scheme, link reliability, energy efficiency and path loss are added based on EMSMO so that optimized routing is performed.

In the paper (Amjad et al. 2020), researchers have tried to optimize energy efficiency (EE) of wireless body area sensor networks. They have formulated a problem. This problem is to optimize EE of the network from sensor nodes acting as transmitters to the aggregator sensor nodes. In this formulation channel state information (CSI) is not considered.

In (Archasantisuk and Aoyagi 2019), authors have proposed transmission power control. They have done this due to variations in path loss. The proposed algorithm is based on motion-aware temporal correlation model. Their proposed algorithm is the combination of transmission power control and human motion classification. This combination has been performed to provide energy-efficient and reliable communication. Human movement is one of the major factors which have significane on the characteristics of WBASN communication. This movement is classified on the basis of identification of the motion which is performed using received signal strength.

Zahid et el in (Ullah et al. 2019), have proposed energy protocol named as "Energy-efficient Harvested-Aware clustering and cooperative Routing Protocol for WBAN (E-HARP)". The proposed algorithm uses a selection of Cluster Head (CH) based on multiple attribute technique. This is the main aim of the proposed algorithm in order to save the energy in the sensor nodes. As the round starts the proposed scheme selects CH. This selection is from all the sensor nodes that form a cluster. The CH selection criteria are based on Cost Factor (CF). CF is to be calculated by various parameters like communication link signal-to-noise ratio (SNR), cumulative energy loss of the network, transmission power needed, and residual energy of the sensor node. The CF is to be calculated in every round so that the every CH should have equal load distribution and the network may remain stable. They have compared their results with several existing algorithms like (i) Energy Harvested-aware Routing protocol with Clustering approach in Body area networks (EH-RCB), (ii) Cooperative Link-Aware and Energy Efficient protocol for Wireless Body Area networks (Co-LAEEBA), (iii) Even Energy Consumption and Backside Routing, and (iv) energy aware link efficient routing approach for WBANs (ELR-W).

A Soft computing approach based energy-efficient algorithm has been proposed by (Rakhee 2018). They have performed the balancing of data packets of all sensor nodes based on clustering with Ant Colony algorithm. They have used ant colony probabilistic function for data routing. They have used Objective Modular Network Testbed in C++ (OMNeT++) for simulation and have found their proposed system performing better as compared to conventional schemes.

3 Mathematical Modeling of Energy

In this section, the mathematical formulations are discussed to justify energy is the most important parameter in WBASN. There are two modes of communication in WBASN as either Single Hop (SH) or Multi Hop (MH). Energy that is consumed during SH communication is mathematically represented as shown in Eq. 1

$$E_{SH} = E_{TX} \qquad (1)$$

E_{TX} is energy consumed by the transmission. As observed in equation 1 in Single Hop communication there is only parameter i.e. transmission that consumes energy.

The second mode of communication is the MH which is mathematically represented as shown in Eq. 2

$$E_{MH} = S * N * \left[E_{TX} + (E_{DA} + E_{RX}) * \frac{N-1}{N} \right] \qquad (2)$$

In Eq. 2, S is the packet size, N being number of nodes used in multi hop communication, energy used in data aggregation is represented by E_{DA}, and E_{RX} is the energy that is used in receiving of the signal.

The power that is consumed in transmission is mathematically represented as shown in Eq. 3

$$E_{TX} = (E_{AMP} + E_{ELEC}) \times S \times D^2 \qquad (3)$$

The cumulative energy that a sensor node consumes is given in Eq. 4

$$E_{NODE} = E_{TX} + E_{RETX} + E_{ACK} + E_{ACC} \qquad (4)$$

E_{RETX} represents the energy being consumed when the signal needs to be retransmitted, channel processing energy is represented as E_{ACC} and the last parameter is E_{ACK} which is the power consumed when an acknowledgement packet is transmitted.

From Eq. 4, it can be observed that out of 4 parameters, 3 are related to transmission. It makes communication more important in terms of energy consumption. Sensors in WBASN have capabilities to perform functions like sensing, processing, and to set the observed data wirelessly so that these functionalities consume energy. In Fig. 3, it can be observed that energy consumed in communication is more than that for the other two parameters (Khan et al. 2017). So communication needs to be accurate and reliable. If for instance, communication is not accurate, then the data needs to be retransmitted. This retransmission again consumes energy again. This will drain the battery of the sensor much faster. The sensors that are used in WBASN are of tiny size and need to be efficient enough so that they may operate for a longer span of time. The data transmission is wireless in nature so there are several parameters which affect the efficiency of these sensors.

In wireless communication, the path loss is a parameter which has a significant impact on communication. It is a parameter that makes transmission weaker and is denoted mathematically as

$$PL_{(f,D)} = PL_o + 10 \times n \times \log_{10} \frac{D}{d_o} + S \qquad (5)$$

Fig. 3 Comparison of power consumption

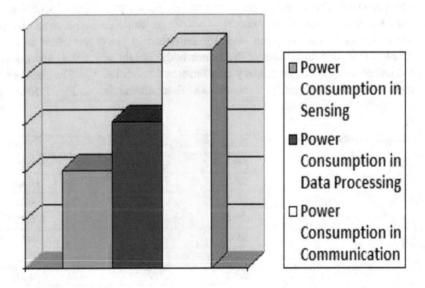

$$PL_o = 10 \times \log_{10} \left[\frac{4 \times \pi \times D \times f}{c} \right]^2 \qquad (6)$$

$$PL_{(f,D)} = 10\log_{10} \left[\frac{4\pi Df}{c} \right]^2 + 10n\log_{10} \frac{D}{d_o} + S \qquad (7)$$

4 Proposed Algorithm

In this section, the proposed algorithm is defined. In the proposed scheme, eight sensor nodes are to be used on the human body and one sink node that has to be placed on the center of the human body.

The proposed scheme is to be compared with one already existing scheme M-ATTEMPT (Javaid et al. 2013). Both algorithms are compared to find out which algorithm performs better. Path loss is based on two parameters (i) frequency and (ii) distance of the sensors and base station. In the proposed scheme the main focus is on the distance parameter in order to reduce path loss in wireless communication.

$$D_i = \sqrt{\left(N_{xi} - S_x \right)^2 + \left(N_{yi} - S_y \right)^2} \qquad (8)$$

Equation 8 represents the EDA of the ith sensor node. N_{xi} represents the ith node's x coordinate, and N_{yi} is the ith node's y coordinate value. X and y coordinates of base station are given as S_x and S_y.

4.1 Sensor Position Deployment

The proposed scheme uses eight sensor nodes. So we propose the coordinated of the sensors as shown in Table 2. These sensor nodes are given distances described in centimeters (cm) in x and y axes of the human body. Height of the human body is considered as y coordinate and width i.e. right hand to left hand as x coordinate. Base station is

proposed to be at 0.40 m and 0.87 m as x and y coordinates, respectively.

5 Simulation Results

In this section, the results of both schemes are given and described in detail. The simulation tool used here is MATLAB. All the required parameters are written into it as a code and the results are discussed.

5.1 Distance Comparison

First, the distance of both the schemes is computed. In Table 3, individual distances of entire sensor nodes of M-ATTEMPT (Javaid et al. 2013), scheme are presented, d1 is the distance of sensor node 1 and the series goes on to d8.

The distances of the proposed scheme are presented in Table 4.

Distance comparison of individual sensor nodes is presented in Fig. 4. Distance values of sensor node 1, 2, 3, and 5 of M-ATTEMPT (Javaid et al. 2013), scheme are lesser than the proposed scheme. The proposed scheme has lesser distance values at sensor nodes 4, 6, 7, and 8. In Fig. 5, the cumulative distance comparison is presented in which it can be observed that the proposed scheme has obtained far lesser distance as compared to M-ATTEMPT (Javaid et al. 2013). The total distance of M-ATTEMPT (Javaid et al. 2013), is 341.8895 cm, while the total distance of the proposed scheme is 336.3679 cm.

The simulation results of ATTEMPT (Javaid et al. 2013), are presented in Fig. 6. The simulation is distance versus path loss.SN-8 is denoted by black colored dot, SN-7 is represented by yellow color, SN-2 by green color, SN-1 is presented in red, magenta represents SN-6, and blue is for SN-3, SN-4, and SN-5.

The simulation results of the proposed scheme are shown in Fig. 7. SN-8 and SN-4 are in black colored dot, SN-1 is in

Table 2 Sensor coordinates

Sensor node (SN) number	X coordinate (m)	Y coordinate (m)
SN-1	0.25	0.10
SN-2	0.40	0.15
SN-3	0.27	0.55
SN-4	0.45	0.56
SN-5	0.60	0.56
SN-6	0.17	0.56
SN-7	0.24	0.50
SN-8	0.40	0.80

Table 3 Individual distances of M-ATTEMPT scheme

Distance							
d1	d2	d3	d4	d5	d6	d7	d8
36.0555 cm	28.2843 cm	31.6228 cm	31.6228 cm	31.6228 cm	41.2311 cm	60.8276 cm	80.6226 cm

Table 4 Individual distances of proposed scheme

Distance							
d1	d2	d3	d4	d5	d6	d7	d8
78.4474 cm	72.0000 cm	34.5398 cm	31.4006 cm	36.8917 cm	40.3113 cm	35.7771 cm	7.000 cm

Fig. 4 Comparison of individual distances

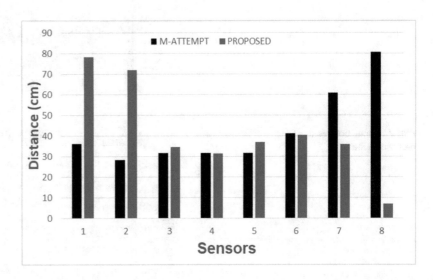

Fig. 5 Comparison of cumulative distance

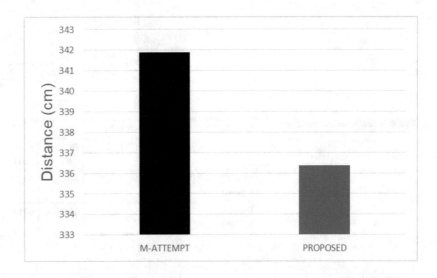

red color, SN-2 is given in green color, blue color represents SN-3, Yellow is for SN-5, SN-6 is represented by magenta color, and SN-7 is denoted by cyan.

The path loss results of both the schemes are presented in Table 5.

It is observed that path losses PL1, PL2, PL3, and PL5 of ATTEMPT (Javaid et al. 2013), are lesser as compared and path loss values PL4, PL6, PL7, and PL8 of the proposed scheme are lesser than that of the ATTEMPT (Javaid et al. 2013), scheme. The cumulative path loss of

Fig. 6 Simulation result of M-ATTEMPT scheme

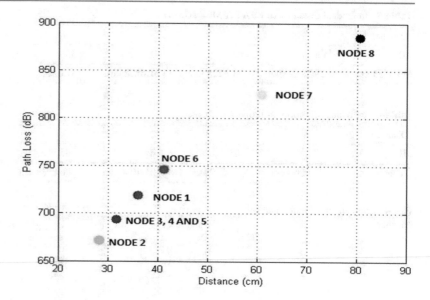

Fig. 7 Simulation result of the proposed scheme

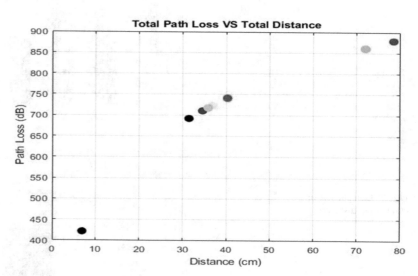

Table 5 Path loss comparison

Scheme	Path loss values								
	PL1 (dB)	PL2 (dB)	PL3 (dB)	PL4 (dB)	PL5 (dB)	PL6 (dB)	PL7 (dB)	PL8 (dB)	PLT (dB)
M-ATTEMPT scheme	719.1	671.7	693.3	693.3	693.3	745.8	825.2	884.6	5926
Proposed scheme	878.5	860.6	710.6	691.9	723.6	741.2	717.5	421.9	5746

ATTEMPT (Javaid et al. 2013), scheme is 5926 decibels (dB) and 5726 dB is the cumulative path loss is of the proposed scheme. There is a difference of 180 dB. The proposed scheme performs better by the transposition suggested making the sensors perform for a longer amount of time and making them consume less energy as the distance between the base station and the sensors has been reduced.

6 Conclusion

In this work the transposition scheme has been described to be used in Wireless Body Area Sensor Networks is proposed. WBASNs are networks related to physiological parameters and their monitoring on a continuous basis and the sensor need to be wireless so that there is no complexity in carrying them. So this sensor needs to perform for a much longer span of time durations so that continuous monitoring is uninterrupted. The main aim of this research is to focus on the sensing points so that if sensors are placed on these positions then there should be two things that need to be achieved (i) No comprise on sensing and (ii) least distance between the sensor and the base station. In the proposed scheme the selection of the sensors has been proposed to be used in any WBASN architecture so that sensors will perform better. The proposed scheme uses Euclidean Distance Algorithm for observing the sensor distances. After the successful implementation of the sensor positions, the simulations were performed. The distance, as well as the path loss comparison, were performed on MATLAB. The simulation results were obtained and it was observed that the proposed scheme was able to achieve lesser distance when compared to the existing scheme making the proposed scheme better. The decrease is distance did allow the proposed scheme to obtain lesser path loss because the path loss is a function of distance.

References

Ahmed, O., Ren, F., Hawbani, A., Al-sharabi, Y.: Energy Optimized Congestion Control-Based Temperature Aware Routing Algorithm for Software Defined Wireless Body Area Networks. IEEE Access **8**, 41085–41099 (2020)

Al Rasyid, M.U.H., Lee, B.H., Sudarsono, A.: Implementation of body temperature and pulse oximeter sensors for wireless body area network. Sens. Mater. **27**(8), 727–732 (2015)

Amjad, O., Bedeer, E., Ali, N.A., Ikki, S.: Robust energy efficiency optimization algorithm for health monitoring system with wireless body area networks. IEEE Commun. Lett. (Early Access) (2020)

Archasantisuk, S., Aoyagi, T.: Transmission power control using human motion classification for reliable and energy-efficient communication in WBAN. IEICE Trans. Commun. **102**(6), 1104–1112 (2019)

Figueiredo, I.N., Leal, C., Pinto, L., Figueiredo, P.N., Tsai, R.: An elastic image registration approach for wireless capsule endoscope localization. arXiv preprint arXiv:1504.06206 (2015)

Hess, P.L., Al-Khatib, S.M., Han, J.Y., Edwards, R., Bardy, G.H., Bigger, J.T., Buxton, A., Cappato, R., Dorian, P., Hallstrom, A., Kadish, A.H.: Survival benefit of the primary prevention implantable cardioverter-defibrillator among older patients: does age matter? An analysis of pooled data from 5 clinical trials. Cir.: Cardiovasc. Qual. Outcomes **8**(2), 179–186 (2015)

Jafri, I.H., Soomro, M.S., Khan, R.A.: ICT in distance education: improving literacy in the province of Sindh Pakistan the Sindh University. J. Educ.-SUJE **40**, 37–48 (2011)

Javaid, N., Abbas, Z., Fareed, M.S., Khan, Z.A., Alrajeh, N.: M-ATTEMPT: A new energy-efficient routing protocol for wireless body area sensor networks. Procedia Comput. Sci. **19**, 224–231 (2013)

Jung, S.J., Shin, H.S., Chung, W.Y.: Driver fatigue and drowsiness monitoring system with embedded electrocardiogram sensor on steering wheel. IET Intel. Transport Syst. **8**(1), 43–50 (2014)

Khan, R.A., Memon, S., Awan, J.H., Zafar, H., Mohammadani, K.H.: Enhancement of transmission efficiency in wireless on-body medical sensors. Eng. Sci. Technol. Int. Res. J. **1**(2), 16–21 (2017)

Khan, R.A., Mohammadani, K.H., Soomro, A.A., Hussain, J., Khan, S., Arain, T.H., Zafar, H.: An energy efficient routing protocol for wireless body area sensor networks. Wireless Pers. Commun. **99**(4), 1443–1454 (2018)

Khan, R.A., Awan, J.H. Memon, Shahzad and Zafar, H.: Influence of cell phone in ICT sector of Pakistan towards advancement. Pak. J. Comput. Inf. Syst. (PJCIS) **2**(1), 17–30 (2017)

Khan, R.A., Memon, S., Zardari, S., Dhomeja, L.D., Usman, M.: Transposition technique for minimization of path loss in wireless on-body medical sensors. Sindh Univ. Res. J.-SURJ (Science Series), 48(4) (2016)

Kumari, R., Nand, P., Astya, R., Chaudhary, S.: Implementation of Square-Odd Scanning Technique in WBAN for Energy Conservation. In: Khanna, A., Gupta, D., Bhattacharyya, S., Snasel, V., Platos, J., Hassanien, A. (eds.) International Conference on Innovative Computing and Communications, vol. 1059, pp. 1–10. Advances in Intelligent Systems and Computing 2019. Springer, Singapore (2019)

Kölbl, F., N'Kaoua, G., Naudet, F., Berthier, F., Faggiani, E., Renaud, S., Benazzouz, A., Lewis, N.: An embedded deep brain stimulator for biphasic chronic experiments in freely moving rodents. IEEE Trans. Biomed. Circuits Syst. **10**(1), 72–84 (2014)

Liu, B., Yan, Z., Chen, C.W.: Medium Access Control for Wireless Body Area Networks with QoS Provisioning and Energy Efficient Design. IEEE Trans. Mob. Comput. **16**(2), 422–434 (2017)

Muramatsu, D., Koshiji, F., Koshiji, K., Sasaki, K.: Effect of user's posture and device's position on human body communication with multiple devices. In: 2015 International Conference on Electronics Packaging and iMAPS All Asia Conference (ICEP-IAAC), pp. 124–127. IEEE, Japan (2015)

Pathan, A.S.K., Lee, H.W., Hong, C.S.: In 2006 8th International Conference Advanced Communication Technology 2006, vol. 2, pp. 1043–1048. IEEE, Korea (2006)

Peter, S., Pratap Reddy, B., Momtaz, F., Givargis, T.: Design of secure ECG-based biometric authentication in body area sensor networks. Sensors **16**(4), 570 (2016)

Rakhee, Srinivas M.B.: Energy Efficiency in Load Balancing of Nodes Using Soft Computing Approach in WBAN. In: Yadav N., Yadav A., Bansal J., Deep K., Kim J. (eds) Harmony Search and Nature Inspired Optimization Algorithms. Advances in Intelligent Systems and Computing 2018, vol. 741, pp. 423–430. Springer, Singapore (2018)

Rao, S., Chiao, J.C.: Body electric: Wireless power transfer for implant applications. IEEE Microwave Mag. **16**(2), 54–64 (2015)

Rathbun, D.L., Jalligampala, A., Stingl, K., Zrenner, E.: To what extent can retinal prostheses restore vision? In: 2015 7th International IEEE/EMBS Conference on Neural Engineering (NER). pp. 244–247. IEEE. France (2015)

Sahndhu, M.M., Javaid, N., Imran, M., Guizani, M., Khan, Z.A., Qasim, U.: BEC: A novel routing protocol for balanced energy consumption in Wireless Body Area Networks. In: International Wireless Communications and Mobile Computing Conference (IWCMC) 2015, pp. 653–658. IEEE, Dubrovnik (2015)

Sajatovic, M., Levin, J.B., Sams, J., Cassidy, K.A., Akagi, K., Aebi, M. E., Ramirez, L.F., Safren, S.A., Tatsuoka, C.: Symptom severity, self-reported adherence, and electronic pill monitoring in poorly adherent patients with bipolar disorder. Bipolar Disord. **17**(6), 653–661 (2015)

Soontornpipit, P.: Study of 403.5 MHz path loss models for indoor wireless communications with implanted medical devices on the human body. ECTI Trans. Electr. Eng. Electron. Commun. **10**(2), 173–178 (2014)

Sugiura, T., Imai, M., Yu, J., Takeuchi, Y.: A low-energy application specific instruction-set processor towards a low-computational lossless compression method for stimuli position data of artificial vision systems. J. Inf. Process. **25**, 210–219 (2017)

Ullah, Z., Ahmed, I., Khan, F.A., Asif, M., Nawaz, M., Ali, T., Khalid, M., Niaz, F.: Energy-efficient harvested-aware clustering and cooperative routing protocol for WBAN (E-HARP). IEEE Access **7**, 100036–100050 (2019)

Van Dam, K., Pitchers, S., Barnard, M.: Body area networks: Towards a wearable future. In: Proc. WWRF Kick Off Meeting, pp. 6–7. Germany (2001)

Xu H and Hua K.: Secured ECG signal transmission for human emotional stress classification in wireless body area networks. EURASIP J. Inf. Secur. **5** (2016)

Yang, G., Wu, X.W., Li, Y., Ye, Q.: Energy efficient protocol for routing and scheduling in wireless body area networks. Wireless Netw. **26**(2), 1265–1273 (2020)

Yang Y, Chae S, Shim J, et al.: EMG sensor-based two hand smart watch interaction. In: Adjunct Proceedings of the 28th Annual ACM Symposium on User Interface Software & Technology, pp. 73–74. ACM, New York (2015)

Zheng, G., Fang, G., Shankaran, R., Orgun, M.A., Zhou, J., Qiao, L., Saleem, K.: Multiple ECG fiducial points-based random binary sequence generation for securing wireless body area networks. IEEE J. Biomed. Health Inf. **21**(3), 655–663 (2016)

Study of Websocket Parent-Teachers/Qualified Teachers in Rural Areas: Case of Central African Republic

Ghislain Mervyl Saint-Juste Kossingou, Nadege Gladys Ndassimba, Edgard Ndassimba, Kéba Gueye, and Samuel Ouya

Abstract

Over the last twenty years, political and military crises have destroyed the Central African education system. Due to insecurity, most of the qualified teachers have left the at-risk provinces to stay in Bangui, the country's capital. Children have no right to education. This constitutes a danger to peace, security, and the socioeconomic development of the country. Parents in at-risk areas are forced to group together to create training for children, but their parent-teachers have not had basic training to teach. In this article, we will set up a Chat and Videoconference platform that allows parents of students who want to replace teachers who are on the run in areas of conflict in the provinces to be qualified and well equipped, enabling children and young people in conflict areas to have the right to education. The methodological approach of our solution is based on the deployment of a common platform using the Node.js server, the MySQL server. The contribution of master-parents to do the rehearsals for children in conflict areas. This solution contributes to the improvement of the schooling rate, the consolidation of peace, security, and socioeconomic development in the Central African Republic.

G. M. S.-J. Kossingou (✉) · N. G. Ndassimba · E. Ndassimba · K. Gueye · S. Ouya
Laboratory LIRT, Higher Polytechnic School University Cheikh Anta Diop of Dakar, Dakar, Senegal
e-mail: skossingou@gmail.com

N. G. Ndassimba
e-mail: nadegegladys.ndassimba@gmail.com

E. Ndassimba
e-mail: edgard.ndassimba@gmail.com

K. Gueye
e-mail: keba.gueye@esp.sn

S. Ouya
e-mail: samuel.ouya@gmail.com

Keywords

Conflict zones • Chat and videoconference platform • Qualified teachers • Abandoned at-risk provinces • Node.js server

1 Introduction

In the Central African Republic, sociopolitical instability for more than 15 years has had extremely negative effects on the education system. In 2015, according to the Education Cluster's evaluation report, enrollment rates fell by 6% compared to 2012, due to the fact that schools were not functional. Among the reasons for this nonfunctioning were mainly the absence of teachers (49% of cases), population displacement (31%), destruction of premises (21%), and insecurity (26%). Qualified teachers fleeing the fighting in rural areas are often replaced by poorly qualified or unqualified "parent-teachers", with 61 percent of these "parent-teachers" in public schools (Education in the Central African Republic Online). The right to education is violated in conflict-affected communities. These factors have a negative impact on access to education, which remains relatively nonexistent among Central African children of school-age, and thus hampers the country's socioeconomic development.

Several articles proposed e-learning solutions to improve children's living conditions. This is the case of a mobile e-learning support system that helps to achieve good teaching and learning outcomes by providing quality and accessible educational content for secondary schools (Akeem and Sun 2018). Another presents the main advantages of using tablet computers in teaching and learning physics in primary schools (Grubelnik and Grubelnik 2016). In (Tibor 2018), the authors demonstrated the effectiveness of modern methods of distance education in primary schools. In these articles, the authors make use of Communications and Information Technology (ICT) infrastructure and use e-learning tools, tablets, and mobiles to maximize the effectiveness of e-learning. However, in the context of countries in conflict such as the Central African Republic, equipment allowing access to the Internet has been destroyed in conflict zones, it is necessary to rely on humanitarian networks to have access to the Internet.

Real-time communication is an evolving technology today. Real-time communication allows users to exchange data and multimedia streams in real time.

Today, many companies are eager to hold their meetings in real time.

This article proposes a solution to help parents who want to replace the teacher on the run in areas of conflict provinces to train remotely. It allows qualified teachers in the capital to assist the parent-teachers at a distance.

The rest of this article is organized as follows. Section 2 describes the state of the art. Section 3 proposes a network architecture based on TVWS and a platform architecture based on WebRTC. Section 4 presents the results and discussion, and finally, Section 5 provides the conclusion.

2 State of Art

2.1 The Situation of Education in Central African Republic

Thirty years after governments around the world adopted the Convention on the Rights of the Child, the right to education is being violated in conflict-affected communities in West and Central Africa (UNICEF Child Alert 2019).

The situation of education in conflict zones in the Central African Republic remains dire. Intercommunity violence is forcing the civilian population to set up fortunate camps protected by UN troops, leaving schools empty.

The conflict is having an impact on children's education, with significant numbers of children unable to attend school, either because of displacement or because schools have been closed, attacked or occupied by armed groups. It is estimated that 31% of school-age children are no longer in school and only 49% of student's complete primary school (European civil protection and humanitarian aid operations Online).

The challenges to be overcome in order for the country's children to have universal access to education are immense. Education is the key that can enable them to draw on this strength to realize their potential and become agents of positive change. Without education, there can be no sustainable recovery, reconciliation, and peace (The government of Central African Republic, Education Cannot Wait Online).

2.2 EasyRTC

EasyRTC is a real-time communication technology on the web. WebRTC is a set of standards that allows video conferencing, text transmission, and data sharing between client browsers (peers) (Cola and Valean 2014). Unlike many web services that already use PSTN, but require downloads, native applications or plug-ins (Pandey et al. 2018), WebRTC provides any browser with the ability to share application data and conduct peer-to-peer teleconferences without having to install a few plug-ins or client software.

EasyRTC OpenSource (Started and with WebRTC using EasyRTC Open Source Online) is:

- A browser client library written in JavaScript. This client manages signaling and largely isolates applications from changes in the WebRTC API.
- A signaling server based on Node.js. Node.js runs on platforms as small as a single-core Raspberry Pi (first edition) on servers in the cloud.

2.3 WebRTC

In this article, the authors set up a system that provides users with a high availability of communication through the convergence of conventional telephony. The system is a web-based application that provides a videoconferencing room for multiple users in real time. The system was implemented using WebRTC for audio and video transmission over Node.js, the Asterisk software (Rosas and Martínez 2016). In another study, a comparative has been made on the different web communication technologies, the strengths, weaknesses, best hardware architectures, processor performance, and storage level studies had been made in this study, allowing us to take into account in relation to our project. Its authors proved that the video conferencing application with WebRTC, nodejs and its components is a real-time communication system, it uses the processor in decoding, encoding and distribution of video and audio in the same period of time (André et al. 2018). The authors show the importance of video conferencing applications in today's society (Pasha et al. 2017).

2.4 Active Pedagogy

The learning methodology for these primary schools is active pedagogy. Active pedagogy has become a

Fig. 1 Proposed deployment architecture

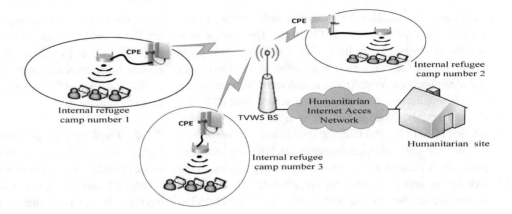

privileged means of transforming the traditional teacher-centered classroom into a new student-centered approach (Learning and Pedagogy Online). (Learning and Pedagogy Online) It is one of the methods that belong to what is called experiential learning, i.e., "learning by doing". The results show that active learning pedagogy activities are important factors that increase student performance compared to students in traditional classrooms (Javed and Odhabi 2018). Learning objects are consolidated as open digital resources to encourage active learning (Salazar and Durán 2017).

3 Solution Architecture

3.1 Simulation Architecture

Figure 1 describes the architecture of the broadband network deployed in conflict zones using TVWS technology.

A TVWS base station is deployed in a humanitarians-controlled area allowing nearby villages to be interconnected to the distance learning facility based in the capital.

CPEs installed in refugee camps allow users connected to the local WIFI network to join the e-learning platform located in Bangui.

3.2 Chat Architecture

To have a solid and reliable real time chat system, we must add to Php, MySQL, and the Node.js server.

The principle is the following:

Send a message from PHP to MySQL server (through a channel), while the Node.js Socket.io module will listen on the same channel, retrieve the message and finally execute an action. So it is our MySQL server that will ensure the communication between our PHP server and Node.js (Fig. 2).

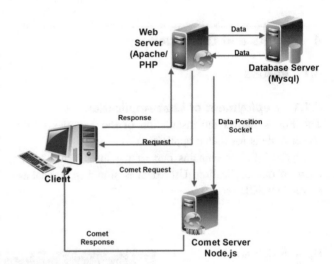

Fig. 2 Chat architecture of Real-time

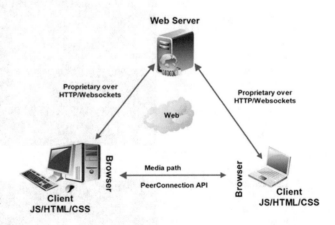

Fig. 3 Video conferencing architecture of Real-time (Loreto and Romano 2014)

3.3 Video Conferencing Architecture

The Websocket protocol provides a bridge between the http protocol and the TCP/IP protocol. This facilitates better real-time communication. Websocket uses full duplex mode, allowing the server to push to its clients. The Webrtc connection is Peer-to-Peer; it is established between peers directly. Once the connection is established, the server will exit the connection loop. Traffic is encrypted between peers.

This architecture consists of Web server (WebRTC and its components, Node.js and its components), client-side JavaScript, HTML, and CSS for the application views.

The real-time videoconferencing architecture provides precision, it showed us that peers send an http request to the web server and these requests are identified in the Websockets server, then the response is sent back to the identified peers.

4 Results and Discussion

4.1 Results

4.1.1 Deployment of Chat Application
The Fig. 4, shows the installation of the modules of our server node.js for a chat application.

Figure 5 below contains part of the information of the users of our application. User data is stored in a database server (MySQL server).

Figure 6 shows us the authentication page, after accessing the normal user account, the user will access the home page (Fig. 7), and the home page allows the user to make a choice, or activate access to videoconferencing or chat.

The Figs. 8 and 9, below allow real-time communication between remote users.

4.1.2 Deployment of Video Conferencing
Figures 10 and 11 show us the different processes of access to videoconference by the user who owns an android device. Figures 12 and 13 similarly show us access to videoconferencing by the user who owns a computer.

Figures 14 and 15 show us the different results of real-time communications by videoconferences.

4.2 Discussions

The client-side JavaScript with html and CSS allowed us to have the client interface active, the server-side JavaScript is node.js.

We implemented the chat client interface using client-side HTML, CSS, and JavaScript. The server-side chat implementation uses the Node.js server and the express, http, and socket.io modules. We used the npm module manager for the installation of the node.js modules.

Fig. 4 Node.js module installation

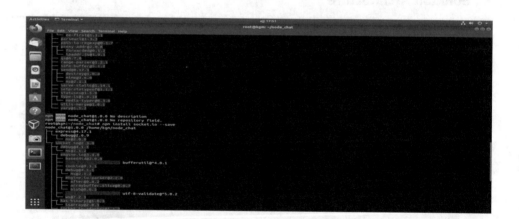

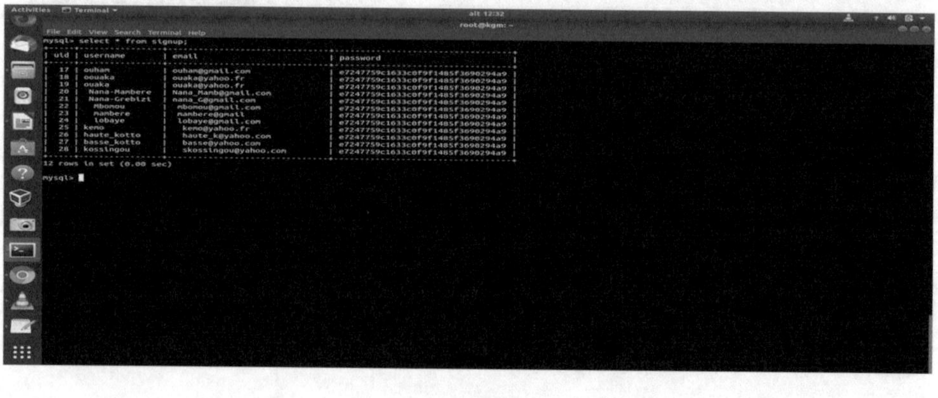

Fig. 5 User database

Fig. 6 Interface 1

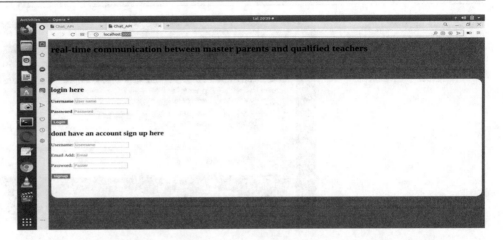

Fig. 7 Interface 2

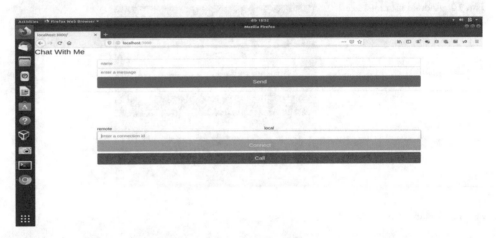

Fig. 8 User interface 1

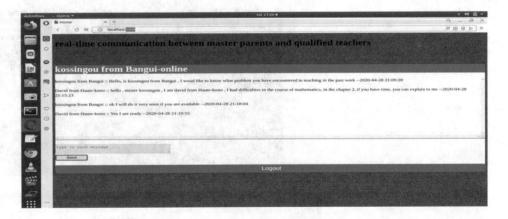

The express module of Node.js allows a good interaction between the http client/server.

The socket.io module allows real-time, threadable interaction.

The database server has been configured using MySQL-Server, a PHP language for database connection and SQL queries.

The MySQL database is designed to store our users, their authentication information, and personal data in real time.

Rooms, chats are also stored on the server level of our MySQL database. To avoid data saturation in the database, we considered sharing the data.

Our video conferencing application WebRTC uses the Node.js server and its modules, the view was realized with HTML, CSS, and client-side JavaScript.

The client-side (HTML, CSS, and JavaScript) interacts with web browsers via the standardized WebRTC API, allowing real-time control of the browsers.

Fig. 9 User interface 2

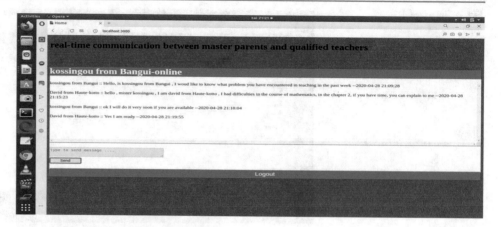

Fig. 10 Android user interface 1

Fig. 11 Android user interface 2

Fig. 12 Computer user interface 1

Fig. 13 Computer user interface 2

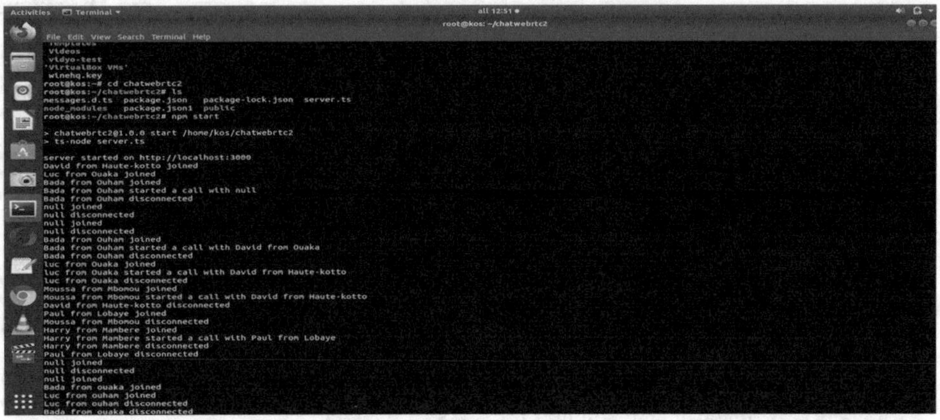

Fig. 14 Result of the calls made between users (master-parent) from different unsecured rural areas

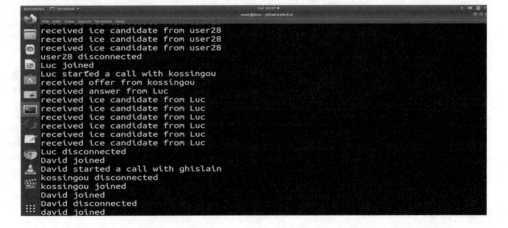

Fig. 15 Result of the calls made between the qualified masters and the users (master-parent) of the different unsecured rural areas

The WebRTC API enables:

Real-time peer-to-peer connection management, negotiation of encoding and decoding capabilities, media selection and control, firewall traversal.

In our platform chat and videoconferencing:

We will have the ability to store hundreds of messages per hour,

Ability to retrieve thousands of messages per hour.

We have taken into account the ability to retrieve messages in chronological order (most recent at the beginning).

Members are likely to read several times, so we need to make the pattern that encourages reading rather than writing.

Possibility to read all messages at once (we think it is difficult).

According to the literature, we believe that our working environment is not the best, but this application could reduce a number of problems in countries in conflict such as the CAR, as most of the provinces of the CAR before the sociopolitical crises we experienced problems of IT infrastructure. This situation has been aggravated by repeated socio-military crises. Our application will enable teacher-parents in conflict zones to train and be in permanent contact with qualified teachers who have given up their jobs due to insecurity problems.

5 Conclusion

In this article, we have shown that the remote support and training solution for master-parents based on WebRTC using the Node.js and MySQL server servers made it possible to fill the lack of qualified teachers in the conflict zones and promotes the sustainable development of its conflict zones.

The Chat and Videoconference platform allows real-time communication between qualified teachers in Bangui and unqualified teachers who are in conflict zones. The interest is to give the right to education to several children who have been deprived of their education in rural areas in conflict. Qualified teachers refuse to return to their posts due to increased insecurity. The lack of teachers in the conflict provinces has made it possible for master-parents to teach in place of qualified teachers.

The data flow management solution for users of the chat and videoconferencing platform proposed to avoid data saturation in the database is not generally sufficient. In the near future, we will have to use data management by collections with the MongoDB database in order to avoid overloading.

We will consider finding an eberger at a lower cost in order to avoid the oversizing of the data storage space, to increase the processing time of data in inputs/outputs.

References

Akeem, O., Sun, Y.: Mobile e-learning support system for secondary schools in Nigeria. In: 2018 1st IEEE International Conference on Knowledge Innovation and Invention (ICKII), pp. 262–265. Jeju (2018)

André, E., Le Breton, N., Lemesle, A., Roux, L., Gouaillard, A.: Comparative study of WebRTC open source SFUs for video conferencing. In: 2018 Principles, Systems and Applications of IP Telecommunications (IPTComm), pp. 1–8. IEEE (2018)

Cola, C., Valean, H.: On multi-user web conference using WebRTC. In: 2014 18th International Conference on System Theory, Control and Computing (ICSTCC), pp. 430–433. Sinaia (2014)

Education in the Central African Republic. [Online]. https://www.globalpartnership.org/where-we-work/central-african-republic

Education under threat in West and Central Africa, UNICEF Child Alert August 2019. [Online]. https://www.unicef.org/child-alert/education-threat-west-central-africa

European civil protection and humanitarian aid operations. [Online]. https://ec.europa.eu/echo/where/africa/central-african-republic_en

Grubelnik, V., Grubelnik, L.: Teaching physics in primary schools with tablet computers: key advantages. In: 2016 39th International Convention on Information and Communication Technology, Electronics and Microelectronics (MIPRO), pp. 847–851. Opatija (2016)

Javed, Y., Odhabi, H.: Active learning in classrooms using online tools: evaluating pear-deck for students' engagement. In: 2018 Fifth HCT Information Technology Trends (ITT), pp. 126–131. Dubai, United Arab Emirates (2018)

Active Learning Pedagogy. [Online]. https://www.researchgate.net/publication/303881379_Active_Learning_Pedagogy

Loreto, S., Romano, S.P.: Real-time Communication with WebRTC: Peer-to-Peer in the Browser. O'Reilly Media, Inc. (2014)

Pandey, N., Bein, D.: Web application for social networking using RTC. In: 2018 IEEE 8th Annual Computing and Communication Workshop and Conference (CCWC), pp. 336–340. Las Vegas, NV (2018)

Pasha, M., Shahzad, F., Ahmad, A.: Analysis of challenges faced by WebRTC videoconferencing and a remedial architecture. arXiv preprint arXiv:1701.09182 (2017)

Rosas, A.S., Martínez, J.L.: Videoconference system based on WebRTC with access to the PSTN. Electron. Notes Theoret. Comput. Sci. **329**, 105–121 (2016)

Salazar, N.I., Durán, E.B.: Design of a learning object for simulation based on emerging pedagogies. In: 2017 Twelfth Latin American Conference on Learning Technologies (LACLO), pp. 1–4. La Plata (2017)

Getting Started with WebRTC using EasyRTC Open Source. [Online]. https://easyrtc.com/docs/easyrtc_gettingStarted.php

The government of Central African Republic, Education Cannot Wait, and a wide coalition of donors and partners launch US $77.6 million education programme for 900,000 children. [Online]. https://reliefweb.int/report/central-african-republic/government-central-african-republic-education-cannot-wait-and-wide

Tibor, H.Z.: Another e-learning method in upper primary school: 3D spaces. In: 2018 9th IEEE International Conference on Cognitive Infocommunications (CogInfoCom), pp. 000405–000410. Budapest, Hungary (2018)

A Comparison of QoS-Based Architecture Solutions for IoT/Edge Computing Environment

Nogaye Lo and Ibrahima Niang

Abstract

Edge computing is a beneficial concept for supporting sensitive IoT applications that now permeate our daily lives. It makes the possibility to migrate the calculation and/or the storage of data towards the "periphery" of the network, close to the end users. However, the onboard calculation method is advantageous for real-time IoT applications that have QoS requirements, for example, the difficulties of high latency and network congestion in IoT infrastructures. To provide efficient and reliable services to end users, it is important to refer to QoS criteria, which implies an effective management of the quality of service in IoT/Edge computing systems. In this paper, we propose an analysis of QoS solutions and architectures for effective quality of service management in the IoT/Edge computing environment. We analyze and propose a classification of QoS mechanisms in relation to the different layers of the IoT architecture. We analyze and propose a classification. This study opens up new research perspectives concerning QoS dynamic management for real-time applications in IoT/Edge computing systems.

Keywords

Internet of thing • Quality of service • Edge computing • Real-time application

N. Lo (✉) · I. Niang
Laboratoire Informatique de Dakar (LID), Cheikh Anta Diop University, Dakar, Senegal
e-mail: nogayelo7@gmail.com; nogaye2.lo@ucad.edu.sn

I. Niang
e-mail: ibrahima1.niang@ucad.edu.sn

1 Introduction

The main idea of internet of things (IoT) is the interconnection of a large number of physical objects via internet. IoT can improve the quality of life in several areas such as transportation, industrial automation, health, and emergency services in natural incidents where humans cannot make decisions directly. In short, IoT allows physical objects to react, perform tasks, share information, and make a decision. To meet market demands and customer needs, emerging service technologies and applications need to evolve proportionally to achieve this exponential growth. In terms of availability everywhere and at any time, devices must be developed according to everywhere and at any time, devices must be developed according to customer requirements via communication technologies, sensor networks, protocols, and internet applications.

With the increase in IoT applications, many of which involve big data, including analysis of resource intensive operations and real-time processing applications, the situation has become exacerbated. Unloading tasks from IoT applications on computer systems with sufficient IT resources is the common way to overcome the small computing resources of IoT devices in various applications (Kumar et al. 2013; Kumar and Lu 2010). Therefore, the unloading method can lead to problems related to the additional overload of the data transmission over networks. This increases network congestion, the latency of IoT applications implying resource intensive operations (Satyanarayanan et al. 2009b; Zeng et al. 2016). These problems of offloading tasks to the remote cloud motivate the recent development of Edge computing (Satyanarayanan et al. 2009b; Beck et al. 2014). To support IoT applications, edge computing paradigm uses network devices with larger computing resources, such as computer switches and computer base stations, verses at the edge of the network.

In addition, to meet the expectations of suppliers and end users, IoT must at least guarantee certain quality properties.

These expectations can be expressed in terms of cost, throughput, bandwidth, response time, accuracy, and efficiency. These needs require a well-defined quality of service (QoS) which is a nonfunctional requirement of IoT. As a very important aspect in the field of Internet of Things, the main role of quality of service is to ensure that services will be provided to end users by meeting their quality of service requirements. Quality of service attributes includes the efficient use of network resources, provision of quality services, high availability, security, trust, resource allocation, etc. (Radovanovic et al. 2013). One of the most challenging aspects of quality of service management is monitoring it in an IoT environment. QoS faces several issues related to scalability, power consumption, resource scarcity, network bandwidth, availability, response time, etc. These issues have a negative impact on the performance of IoT applications.

In this paper, we present an analysis of QoS management in IoT/Edge computing environment. We summarize the QoS requirements through an analysis based on the characteristics of the application layer, network layer, and perception layer services. Our study offers an extended study of QoS management in an IoT/Edge computing environment. This paper is organized as follows: in Sect. 2, we propose an overview of concepts, challenges, and architectures of IoT and edge computing, respectively. A review of QoS architectures in IoT Environment is presented in Sect. 3. Section 4 presents a Qos-aware of Edge computing services to the Internet of things. The results are presented in Sect. 5. Open questions and future direction are proposed in Sect. 6. Finally, we present a conclusion in Sect. 7.

2 Background

Internet of Things (IoT) has become a companion in our daily lives, millions of sensors and devices have supported machine-to-machine communications and permanently produce data and exchange important messages over complex networks. To solve the needs in IoT and localized computing, edge computing has emerged as a new paradigm as a strategy to mitigate the escalation of resource congestion. Edge computing will migrate computing or data storage to the "edge" of the network, close to end users. Thus, in this section, we will discuss the two systems IoT and edge computing and the advantages of IoT compared to edge computing.

2.1 IoT Systems

2.1.1 IoT-Concepts and Challenges

The Term "Internet of Things" was first used in 1999, by Kevin Ashton in a presentation by Procter and Gamble. This term refers to the world of objects, devices, sensors, and actuators linked by the internet (Saleh 2017). The authors of (Manyika et al. 2013), define IoT by referring to two aspects: spatial and temporal illustrated in Fig. 1, which allow people to connect anywhere and at any time through connected objects (smartphones, tablets, sensors, etc.). The security of the Internet of Things is a key phenomenon and is designed to avoid potential threats and risks while masking the underlying technological complexity.

As a topic of current research, IoT must address many hardware and software challenges (e g., reliability, mobility, availability, scalability, and interoperability). These challenges allow for better management of service quality in several areas: technical, social, environmental, and others.

2.1.2 IoT Architecture-Applications and Future Vision

The main objective of IoT is to interconnect a large number of heterogeneous objects via internet. These objects can be used in many areas to make life easier for people. IoT applications now affect almost sectors such as:

- Smart agriculture to optimize the use of water;
- Connected vehicles to optimize urban traffic management;

Fig. 1 A new dimension for IoT (Saleh 2017)

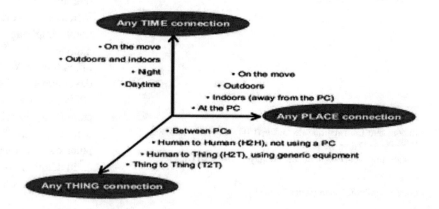

- Connected devices to optimize the consumption distribution of electrical energy;
- Digital Arts.

Given the need of customers and the challenges faced by the IoT, it is, therefore, essential to have a flexible layered architecture. Several researches have been done on the architecture of IoT but there is still no common model accepted by everyone. The majority of the models proposed that the reference model is a 3-layer architecture (Duan et al. 2011; Yang et al. 2011), the application layer, the network layer, and the perception layer.

2.2 Edge Computing Systems

2.2.1 Edge Computing Concepts and Challenges

With the massive growth of millions of sensors and mobile devices, conventional centralized cloud computing is struggling to satisfy the quality of service of many applications. These devices are constantly generating data and exchanging important messages over complex networks that support machine-to-machine communications. Data is increasingly being generated at the edge of the network, so it would be more efficient to process it at the edge of the network where it is generated. To get around this problem, the concept of Edge Computing is introduced to process IoT data with a server at the edge of the network instead of the centralized cloud server. Edge computing (Bonomi et al. 2012), emerged because cloud computing is not always efficient for processing data produced at the edge of the network. Edge computing or data processing at the edge of the network can be defined as an open distributed computing architecture that provides decentralized data processing power. Others define it as a technology that allows computations to be per-formed at the edge of the network, on downstream data on behalf of cloud services and upstream data on behalf of Internet of Things services (Satyanarayanan et al. 2009a). Figure 3 shows a difference in data production and consumption between edge and cloud computing discussed in (Shi et al. 2016).

There are a number of needs that motivate the presence of edge computing in the field of cloud computing. In (Varghese et al. 2016), the authors use network performance and quality of service (QoS) requirements to classify edge computing challenges and opportunities. Figure 4 summarizes the motivations, challenges, and opportunities of edge computing.

2.2.2 Edge Computing Architecture and Application

Edge servers are closer to end users than cloud servers. Edge servers typically offer better quality of service (QoS) and lower latency to end users even though they have lower computing power than cloud servers. To better understand this technology, we will focus on the architectures of the two types of servers and compare them afterwards. Edge computing integrates edge nodes into the network as opposed to the cloud where all servers are centralized. In this document, edge computing nodes are called Edge/Cloud servers. The edge computing structure is generally divided into three aspects (Wei et al. 2017): Front-end (or edge), Near-end (or fog), and Far-end (cloud), as shown in Fig. 5. The differences between these domains are described in detail below.

Front-end (or edge): In this part is the terminal devices typically used in IoT (e.g. sensors, actuators) that are deployed at the front-end of the edge computing structure. It is called the perception layer in the IoT domain. Edge computing can provide real-time services for certain applications using the computing capacity provided by nearby end devices. However, most requirements cannot be met in the edge environment due to the limited capacity of the end devices (Wei et al. 2017).

Near-end (or fog): Most traffic flows in networks are supported by gateways deployed in the near-end environment. Peripheral/cloud servers may also have many resources such as real-time data processing, data caching, and compute offloading. Most of the computation and data storage will be migrated to this near-end environment for quick and easy processing. However, end users will gain in terms of latency and can get much better performance on computation and data storage (Shi et al. 2016).

Far-end: These servers in the cloud are deployed further away from the end devices. Transmission latency is high between these servers and the end devices. Nevertheless, cloud servers can provide more computing power and more data storage. Their storage capacities are generally larger (Bonomi et al. 2012; Wei et al. 2017).

2.3 Benefits of Edge Computing for IoT Environment

Edge computing has solved many IoT issues despite their independence. In this section, we analyze the benefits of edge computing over the Internet of Things. To illustrate how edge computing improves IoT performance, we focus

Table 1 Benefits of edge computing for IoT

Applications	
Data transmission	Data storage and processing
Benefits of edge computing for IoT	
Latency/bandwidth: Low latency allows real time communication, which leads to better decision-making in edge computing based on IoT. IoT applications generate millions of data per second, edge computing is used to reduce the amount of data sent to the cloud in order to reduce bandwidth consumption and response time (Radovanovic et al. 2013)	**Data analysis/decision-making**: Edge computing allows local data analysis for IoT applications in real time. For example, traffic light cameras can analyze the data for themselves, and make immediate decisions to complete the required tasks instead of sending them to the cloud (Bonomi et al. 2012)
Mobility: mobility support for certain edge applications thanks to the switchover from one edge server to another	**Security**: data security can be ensured because customer data is aggregated at certain access points located close to the end user
Real time interactions: IoT uses edge computing for certain advanced applications sensitive to latency and requiring interaction in real time. (Health, augmented reality)	**Interoperability**: Applications and services developed for edge computing based on IoT are interoperable to ensure compatibility with other applications and hardware components (Radovanovic et al. 2013)
Energy Consumption: For devices with a limited battery, it is important to maximize their lifespans. To achieve this goal, Edge computing can integrate a flexible task offload scheme that takes into account the energy resources of each device (Bonomi et al. 2012)	**Data acquisition**: Onboard devices that can capture data for rapid analysis (reducing latency in accessing the service) and immediate actions related to data processing
Data loss: With the edge storage distributed storage system, loss-sensitive IoT data can be reproduced and stored in different storage locations. This can remarkably reduce data loss	**Lower management costs**: Edge Computing is a good way for companies to reduce the operating cost of their data storage infrastructures. Sending data to the cloud costs more than the enterprise produces and collects more and more data with IoT

on data transmission, storage, and processing. Table 1 describes the advantages of edge computing over IoT.

3 Review of QoS Architectures in IoT Environment

3.1 QoS-Based IoT Layers

Quality of service of each layer is strongly related to its position and operation of this layer. As defined in Fig. 2, the application layer responds directly to needs of end users, the network layer manages the transmission of data between the two layers and the perception layer is responsible for data collection. In (Naik et al. 2016), authors propose a guarantee of QoS in all layers of the IoT system. Several QoS attributes have been proposed by researchers to meet the needs of IoT applications in terms of QoS. We classified these attributes based on the three IoT layers discussed in Sect. 1. In (Duan et al. 2011), authors specified many QoS attributes that include delay time, packet loss rate, transmission capacity, service level, and the accuracy of the data. They relied on application types to classify the quality of service attributes.

Authors of (Li et al. 2014), made their choices on the functionality of application service that is delivered directly to end users. In (Singh and Baranwal 2018), authors are interested in cloud and data processing to define their QoS criteria. Table 2 describes a set of QoS attributes based on (Li et al. 2014; Duan et al. 2011; Singh and Baranwal 2018) according to three IoT dimensions which are perception layer, the network layer, and the application layer. This table makes a specific ranking for each layer aimed at end-to-end QoS management.

3.2 QoS Architectures in IoT Environment

Quality of service must be guaranteed in each layer of IoT as shown in the table above. The quality of service requirements must be from one layer to another. For example, from the upper layer to the lower layer, while feedback on the quality of service is transmitted from the lower layer to the upper layer. It is, therefore, necessary to find a way to manage the common QoS of the three layers of IoT.

In (Duan et al. 2011), authors propose an architecture for managing quality of service in the IoT environment. This

Fig. 2 IoT architecture

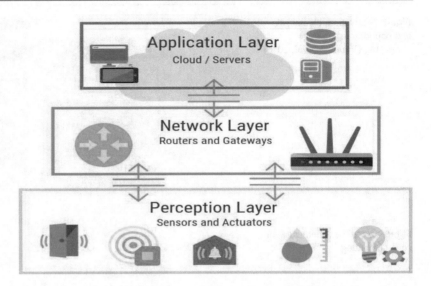

Table 2 QoS attributes for IoT system

Dimensions	QoS Attributes based on (Naik et al. 2016)	QoS Attributes based on (Maalel et al. 2013)	QoS Attributes based on (Rausch et al. 2018)
Perception layer	Accuracy, Availability, Stability	Sampling parameters time, Synchronization Coverage and location/mobility	Weight interoperability, Overall Accuracy, Response Time, range
Network layer	Transmission time, Storage capacity, Reliability	Bandwidth, Delay, packet loss rate, Jitter	Jitter, Bandwidth, Network Connection Time, Throughput and Availability
Application layer	Functionality, Normative, Robustness	Service Time, Service delay, service accuracy, service load and service vice priority	Response Time, Pricing, Reliability, Dynamic Availability

Fig. 3 Edge/cloud paradigm

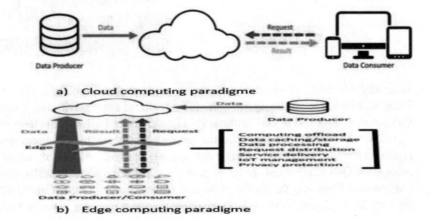

Fig. 4 Motivation, challenges, and opportunities in edge computing (Varghese et al. 2016)

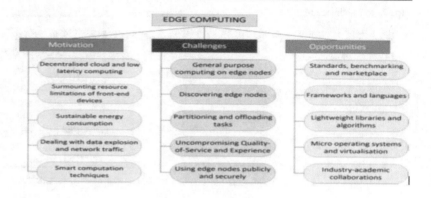

Fig. 5 A typical architecture for edge computing

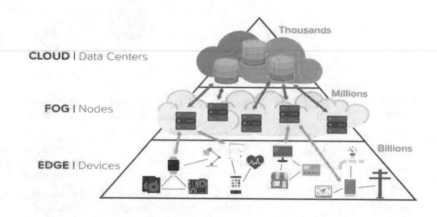

Fig. 6 QoS architecture for IoT (Duan et al. 2011)

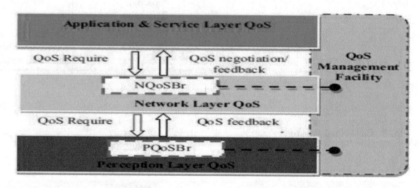

architecture is made up of quality of service management function common to all the layers of the IoT system, of the QoS for each layer and of QoS brokers. Two brokers named NQoSBr and PQoSBr are placed in the network and perception layers, respectively. Their roles are to analyze, process the QoS requirements received from the upper layer and translate them into the quality of service requests from the local layer. Figure 6 shows the architecture proposed by these authors.

Application layer communicates with the NQoSBr network layer broker with an interface called CallNQoSBr to send the quality of service requirements to the lower layer. Depending on the QoS mechanism applied in the network layer, NQoSBr translates the requirements into the quality of service requests and calls the QoS service. The NQoSBr

broker is also responsible for maintaining the quality of service. NQoSBr would negotiate with the upper layer to lower the level of quality of service or change its decision if the resources are not sufficient to satisfy the quality of service. QoS requests from the network layer are transmitted to the PQoSBr perception layer broker through an interface called CallPQoSBr. It monitors the information returned by the network layer and also the performance of QoS in the perception layer.

Authors of (Peros et al. 2018), focus on SDN to make the quality of service dynamic in the IoT system. An architecture is proposed by integrating a centralized SDN controller. The non-IP network closes non-IP devices that IP send data to client applications that operate in the IP network via the gateway. The IP network, on the other hand, closes the IP equipment, the data plan, and the network hosts. The data plane site is made up of a set of OpenFlow switches managed by the SDN controller in the control layer. The data plan is administered by the API control layer. This functionality is made subject to the implementation of an SDN controller. The management layer, for its part, contains IoT-QoS applications which allow to dynamically reconfigure the network to ensure QoS, to calculate the quality of service paths between the devices, to determine the assignable quality of service, the detecting new devices joining the network and specifying device types of users. The authors use the network reconfiguration time following an intrusion captured by a motion detector. If the quality of service after motion detection is high, the experience is successful and the performance of the framework is evaluated by measuring the time required to reconfigure the network while respecting the constraints of QoS. Several attempts show results allowing to say that QoS is dynamic.

A QoS routing algorithm (AQRA) based on SDN and compatible with IoT applications is proposed by the authors of (Deng 2018). The authors of this article focus their solutions on the IoT/SDN system with an emphasis on quality of service (QoS). It makes a dynamic classification of flows according to their priorities. Based on the QoS requirements (i.e., delay, jitter, and packet loss rate) of each IoT application, AQRA calculates multiple routing paths. To achieve load balancing between paths, the available bandwidth of each path is taken into account when making the final decision on the routing path. It guarantees the QoS requirements of high priority IoT applications. These works were compared with others to assess their results. The result shows that several QoS requirements of critical applications can be guaranteed. The following table analyzes the papers examined and deduces a classification of the results obtained for QoS in IoT.

4 QoS-Aware of Edge Computing Services to Internet of Things

Edge computing is the extension of cloud computing at the edge of the network that allows fast data access and processing. Edge computing does not yet have a standard architecture. Several sectors are open to research for better management of this new paradigm. As a result, edge computing has solved many problems related to the Internet of Things. Many researchers are digging for effective quality of service management in the IoT/edge environment.

The authors of (Yin et al. 2019), propose a method of predicting quality of service (QoS) in an IoT/edge environment for recommending services, with a thorough learning of the functionalities. With the use of CNN (Convolutional

Neural Network), both on the user side and on the service side, the proposed model learned the deep latent characteristics of the neighbors based on QoS records. The authors focus on a new calculation method based on IIFU (reverse invocation frequency of users) and IIFS (reverse invocation frequency of services) for improving the qualitative selection of neighbors. Concerning the learning of functionalities, to learn the deep latent characteristics of user neighbors and service neighbors, they propose a model which extends the convolutional neural network. An evaluation of the results shows that the proposed model improves the capacity for low data density, as well as the accuracy of predictions.

With the emergence of new paradigms such as edge computing and rapid changes in quality of service requirements of the Internet of Things, several measures of QoS have been identified in (Sharghivand et al. 2018). To match cloudlets to IoT applications when providing edge computing services, you need to have efficient QoS compatible matching algorithms.

In (Sharghivand et al. 2018), the authors propose new bilateral matching techniques, QMECS (QoS-aware Matching of Edge Computing Services), and M-QMECS (Modified QoS-aware Matching of Edge Computing Services), for edge services taking into account QoS requirements in terms of service response time. Other solutions for determining payments taking into account incentives from cloudlets and preferences, IoT users, and the system are offered. The proposed matching techniques are efficient for Pareto and efficient in terms of calculations, compatible with incentives and less balanced in terms of plan. The evaluation of the results obtained shows us the suitability of the mechanisms for edge services by providing optimal correspondence, pricing efficiency, and a guarantee of quality of service.

In (Song et al. 2017), authors emphasize the quality of service management by proposing an approach for the periodic distribution of incoming tasks in the edge computing system. This allows a massive increase in the number of spots to be processed in the edge network. This approach generates a task distribution to meet all the quality of service (QoS) specifications of the tasks hosted in the IoT applications. Several hypotheses have been developed:

- The priority of the tasks to be treated in the edge system is different.
- The tasks to be processed in edge computing are independent.
- Processing can be done on any VM with sufficient resources.

In edge computing, the task distribution process must be applied periodically since incoming jobs are sent continuously from IoT devices. With the dynamic change of incoming tasks for each application in terms of arrival rate and specifications (workload, data size, deadline, etc.), they must have a dynamic adjustment to maintain acceptable performance of the distribution process stain. For effective stain treatment, the authors have established a stain distribution process. First a formulation of the task distribution problem in edge computing (QoS requirement, resource constraints etc.), then go to the linearization of the formulation and finally generate the task distribution using their own algorithm.

The Internet of Things works on a subscription-to-subscription middleware basis for easy communication between devices. IoT applications impose QoS requirements which generally cannot be managed by the cloud. Edge computing has emerged for customer mobility and dynamic availability of resources. It is in this context that the paper authors (Rausch et al. 2018), present an EMMA solution which is MQTT middleware which aims to meet the strict QoS requirements for edge computing applications. EMMA allows messages to be delivered to geographically different locations with minimal overhead, and can also provide low latency for devices near the network. EMMA is based on the MQTT pub/sub protocol used in IoT systems and which works with a broker distributed transparently. EMMA allows existing IoT software that uses MQTT to be used transparently. EMMA has taken the functionality of the MQTT protocols with the three QoS levels and users have the privilege of choosing their QoS levels.

5   Results and Comparison

In this part, we analyze and classify the different papers examining the management of QoS in IoT and edge computing systems. The results of these analyzes show:

– For the IoT system, most of the authors present a QoS management based on SDN by taking into account its

centralized controller in order to make QoS dynamic. QoS is evaluated against metrics such as latency, packet loss rate, availability, and bandwidth
– For the IoT/edge computing system, several algorithms are created to support QoS, most of which focus on the distribution of tasks in the edge computing environment by evaluating the service response time and latency.

The following Tables 3 and 4, show a classification of the results obtained, respectively, for QoS in the IoT system and edge computing

6   Open Research Issues and Future Direction

Several quality of service indicators have been proposed to meet the requirements of IoT applications. In this contribution, we have analyzed and classified the QoS attributes on each IoT layer: application, network and perception. Further analysis is performed on recently proposed architectures and solutions for QoS management in the IoT/edge computing environment. Our studies show that most of the proposed architectures are based on a centralized controller that will have a global vision of the network. Some works proposing solutions integrating QoS do not present efficient results. This is often due to the fact that the proposed simulation environments do not release all the functionalities of the IoT. This suggests new research challenges: an optimal management of QoS in IoT communication protocols. It should be noted that the solutions to be proposed must imperatively take into account the management of a dynamic QoS. This dynamic QoS in IoT systems will be the subject of our next contribution, which will be an analysis following an experimental validation of a new concept of optimal QoS management in IoT. To do so, it is imperative to analyze existing solutions and techniques in order to identify their advantages and limitations while taking into account the QoS parameters discussed in the previous section.

Table 3 QoS IoT system: results and comparison

Approach	Architecture	Algorithm/framework	QoS metric
AQRA (Deng 2018)	SDN-based IoT	Application-aware SA-based QoS routing with adaptive weights, load balancing, admission control and starvation avoidance	Delay, jitter, packet loss rate, and available bandwidth
Peros et al. (2018)	SDN-based IoT	Framework built with SDN, to support dynamic QoS for the IoT	Delay, jitter and packet loss rate
Duan et al. (2011)	IoT architecture-Based on NQoSBr and PQoSBr brokers	A common cross-layer facility for supporting QoS operations in all three layers	Delay, jitter, packet loss rate, and available bandwidth

Table 4 QoS-aware of IoT/edge system: results and comparison

Approach	Architecture	Algorithm/framework	QoS metric
Yin et al. (2019) (CNN-MF (JCM)	IoT-edge computing	A novel prediction model with neighbors future learning based QoS on CNN and MF, named Joint CNN-MF(JCM)	The methods are evaluated using two measures inclusing the mean absolute error (MAE) and the mean square error (RMSE) smaller the MAE, higher the accuracy of prediction and higher the QoS
Sharghivand et al. (2018)	IoT-edge computing	Bilateral matching mechanism based on QMECS and M-QMECS for edge services to manage QoS requirements	Service response time
Song et al. (2017)	IoT-edge computing	An approach to QoS-based task distribution in edge computing networks for IoT applicatrion	Resource reservation
Rausch et al. (2018)	IoT-edge based Publish–subscribe	Distributed QoS-Aware MQTT Middleware for Edge computing Application	Latency

Indeed, interactive real time online applications (ROIA) are increasingly present in Cloud/Edge computer systems. The particularity of these applications means that expectations in QoS change during their execution. This means that existing static QoS management solutions will not be functional. Indeed, with the emergence of software defined networking (SDN), a new dynamic operation and control of the network is possible. These operations and controls require the realization of well-established quality of service (QoS) concepts using a centralized controller. An evaluation of the IT architecture is necessary for an interaction between the centralized controller and the IoT system, going from a cloud-centric architecture to a distributed and reinforced computing architecture at the level of the equipment at the edge of the network located at closer to the sensors (edge computing). This dynamism of QoS can be achieved by relying on these three paradigms: SDN, IoT, and edge computing.

7 Conclusion

Given the importance of IoT in everyday life, effective IoT applications require quality of service management. The use of this QoS in the IoT/edge computing environment is due to the satisfaction of end users, as well as service providers. We presented the most recent solutions and architectures in IoT/edge systems from a quality of service perspective. For this, we studied in detail several architectures and techniques based on a centralized integrated QoS controller. Classification was made to have the most appropriate QoS attributes for each layer of the IoT architecture. A logical continuation of this work is to study solutions based on the SDN

architecture with an optimization of QoS management for interactive applications in real time.

References

Al-Fuqaha, A., Guizani, M., Mohammadi, M., Aledhari, M., Ayyash, M.: Internet of things: a survey on enabling technologies, protocols, and applications. IEEE Commun. Surv. Tutor. **17**, 4, 2347–2376 (2015)

Beck, T., Werner, M., Feld, S., Schimper, S.: Mobile edge computing: a taxonomy. In: Proc. Intl. Conf. on Advances in Future Internet. Citeseer (2014)

Bonomi, F., Milito, R., Zhu, J., Addepalli, S.: Fog computing and its role in the internet of things. In: Proceedings of the First Edition of the MCC Workshop on Mobile Cloud Computing, pp. 13–16. ACM (2012)

Bonomi, F., Milito, R.A., Zhu, J., Addepalli, S.: Fog computing and its role in the internet of things.In: Proc. Workshop on Mobile Cloud Computing, pp. 13–16 (2012)

Daniel Macedo, Luiz A onso Guedes, and Ivanovitch Silva. 2014. A dependability evaluation for Internet of Things incorporating redundancy aspects. In Proceedings of the 11th IEEE International Conference on Networking, Sensing and Control. IEEE, 417–422.

Deng, G.C., Wang, K.: An application-aware QoS routing algorithm for SDN-based IoT networking. In: 2018 IEEE Symposium on Computers and Communications (ISCC) (2018)

Duan, R., Chen, X. and Xing, T.: A QoS architecture for IOT. In: 2011 International Conference on Internet of Things and 4th International Conference on Cyber, Physical and Social Computing, pp. 717–720. IEEE (2011)

Elsaleh, T., Gluhak, A., Moessner, K.: Service continuity for subscribers of the mobile real world Internet. In: 2011 IEEE International Conference on Communications Workshops (ICC), pp. 1–5. IEEE (2011)

Ganz, F., Li, R., Barnaghi, P., Harai, H.: A resource mobility scheme for service-continuity in the Internet of Things. In: 2012 IEEE International Conference on Green Computing and Communications, pp. 261–264. IEEE (2012)

Kempf, J., Arkko, J., Beheshti, N., Yedavalli, K.: Thoughts on reliability in the internet of things. In: Interconnecting Smart Objects with the Internet Workshop, vol. 1, pp. 1–4 (2011)

Kumar, K., Liu, J., Lu, Y., Bhargava, B.K.: A survey of computation o oading for mobile sys-tems. Mobile Netw. Appl. **18**(1), 129–140 (2013)

Kumar, K., Lu, Y.: Cloud computing for mobile users: can o oading computation save energy? IEEE Comput. **43**(4), 51–56 (2010)

Li, L., Rong, M., Zhang, G.: An Internet of things QoS estimate approach based on multidimension QoS. In: 2014 9th International Conference on Computer Science Education, pp. 998–1002. IEEE (2014)

Li, L., Jin, Z., Li, G., Zheng, L., Wei, Q.: Modeling and analyzing the reliability and cost of service composition in the IoT: a probabilistic approach. In: 2012 IEEE 19th International Conference on Web Services, pp. 584–591. IEEE (2012)

Maalel, N., Natalizio, E., Bouabdallah, A., Roux, P., Kellil, M.: Reliability for emergency applications in internet of things. In: 2013 IEEE International Conference on Distributed Computing in Sensor Systems. IEEE, 361–366 (2013)

Manyika, J., Chui, M., Bughin, J., Dobbs, R., Bisson, P., Marrs, A.: Disruptive Technologies: Advances that Will Transform Life, Business, and the Global Economy, vol. 180. McKinsey Global Institute San Francisco, CA (2013)

Naik, N., Jenkins, P., Davies, P., Newell, D.: Native web communication pro-tocols and their e ects on the performance of web services and systems. In: 2016 IEEE International Conference on Computer and Information Technology (CIT), pp. 219–225. IEEE (2016)

Peros, S., Janjua, H., Akkermans, S., Joosen, W., Hughes, D.: Dynamic QoS support for IoT backhaul networks through SDN. 978-1-5386-5896-3/18/2018 IEEE

Radovanović, S., Nemet, N., Ćetković, M., Bjelica, M. Z., Teslić, N.: Cloud-based framework for QoS monitoring and provisioning in consumer devices. In: 2013 IEEE Third International Conference on Consumer Electronics> Berlin (ICCE-Berlin), pp. 1–3. IEEE (2013)

Rajan, M.A., Balamuralidhar, P., Chethan, K.P., Swarnahpriyaah, M.: A self-recon gurable sensor network management system for internet of things paradigm. In: 2011 International Confer-ence on Devices and Communications (ICDe-Com), pp. 1–5. IEEE (2011)

Rausch, T., Nastic, S., Dustdar, S.: Emma: Distributed QoS-aware mqtt middleware for edge computing applications. In: 2018 IEEE International Conference on Cloud Engineering (IC2E), pp. 191–197. IEEE (2018)

Saleh, I.: Les enjeux et les de s de l'Internet des Objets (IdO). Int. Des Objets **1**, 1 (2017)

Satyanarayanan, M., Bahl, P., Caceres, R., Davies, N.: The case for vm-based cloudlets in mobile computing. IEEE Pervasive Comput. **8**(4), 14–23 (2009)

Satyanarayanan, M., Bahl, P., Caceres, R., Davies, N.: The case for vm-based cloudlets in mobile computting. Pervasive Comput. IEEE **8**(4), 14–23 (2009)

Sharghivand, N., Derakhshan, F., Mashayekhy, L.: QoS-aware matching of edge computing services to internet of things. In: 2018 IEEE 37th International Performance Computing and Communications Conference (IPCCC), pp. 1–8. IEEE (2018)

Shi, W., Cao, J., Zhang, Q., Li, Y., Xu, L.: Wayne State University Edge computing: vision and challenges, 2327–4662 (c) 2016 IEEE. Personal use is permitted, but repub-lication/redistribution requires IEEE permission. See https://www.ieee.org/publications-standards/publications/rights/index.html for more information

Singh, M., Baranwal, G.: Quality of service (QoS) in internet of things. In 2018 3rd International Conference on Internet of Things: Smart Innovation and Usages (IoT-SIU), pp. 1–6. IEEE (2018)

Song, Y., Yau, S. S., Yu, R., Zhang, X., Xue, G.: An approach to QoS-based task dis-tribution in edge computing networks for IoT applications. In: 2017 IEEE International Conference on Edge Computing (EDGE), pp. 32–39. IEEE (2017)

Varghese, B., Wang, N., Barbhuiya, S., Kilpatrick, P., Nikolopoulos, D.S.: Challenges and opportunities in edge computing. In: Proceedings of the IEEE International Conference on Smart Cloud (IEEESmart-Cloud). IEEE (2016). https://doi.org/10.1109/SmartCloud.2016.18

Yang, Z., Yue, Y., Yang, Y., Peng, Y., Wang, X., Liu, W.: Study and application on the architecture and key technologies for IOT. In: 2011 International Conference on Multimedia Technology, pp. 747–751. IEEE (2011)

Yin, Y., Chen, L., Xu, Y., Wan, J., Zhang, H., Mai, Z.: QoS Prediction for service recommendation with deep featureLearning in edge computing environment. Springer Science+Business Media, LLC, part of Springer Nature (2019)

Yu, W., Liang, F., He, X., Hatcher, W.G., Lu, C., Lin, J. and Yang, X.: A survey on the edge computing for the internet of things, 2169–3536 (c) 2017 IEEE. Translations and content mining are permitted for academic research only. Personal use is also permitted, but republication/redistribution requires IEEE permission. See https://www.ieee.org/publications-standards/publications/rights/index.html for more information

Zeng, D., Gu, L., Guo, S., Cheng, Z., Yu, S.: Joint optimization of task scheduling and image placement in fog computing supported software-de ned embedded system. IEEE Trans. Comput. **65**(12), 3702–3712 (2016)

Towards Sustainable e-Learning Systems Using an Adaptive Learning Approach

El Miloud Smaili⬥, Soukaina Sraidi⬥, Salma Azzouzi⬥, and My El Hassan Charaf⬥

Abstract

Over the last years, sustainable learning has adopted various dimensions aimed at limiting negative impacts at the individual, social, and technological levels. For this reason, formal classrooms have been partially replaced by e-learning systems in order to ensure the sustainability of these systems. In addition, the MOOC (Massive Open Online Course) revolution is gaining an increasing popularity since everyone can improve their knowledge and skills by using these courses. Moreover, MOOCs are, especially interesting, as they handle not only a large amount of teaching data, but also a large number of users with diverse educational and cultural backgrounds. However, the retention rate of learners, which is generally around 10%, raises the question of the sustainability of this mode of teaching. In this paper, we suggest a sustainable e-learning system to tackle school dropout. The idea is to provide courses that correspond to the way in which learners could complete adequately their learning process. To this end, the model will be realized through an adaptive e-learning system by exploiting the traces left by users' interactions with their learning environment. Depending on the learners' profile, the system will automatically determine the path and recommend the appropriate courses for them using the ant colony algorithm.

El Miloud Smaili (✉) · S. Sraidi · S. Azzouzi · My El Hassan Charaf
Informatics Systems and Optimization Laboratory (ISOLab), Ibn Tofail University-Kenitra, Kenitra, Morocco
e-mail: smaili.miloud@gmail.com

S. Sraidi
e-mail: sraidi.soukaina@gmail.com

S. Azzouzi
e-mail: salma.azzouzi@gmail.com

My El Hassan Charaf
e-mail: charaf@gmail.com

Keywords

Sustainability • Ant colony algorithm • MOOC • Adaptive learning

1 Introduction

Nowadays, the rapid growth of information and communication technologies has led to a learner-centered approach instead of a teacher-centered approach using popular e-learning platforms (Alharthi et al. 2019). However, recent studies have shown that if we developed such systems without taking into account the requirements of sustainability, we will be faced with the problem of the effectiveness of this model of education.

In this context, the sustainability is still an important challenge for e-learning systems as they process not only a huge amount of knowledge, but also various participants of different ages, learning habits, and backgrounds with only a single proposed path that cannot meet everyone's needs.

One popular approach to disseminate knowledge to a large number of people through such e-learning systems is the use of MOOCs (Massive Open Online Course). The revolution of MOOCs has emerged as these new online education systems offer free courses to anyone with access to the Internet. In addition, these systems manage a diverse set of course materials, such as course videos, written lessons, and assessment exercises (Khalil 2018). Moreover, these courses enable teachers to innovate easily in their pedagogical actions and provide opportunities for people to reinforce their social inclusion in education (Gulati 2013). However, the retention rate of learners using the MOOCs raises the question of the sustainability of this mode of teaching.

Several works have associated this dropout to the platform management, the course content or even the lack of motivation of the learners or a lack of their commitment.

M. Ben Ahmed et al. (eds.), *Emerging Trends in ICT for Sustainable Development*,
Advances in Science, Technology & Innovation, https://doi.org/10.1007/978-3-030-53440-0_38

However, the heterogeneity of learner profiles is widely omitted as one of the main causes of the low completion rate in these platforms.

In order to deal with this situation, we suggest an adaptive approach to ensure that the proposed course corresponds adequately to the way the learners could complete their learning process. Therefore, our objective in this paper is to improve course completion rates and to provide a basis for building sustainable e-learning systems. Consequently, we aim to address the sustainability of e-learning systems by designing a new adaptive learning system that tackles the dropout problem from the MOOCs.

The basic idea of this paper is to collect the traces left during the users' interactions with their learning environment. By using these traces, we obtain all relevant information related to the learner's profiles. Hence, the prototype as designed set the learners' characteristics, such as goals, preferences, level of knowledge, learning styles, and academic motivations into predefined profiles. Furthermore, we suggest generating via ant colony algorithm, recommendations adapted to each learner.

To this end, we generate the learner's profiles by collecting firstly static data predefined by the learner (i.e., the learner identification is provided during the learner's first interaction with the learning environment even before the beginning of the learning process). Then, we gather dynamic data that are updated in real time through the use of learners' traces during their interaction with MOOC forums, as well as their associated social media groups. Based on the learner's profile, the system will automatically determine the path to be followed by each learner and recommend the appropriate courses for them using the ant colony algorithm.

The paper is structured as follows: The second section provides some preliminaries. In Sect. 3, we present the problematic statement. Then, we discuss some work related to the adaptive e-learning system in Sect. 4. The Sect. 5, gives an overview of our adaptive learning model by highlighting the basic concepts of our architecture. In the Sect. 6, we explain and discuss the implementation of the proposed solution. Finally, Sect. 7 gives some conclusions and identifies future works.

2 Preliminaries

2.1 Sustainable E-learning

Sustainable software is defined as software that has a positive effect on sustainable development and whose direct and indirect negative impacts resulting from its development, deployment and use are minimal (Naumann et al. 2011). In particular, sustainable e-learning is seen as a system where organizational, technological and pedagogical activities interact to ensure sustainability (Robertson 2008).

2.2 Adaptive Learning

Adaptive learning consists of providing a personalized learning resource for students. It is also defined as the automatic structuring of learning paths to meet the needs of the learner. Consequently, the adaptive learning environments aims to change the traditional perspective of teaching by transforming education systems from their static form to a dynamic form and to provide learners with learning content that meets their requirements (Muhammad et al. 2016).

2.3 Swarm Intelligence

The main idea behind the swarm intelligence concept is to imitate the behavior of social insects swarming and flocking. The objective is to probe the search space in order to find the more convenient. We mentioned here the most swarm intelligence techniques: Particle Swarm Optimization (PSO) and Ant Colony Optimization (ACO). PSO is a stochastic global optimization method based on the social behavior (bird flocking or fish schooling) and intelligence of swarm searching for the global optimal whereas ACO is based on ants' behavior looking for a way between their state and a source of nourishment (Rozenberg et al. 2012).

3 Problematic Statement

One main objective of developing MOOCs is to reinforce the principles of open pedagogy by promoting free access to the education quality through online courses and using several types of resources such as videos, quizzes or texts. However, the sustainability for these e-learning systems is still an important challenge as most research points to a success rate of less than 10%.

Many works have coupled this low rate to the management of the platform or the content of courses or even the lack of motivation and commitment of the learners. However, the factor of learners' heterogeneity is widely omitted while addressing the dropping out problem.

We address in this article an adaptive approach to take into account the diversity and variability of learners. Hence, we seek to apply personalization techniques to adapt the MOOC content to the specificities of each learner by taking into account the expectations, motivation, learning styles, habits, and needs of each learner. To this end, we suggest in this paper an adaptive learning system based on ant colony optimization algorithm to meet the diversified demands of most learners.

4 Related Works

A number of studies showed that if an e-learning system is developed without taking into account sustainability requirements, this system could have negative impacts particularly on individual and social levels (Penzenstadler and Femmer 2013; Stepanyan et al. 2013).

Therefore, the authors suggest in (Alharthi et al. 2019), to cover some requirements related to human, technical, economic, and environmental aspects in order to ensure the sustainability of an e-learning system. Hence, they provide a systematic literature review of the sustainability meta-requirements for e-learning systems to identify open problems and to present the state of the art of this research area.

Another work (Farid et al. 2017), suggests a sustainable quality assessment model for the e-learning systems keeping software perspective under consideration. The authors of (Demirkan et al. 2010), address the basic and advanced sustainability capabilities that integrate partner, application, faculty, student, and e-learning service system issues.

However, the factor of learners' heterogeneity is widely omitted while addressing the sustainability issues for these eLearning systems. In this context, the authors in (Tlili et al. 2017), refer to Learning Analytics (LA) to implicitly model learners' personalities based on their traces generated during the learning-playing process. In (Tashtoush et al. 2017), the paper proposes a new adaptive e-learning system that integrates a well known intelligent web-based English e-learning tutor with data mining techniques. The work (Kotova and Pisarev 2017), proposes a method developed for the sequential refinement of the prediction of students' learning, taking into account individual cognitive characteristics.

The objective of the paper (Ennouamani and Mahani 2017), is to present the state of the art in adaptive e-learning systems as an alternative to the traditional learning by describing its dimensions, design, architecture, and theoretical approaches. A new learner model is proposed in the paper (DIng et al. 2018), including four basic features: basic information, learning style, knowledge state, and cognitive ability. The authors give a formal representation of these four feature elements and a method for initializing and updating the values.

Furthermore, an adaptive framework for learning is given in (Pandit and Bansal 2019), where groups, students plan with the end goal of helping the learner complete all topics in the course. The proposed model in (Radosavljevic et al.

2019), takes advantage of a smart classroom environment for the realization of adaptive learning. As adaptation criteria, the authors use parameters of motivation, student's prior knowledge, cognitive load, and a dynamic environmental parameter.

The work (Li 2019), is based on the improved ant colony algorithm, an adaptive learning system model that can satisfy learners' demands is built herein with reference to the foraging approach of ants to traverse the paths, thereby to find the best learning path. The paper (Šumak et al. 2019), reports the design and development architecture of an AIAES (Autonomous Intelligent and Adaptive E-learning Systems), the main objective of the work is to improve the information literacy of adolescents. The authors in (Rosen et al. 2018), report an experimental implementation of adaptive learning functionality in a self-paced Microsoft MOOC (massive open online course) on edX.

Finally, we focus in this article on sustainable learning and we suggest an adaptive approach that takes into account the learners' heterogeneity. To this end, we describe in the next section our model to deal with the dropout problem from the MOOCs.

5 Our Sustainable Model-Based ACO Algorithm

5.1 Architecture

We describe in the figure (Fig. 1), the architecture of our adaptive learning model. We aim to build an appropriate solution based on ant colony optimization to meet the diversified needs of the learners and by the way to face the dropout problem.

In what follows, we highlight the basic concepts of our adaptive architecture.

5.2 Data Collection Approach

Static Data. The personal information of learners will be collected and stored in our database. We used such information to identify the learners throughout the training and at each interaction with the system. This information is static and cannot be changed during the learning process. It will be provided at the beginning of the learning process.

Dynamic Data. Although the learning platforms have forums, these forums often don't respond to students' social needs. Recent studies show the social media use by students compared to embedded MOOC forums and also the engagement from both instructors and students. In fact, the

Fig. 1 Adaptive learning
Architecture

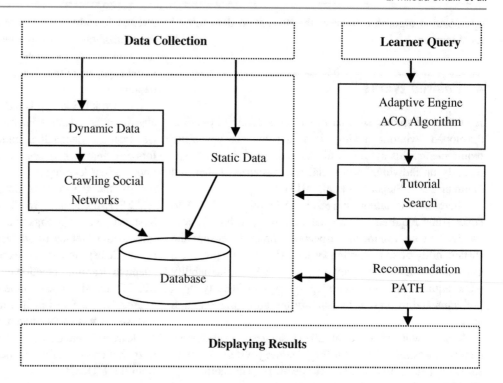

authors in (Zheng et al. 2016), have found indications that Facebook groups for a course can be more active than the official forums of the course. Generally, students may create online support communities on social networking sites and previous participants of a MOOC may independently volunteer to support participants in new courses. For instance, some participants can set up a Facebook group at the beginning of the course to create their own space outside the Coursera platform for increasing interactions: sharing questions, concerns, and successes and failures.

Crawling Online Social Networks. In order to collect dynamic data described above, we need to identify and download a useful set of web pages within time and bandwidth constraints. Generally, web crawlers are used for social media and general web. One important difference is the opportunity for a social media crawler to use information about how a site is organized to crawl it more efficiently.

Learner's Profiles Generation. The work on adaptive learning focuses on the learner and the way in which can be observed. The learner's profile generation consists in collecting the characteristics of the learners, namely the objectives, preferences, level of knowledge, learning styles, and academic motivations. In fact, the learners are identified by their profiles and each user needs to create a complete profile with current learning level (beginners, intermediate or advanced expertise) and learning style. These characteristics will be grouped into profiles and the system will then determine automatically for each learner, based on the profile, the activities that should be recommended.

5.3 Ant Colony Optimization Algorithm: Learners as Ants

Overview. The ant colony optimization (ACO) is an optimization technique introduced in the early 90 s. The initial idea comes from the observation of ants. They have the ability to find the shortest path between their nest and a food source by bypassing the obstacles in their path (Dorigo et al. 1999).

Principle. Ants have developed very elaborate communication mechanisms; they deposit by walking chemical markers called pheromones. The other ants follow the path traced by these pheromones. Consequently, the presence of the pheromone in a track plays the role of a means allowing the attraction of ants to the reinforced track. When an ant has to make a decision on the direction to take, it must choose the path with the highest pheromone concentration (Fig. 2).

Algorithm Description. The ant colony optimization (ACO) is a commonly used algorithm to find the optimal path using heuristic algorithms within an acceptable time. As shown below (Algorithm1), the ACO approach attempts to solve an optimization problem by repeating the following two steps:

- The candidate solutions are built using a pheromone model, that is, a parameterized probability distribution over the solution space;
- The candidate solutions are used to modify pheromone values in a way that is deemed to bias future sampling toward high quality solutions.

Fig. 2 Learners as Ants using ACO approach

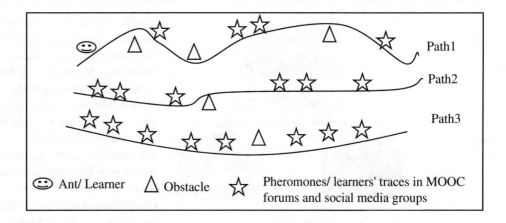

Path1

Path2

Path3

☺ Ant/ Learner △ Obstacle ☆ Pheromones/ learners' traces in MOOC forums and social media groups

---

Algorithm 1
Set parameters, initialize pheromone trails
BEGIN_SCHEDULE_ACTIVITIES
    Construct Ant Solutions
    Daemon Actions   {optional}
    Update Pheromones
END_SCHEDULE_ACTIVITIES

Mathematical Model. In order to determine the next node to be accessed, the algorithm applies the probabilistic behaviors as the ants randomly select the vertex to be visited. According to (Dorigo 2006), when the ant k is in vertex i, the probability of going to vertex j is given by the formula

$$p_{ij}^k = \begin{cases} \frac{(\tau_{ij})^\alpha \cdot (\eta_{ij})^\beta}{\sum_{l \in J_i^k} (\tau_{il})^\alpha \cdot (\eta_{il})^\beta} & \text{If } j \in J_i^k \\ 0 & \text{If } j \notin J_i^k \end{cases}$$

- J_i^k is the set of neighbors of vertex i of the kth ant,
- τ_{ij} is the amount of pheromone trail on edge (i, j),
- α and β are weightings that control the pheromone trail
- The visibility value η_{ij} is given by: $\eta_{ij} = \frac{1}{d_{ij}}$
- d_{ij} is the distance between vertices i and j.

The pheromone values are updated by all the m ants that have completed the tour. The pheromone τ_{ij}, which is associated with the edge joining vertices i and j, is updated as follows:

$$(\tau_{ij}) = (1 - \rho)\tau_{ij} + \sum_{k=1}^m \Delta\tau_{ij}^k$$

where

- ρ is the pheromone evaporation rate,
- m is the number of ants,
- $\Delta\tau_{ij}^k(t)$ is the quantity of pheromone laid on edges (i, j) by the ant k

$$\Delta\tau_{ij}^k = \begin{cases} \frac{Q}{L^k} & \text{If the ant k used edge } (i, j) \text{ in its tours,} \\ 0 & \text{Otherwis} \end{cases}$$

- Q is a constant and L^k is the length of the tour constructed by the ant k.

5.4 Processing Data

In this experiment, we suggest to simulate the behavior of ants (learners) using the Matrix Laboratory (MATLAB) software. Furthermore, the choice of input data is done accordingly to the approach described in Sect. 5.2.

Therefore, we choose to implement our algorithm using the MATLAB (version: R2010b) language, and the hardware configuration used for our experiment is an Intel® Core™ i5-6500 3.19 GHz, RAM 8 GB running the OS: Windows 7.

Thus, the method as implemented will allow us to accurately predict the learner profile at an early stage. Consequently, it will automatically determine the path to be followed by each learner and recommend the appropriate courses for them.

6 Adaptive Learning Application

6.1 Prototype

The figure below (Fig. 3), gives an overview of our prototype. The figure describes the prototype adopted in our context from the first contact of the learner with the learning system to the completion phase of the course.

The proposed approach can serve as a decision support system to set goals and priorities. We discuss in the next section the different steps of our model.

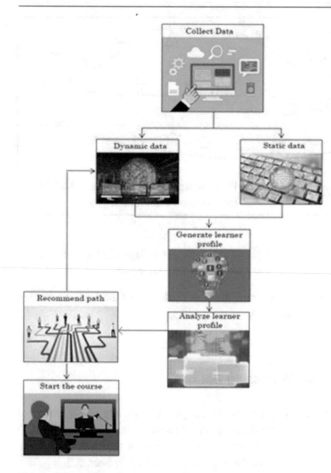

Fig. 3 Adaptive learning prototype

6.2 Model Description

The purpose of the model is to give a comprehensive view to overcome the MOOCs low completion rate.

The learner identification will be provided during the first interaction of the learner with the learning environment even before the start of the learning process, some information is static and cannot be changed during the learning process, others are dynamic and both serve to generate the learner's profile. All learner interactions are collected including the learners' personal information used for identification during the training period. However, if the user logs in the system for the first time, after the registration, the current system has no history of his or her searches.

The other category of traces concerns the characteristics of the learners: the objectives, preferences, level of knowledge, learning styles, and academic motivations, or even the learners' expectations of the course. We address such data by the collection and analysis of real data from MOOC forums, as well as their associated social media groups. We

plan to collect data (forum posts) from Facebook groups organized around some Coursera courses. In such context, the dataset can be obtained throughout application programming interfaces (APIs) or obtained automatically through web crawling tools.

After the acquisition of data text, we proceed on data preprocessing through replacing all the emoticons with their sentiment and removal of hashtags, URLs, stop words, non-English words, punctuation, symbols, and numbers. Afterwards, the traces are stored in a Mango database so that we can analyze these traces to generate learner profiles based on the characteristics collected. Hence, the generated profiles will allow the system to suggest automatically the activities that should be recommended for each learner.

6.3 Discussion

In this paper, we suggest a self-adaptive approach based on an analysis of learners' interactions within the e-learning platform. The aim is to build a sustainable e-learning system in order to raise learners' motivation and reduce dropout rates.

Besides, several works have conducted to tackle the retention rate of learners. However, the heterogeneity of learner profiles is widely omitted as a cause of the low completion rate in these platforms.

According to the proposed approach, the "data collection" will be done automatically from the main MOOC dissemination platforms, and also manually when the automated approaches fail.

Furthermore, the learners' paths selection is made using the ant colony optimization algorithm. In this case, the learners are considered as ants and the pheromones (Chemical markers generated by ants while walking) are reflected on as learners' traces in MOOC forums and social media groups. Moreover, the learners will also have learning objects that have been accessed most often recently. As a result, ant colony optimization can use the same approach to recommend the appropriate course for learners. In this context, adaptive learning based on the ant colony algorithm exploits the characteristics of the learners and the learning methods to provide the most appropriate courses in the learning system.

Moreover, while the learner chooses the proposed course and begins the learning process, the system takes care of the supervision of all its interactions with the system by updating the learner profile to regenerate the course and guide the learner's choices. At each interaction with the system, the learner's profile will be analyzed by the proposed algorithm, so that the learning materials can easily be recommended to the learner. Consequently, the proposed system guesses the knowledge and preferences based on both user's own data and other users' desire collected from online social networks.

Finally, the use of ant colony optimization approach will help users to quickly get the resources they desire and respond to their educational demands.

7 Conclusion

As a conclusion, we deal in this paper with the sustainability of e-learning systems by providing learners the necessary support to tackle the low completion rate of MOOC's. To this end, we suggest an adaptive learning framework based on the ant colony optimization algorithm.

The development of this framework is based on the learners' profiles. Hence, the prototype as designed set the learners' characteristics into predefined profiles. These profiles are, therefore, generated by collecting: static data introduced at the first time by the learners and dynamic data that are updated in real time through the use of learners' traces during their interaction with social media and forums associated to the MOOC. Based on the learner's profile, the system will automatically determine the path to be followed by each learner and recommend the appropriate courses for them using the ant colony algorithm. Finally, our prototype's realization of this model is under experimentation and we conduct actually empirical studies to evaluate the efficiency of our approach using MATLAB software.

In the future works, we will pay greater attention to other characteristics such as reasoning and cognitive styles. In addition, the social context should be investigated more in order to improve the efficiency of our model.

References

Alharthi, A.D., Spichkova, M., Hamilton, M.: Sustainability requirements for eLearning systems: a systematic literature review and analysis. Requir. Eng. **24**, 523–543 (2019). https://doi.org/10.1007/s00766-018-0299-9

Demirkan, H., Goul, M., Gros, M.: A reference model for sustainable e-learning service systems: experiences with the joint university/teradata consortium. Decis. Sci. J. Innov. Educ. **8**, 151–189 (2010). https://doi.org/10.1111/j.1540-4609.2009.00250.x

DIng, W., Zhu, Z., Guo, Q.: A new learner model in adaptive learning system. In: 2018 3rd Int. Conf. Comput. Commun. Syst. ICCCS 2018, pp. 472–476 (2018). https://doi.org/10.1109/CCOMS.2018.8463316

Dorigo, M.: An introduction to ant colony optimization. In: IRIDIA – Tech. Rep. Ser. Tech. Rep. No. TR/IRIDIA/2006-010 (2006)

Dorigo, M., Di Caro, G., Gambardella, L.M.: Ant algorithms for discrete optimization. Artif. Life. **5**, 137–172 (1999). https://doi.org/10.1162/106454699568728

Ennouamani, S., Mahani, Z.: An overview of adaptive e-learning systems. In: 2017 IEEE 8th Int. Conf. Intell. Comput. Inf. Syst. ICICIS 2017. 2018-Janua, pp. 342–347 (2018). https://doi.org/10.1109/INTELCIS.2017.8260060

Farid, S., Ahmad, R., Alam, M.: A sustainable quality assessment model for the information delivery in e-learning systems. Inf. Discov. Deliv. (2017)

Gulati, A.: An overview of massive open online courses (MOOCs): some reflections. Int. J. Digit. Libr. Serv. **3**, 37–46 (2013)

Khalil, M.: Learning analytics in massive open online courses (2018)

Kotova, E.E., Pisarev, A.S.: Adaptive prediction of student learning out-comes in online mode. In: Proc. 2017 IEEE 2nd Int. Conf. Control Tech. Syst. CTS 2017, pp. 138–141 (2017). https://doi.org/10.1109/CTSYS.2017.8109509

Li, R.: Adaptive learning model based on ant colony algorithm. Int. J. Emerg. Technol. Learn. **14**, 49–57 (2019). https://doi.org/10.3991/ijet.v14i01.9487

Muhammad, A., Zhou, Q., Beydoun, G., Xu, D., Shen, J.: Learning path adaptation in online learning systems. In: Proc. 2016 IEEE 20th Int. Conf. Comput. Support. Coop. Work Des. CSCWD 2016, pp. 421–426 (2016). https://doi.org/10.1109/CSCWD.2016.7566026

Naumann, S., Dick, M., Kern, E., Johann, T.: The GREENSOFT model: a reference model for green and sustainable software and its engineering. Sustain. Comput. Inform. Syst. **1**, 294–304 (2011). https://doi.org/10.1016/j.suscom.2011.06.004

Pandit, D., Bansal, A.: A declarative approach for an adaptive frame-work for learning in online courses. In: Proc. - Int. Comput. Softw. Appl. Conf. 1, pp. 212–215 (2019). https://doi.org/10.1109/COMPSAC.2019.00039

Penzenstadler, B., Femmer, H.: A generic model for sustainability with process- and product-specific instances. In: GIBSE 2013 - Proc. 2013 Work. Green Softw. Eng. Green by Softw. Eng., pp. 3–7 (2013). https://doi.org/10.1145/2451605.2451609

Radosavljevic, V., Radosavljevic, S., Jelic, G.: A model of adaptive learning in smart classrooms based on the learning strategies. PEOPLE Int. J. Soc. Sci. **5**, 662–679 (2019). https://doi.org/10.20319/pijss.2019.52.662679

Robertson, I.: Sustainable e-learning, activity theory and professional development. In: ASCILITE 2008 - Australas. Soc. Comput. Learn. Tert. Educ, pp. 819–826 (2008)

Rosen, Y., Munson, L., Lopez, G., Rushkin, I., Ang, A., Tingley, D., Rubin, R., Weber, G.: The effects of adaptive learning in a massive open online course on learners' skill development. In: Proc. 5th Annu. ACM Conf. Learn. Scale, L S 2018 (2018). https://doi.org/10.1145/3231644.3231651

Rozenberg, G., Bäck, T., Kok, J.N.: Handbook of Natural Computing. Springer, Berlin, Heidelberg (2012)

Stepanyan, K., Littlejohn, A., Margaryan, A.: Sustainable e-learning: toward a coherent body of knowledge. J. Educ. Technol. Soc. **16**, 91–102

Šumak, B., Podgorelec, V., Karakatic, S., Dolenc, K., Šorgo, A.: Development of an autonomous, intelligent and adaptive e-learning system. In: 2019 42nd Int. Conv. Inf. Commun. Technol. Electron. Microelectron. MIPRO 2019 - Proc., pp. 1492–1497 (2019). https://doi.org/10.23919/MIPRO.2019.8756889

Tashtoush, Y.M., Al-Soud, M., Fraihat, M., Al-Sarayrah, W., Alsmirat, M.A.: Adaptive e-learning web-based English tutor using data mining techniques and Jackson's learning styles. In: 2017 8th Int.

Conf. Inf. Commun. Syst. ICICS 2017. 86–91 (2017). https://doi.org/10.1109/IACS.2017.7921951

Tlili, A., Essalmi, F., Ayed, L.J. Ben, Jemni, M., Kinshuk: A smart educational game to model personality using learning analytics. In: Proc. - IEEE 17th Int. Conf. Adv. Learn. Technol. ICALT 2017, pp. 131–135 (2017). https://doi.org/10.1109/ICALT.2017.65

Zheng, S., Han, K., Rosson, M.B., Carroll, J.M.: The role of social media in MOOCs: how to use social media to enhance student retention. In: L@S 2016 - Proc. 3rd 2016 ACM Conf. Learn. Scale, pp. 419–428 (2016). https://doi.org/10.1145/2876034.2876047

Toward a Mobile Remote Controlled Robot for Early Childhood in Algeria

Ehlem Zigh, Ayoub Elhoucine, Abderrahmane Mallek, and Belcacem Kouninef

Abstract

The mobile phone has made the learning process for children interesting. It will be certainly much more fascinating if it is combined with a hardware conception. For that, we propose in this paper a pedagogical package including a robot car and an android application. We describe all its components and presume that it will be useful for early childhood and first primary education. It is an inexpensive interactive pedagogical package, where a child solves three games quiz related to three levels consecutively. Therefore, the further the kid moves into the game, the more control he has over the robot car. He could play alone or with his friends into group game mode. We have proposed another software implementation including some educational videos that a child could watch if he wants to use a smartphone without a robot. Also, for more flexibility use and for disabled children, we have implemented a speech recognition technique, therefore, he could use his voice commands as forward, backward, right, left, stop to interact easily with a robot. The proposed pedagogical package constructs the high-level cognitive abilities of children. It is designed for Algerian early childhood, it could be modified according to the educational needs of other children from other countries around the world.

Keywords

Internet of Things • Early childhood education in Algeria • Arduino card • Remote controlled robot • Android application • Mobile communication

1 Background

The current pedagogical approaches for early childhood education support the ability of the child to have control of his activities, while the active participation constitutes one of the basic parameters for the construction of knowledge (Komis and Misirli 2016). This participation could be ensured by the use of robots or mobile applications. Recent works have shown how the field of robotics applications, in particular, is a suitable educational framework within which it is possible to develop high-level cognitive skills for early childhood, such as critical thinking, problem-solving, team-work, logical, and linguistics abilities (Bers et al. 2002, 2013; Cejka et al. 2006; Perlman 1976; Wyeth 2008; Sullivan et al. 2013). In this context, we can distinguish two major types of contexts for early childhood and first school age: the robotic construction kits, Lego-logo like environments (Kibo, LEGO®-WeDo™) allows construction and programming of the robot and programmable robots Logo-like environments (Bee-Bot™, Pro-Bot™, Constructa-Bot™).

The educational robotic provides attractive learning environment (Alimisis 2013). To show its importance, we can cite, for example, the research of (Cheng et al. 2017) which indicates that the educational robotics is classified as the second large demand (Fig. 1).

On the other hand, children are exposed to handheld mobile technology or devices like smartphones and tablets. Many studies showed that there are improvements in terms of children performance before and after they used the multimedia mobile application as a learning tool in the area

E. Zigh (✉) · B. Kouninef
Laboratoire LaRATIC. Institut National des Télécommunications et des TIC D'Oran, BP 1518 Oran El M'nouer, 31000 Oran, Algérie
e-mail: ezigh@inttic.dz

A. Elhoucine · A. Mallek
Institut National des Télécommunications et des TIC D'Oran, BP 1518 Oran El M'nouer, 31000 Oran, Algérie

Fig. 1 The ranking of the needs for educational robots (Cheng et al. 2017)

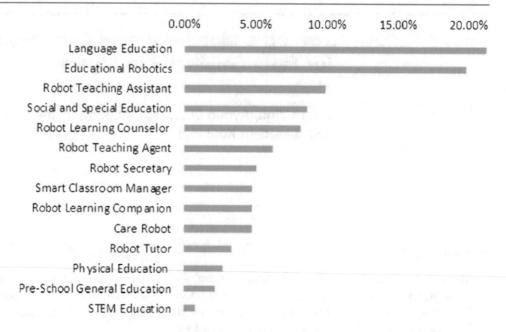

of language information, learning skills, cognitive strategy, attitude, emotion, and motor skills (Kokkalia and Drigas 2016).

We can cite, for example, (Yang et al. 2014) who proposed a hand gesture recognition method for a smartphone-based robot and applied it to early childhood mathematics education. (Hadron Solutions India Pvt. Ltd. 2013) have developed "Alphabet find application" that includes letters, numbers and some colors to be helpful and attractive for children. Kids number and math is a package of brightly designed, pleasantly narrated number recognition and math games which is made with little hands and budding mathematicians in mind.

Thus, we found that it will be even more interesting if we can benefit from both robotic and smartphone fields to build an educational kit for Algerian early childhood as an example.

The first motivating element is that the number of smartphone users in Algeria is increasing from one year to another. According to the annual Global Mobile Market Report, which shows the top 50 countries in terms of smartphone users in 2017, Algeria has been classified at the 32 position over the world, and it is ranked as the third Arabic country. Also, Algeria has a high number of mobile subscriptions according to Hootsuite statistics done in January 2019 (Fig. 2).

The second motivating element is that the school accompaniment of disabled early childhood in Algeria is not often suitable for this category of pupils (Bessai 2018).

Furthermore, to acquire an educational robot for Algeria country is still expensive.

To overcome these issues, we propose in this paper, a combination of a robot car and an android application to obtain a suitable inexpensive pedagogical package for early childhood. We aim via this package:

- To contribute to childhood education in third world countries like Algeria.
- To help parents for keeping their kids away from violent games.
- To provide a deal of support for disabled kids by allowing them the same games like other kids.
- To offer Algeria childhood education an inexpensive package including both hardware and software parts.

2 The Overall Implementation of the Proposed Package

The proposed package is a mobile remote controlled robot car. Its implementation is illustrated in Fig. 3. The mobile phone retrieves the needed data from Firebase database via the Internet and communicates with the robot via Wi-Fi module in local area network (LAN).

The Robot acts as a server and the Smartphone as a client during the communication process. This last was established using sockets, in which the client needs the port number and the IP address of the Server, whereas the server needs only the port number. The Wi-Fi module receives the messages and transfers it to the Arduino microcontroller via serial communication (Rx/Tx). The Arduino microcontroller does the processing operation and gives the orders to the motor driver to control the spin direction of the actuators.

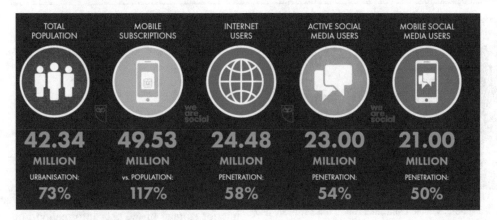

Fig. 2 Mobile use compared to Internet and social media use in Algeria *Source* United nations; U.S.Census Bureau. Mobile: QSMA intelligence. Internet: Internetworldstats; ITU; world bank; CIA world facebook, Eurosat, Local goverment bodies and regulatory authorities; Mideastmediaorg;reports in regular media. Social media: platforms'self-serve advertising tools, press releases and investor earnings announcements, arab social media report, techrasa, niki aghael, rose RU

Fig. 3 The overall system block diagram

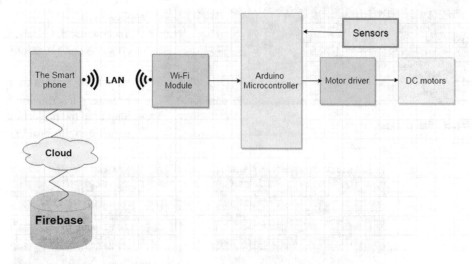

3   The Hardware and Software Parts of the Proposed Mobile Remote-Controlled Robot

We show in Fig. 4, the implemented mobile remote controlled robot. It includes hardware and software parts.

3.1   The Hardware Part

The hardware part could be divided into three main parts:

- The hardware controller (Fig. 5): it includes Arduino Uno ATmega 328, motor driver L293D and Wi-Fi module ESP8266.
- The movement mechanism: it is front and back wheels of the robot car. We have increased the wheel torque using the gear system on the back wheels (Fig. 6).

- The energy: the power output delivered from the Arduino Card (5 V) is used to supply the integrated circuit of the motors' driver. To feed the two DC motors, we have used 6 V power supply obtained from 4 × 1.5 V batteries mounted in series.

3.2   The Software Part

The Application Architecture (the Model View Controller)

The application architecture called the model view controller divides an interactive application into three components (Fig. 7): model, views, and controller. The model contains the core functionality and data. Views display information to the user. Controllers handle user input. Views and controllers together comprise the user interface. A change-propagation mechanism ensures consistency between the user interface and the model.

Fig. 4 A proposed mobile remote controlled robot: the hardware and software parts

Fig. 5 The hardware controller

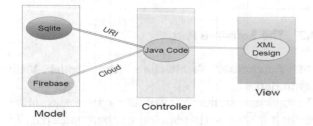

Fig. 6 The front and back wheels of the robot car

Fig. 7 The application architecture

```
dependencies {
    classpath 'com.android.tools.build:gradle:1.5.0'
    classpath 'com.google.gms:google-services:3.0.0'
}
```

Fig. 8 Adding the classpaths

```
apply plugin: 'com.google.gms.google-services'
```

Fig. 9 Applying the file as plugin

In our application, we have used Sqlite database for the interior storage of the user information in the phone. These informations are the user name, the user family name, the user age, the game level, the level score.

To (read/write) (from/to) the Sqlite database, we have created a java class (User helper) to do this work.

We have used the Firebase as a backend to the application.

To connect the Firebase with the application, we needed to do some configurations in Android studio SDK:

- Step 1: we needed to upload JSON file that holds all the necessary information like Firebase project information
- Step 2: in the project file we needed to add two classpaths variable to the **build.gradle** file inside the dependencies

The gear system

- Step 3: in the app file, we needed to apply the file that we have uploaded as plugin inside the **build.gradle** file.

In our case, the controller is a bunch of java classes as illustrated in Fig. 10.

The view is a bunch of layouts, menus and drawable that reside in res file.

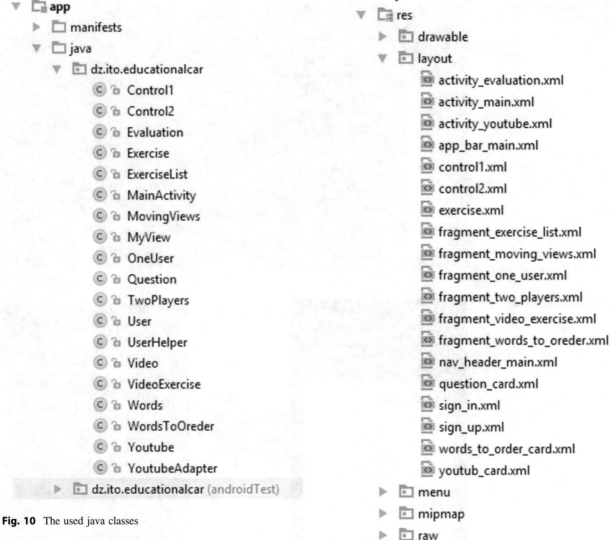

Fig. 10 The used java classes

Fig. 11 The **res** file

The Application Components

The application components are the essential building blocks of an Android application. Each component is an input point through which a system or a user can enter the application. Some components depend on others. In our application, we have used three components which are activity, content provider, and broadcast receiver. Each type serves a distinct purpose and has a distinct lifecycle that defines how the component is created and destroyed.

The Code Simulation

To simulate the code of the Robot execution, we have used PROTEUS virtual System Modeling (VSM). The circuit is created in PROTEUS using the previously described components in order to test and debug the final program before uploading it to the Arduino (Fig. 12).

The Speech Recognition Implementation

At the present time, mobile products of speech recognition are pervasive. There are numerous speech recognition applications that are supported in android products.

We have chosen Google Voice Recognition, which is preinstalled in many android devices as our recognition engine. It has been added as an additional option, especially for disabled children. We have designed the application with five meaningful voice commands which are Forward, Backward, Right, Left and Stop.

As the voice commands to the robot are processed via an android application and transferred via Wi-Fi, a decision to

Fig. 12 A screenshot of the proteus simulation

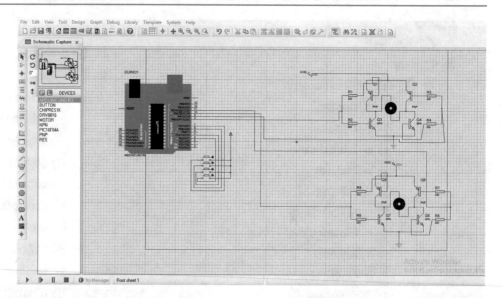

Fig. 13 A new user into the game

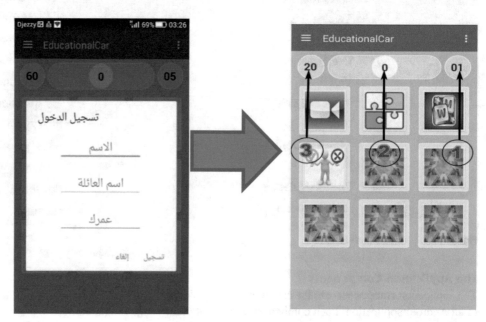

use an Android Operating System interface as the speech processing platform was made due to its flexibility and numerous features. As of May 2017, Android has over two billion monthly active users, the largest installed base of any operating system, and as of December 2018, the Google Play store features over 2.6 million applications (14). An android operating system allows an easy and reliable connection with the Google Speech processing libraries for smooth and accurate speech recognition.

The Libraries .

Google voice recognizer is an application developed by Google Company. This application enables a child to type with his voice, meaning by dictating the words, the application will receive the input, analyze it, and convert it to text.

To make it easier for the application developers, android added the pre-defined APR for speech recognizer into its library. In this way, we have only needed to add the library into the application and call the right function and method in the java class.

In order to write an offline voice recognizer which recognize the dictated word and convert it to text, importing below JAR file into the java class is recommended:

Import Android.Speech.RecognitionListener;

Import Android.Speech.RecognizerIntent;
Import Android.Speech.SpeechRecognizer;
Import Android.App.Activity;
Import Android.Content.Intent;
Import Android.View.View;

Fig. 14 First Level: Example of an educational video

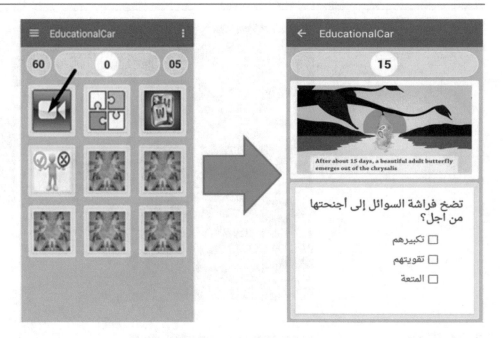

Import Android.Widget.CompoundButton;
Import   Android.Widget.CompoundButton.
OnCheckedChangeListener;
Import Android.Widget.ProgressBar;
Import Android.Widget.TextView;
Import Android.Widget.ToggleButton;

The Package Definitions

- Android.speech package contains one interface (Recognition Listener) and five classes (Recognition Service, Callback, Recognizer Intent, Recognizer Results Intent and Speech Recognizer)
- Android.app package contains high-level classes encapsulating the overall Android application model.
- Android.Widget package, to create our own widget, we have extended View.

4   Results

4.1   The New User Interface

For a new user, firstly he needs to create a new account, the application will save his information and generate two variables which are level and score, the level variable equal to one and the score equal to zero for a new account, the more the user answers correctly, the more score he wins and the more advanced level he moves on.

According to Fig. 36, we have

1. The games level.
2. The user score.
3. The minimum required score to move to the next level.

4.2   The Levels of the Proposed Educational Games

After creating an account, the user (a child) will find a list of games for each level. We have proposed three levels. The complexity of games will increase according to the increase in levels. Also, the content of the games could be modified according to the educational needs (the educational system of each country, the age of the child, etc.).

The First Level
This first level includes a list of games related to educational videos taken from ausum (15). The longest proposed video is around 6 min. For that, a child watches the video, after that, he has to choose a correct answer from three proposed responses (Fig. 14).

The Second Level
The games of this level consist of ordering the words that matches with the content of the image in order to obtain a correct sentence (Fig. 15).

The Third Level
In this level, a child has to intersect the right word with the right image (Fig. 16).

Fig. 15 Second level: example of putting words in order

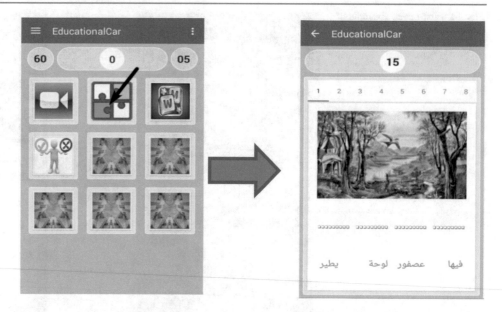

Fig. 16 Third level: example of intersection word to image

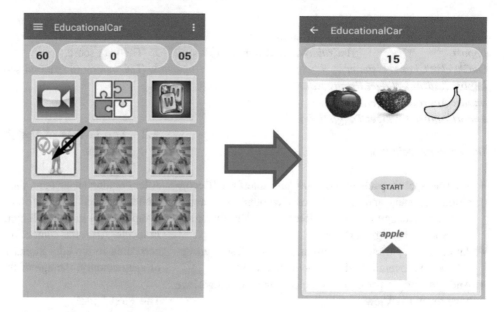

4.3 The Robot Page Controller

It is the page where the kid finds the available robot control. Each allowed robot control depends on the reached level (Fig. 17).

- (a): First robot control is related to the first level of games, so, if the child wins the first level, he can move the robot car forward and back.
- (b): Second robot control is related to the second level, so, if the child wins the second level, he can turn the robot on the right and the left.
- (c): Third robot is control related to the third game level, so, if the child wins the third level, he can put on/put off the lights of the robot car.

We have added a voice control option for disabled kids or for those who want to use their vice to remote control a robot car (Fig. 18).

4.4 Other Software Implementation: Educational Videos

We have added another software implementation. It is optional and consists of using a smartphone without a robot. For that, we have proposed an automatic link from the mobile phone to some educational videos. Therefore, the child is enjoying another kind of edutainment (Fig. 19). For that, we have selected beforehand, some videos from youtube and we have saved them in the Firebase database. We

Fig. 17 The robot page controller. **a** first control. **b** second control. **c**. third control

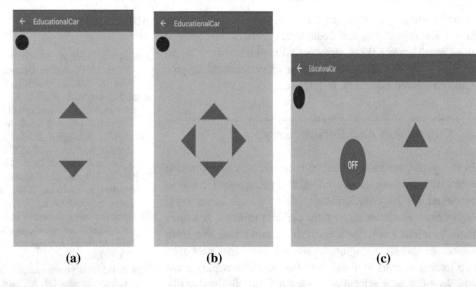

(a) (b) (c)

(a) (b)

Fig. 18 Example of a robot voice control. **a** First interface, **b** Forward control

Fig. 19 Some selected vidéos

have organized these videos into subjects like human stories in Quran (16), women stories in Quran (17), and animal stories in Quran (18). They exist in Arabic version and English subtitles.

5 Discussion

We have described a proposed educational application to remote control a robot. The proposed software part of our prototype includes three examples of games related to three levels, respectively. The first one consists in multiple choice questions from a short video introduced in advance, the second game is the ordering of a set of words according to a

content of the illustrated image and in the third game, a kid should intersect the right word with the right displayed image.

To ensure an attractiveness for the proposed application, we have tried to set for each level his own design.

The success of each level allows a child a remote control of his robot in easy and affordable way.

Also, for disabled kids, we have added a speech remote control option, therefore, a child could use his voice commands as forward, backward, right, left, stop, to interact easily with a robot. Furthermore, we have taken into consideration the possibility of using a smartphone without a robot, for that some selected educational videos are proposed for a child. Although we did not use standardized methods to evaluate the proposed package in term of social acceptance and usability, the oral and behavioral feedback we received from some people (kids, parents and engineers) may be an indicator that the proposed package can contribute to promote childhood education in Algeria.

6 Conclusion and Perspectives

In this paper, we have described a mobile remote controlled robot. It is an inexpensive interactive pedagogical package proposed for Algerian children. This package is an easy, enjoyable and educative way for children to learn. It allows them to interact with both smartphone and robot. We have added a speech recognition control into the proposed application in order to allow child free hands smartphone use and to offer this educational package for disabled early childhood. For that, we have chosen Google voice recognition-based system, this implementation allows the child to control the robot using his voice commands as forward, backward, right, left, stop. Furthermore, we have added another software option. It is an automatic link to some educational videos which could be accessible by a child when he wants to use a smartphone without a robot car.

Therefore, this pedagogical package could be used by children in order to learn in an attractive and interactive environment. We are sure that it will bridge the gap between education and engineering in the future.

A proposed implementation offers significant flexibility and extension possibilities for the field of robotics education. In one hand, a hardware part could be improved by using ultrasonic sensors, for example. In the other hand, we can plane to share the software part via social media to allow the robot control to two players (two kids) using two smartphones in order to raise much more a competitive aspect of the game.

We can also change the kind of games on the Android platform according to a child's need and an education system of the country.

For future, we plan to use standardized methods to evaluate the proposed package in term of social acceptance and usability in Algeria.

References

Komis, V., Misirli, A.: The environments of educational robotics in Early Childhood Education: towards a didactical analysis. Educ. J. Univ. Patras UNESCO Chair **3**(2), 238–246 (2016). ISSN: 2241–9152

Bers, M., Ponte, I., Juelich, K., Viera, A., Schenker, J.: Teachers as designers: integrating robotics in early childhood education. Inf. Technol. Childhood Educ. AACE, 123–145 (2002)

Bers, M.U., Seddighin, S., Sullivan, A.: Ready for robotics: bringing together the T and E of STEM in early childhood teacher education. J. Technol. Teacher Educ. **21**(3), 355–377 (2013)

Cejka, E., Rogers, C., Portsmore, M.: Kindergarten robotics: using robotics to motivate math, science, and engineering literacy in elementary school. Int. J. Eng. Educ. **22**(4), 711–722 (2006)

Perlman, R.: Using computer technology to provide a creative learning environment for preschool children. Logo memo No. 24, p. 260. MIT Artificial Intelligence Laboratory Publications, Cambridge, MA (1976)

Wyeth, P.: How young children learn to program with sensor, action, and logic blocks. Int. J. Learn. Sci. **17**(4), 517–550 (2008)

Sullivan, A., Kazakoff, E.R., Bers, M.U.: The wheels on the bot go round and round: robotics curriculum in pre-kindergarten. J. Inf. Technol. Educ.: Innovations Pract. **12**, 203–219 (2013). Retrieved from https://www.jite.org/documents/Vol12/JITEv12IIPp203219Sullivan1257.pdf

Alimisis, D.: Educational robotics: open questions and new challenges. Themes Sci. Technol. Educ. **6**(1), 63–71 (2013)

Cheng, Y.W., Sun, P.C., Chen, N.S.: An investigation of the needs on educational robots. In: IEEE 17th International Conference on Advanced Learning Technologies (ICALT). Timisoara, Romania, 3–7 July 2017

Kokkalia, G.K., Drigas, A.S.: Mobile learning for special preschool education. Int. J. Interact. Mobile Technol. **10**(1) (2016). https://doi.org/10.3991/ijim.v10i1.5288

Yang, D., Lim, J.K., Choi. Y.: Early childhood education by hand gesture recognition using a smartphone based robot. In: The 23rd IEEE International Symposium on Robot and Human Interactive Communication, Edinburgh, Scotland, UK, 25–29 August 2014

Hadron Solutions India Pvt. Ltd.: "Alphabet Find". https://www.amazon.com/Hadron-Solutions-India-Pvt-Ltd/dp/B005MJITZM. Latest Developer Update: 10 April 2013

Bessai, R.: Access to schooling for people with special needs in Algeria. Sociol. Int. J. **2**(5), 371–375 (2018)

https://www.statista.com/statistics/266210/number-of-available-applications-in-the-google-play-store/. (2019). Accessed Feb 2019

https://www.youtube.com/watch?v=O1S8WzwLPlM

https://www.youtube.com/watch?v=kCImYSW49u0. (2020). Accessed 10 March 2020

https://www.youtube.com/watch?v=hSrxF3NgCw&list=PLL3q6xOkK5CkxjlnpETPU5z1bS2es_9Wh. (2018). Accessed 05 Dec 2018

https://www.youtube.com/watch?v=-ab2d4Xabuo. (2018). Accessed 01 Dec 2018

Multi-Directional Total Variation and Wavelet Transform Based Methods: Application for Correlation Fringe Patterns Denoising and Demodulation

Mustapha Bahich and Mohammed Bailich

Abstract

In this work, we present a multi-directional total variation method to reduce the high frequency speckle noise in order to prepare the digital speckle pattern interferometry fringes for demodulation. In this method, the local fringe directional information is involved to increase the sensitivity of the variational regularization to the spatially varying directional intensity changes. Once the speckle noise is removed from the fringe pattern, the fringes are demodulated using directional wavelet-based phase retrieval algorithm. The extracted phase can be used for the three-dimensional defect inspection. Through simulation works, we demonstrate the ability of the denoising method to reduce the speckle noise, in terms of accuracy and computational efficiency. A comparison with the Stationary Wavelet Transform based denoising results is made.

Keywords

Image processing • Computer vision • Speckle noise • Fringe denoising • Multi-directional total variation • Directional wavelet transform • Phase retrieval

1 Introduction

Many industrial fields need more and more potent quality control systems. It's appearing that optical metrology techniques are effective non destructive methods that are capable to conduct several dimensional and physical measurements (shape, size, surface, thickness, defects, refractive index, heat, …) with a resolution going to nanometer scale (Zhu et al. 2011). Digital speckle pattern interferometry (DSPI) is an interferometry based technique that allows achieving whole-field, non-contacting 3D measurements of microscopic deformations. It forms an intensity images calling correlation fringe patterns or correlograms that are coded by the studied deformation (Kaufmann 2011). Due to the under test object roughness, a multiplicative measurement noise, called speckle noise, is added to the fringes. Furthermore, fringe demodulation methods are noise sensitive and have been developed to acquire only the optical phase and then the needed information of deformation from a correct intensity data. Therefore, a denoising algorithm is required for removing the speckle noise in DSPI fringe patterns to bring improvement to fringe demodulation performance (Barj et al. 2006). The information about the deformation to be measured is related to the fringes coding phase that needs to be extracted using fringes demodulation algorithm (Rastogi 1997).

In this work, we propose a multi-directional total variation (MDTV) based denoising method. Our algorithm is inspired by the total variation (TV) minimization principle. In the other hand, we use the directional wavelet transform based algorithm for fringes demodulation and phase retrieval (Bayram and Kamasak 2012; Bahich et al. 2013).

2 Digital Speckle Pattern Interferometry

Digital Speckle Pattern Interferometry is a metrology technique that allows the measurement of object deformations through the analysis of interference fringes and their evolution when the object is been deformed. In the classical interferometry where optically smooth surfaces are studied, the information of deformation is obtained by comparing the deformed reflected wave with a plane reference wave, no speckles are present. The difference between these two waves will give rise to interference fringes describing the shape of the object. However, when we test an object with a rough diffusing surface, the wave front coming from it will

M. Bahich (✉)
Physics Department, Faculty of Sciences, Moulay Ismail University, Meknes, Morocco
e-mail: mbahich@gmail.com

M. Bailich
EIT Department, ENSAO, Mohammed First University, Oujda, Morocco

M. Ben Ahmed et al. (eds.), *Emerging Trends in ICT for Sustainable Development*,
Advances in Science, Technology & Innovation, https://doi.org/10.1007/978-3-030-53440-0_40

Fig. 1 Subjective speckle pattern formation

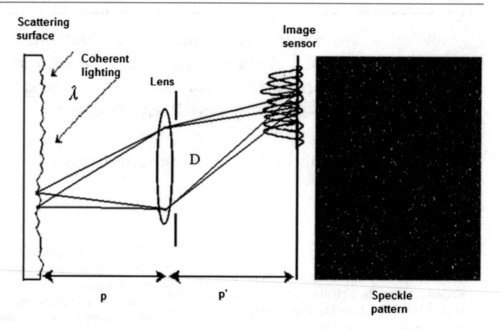

be very deformed and the interferences image will be embedded in a structure of random interference so-called speckle (Fig. 1). The fringes embedded in this random phenomenon are difficult to be distinguished, and therefore, the direct analysis becomes a challenge to demodulate the fringes and extract the phase and then the object deformation (Fujieda et al. 2009).

The basic principle of the DSPI technique is that the speckled pattern intensity distribution is a function of the relative phases of two interfering plane waves inside each resolution cell of an imaging setup (Fig. 2).

Generally, the fringe patterns are produced by the interference of reflected light beams by two surfaces in an interferometer (Slangen et al. 1996).

The intensity of the primary specklogram is written as follows:

$$I_1(x, y) = I_0(x, y)[1 + V(x, y) \cos(\phi_s(x, y))] \quad (1)$$

where (x, y), I_0, and V are the pixel coordinates, the average intensity and, the visibility, respectively. ϕ_s is the speckle phase.

The intensity distribution modeling the secondary specklogram is expressed as follows:

$$I_2(x, y) = I_0(x, y)[1 + V(x, y) \cos(\phi_s(x, y) + \varphi(x, y))] \quad (2)$$

where φ represents the phase change due to the deformation.

The specklograms, according to the principle of statistical optics, are the noise-carrier patterns. The correlation fringes of deformation cannot be directly distinguished in the secondary specklogram because of its both low signal to noise

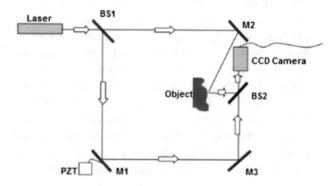

Fig. 2 Experimental DSPI setup

ratio (SNR) and contrast that are caused by the presence of random rapid varying speckle noise.

After the subtraction, the final correlogram intensity distribution can be modeled as follows:

$$I(x, y) = I_2 - I_1 = 2I_0(x, y)V(x, y) \sin(\frac{\varphi(x, y)}{2}) \sin(\phi_s(x, y) + \frac{\varphi(x, y)}{2})$$

$$(3)$$

Now, our goal is to extract the phase map φ, and from it find the deformation. Unfortunately, the interpretation of the speckle correlation fringes is difficult for the reason of the presence of a high frequency multiplicative noise $\sin(\phi_s + \varphi/2)$ (called speckle noise) that need to be reduced before demodulating the fringes. In this work, we use a new variant of the total variation based denoising method and with the logarithm transform; we convert the multiplicative noise into additive noise.

3 Multi-Directional Total Variation Denoising

The classical variational denoising algorithm is proposed by Rudin-Osher-Fatemi (ROF) and it is formulated as a total variation minimization problem (Rudin et al. 1992):

$$J(u) = \int \|\nabla u\|_2 d + \lambda \int (u - I)^2 d \qquad (4)$$

where I is the noisy image, u is the denoised image, Ω is the image domain, ∇ is the gradient operator, and λ is a positive regularization parameter.

The classical TV of u can be defined as the L_2 norm of the gradient

$$\sum_{i,j} \|\nabla u(i,j)\|_2 = \sum_{i,j} \sup_{t \in B_2} \langle \nabla u(i,j), t \rangle \qquad (5)$$

where B_2 is the unit ball.

For this, TV measure can be viewed as a support function of the set B_2, making it rotation invariant. This TV minimization tends to yield a piecewise constant image, and thus can give rise to the staircase effect in regions with a smooth change in its intensity value. To reduce the staircase effect, the MDTV is proposed while using a support function of the set $E_{a,\theta}$, which denotes an ellipse that is oriented along θ and has major axis length of $a > 1$ and minor axis is 1 (Zhang and Wang 2013). An ellipse can be formulated in terms of a rescaled and rotated unit ball as $E_{a,\theta} = R_\theta \Lambda_a B_2$,

where $R_\theta = \begin{bmatrix} \cos\theta & -\sin\theta \\ \sin\theta & \cos\theta \end{bmatrix}$ and $\Lambda_a = \begin{bmatrix} a & 0 \\ 0 & 1 \end{bmatrix}$.

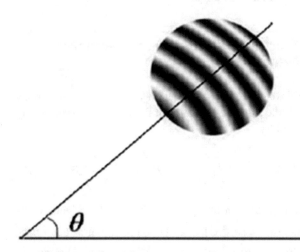

Fig. 3 Fringe direction angle

Hence, MDTV measure can also be expressed as follows:

$$\sum_{i,j} \|\nabla u(i,j)\|_2 = \sum_{i,j} \sup_{t \in E_{a,\theta}} \langle \nabla u(i,j), t \rangle$$
$$= \sum_{i,j} \sup_{t \in B_2} \langle \nabla u(i,j), R_\theta \Lambda_a t \rangle \qquad (6)$$

Notice that, the above described MDTV can be effective for denoising arbitrary images with no single dominant direction, which is the case of our correlation fringe patterns. Therefore, the directions of the ellipses have to be locally changed depending on the local fringe direction. To have prior knowledge on θ, we proposed a pre-estimation of the spatial fringe orientation angle to give a range of θ values (while keeping $a > 1$ constant) and picking the one which yields the smallest TV. Due to its accuracy and computation time, we use in this work the gradient-based method to compute the fringe direction map. Figure 3 shows an illustration of a fringe direction angle. Solving this minimization problem can be carried out through Chambolle's projection algorithm (Chambolle 2004), which is a recent and fast method to solve this TV model.

4 Directional Wavelet Transform

The wavelet technique is a powerful tool to locally analyze a signal and obtain a space–frequency description of it. A wavelet is a small wave or pulse which can be compressed and stretched at different scales (Mallat 1999). The basic wavelet function ψ is called "mother wavelet" with zero mean and satisfying the following admissibility condition:

$$C_\Psi = \int_{-\infty}^{+\infty} \frac{|\hat{\Psi}(k)|^2}{k} \, dk < +\infty \qquad (7)$$

One of the wavelets having a strong presence in optical metrology is "Morlet wavelet". It is defined as

$$\psi(x) = \frac{1}{\sqrt{\pi}} \exp(2i\pi k_c x) \exp(-x^2) \qquad (8)$$

where k_c is the wavelet center frequency.

A family of analyzing wavelets are generated by dilating the mother wavelet using the scale parameter $s > 0$ and translating it using the location parameter $\xi \in R$. It can be expressed as

$$\psi_{s,\xi}(x) = \frac{1}{\sqrt{s}} \psi\left(\frac{x - \xi}{s}\right) \qquad (9)$$

We note that the scale parameter s is related to the frequency concept. The wavelets with small values of s have

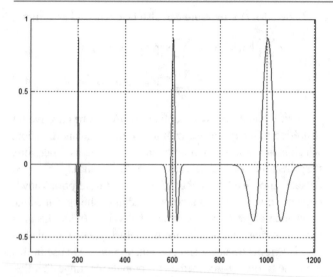

Fig. 4 Mexican hat wavelet with scale values 2, 1, and 0.5 at the locations 200, 600, and 1000

narrow spatial support and consequently rapid oscillations, making them well adapted to selecting high frequency components of a signal, and the converse is true for wavelets with big values of s.

Figure 4 shows another commonly used wavelet (Mexican hat) with three scale values located at three different positions.

The one-dimensional continuous wavelet transform (1DCWT) of a signal $f(x)$ is given by

$$
\begin{aligned}
W_f(s, \xi) &= \int_{-\infty}^{+\infty} f(x) \psi_{s,\xi}^*(x) dx \\
&= \int_{-\infty}^{+\infty} f(x) \frac{1}{\sqrt{s}} \psi^*\left(\frac{x - \xi}{s}\right) dx
\end{aligned}
\tag{10}
$$

where x is the 1D space coordinate and $*$ is the complex conjugate symbol.

The 1DCWT yields a series of coefficients $W_f(s, \xi)$ measuring the degree of similarity between the signal f and the different analyzing wavelets.

The inverse wavelet transform allows the reconstruction of the original signal as follows:

$$
f(x) = \frac{1}{C_\Psi} \int_0^{+\infty} \int_{-\infty}^{+\infty} W_f(s, \xi) \, \Psi_{s,\xi}(x) \, \frac{ds \, d\xi}{s^2}
\tag{11}
$$

In two-dimensional space, the directional wavelet transform of a given image I is expressed as (Yarlagadda and Kim 2012):

$$
W_I\left(s, \vec{\xi}, \alpha\right) == \iint_{R^2} I(\vec{x}) \psi_{s,\vec{\xi},\alpha}^*(\vec{x}) d\vec{x}
\tag{12}
$$

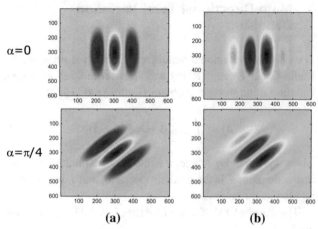

Fig. 5 Complex Morlet with two different orientation angles: **a** Real parts, **b** imaginary parts

where $s > 0$ is the scale parameter, $\vec{\xi} \in R^2$ is the translation parameter, and $\alpha \in [0, 2\pi]$ is the orientation parameter.

In the directional case, the analyzing wavelet is a small orientable function that can be defined as follows:

$$
\psi_{s,\vec{\xi},\alpha}(\vec{x}) = \frac{1}{s} \psi\left(R^{-\alpha} \frac{\vec{x} - \vec{\xi}}{s}\right)
\tag{13}
$$

R^α designates the orientation matrix with the angle α

$$
R^\alpha = \begin{pmatrix} \cos\alpha & -\sin\alpha \\ \sin\alpha & \cos\alpha \end{pmatrix}
\tag{14}
$$

Thus, these analyzing wavelets are resulted from the wavelet by dilation, translation, and orientation processes.

Figure 5 shows the 2D complex Morlet wavelet with two different orientation angles.

5   Directional Wavelet Phase Recovery Method

To retrieve the phase information using the ridgelet based method, the ridgelet transform of the fringe pattern is firstly computed by

$$
R\left(s, \vec{\xi}, \alpha\right) = \iint_{R^2} I(\vec{x}) \frac{1}{s} \psi^*\left(R^{-\alpha} \frac{\vec{x} - \vec{\xi}}{s}\right) d\vec{x}
\tag{15}
$$

The idea of data fitting is as follows. The ridgelet coefficients array $R\left(s, \vec{\xi}, \alpha\right)$ is a 4D complex matrix that quantifies the local resemblance degree between the fringe pattern and the ridgelets for the different values of scale, location, and orientation (Watkins 2012).

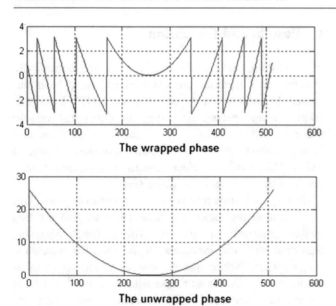

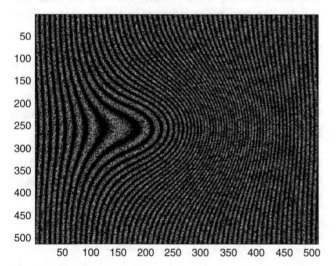

Fig. 6 Phase unwrapping principle

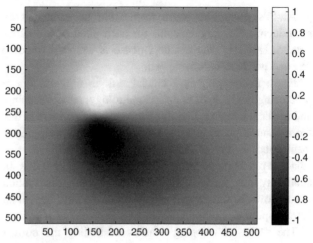

Fig. 8 Fringe direction angle map (in radian)

Fig. 7 Simulated speckled correlation fringe pattern

The ridgelet coefficients, for the same location parameter value $\vec{\xi}$, form a 2D complex array $R(s, \alpha)$ that will be used to compute the phase of the fringe pattern's pixel with coordinates $\vec{\xi}$.

To retrieve the phase of this pixel, the ridge of coefficients array is firstly determined and then the phase will be its corresponding angle.

$$\varphi(\vec{\xi}) = angle(R(s_m, \alpha_m)) \qquad (16)$$

With

$$|R(s_m, \alpha_m)| = \max(|R(s, \alpha)|) \qquad (17)$$

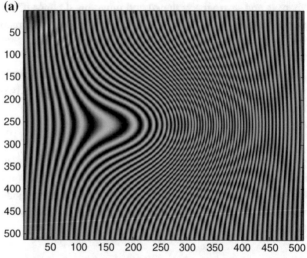

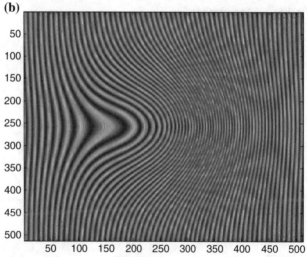

Fig.9 **a** Theoretical correlation fringe pattern and **b** denoised correlation fringe pattern

At this point, the phase map could be estimated by reproducing the same calculus for each pixel of the fringe pattern. However, the resultant phase is wrapped (modulo 2π) and needs to be unwrapped.

Phase unwrapping is a crucial task in different image processing fields, such as optical metrology, remote sensing, and medical imaging. For all these areas, the problem is that the measured phase is always modulo 2π. This wrapping is manifested in the phase map by sudden jumps of two adjacent pixels phase value. Thus, we must remove this 2π ambiguity and recover the continuous phase distribution (also known as unwrapped or absolute), we must appeal an "unwrapping algorithm".

The phase unwrapping consists of finding the correct multiple of 2π as follows:

$$\phi(p) = \varphi(p) + 2k(p)\pi \qquad (18)$$

where φ and ϕ are the wrapped and unwrapped phase of the pixel p, respectively, and k is an integer. Along a phase map row (or a column), the wrapped phase appears as a saw teeth profile (Fig. 6).

The correct phase difference between two adjacent pixels must, therefore, be less than π. If p_1 and p_2 denote two neighboring pixels, then we have

$$|\phi(p_1) - \phi(p_2)| < \pi \qquad (19)$$

Several phase unwrapping methods have been conceived, which are differentiated by their noise sensitivity, computational approaches, and complexity, etc. These methods could be grouped into two main categories: local and global methods.

6 Results and Discussion

In this section, the simulation results of the above mentioned and proposed methods are obtained and have been discussed. In order to test the ability of the MDTV based correlation fringe pattern denoising for directional wavelet phase extraction algorithm, the simulation algorithms have been implemented in MATLAB software. Firstly, a 512×512 simulated fringe pattern is generated with a gaussian elevation as a test phase (Fig. 7).

The regularization parameter was chosen to be $\lambda = 9$. This regularization parameter was hand tuned to give the best quality improvement. We use MDTV based denoising algorithm with 150 iterations on the speckled correlation fringe pattern. Notice that the convex closed set of minimization is chosen for an ellipse with a = 8 and b = 1 as its major and minor axis lengths respectively and locally oriented using the fringe direction map of Fig. 8. The performance of this denoising algorithm is evaluated using the structural similarity index (SSIM), between the theoretical and the denoised fringe patterns, as a quantitative performance measure. Considering the intensity distortion, the obtained SSIM value is 0.7.

Figure 9a and b show the theoretical and the denoised fringe patterns, respectively. As shown in Fig. 10, the proposed denoising method showed a good visual result with not only removing speckle but also preserves the fringes properties.

The directional wavelet-based phase retrieval algorithm was applied to the denoised fringe pattern of Fig. 9b, and the resulted phase distribution is illustrated in Fig. 11a. Notice that the resulted phase is wrapped (modulo 2π) and an

Fig.10. 256-line intensity distribution: **a** speckled fringe pattern, **b** theoretical fringe pattern, and **c** denoised fringe pattern

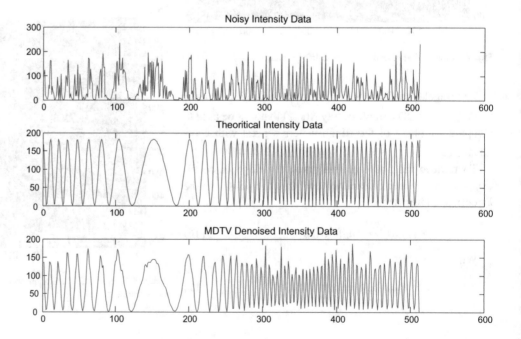

(a)

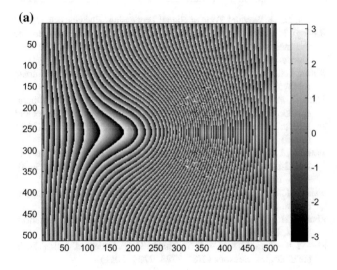

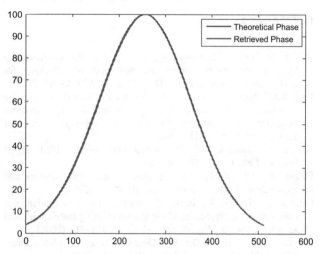

Fig. 12 256-line phase distributions

(b)

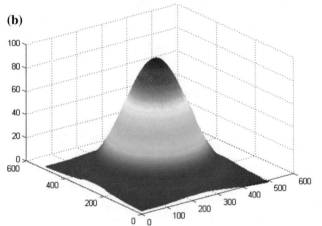

(c)

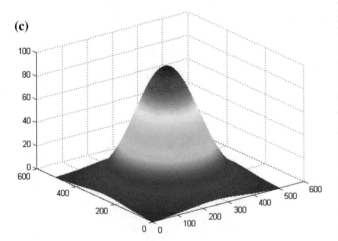

Fig. 11 **a** Wrapped retrieved phase, **b** unwrapped retrieved phase, and **c** the theoretical phase (in radians)

unwrapping method has to be chosen to obtain the final unwrapped phase (Ghiglia and Pritt 1998). It is clear that the phase of interest is well estimated (Fig. 11b and c).

Indeed, theoretical and retrieved phase maps are perfectly superimposed as illustrated in Fig. 12, and the given RMS error (Root Mean Square) between them is 1.5 rad.

7 Conclusion

In this chapter, we show that the MDTV approach is a very efficient technique to reduce the speckle noise in DSPI fringe patterns. It is also shown that the directional wavelet transform approach is a very useful tool and can be used to retrieve phase distributions from restored correlation fringes with good accuracy. The MDTV based algorithm is based on the variational model with the involvement of the acquired local fringes orientation information that increases the sensitivity of the method to the intensity image directions enabling an efficient and stable recovery of the DSPI correlation fringes. Moreover, wavelet-based fringes demodulation method guarantees the retrieval of the optical phase distribution using only one single fringe pattern to be acquired and making it much suitable for the real-time dynamic measurements. Nevertheless, and due to the high sensitivity of intensity gradients to the speckle noise, there is more work to be done for a correct local fringe orientation pre-estimation of real DSPI fringe patterns.

References

Bahich, M., Bailich, M., Imloul, A., Afifi, M., Barj, E.: A comparative study of one and two-dimensional wavelet-based techniques for noisy fringe patterns analysis. Opt. Commun. **290**, 43–48 (2013)

Barj, E.M., Afifi, M., Idrissi, A.A., Nassim, K., Rachafi, S.: Speckle correlation fringes denoising using stationary wavelet transform. Application in the wavelet phase evaluation technique. Opt. Laser Technol. **38**(7), 506–511 (2006)

Bayram, İ., Kamasak, M.E.: Directional total variation. IEEE Signal Process. Lett. **19**(12), 781–784 (2012)

Chambolle, A.: An algorithm for total variation minimization and applications. J. Math. Imaging Vis. **20**, 89–97 (2004)

Fujieda, I., Kosugi, T., Inaba, Y.: Speckle noise evaluation and reduction of an edge-lit backlight system utilizing laser diodes and an optical fiber. J. Display Technol. **5**(11), 414–417 (2009)

Ghiglia, D.C., Pritt, M.D.: Two-Dimensional Phase Unwrapping: Theory, Algorithms and Software. Wiley Eds (1998)

Kaufmann, G.H.: Advances in Speckle Metrology and Related Techniques. Wiley Eds (2011)

Mallat, S.: A Wavelet Tour of Signal Processing. Elsevier Eds (1999)

Rastogi, P.K.: An electronic pattern speckle shearing interferometer for the measurement of surface slope variations of three-dimensional objects. Opt. Lasers Eng. **26**(2), 93–100 (1997)

Rudin, L.I., Osher, S., Fatemi, E.: Nonlinear total variation based noise removal algorithms. Physica D **60**(1), 259–268 (1992)

Slangen, P., Berwart, L., Veuster, C., Golinval, J.C., Lion, Y.: Digital speckle pattern interferometry (DSPI): a fast procedure to detect and measure vibration mode shapes. Opt. Lasers Eng. **25**(4), 311–321 (1996)

Watkins, L.R.: Review of fringe pattern phase recovery using the 1-D and 2-D continuous wavelet transforms. Opt. Lasers Eng. **50**(8), 1015–1022 (2012)

Yarlagadda, P., Kim, Y.H.: Measurement Technology and its Application. Trans Tech Publications Ltd. (2012)

Zhang, H., Wang, Y.: Edge adaptive directional total variation. The J. Eng. **61**, 1–2 (2013)

Zhu, Y.K., Tian, G.Y., Lu, R.S., Zhang, H.: A review of optical NDT technologies. Sensors **11**(8), 7773–7798 (2011)

A Multi-Agent System for Color Video Decomposition

Insaf Bellamine

Abstract

Dynamic texture (DT) analysis is an important research area in the field of computer vision. Actually, several methods are used to recognize the dynamic texture, but a good compromise between computational cost and performance is hard to achieve. For color sequences, we propose a fast color decomposition model which splits a video into two components, a first one containing geometrical information, the structure of the sequence and a second one dynamic color texture (DCT). This work presents a combined color video decomposition with multi-agent system (MAS). The combination is used to increase the robustness of the dynamic texture recognition and to decompose a color video in a reasonable time. Experimental results are obtained from Dyntex database which offers a wide variety of natural sequences all presenting dynamic textures.

Keywords

Dynamic texture • Multi-agent system • Color video decomposition

1 Introduction

The extracted features from dynamic texture can be utilized for tasks of images sequences segmentation, classification, recognition, and retrieval. They combine appearance description and motion, and also be invariance to some transformation such as rotation, translation, and illumination (Zhao and Matti 2007). Comparing with texture found within static images, analyzing dynamic texture is a challenging problem.

Dynamic texture recognition is an essential task in a number of applications such as motion detection (Lloyd 2017), segmenting the sequence images of natural scenes (Doretto 2003), video indexing, video classification, and video retrieval (Yan 2020; Yan et al. 2019).

In the literature, there are many methods to recognize dynamic textures, which are based on optical flow (Chetverikov and Péteri 2005), local spatio-temporal filtering (Wildes and James 2000), global spatio-temporal transform (Smith et al. 2002), computing geometric properties (Otsuka, et al. 1998), and color video decomposition using bounded variation and oscillatory functions (Mathieu et al. 2008).

In order to extract the pure dynamic texture component and the geometric structure component, Mathieu et al. (2008) proposed to extend the Osher–Vese color decomposition model (Vese and Stanley 2006).

Mathieu et al. (2008) considered a video as a 3-D image (Aubert and Pierre 2006) in order to take into account the spatio-temporal structure, i.e., a volume, in order to be able to apply image algorithms 2-D extended to 3-D case.

To decompose all image sequences of the video, the color video decomposition performs slowly because of the big size of data. To get over this problem and provide a sophisticated decomposition system, a multi-agent system (MAS) based on the cooperative agents work to achieve the global decomposition of the color video is proposed.

This paper is organized as follows: Sect. 2 presents the Multi-Agent System, Sect. 3 presents the Color video decomposition, Sect. 4 presents our proposed approach, and finally, Sect. 5 shows our experimental results.

I. Bellamine (✉)
LAROSERI, Department of Computer Science, Faculty of Sciences, Chouaïb Doukkali University, El Jadida, Morocco
e-mail: insafbellamine20@gmail.com

© The Editor(s) (if applicable) and The Author(s), under exclusive license to Springer Nature Switzerland AG 2021
M. Ben Ahmed et al. (eds.), *Emerging Trends in ICT for Sustainable Development*,
Advances in Science, Technology & Innovation, https://doi.org/10.1007/978-3-030-53440-0_41

2 Related Work

In general, a multi-agent system (MAS) consists of several interacting intelligent agents. Since the late 1980s, agent technology has received considerable attention in many fields of application such as industry, medicine, and computer science.

Within a multi-agent system (Jennings et al. 1998), Jennings et al. emphasized interactions (cooperation, coordination, and negotiation). Wooldridge and Nicholas (1995) put the three main elements necessary for the implementation and the design of intelligent agents (theories, architectures, and languages).

In the MAS field, there is no single definition of an agent. Wooldridge (1992) proposed a definition of agent based on the concept of autonomy. According to Wooldridge (1992), a multi-agent system can be defined as a network of autonomous agents which interact with each other and with their environment to achieve a common objective (Niazi and Hussain 2011). There are several applications on multi-agent systems (online commerce, (Rogers 2007) disaster response (Schurr 2005; Genc 2013) and modeling of social structure (Sun and Isaac 2004).

A detailed list of potential applications of the multi-agent system associated with many fields of application (logistics, manufacturing control, production planning) was presented by Pechoucek and Marik (Pechoucek and Vladimir 2008). Stonedahl et al. (2011) implemented the MAgICS framework as a coherent introductory computer program based on an agent-based model (ABM) and multi-agent system. Fabregues and Sierra (Fabregues and Carles 2014) proposed modular software architecture based on an innovative research and negotiation method and which includes the BDI model (belief, desire, and intention).

Agents must be able to interact with each other and with their environment in order to meet the different purposes for which MAS is designed. In literature, three agent architectures have been proposed: reactive (Brooks 2014), deliberative (Rao and Georgeff 1995), and hybrid (Ferguson 1992) architectures.

3 Color Video Decomposition

There is a great deal of literature on image decomposition models. Thus, among many others, we cite only the most relevant works which appear to us to be the most useful papers. In this way, the reader can refer to the work of Rudin et al. (Rudin et al. 1992) and Mayer et al. (2001).

Many algorithms have been proposed to solve the Mayer model numerically. The most popular algorithms are those of the Aujol algorithm (Meyer 2006; Aujol 2005; Aujol and

Antonin 2005; Aujol and Sung 2006) and Vese and Osher (2003), (2004) to cover the most recent and relevant advances.

In the following work, the decomposition of color images (Vese and Stanley 2006) into geometrical and texture components appeared as a good means of extracting the significant information, i.e., the color texture component, independently of the color geometric information.

In order to extract the pure dynamic texture component and the geometric structure component, Mathieu et al. (2008) proposed to extend the Osher–Vese color decomposition model (Vese and Stanley 2006).

Mathieu et al. (2008) considered a video as a 3-D image (Aubert and Pierre 2006) in order to take into account the spatio-temporal structure, i.e., a volume, in order to be able to apply image algorithms 2-D extended to 3-D case.

Let an original image sequence $f \in L^2(\Omega)$, where Ω is a bounded open domain of R^3, with Lipschitz boundary conditions. In order to extract u (structure component) and v (texture component) from f, in order to recover u and v from f, Mathieu et al. (Aujol et al. 2006) proposed

- a spatio-temporal version of the gradient vector $|\nabla u_{\mathrm{xyt}}|$ discretized (Mathieu et al. 2008):

$$(\nabla \mathrm{u})_{\mathrm{i,j,k}} = ((\nabla u)^x_{i,j,k}, (\nabla u)^y_{i,j,k}, (\nabla u)^t_{i,j,k})) \qquad (1)$$

$$(\nabla u)^x_{i,j,k} = \begin{cases} u_{i+1,j,k} - u_{i,j,k} & \text{if } i < N \\ 0 & \text{if } i = N \end{cases}$$

$$(\nabla u)^y_{i,j,k} = \begin{cases} u_{i,j+1,k} - u_{i,j,k} & \text{if } j < N \\ 0 & \text{if } j = N \end{cases}$$

$$(\nabla u)^t_{i,j,k} = \begin{cases} u_{i,j,k} - u_{i,j,k-1} & \text{if } k < N \\ 0 & \text{if } k = N \end{cases}$$

- The redefinition of the total discrete variation taking into account the joint evolution of color planes inspired by (Duval et al. 2008)

$$J(u) = \int_\Omega \sqrt{\sum_{c=R,G,B} \sum_{i=x,y,z} \left| ((\nabla u)^i_{i,j,k})_c \right|^2} \, \mathrm{dxdydt} \qquad (2)$$

- The vector version of the space G adapted to the RGB color space and extended to the temporal dimension.

Let G be the Banach space composed of all the vectorial functions.

$$\vec{v}(x,y,t) = 0.17em(v_R(x,y,t), v_G(x,y,t), v_B(x,y,t))$$

which can be written as follows:

$$\vec{v}(x,y,t) = \left(\mathrm{div}\,\overrightarrow{g_R},\,\mathrm{div}\,\overrightarrow{g_G},\,\mathrm{div}\,\overrightarrow{g_B}\right) \quad (3)$$

with $g_{x,c}, g_{y,c}, g_{t,c} \in L(R^3), c = R, G, B$

Induced by the standard $\|v\|$ defined as the smallest bound of all L^∞ norms of functions $|\vec{g}|$, where

$$|\vec{g}| = \sqrt{|\overrightarrow{g_R}|^2 + |\overrightarrow{g_G}|^2 + |\overrightarrow{g_B}|^2} \quad (4)$$

Mathieu et al. (Aujol et al. 2006) proposed the following minimization problem inspired by (3) and (Aujol 2005), for each color channel:

$$\inf_{u,g1,g2,g3}\left\{ G_p(u, g1, g2, g3) = \int \lfloor \nabla u \rfloor + \lambda \int |f - u - \partial_x g1 \right.$$

$$+ \partial_y g2 + \partial_z g3|^2 \mathrm{dxdydt} +$$

$$\left. \mu\left[\int \left(\sqrt{g1^2 + g2^2 + g3^2}^2 \right) \mathrm{dxdydt} \right]^{\frac{1}{p}} \right\} \quad (5)$$

where λ and μ are tuning parameters.

Let reintroduce that $\vec{u} = (u_R, u_G, u_B)$, and $\vec{g_i} = (\vec{g_i}, R, \vec{g_i}, G, \vec{g_i}, B)$, $i \in \{1,2,3\}$.

Formally minimizing the above energy equation with respect to $u, g1, g2, g3$ yields the following Euler–Lagrange equation for each color channel:

$$u = f - \partial_x g1 - \partial_y g2 - \partial_x g3 + \frac{1}{2\lambda}\mathrm{div}\left(\frac{\nabla u}{|\nabla u|}\right) \quad (6)$$

$$\cdot\mu\left(\left\|\sqrt{g1^2 + g2^2 + g3^2}\right\|\right)^{1-p}\left(\sqrt{g1^2 + g2^2 + g3^2}\right)^{p-2} g1$$

$$= 2\lambda\left[\frac{\partial}{\partial_x}(u - f) + \partial_{yx}^2 g1 + \partial_{yy}^2 g2 + \partial_{yt}^2 g3\right] \quad (7)$$

$$\mu(\|\sqrt{g1^2 + g2^2 + g3^2}\|)^{1-p}\left(\sqrt{g1^2 + g2^2 + g3^2}\right)^{p-2} g2$$

$$= 2\lambda\left[\frac{\partial}{\partial_y}(u - f) + \partial^2_{yx} g1 + \partial^2_{yy} g2 + \partial^2_{yt} g3\right] \quad (8)$$

$$\mu(\|\sqrt{g1^2 + g2^2 + g3^2}\|)^{1-p}\left(\sqrt{g1^2 + g2^2 + g3^2}\right)^{p-2} g3$$

$$= 2\lambda\left[\frac{\partial}{\partial_t}(u - f) + \partial^2_{xt} g1 + \partial^2_{yt} g2 + \partial^2_{tt} g3\right] \quad (9)$$

Dynamic decomposition has been applied to traffic video from the Dyntex database (Péteri et al. xxxx). The result of the decomposition using the parameters ($\mu = 1000, \lambda = 0.1$) is shown in Fig. 1.

(a)

(b) **(c)**

Fig. 1 The dynamic decomposition with an adapted Aujol Algorithm: **a** Fountain image **b** the dynamic texture component **c** the dynamic structure component

The wavelets located near the fountain are completely integrated into the dynamic texture sequence, and the circumference of the fountain, therefore, seems completely frozen in the u sequence, dynamically regularized. The wavelet size taken into account is linked to the scale parameter λ.

4 Proposed Approach

To decompose a single image of a video, an image decomposition model (Vese and Stanley 2006) is sufficient, where it provides good results within a reasonable time, thanks to the spatial information used. To decompose all the image sequences of the video, the decomposition of the color video is carried out slowly due to the large size of the data. To overcome this problem and provide a sophisticated decomposition system, a multi-agent system based on the work of cooperative agents to achieve the overall decomposition of color video is proposed.

In fact, we proposed to divide each image into parts $c \times c$ (Fig. 2), so that the image is divided horizontally into sub-images and vertically also. Each agent manages a part of the image different from the other parts managed by other

Fig. 2 The Fountain image is splitted into 4*4 parts ($c = 4$)

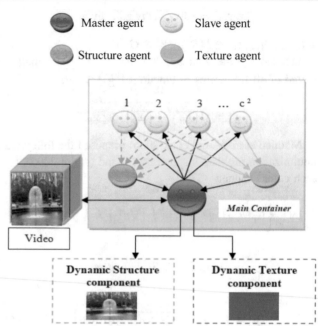

Fig. 3 The conceptual architecture of the color video decomposition

agents. Each agent breaks down part of the image separately and in parallel, which can significantly reduce processing time.

The proposed multi-agent system takes a series of image sequences as input and returns the corresponding decomposed images by its output. It is composed of four different types of agents: a Master agent, a texture agent, a structure agent, and c^2 slave agents.

In general, the master agent is the first agent created in the platform. He takes control of the system and has $c*c$ slave agents who perform partial decompositions; share memory space and communicate via the exchange of messages agent communication language (ACL); in order to achieve the overall breakdown of the video.

The conceptual architecture of the multi-agent approach illustrating the types of agents proposed and their interactions is illustrated in Fig. 3.

To learn more about this question, explain a usage scenario below, after a sequence that first describes how the main ideas of our approach work:

1. The Master agent initiates the interaction between a user and the MAS and obtains the name of the image and the decomposition parameter as input.
2. To start the decomposition process, the Agent Master divides the image into $c * c$ sub-images. Then, it creates c^2 slave agents and sends requests to them. When all the decomposition conditions are satisfied, the Master agent warns the slave agents in order to start their tasks and waits for them to respond.

3. Since a slave agent has received the Master's notification message, it performs the decomposition process; using the color video decomposition presented in the previous subsection. When it has finished decomposing its specific part of a given image, it sends the structure component to the structure agent and the dynamic texture component to the texture agent.
4. The texture agent combines all the textured sub-images to obtain the whole, and then sends the textured component to the managing agent.
5. The structural agent combines all the structured sub-images to obtain the whole, and then sends the textured component to the master agent.

   Each slave agent automatically goes to the next image without waiting for the end of the other agents and when it is done; there are no more image sequences to process; he informs the master agent that his mission has been successfully completed.
6. The master agent notifies the user that the decomposition process has ended.

5   Experimental Results and Discussion

5.1   Experimental Results

The different sequences proposed in this section come from the Dyntex database (Péteri et al. xxxx) which offers a wide variety of natural sequences all presenting dynamic textures.

DynTex (Péteri et al. xxxx) provides a large and diverse database of high-quality dynamic textures that can be used for a wide variety of research.

The total contents of the DynTex database consist of 679 sequences for each following directories:

- pr1: Contains processed sequences that are down sampled to 352 × 288 and encoded to DivX Mpeg4.
- raw: Contains original DV compressed sequences in PAL resolution (720 × 576, 25 fps). The avi-files also contain a sound track. Total size: 52.0 Gb.
- pr0: Contains processed sequences that are still DV compressed (PAL resolution), but have been interlaced with a spatio-temporal median filter. The files do not contain sound tracks. Total size: 52.0 Gb.

Color video decomposition was applied to part of the Dyntex database (Péteri et al. xxxx). The result of the decomposition using the parameters ($\mu = 1000$, $\lambda = 0.1$) is shown in Fig. 4.

The pure dynamic texture component, the structural elements of the regularized space–time sequence as well as the noise are obtained with the approach presented above. Figure 4 clearly shows the contribution and the temporal impact of the decomposition:

As in the static case, these results are a step prior to the characterization into dynamic image sequences of images which, therefore, represents the indexing of video sequences.

Fig. 4 The original image on the left, the dynamic color texture (DCT) component on the right, and the structure component in the center

Moving escalator

Sea-waves

Texure

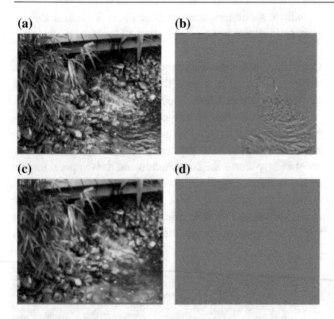

Fig. 5 The color video decomposition: **a** The original image (waterfall's sequence) **b** the dynamic color texture (DCT) component **c** the dynamic color structure (DCS) component **d** texture component with the classic color decomposition (Vese and Stanley 2006)

5.2 Comparison and Discussion

We present a comparison between two methods (the static decomposition and the dynamic decomposition are both calculated with the same classical parameters).

We can easily see that the resulting temporal impact, the water in Fig. 5 is well regularized and the fluid aspect is well represented in the dynamic color texture component.

Thanks to the results illustrated in Fig. 5, we are able to locate moving objects in the cascade sequence (see Fig. 5d).

The performances of the implementation of color video decomposition discussed above are compared to those of the sequential dynamic decomposition model (Mathieu et al. 2008), in terms of cost in time. The program was run on a PC with a 2.4 GHz Intel core (TM) i7 processor with 4 GB of RAM.

Color video decomposition performance was measured for the cascade sequence by increasing the number of slave agents and calculating the time cost. We calculated our decomposition on 10 images treated as a block (Fig. 6).

The more the number of slave agents is increased the more the execution time decreases.

From Fig. 7, we note that as the number of the input images increases, both methods require much more time. In addition, all the running times performed by our proposed method are smaller than those performed by the sequential version of the color video decomposition method (Mathieu et al. 2008).

From Fig. 8, our proposed method requires less time and can effectively split a Grayscale video into two components, a first one containing geometrical information, the structure of the sequence and a second one dynamic color texture (DCT).

6 Conclusion

This paper mainly contributes to the color video decomposition which can split a video into two components: a first one containing geometrical information, the structure of the sequence and a second one dynamic color texture (DCT). To decompose a color video in a reasonable time, our method consists to use the notion of multi-agent system (MAS). Experimental results are obtained from Dyntex database which offers a wide variety of natural sequences all presenting dynamic textures.

Fig. 6 The running times consumed by our Color Video Decomposition on the waterfall's sequence

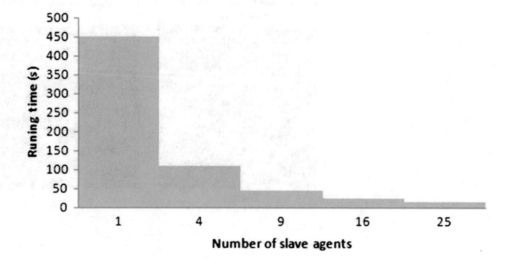

Fig. 7 The running times consumed on a wide variety of natural color sequences from Dyntex database

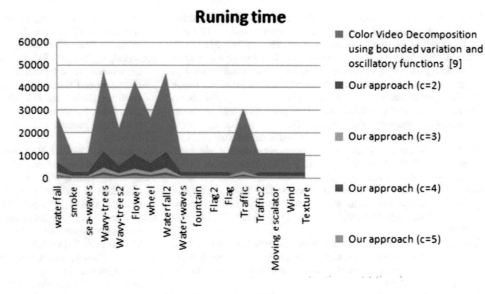

Fig. 8 The running times consumed on a wide variety of natural grayscale sequences

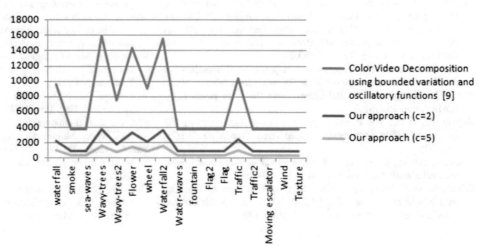

References

Aubert, G., Pierre K.: Mathematical Problems in Image Processing: Partial Differential Equations and the Calculus of Variations, vol. 147. Springer Science & Business Media (2006)

Aujol, J., et al.: Image decomposition into a bounded variation component and an oscillating component. . J. Math. Imaging Vis. **22**(1), 71–88 (2005)

Aujol, J., Antonin, C.: Dual norms and image decomposition models. Int. J. Comput. Vis. **63**(1), 85–104 (2005)

Aujol, J., Sung, H.: Color image decomposition and restoration. . J. Vis. Commun. Image Represent. **17**(4), 916–928 (2006)

Aujol, J., Gilboa, G., Chan, T., Osher, S.: Structure–texture image decomposition-modeling, algorithms, and parameter selection. Int. J. Comput. Vis. **67**(1), 111–136 (2006)

Brooks, R.: How to build complete creatures rather than isolated cognitive simulators. In: Architectures for Intelligence. Psychology Press, pp. 239–254 (2014)

Chambolle, A.: An algorithm for total variation minimization and applications. J. Math. Imaging Vis. **20**(1–2), 89–97 (2004)

Chetverikov, D., Péteri, R.: A brief survey of dynamic texture description and recognition, pp. 17–26. Computer Recognition Systems. Springer, Berlin, Heidelberg (2005)

Doretto, G.: Dynamic texture segmentation. null. IEEE (2003)

Duval, V., Jean-François, A., Luminita, V.: A projected gradient algorithm for color image decomposition. UCLA CAM report (2008)

Fabregues, A., Carles, S.: Hana: a human-aware negotiation architecture. Decis. Support Syst. **60**, 18–28 (2014)

Ferguson, I.A.: TouringMachines: an architecture for dynamic, rational, mobile agents. No. UCAM-CL-TR-273. University of Cambridge, Computer Laboratory, (1992)

Franklin, S., Art, G.: Is it an Agent, or just a Program. A Taxonomy for Autonomous Agents. International Workshop on Agent Theories, Architectures, and Languages. Springer, Berlin, Heidelberg (1996)

Genc, Z.: Agent-based information infrastructure for disaster management. In: Intelligent Systems for Crisis Management. Springer, Berlin, Heidelberg, pp. 349–355 (2013)

Jennings, N., Katia, S., Michael, W.: A roadmap of agent research and development. Auton. Agent. Multi-Agent Syst. **1**(1), 7–38 (1998)

Lloyd, K.: Detecting violent and abnormal crowd activity using temporal analysis of grey level co-occurrence matrix (GLCM)-based texture measures. Mach. Vis. Appl. **28**(3–4), 361–371 (2017)

Mathieu, L., Michel, M., Abdallah, H.: Dynamic color texture modeling and color video decomposition using bounded variation and oscillatory functions. In: 3rd International Conference on Image and Signal Processing, Cherbourg-Octeville, France, pp. 277 (2008)

Meyer, Y.: Oscillating patterns in image processing and nonlinear evolution equations: the fifteenth Dean Jacqueline B. Lewis memorial lectures. Vol. 22. American Mathematical Soc. (2001)

Meyer, Y.: Oscillating patterns in some nonlinear evolution equations, pp. 101–187. Mathematical foundation of turbulent viscous flows. Springer, Berlin, Heidelberg (2006)

Niazi, M., Hussain, A.: Agent-based computing from multi-agent systems to agent-based models: a visual survey (PDF). Scientometrics **89**(2), 479–499 (2011). https://doi.org/10.1007/s11192-011-0468-9

Otsuka, K., et al.: Feature extraction of temporal texture based on spatiotemporal motion trajectory. In: Proceedings. Fourteenth International Conference on Pattern Recognition (Cat. No. 98EX170), vol. 2, IEEE (1998)

Pechoucek, M, Vladimir, M.: Review of industrial deployment of multi-agent systems. Gerstner Laboratory, Agent Technology Group, Department of Cybernetics, Czech Technical University in Prague, Czech Republic and Rockwell Automation Research Center, Prague, Czech Republic (2008)

Péteri, R., Huiskes, M., Fazekas, S.: (Dyntex: A comprehensive database of dynamic textures). https://dyntex.univ-lr.fr/index.html

Rao, A.S., Georgeff, M.P.: BDI agents: from theory to practice. In: ICMAS, vol. 95 (1995)

Rogers, A., et al.: The effects of proxy bidding and minimum bid increments within eBay auctions. ACM Trans. Web (TWEB) **1**(2), 9 (2007)

Rudin, I., Stanley, O., Emad, F.: Nonlinear total variation based noise removal algorithms. Physica D **60**(1–4), 259–268 (1992)

Schurr, N., et al.: The Future of Disaster Response: Humans Working with Multiagent Teams using DEFACTO." . AI technologies for homeland security, AAAI spring symposium (2005)

Smith, J., Ching-Yung, L., Milind N.: Video texture indexing using spatio-temporal wavelets. In: Proceedings. International Conference on Image Processing, vol. 2, IEEE (2002)

Stonedahl, F., Michelle W., Uri, W.: MAgICS: toward a multi-agent introduction to computer science. In: Multi-agent Systems for Education and Interactive Entertainment: Design, Use and Experience. IGI Global, pp. 1–25 (2011)

Sun, R., Isaac, N.: Simulating organizational decision-making using a cognitively realistic agent model. J. Artif. Soc. Soc. Simul. **7**(3) (2004)

Szummer, M., Rosalind, P.: Temporal texture modeling. In: 3rd IEEE International Conference on Image Processing, vol. 3, IEEE (1996)

Vese, L.A., Osher, S.J.: Modeling textures with total variation minimization and oscillating patterns in image processing. J. Sci. Comput. **19**(1–3), 553–572 (2003)

Vese, L.A., Osher, S.J.: Image denoising and decomposition with total variation minimization and oscillatory functions. J. Math. Imaging Vis. **20**(1–2), 7–18 (2004)

Vese, L., Stanley, O.: Color texture modeling and color image decomposition in a variational-PDE approach. In: 2006 Eighth International Symposium on Symbolic and Numeric Algorithms for Scientific Computing. IEEE (2006)

Wildes, R., James B.: Qualitative spatiotemporal analysis using an oriented energy representation. In: European Conference on Computer Vision. Springer, Berlin, Heidelberg (2000)

Wooldridge, M.: The logical modelling of computational multi-agent systems. Diss. University of Manchester, Institute of Science and Technology (1992)

Wooldridge, M., Nicholas, R.: Intelligent agents: theory and practice. The Knowl. Eng. Rev. **10**(2), 115–152 (1995)

Yan, C.: Spatial-temporal attention mechanism for video captioning. IEEE Trans. Multimedia **22**(1), 229–241 (2020)

Yan, C., Gong, B., Wei, Y., Gao, Y.: Deep multi-view enhancement hashing for image retrieval. IEEE Trans. Pattern Anal. Mach. Intell. (2019)

Zhao, G., Matti, P.: Dynamic texture recognition using local binary patterns with an application to facial expressions. IEEE Trans. Pattern Anal. Mach. Intell. **6**, 915–928 (2007)

Serious Games for Sustainable Education in Emerging Countries: An Open-Source Pipeline and Methodology

Younes Alaoui, Lotfi El Achaak, Amine Belahbib, and Mohammed Bouhorma

Abstract

Serious Games start playing an important role in education (Greitzer et al. in J. Educ. Resour. Comput. 7: 2-es, Greitzer, F.L., Kuchar, O.A., Huston, K.: Cognitive science implications for enhancing training effectiveness in a serious gaming context. J. Educ. Resour. Comput. 7 (3), 2-es (2007). 10.1145/1281320.1281322). Video games have transformed how we spend our leisure time, and started to transform how we learn through serious games (Young et al. in Rev. Educ. Res. 82:61–89, Young Mar. Rev. Educ. Res. 82:61–89, 2012). Developed economies have used serious games in multiple domains like military, healthcare, or telecom to train workforce (Liarokapis and de Freitas in A Case Study of Augmented Reality Serious Games, IGI Global, Liarokapis, F., de Freitas, S. (eds.): A Case Study of Augmented Reality Serious Games. IGI Global (2010); Bellotti et al. in Adv. Hum.-Comput. Interact. 2013: 1–11, Bellotti, F., Kapralos, B., Lee, K., Moreno-Ger, P., Berta, R.: Assessment in and of serious games: an overview. Adv. Hum.-Comput. Interact. 2013, 1–11 (2013)). Serious games are used as an eLearning tool also to complement traditional education or for distance learning. Researchers have tested serious games in preschools and primary schools recently, and results proved efficiency (Papanastasiou et al. in Int. J. Emerg. Technol. Learn. 12:44, Papanastasiou et al. Int. J. Emerg. Technol. Learn. 12:44, 2017). Serious games bring additional training activities to the learning process (Zierer and Seel in SpringerPlus 1:15, Zierer and Seel Dec. SpringerPlus 1:15, 2012; Koper in Educational Technology Foundations of Electronic Learning Environments, p. 41, Koper, R.: Educational technology foundations of electronic learning environments, p. 41 (2015)) and train schoolchildren on applying knowledge. Training schoolchildren on applying knowledge makes education sustainable and is a challenge for teachers in some developing countries. Using serious games can help in this area. To create such serious games, we need a process and a set of methodologies able to manage the development process of such serious games. In this paper, we present a pipeline or a process and a methodology to develop serious games for sustainable education. This process is called Gaming and Learning Unified Process to engineer Software, or GLUPS.

Keywords

Serious game • Video game • Development processes • Instructional design • Software pipeline • Domain-specific language • Open-source • Sustainable education

1 Introduction

Video games have transformed how we spend our leisure time, and started to transform how we learn (Young et al. 2012). The video game industry is the sector involved in the development, marketing and monetization of video games. This industry has grown in size as a business. Newzoo estimated the size of the games market to be around $137Bn in 2018. The cost of developing a game depends on many factors, the most important being the complexity of the game. Developing a simple puzzle game can cost $500, while the development of GTA 5 has costed in the $265Mn (https://en.wikipedia.org, w, index.phptitle=List_of_most_-expensive_video_games_to_developoldid=946053757). To manage video game development projects, developers of video games have blended processes from the movie-making industry and from the software development industry to design Game Development Pipelines, or development processes to develop and engineer games (Electronic Arts,

Y. Alaoui (✉) · L. El Achaak · A. Belahbib · M. Bouhorma
PLIST Laboratory UAE University, P.O. Box 416
Tangier, Morocco
e-mail: younes.alaoui@amana.ac.ma

CG-Spectrum). Research studies have formalized some of the game development processes and categorized these processes under the name Game Development Software Engineering (Papanastasiou et al. 2017).

Serious games is a sub-category of video games that uses video gaming to serve training or education purposes. Sande Chen and David Michael define serious games as "games whose first purpose was not mere entertainment" (Alvarez and Djaouti 2011). Alvarez and Djaouti define serious games by a relationship (Alvarez and Djaouti 2011):

Serious Game = UtilitarianFunction(s) + Video Game

Wilkinson defines serious games as games with purpose (Wilkinson 2016), and link them to purposing non-digital games promoted by the work of Plato. From these perspectives, serious games are at the intersection of instructional design, game design, and software engineering.

Researchers started testing serious games to complement education in preschool and primary schools. The results indicated positive effect (Papanastasiou et al. 2017). As education in Morocco is judged lacking efficiency and its improvement has been identified as a nation priority in 2000, 2008, and 2019 (Charte nationale d'éducation et de formation 1999) we have explored how to develop serious games that can help improve schoolchildren skills and teacher competencies in Morocco and in developing countries. The development processes used for video games are not directly applicable to serious games for many reasons: size of projects, available budget, disciplines involved. Serious games for education require usually some of the skills required for video games plus pedagogical skills. Budget wise, serious game for education have a very limited budget, and require less workload than big video games. In this paper, we present a process to follow when developing serious games for schoolchildren by a community.

1.1 Organization of Our Paper

Section 2 lists the main findings and design recommendations related to Serious Games for education and eLearning. Section 3 presents the processes that we have explored to coordinate the serious game development process. Section 4 presents the instructional design and serious game development methodologies that we have explored and blended, to build-up GLUPS. Section 5 presents the artifacts of GLUPS among a domain-specific modeling language. Section 6 presents the application of GLUPS to develop a serious game. Section 7 presents first findings and outlooks.

2 Defining the Scope

Researchers have conducted different experiments to measure the results of game-based learning. Results suggest that game-based learning improve achieving learning objectives (Liarokapis et al. 2010; Bellotti et al. 2013). According to Papanastasiou et al. (2017) a survey of over 1,600 practicing teachers in English state primary and secondary schools, showed that the majority of the teachers believe that digital games can help support children's motor and cognitive development (83%) and their higher order thinking skills (such as logical thinking, planning and strategizing) (65%).

Serious games are effective when they are based on both pedagogical and gaming principals (Liarokapis et al. 2010).

In our research, we work on enabling development of open-source serious games for schoolchildren by a community. This development of open-source serious games is multidisciplinary and involve people with competencies in didactics, game design, or software development.

To reduce the cognitive load on schoolchildren, we need to design games that follow similar logics and metaphors. We need to produce blueprints that can apply to multiple games.

This led us to set up Gaming and Learning Unified Process to engineer Software (GLUPS), a hybrid approach to design and develop serious games blending learning design and software engineering. GLUPS is open-source.

2.1 Related Works

Development of serious games is in the intersection of instructional/learning design, game design, and software engineering.

We reviewed methodologies from instructional and learning design mainly the ADDIE umbrella (Zierer and Seel 2012), serious game design, namely, the Four-Dimensional Framework (Liarokapis et al. 2010), game development software engineering (multiple (Aleem et al. 2016; Ramadan and Widyani 2013)), and the software development process, namely, Unified Process (Kruchten 2005; Scott 2002). To streamline design, we have reviewed visual modeling languages and domain-specific modeling languages for serious games (Marchiori and del Blanco 2011; Prasanna 2012).

Instructional Design methodologies focus on instruction and do not cover software development. Serious game design methodologies focus on the requirement analysis mainly. Their objective is to insure comprehensive analysis. Game Development Software Engineering adapts software engineering processes to games, but do not cover learning design nor the gamification process for learning.

One solution can be to use 2 or 3 methodologies in parallel. But this will require users with serious training and experience to coordinate between these methodologies and push the 2 or 3 tracks in parallel.

We preferred to set up an easy to understand and easy to use unified process that drives through the important steps to follow to develop serious games for schoolchildren by a community.

3   Serious Game Development Process

Current game development software engineering methodologies are based on software engineering methodologies (Aleem et al. 2016).

The functional analysis used when developing serious games is different from the analysis usually used in software development. As in software development, functional analysis specifies the objectives (the why and the what) and decide on the main designs (the how). But the "how" question for serious games has two dimensions: learning design and game design.

The learning dimension selects the didactics or pedagogy to use to achieve the learning objectives: how to design the learning steps to achieve learning objectives. The game design creates a game scenario leveraging game entertainment techniques while staying coherent with the learning design.

3.1   The Unified Process Framework

The Unified Process is a framework for iterative and incremental software development process (Scott 2002). This framework has been instantiated and adapted on multiple flavors, a commercial version called Rational Unified Process (Kruchten 2005) (an open-source version called OpenUP (Scott 2002), and a version for agile development called Agile Unified Process.

Unified Process builds on the following principles (Thillainathan 2014):

- Iterative and incremental development
- Collaborative development
- Development driven by use cases and centered on architecture and design
- Risk management
- Requirement management
- Change control
- Continuous evaluation of quality and efficiency.
- Visual modeling.

Many of these principles are well suited for our project: collaborative development, design-driven development, risk management, and continuous evaluation of quality and efficiency.

The OpenUP version of the Unified Process is open-source and available under the Eclipse license. OpenUP is part of Eclipse Project Framework.

The Unified Process defines four iterative project phases: inception, elaboration, construction, and transition. The Unified Process identifies also the building blocks that will produce the aimed work: the engineering disciplines that produce (the What) and the tasks to execute (the How). For this reason, the engineering disciplines are called the workflow.

UP represents the phases and the engineering disciplines in a matrix: project phases are the columns; the engineering disciplines are the rows (See Table 1).

We have adapted the Unified Process framework (UP) in its version OpenUP. Game Development Software Engineering methodologies usually group phases into pre-production, production, and post-production (Aleem et al. 2016), (Ramadan and Widyani 2013). We have renamed Elaboration into Pre-Production, and Construction into Production.

The following sections describe how we have adapted the engineering disciplines of OpenUP to support learning design and serious game design.

Table 1 The Unified Process Matrix

| | | Phases | | | |
		Inception	Elaboration	Construction	Transition
Engineering disciplines	Business modeling				
	Requirements				
	Analysis				
	Design				
	Implementation				
	Test				
	…				

Fig. 1 Iterative and incremental
development in UP

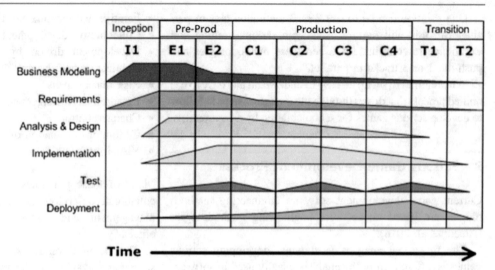

3.2 UP Engineering Disciplines

The engineering disciplines of UP cover the tasks to perform during the project life cycle. One can see them as the ingredients to mix to get an outcome. The phases of UP cover the "quantity" or the "doses" of each activity or ingredient to involve at this phase. First phases (inception or pre-production) involve mainly modeling, requirement, analysis, and design. These first phases can also involve some activities of implementation and test. The production phase involves implementation activities mainly. However, it can also involve some complementary tasks of modeling and design. This structure enables iterative and incremental development of deliverables (see Fig. 1).

4 Designing Learning in Serious Games

Greitzer stated in (Greitzer et al. 2007) that "to be effective, serious games must incorporate sound cognitive, learning, and pedagogical principles into their design and structure." When researching on approaches to help design e-learning activities through serious games, one comes naturally across instructional design and serious game design.

For learning design, we have explored methodologies for Instructional Design (ADDIE and CLT). For Serious Game Design we have explored Four-Dimension Framework (4DF) and the Rapid Prototyping Model for Serious Games Development (Lotfi et al. 2014). We have also explored visual modeling languages to capture game design.

4.1 Learning Design from Instructional Design

Instructional Design is used in the USA to enable systematic application of knowledge to the development of learning

systems. This translation into a "systematic application process" is called "Instructional Systems Development" (ISD).

ISD provides guidelines that instructional designers follow in order to create a workshop, a course, a curriculum, or a training (McGriff 2005). Currently, there are two prevalent models of Instructional Systems Development (Zierer and Seel Dec. 2012): ADDIE (Peterson 2003) is targeting the macro-level of application while the Cognitive Load Theory is targeting the micro-level of Instructional DesignInstructional Systems Development.

ADDIE covers the major processes of the generic Instructional Systems Development process: Analysis, Design, Development, Implementation, and Evaluation (thus the ADDIE acronym). Figure 2 represents the phases of ADDIE.

Each phase covers specific topics:

- The phase Analysis covers needs analysis: learner situation, knowledge available with learner, objectives, targeted performances and cost-benefit.

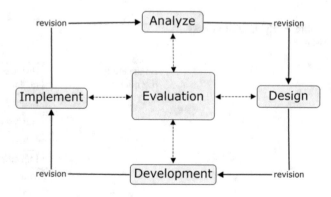

Fig. 2 ADDIE

- The phase Design focuses on the design of a blueprint of the outcome and a storyboard of the instruction.
- The Development phase covers development of instructional material according to decisions made following the designs of the design phase.
- Implementation contains the concrete realization of planned instruction in a real setting.
- The Evaluation phase of ADDIE refers to testing the success of instruction, and improving the instructional design.

The Dick and Carey model complements ADDIE and is one of the most popular models of Instructional Design and Instructional Systems Development (Zierer and Seel Dec. 2012). Dick and Carey published a textbook The Systematic Design of Instruction that was used to train many instructional designers. The Dick and Carey model is task-oriented and bottom-up. These tasks fit within the ADDIE phases as follows:

- Analysis: identify instructional goals (instructional analysis, analyzing learners, and contexts), write performance objectives
- Design: develop assessment instruments, develop instructional strategy
- Development: develop instructional material, design formative evaluation, design summative evaluation
- Evaluation: revise instruction.

The Dick and Carey model takes into consideration the learners and contexts in the analysis phase and emphasizes assessment in the design phase. As we will see in the following sections, frameworks proposed for serious games development use also these criteria (Liarokapis et al. 2010).

El Achaak et al. have adapted the Rapid Prototyping Model to the development of serious games (Lotfi et al. 2014). Rapid Prototyping Model is an instructional design methodology with many similarities to ADDIE (Fig. 3. Process of Rapid Prototyping Model).

ADDIE, Dick and Carey model and the Rapid Prototyping Model provide a framework that help instructional designers to take into consideration important design objectives and constraints for a successful outcome.

4.2 Learning Design from Serious Games Methodologies

The Serious Game Institute has developed the four-dimensional framework (4DF), a model to support design of games (Liarokapis et al. 2010) (Fig. 4). According to Fotis and de Freitas (Liarokapis et al. 2010), 4DF brings together: learner modeling, a consistent consideration of the pedagogic models adopted with the game, levels of required interactivity and immersion, and the context where the game will be used.

Learner Specifics and Pedagogy link to learning design. Representation and Context link to game design.

The 4DF has highlighted the importance of pedagogy and representation in the development of serious games. However, we did not find it as a methodology that is able to support serious game development end-to-end.

The first row dimensions of 4DF relating to learners and pedagogy can merge into ADDIE engineering blocks. The second row dimensions, representation and context, can merge into the requirement building block of the Unified Process.

4.3 Merging ADDIE and 4DF into UP

We found Unified Process and ADDIE methodologies both building on the same principles:

- Phases
- Iterations within phases
- Objective- and constraint-based design: engineering disciplines to identify objectives, analyze requirements, sketch up and test solutions, design blueprints, develop, deploy, collect feedback, and so on.

These common foundations helped us blend elements from both methodologies.

ADDIE is used for generic instructional design. We have taken from ADDIE what is helpful for serious game development.

We have kept the Unified Process managing the iterative process on the horizontal axis: manage iterations and include agility.

Fig. 3 Process of rapid prototyping model

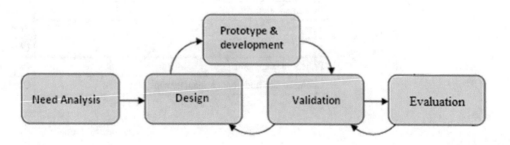

Four Dimensional Framework	
Learner Specifics	**Pedagogy**
Profile, Role, Competences	Associative, Cognitive, Social/Situative
Representation	**Context**
Fidelity, Interactivity, Immersion	Environment, Access to learning, supporting resources

Fig. 4 The four-dimensional framework

We have used ADDIE and 4DF to adapt the workflow of the building blocks of the vertical axis. Table 2 shows the correspondence between the OpenUP disciplines and GLUPS disciplines:

In our work, we have identified two adaptations that were not intuitive but essential to make the model work: mapping "Business Modeling" to "Learning Modeling," and making "Learning Objectives," "Learning Assessment" and "Game Representation" as mandatory pieces of "Requirements."

Learning Modeling refers to the pedagogical approach to follow in the game. We consider Learning Modeling as a given parameter that we should follow. Learning Modeling defines the type of learning objectives to target, the learning steps to follow, and the typical training situations to use to reach these objectives.

Requirements cover both Learning and Serious Game, as presented earlier. To capture learning requirements, our experience showed us that we could focus on capturing how to assess achievement of learning objectives.

We have also divided "Testing" into two disciplines: software and game testing that we kept calling "Testing." We have also added "Learning Evaluation" to testing. Even if the software is working well and the game is entertaining, we still need to test and evaluate the learning outcome.

5 GLUPS Artifacts

Artifact is another important component of the Unified Process. Artifacts are project outputs or by-products expected at milestones. An artifact is produced, modified, or used by a task from a discipline. Artifacts have value and are subject to version control. Examples of artifacts are project plan, risk list, user stories, architecture notebook, wireframes, storyboards, work item list, and so on.

OpenUP defines multiple artifacts to capture product and project-related information. OpenUP identifies which information or design to produce at each project phase to build up project value. There is no obligation to capture information in formal artifacts. The project team can capture information on whiteboards informally or sketch designs. What is important is to explicitly produce and capture that piece of information or design at a given milestone.

Table 3 describes the artifacts that we have introduced for GLUPS:

5.1 ABCD User Stories

We have tested and adopted writing learning objectives following the ABCD approach, inspired by the Educational

Table 2 Comparison between OpenUP and GLUPS disciplines

OpenUP disciplines	GLUPS disciplines
Business modeling	Learning objectives and didactics
Requirements	Learning competencies
	Learning assessment
	Learning tasks
Analysis	Metaphor (immersion, characters)
	Storyline
	Gamification
Architecture	Architecture and design
Implementation	Implementation
Test	Test
	Learning evaluation
Project management	Project management
Configuration and change management	Configuration and change management

Table 3 Artifacts associated with GLUPS specific disciplines

GLUPS Disciplines	Artifacts
Learning objectives and assessment	ABCD user stories
Learning assessment	User stories
Game representation (interactivity, immersion...)	User stories
Architecture and design	Storyline using a domain-specific modeling language

Modeling Languages (Koper 2015). A learning objective following the ABCD approach consists of four elements: (A) Audience, (B) Behavior, (C) Condition, and (D) Degree.

A-Audience: Determine who will achieve the objective.
B-Behavior: Use action verbs to write observable and measurable behavior that shows mastery of the objective.
C-Condition: If any, state the condition under which behavior is to be performed. (Optional)
D-Degree: If possible, state the criterion for acceptable performance, speed, accuracy, quality, etc. (Optional).

Example: Euclidean division, "Given a number of coins (dividend) and a number of pirates (divisor), learners will be able to identify dividend and divisors and divide coins on pirates following the Euclidean division."

Audience: learners
Behavior: will be able to divide coins on pirates, give to all pirates the same number of coins, keep for himself the rest that must be less than the number of pirates. Doing so, he wins the rest, and the pirates do not attack him.
Condition: "Given a number of coins (dividend) and a number of pirates (divisor)."
Degree: all pirates get the same number of coins. The rest that he keeps for him is less than the divisor.

Writing learning objectives following the ABCD user-story artifact has proven to be effective.

5.2 Formal Language for Modeling and Design

The design of serious games involves collaboration between teachers and software developers. We have looked at approaches to enable teachers and developers explicit specifications and elaborate designs understandable by both parties.

Researchers have tested and proposed different modeling approaches. El Achaak et al. propose a set of artifacts to structure the design (game story, scenario flowchart, non-play, mission, immersion, playing character, levels) (Lotfi et al. 2014). Other researchers used modeling

languages based on UML (Thillainathan 2014) or defined Domain-Specific Modeling Languages DSML (Marchiori and del Blanco 2011; Prasanna 2012) to model the entities and the relations of the game. Some researchers have tested generating game code from these models using Model-Driven Software Development approaches (Matallaoui et al. 2015).

We have tested some of the artifacts of El Achaak and tested their modeling languages and approaches in a manual process. Marchiori et al. have defined a Domain-Specific Visual Language (DSVL) to model serious games following the adventure point-and-click paradigm, leveraging state diagrams.

We found the graphical modeling language proposed by Prasanna easy to use and promising, mainly the first part of this language covering storylines. The second part of this language covers the decomposition models that Prasanna uses to define the actions and the detailed behavior. We found the second part incomplete and lacking expression power. We have extended the use-case diagrams of UML to implement the storyline constructs proposed by Prasanna.

5.3 GLUPS Matrix

The matrix crossing the Engineering Disciplines of GLUPS by the phases is illustrated in Table 4.

6 Applying Glups to a Usecase

We have used GLUPS to manage the development of a serious game to help school children apply the Euclidean division. Euclidean division is studied first at the fourth primary grade in Morocco. Schoolchildren take 2 to 3 years to master it and to identify when to apply it. The objective of the game is to help learners apply the Euclidian division and remember its algorithm.

For the development of this game, we have used iterations of 2 weeks time. Inception had 1 iteration, Pre-Production 3 iterations, Production 2 iterations, and Transition 1 iteration.

Table 4 Disciplines × Phases for GLUPS

		Phases			
		Inception	Pre-Production	Production	Transition
Engineering disciplines	Learning objectives and didactics				
	Learning modeling				
	Game modeling				
	Architecture and design				
	Implementation				
	Test				
	Learning evaluation				
	Project management				
	Configuration and change management				

6.1 Game Metaphor

The metaphor of the game is the following: pirates have found chests full of coins and moneybags. The player needs to solve multiple math challenges that require him to divide a set of coins among the pirates (coins, moneybags or a mix of coins and moneybags). After dividing the coins among the pirates, the player keeps the remaining coins for himself and wins additional points. If the pirates do not get the same number of coins, they fight against each other and may attack the player. If the player keeps more coins than the number of pirates, then the pirates will attack him as each pirate can still get one additional coin from the remaining coins. In this case, the player loses some points. If the player gives to each pirate the quotient and keeps for himself the rest of the Euclidian division, then all pirates are happy; they have no reason to fight against each other; they have no reason to attack the player. The player gets a reward. Figure 5. Euclidean Division Challenge shows the game.

6.2 Applying GLUPS Artifacts

The learning model used in this game is "situational" or learning by repeating, learning by doing mainly. We have defined the learning objectives using the ABCD user-story artifact. We have already presented one of the learning objectives of this game when we have presented the ABCD user-story artifact.

An additional learning objective is to divide moneybags and coins on the pirates, where each moneybag contains 10 coins. A typical example is 3 money bags plus 7 coins and 2 pirates.

Fig. 5 Euclidean division challenge

Audience: learners

Behavior: will be able to divide a mix of moneybags and coins on pirates.

Condition: "He can divide the moneybags first. He can open the remaining bags when the number of the remaining bags is less than the number of pirates."

7 Conclusion and Future Work

To develop serious game for education we need a process or a pipeline. Such process is even more important when a community will conduct this development and in a collaborative open-source mode.

In this paper, we describe a process to develop serious games based on instructional design and software development best practices. The proposed process is based on OpenUP, which is part of Eclipse Project Framework. To cover the pedagogical aspect of serious games, we have included in the proposed process activities and building blocks from ADDIE that engineer the development of instructional products. We have named this process GLUPS for Gaming and Learning Unified Process to engineer Software. We have kept GLUPS simple and reduced its artifacts to the minimum requirement to develop serious games for education in preschools and primary schools. Future work will include testing GLUPS on additional projects and on complex projects with participants from different countries. We will work also on enhancing GLUPS with tools to accelerate modeling designs, creating game templates, or generating pieces of code automatically.

References

"Charte nationale d'éducation et de formation," Commission Spéciale Education Formation, Royaume du Maroc (1999)

Conseil Supérieur de l'Education, de la Formation et de la Recherche Scientifique, Une école de justice sociale: contribution à la réflexion sur le modèle de développement (2018)

"GUIDE DE LA PEDAGOGIE DE L'INTEGRATION dans l'école marocaine." Le Centre National d'Innovatin Pédagogique et d'Expérimentation

Laabou, M.: "Pédagogie de l'intégration : la grande imposture ?," Libération. https://www.libe.ma/Pedagogie-de-l-integration-la-grande-imposture_a19544.html

"Nouveau Curricula-1e-2e-3e-4e_AP-VF," Direction des Programmes, May 2019

Liarokapis, F., de Freitas, S. (eds.): A Case Study of Augmented Reality Serious Games. IGI Global (2010)

Bellotti, F., Kapralos, B., Lee, K., Moreno-Ger, P., Berta, R.: Assessment in and of serious games: an overview. Adv. Hum.-Comput. Interact. **2013**, 1–11 (2013)

Greitzer, F.L., Kuchar, O.A., Huston, K.: Cognitive science implications for enhancing training effectiveness in a serious gaming context. J. Educ. Resour. Comput. **7**(3), 2-es (2007). https://doi.org/10.1145/1281320.1281322

Thillainathan, N.: Serious game development for educators—a serious game logic and structure modeling language, p. 12 (2014)

Marchiori, E.J., del Blanco, A´., Torrente, J., Martinez-Ortiz, I., Fernandez-Manjo´, B.: A visual language for the creation of narrative educational games. J. Vis. Lang. Comput. (22), 443–452 (2011)

Prasanna, A.T.: A domain specific modeling language for specifying educational games, p. 109 (2012)

Matallaoui, A., Herzig, P., Zarnekow, R.: Model-driven serious game development integration of the gamification modeling language GaML with unity. In: 2015 48th Hawaii International Conference on System Sciences, HI, USA, pp. 643–651 (2015). https://doi.org/10.1109/HICSS.2015.84

Lotfi, E.A., Belahbib, A., Bouhorma, M.: Adaptation of rapid prototyping model for serious games development. J. Comput. Sci. Inf. Technol. (2014)

Zierer, K., Seel, N.M.: General didactics and instructional design: eyes like twins a transatlantic dialogue about similarities and differences, about the past and the future of two sciences of learning and teaching. SpringerPlus **1**(1), 15. https://doi.org/10.1186/2193-1801-1-15

Koper, R.: Educational technology foundations of electronic learning environments, p. 41 (2015)

Peterson, C.: Bringing ADDIE to life: instructional design at its best, p. 15

Balduino, R.: Introduction to OpenUP (Open Unified Process). https://www.eclipse.org/epf/general/OpenUP.pdf

Kruchten, P.: Le Rational Unified Process®, p. 22 (2005)

Scott, K.: The unified process explained. Addison-Wesley, Boston (2002)

Aleem, S., Capretz, L.F., Ahmed, F.: Game development software engineering process life cycle: a systematic review. J. Softw. Eng. Res. Dev. **4**(1), 6 (2016)

Ramadan, R., Widyani, Y.: Game development life cycle guidelines. In: 2013 International Conference on Advanced Computer Science and Information Systems (ICACSIS), Bali, pp. 95–100 (2013)

McGriff, S.J.: ISD knowledge base/instructional design & development/instructional systems design models, vol. 15, p. 2005 (2001)

Papanastasiou, G., Drigas, A., Skianis, C.: Serious games in preschool and primary education: benefits and impacts on curriculum course syllabus. Int. J. Emerg. Technol. Learn. **12**(01), 44 (2017). https://doi.org/10.3991/ijet.v12i01.6065

Alvarez, J., Djaouti, D.: An introduction to serious game definitions and concepts. Serious Games & Simul. Risks Manag. 11–15 (2011)

Wilkinson, P.: A brief history of serious games. In: Dörner, R., Göbel, S., Kickmeier-Rust, M., Masuch, M., Zweig, K. (eds.) Entertainment Computing and Serious Games, vol. 9970, pp. 17–41. Springer International Publishing, Cham (2016)

"List of most expensive video games to develop," *Wikipedia*. https://en.wikipedia.org/w/index.php?title=List_of_most_expensive_video_games_to_develop&oldid=946053757 (2020). Last accessed 31 March 2020

Young, M.F., et al.: Our princess is in another castle: a review of trends in serious gaming for education. Rev. Educ. Res. **82**(1), 61–89. https://doi.org/10.3102/0034654312436980